Theory and Technology of Rock Excavation for Civil Engineering

Dingxiang Zou

Theory and Technology of Rock Excavation for Civil Engineering

冶金工业出版社
Metallurgical Industry Press

Springer

Dingxiang Zou
Hong Kong
Hong Kong

ISBN 978-981-10-1988-3 ISBN 978-981-10-1989-0 (eBook)
DOI 10.1007/978-981-10-1989-0

Jointly published with Metallurgical Industry Press

Library of Congress Control Number: 2016954905

Printed on acid-free paper

This Springer imprint is published by Springer Nature
The registered company is Springer Nature Singapore Pte Ltd.
The registered company address is: 152 Beach Road, #22-06/08 Gateway East, Singapore 189721, Singapore

Foreword I

Rock excavation is a complex, nuanced, and hard work. The book “Theory and Technology of Rock Excavation for Civil Engineering” not only outlines various theoretic topics on rock excavation, but also introduces various excavation techniques under the guidance of these theories. This book, which combines theory and practice with an abundance of figures, data and illustrations, can be applied as the reference material for the relevant subjects in addition to being an invaluable handbook.

Mr. D. Zou, the author of this book, is one of the standing directors of the China Society of Explosives and Blasting. “A Three-Dimensional Mathematic Model in Calculation the Rock Fragmentation of Bench Blasting in the Open Pit” (BMMC model)—which was developed by the author during the period of working in the Maanshan Institute of Mining Research in early 1980s—is described as “the first complete numerical model for blasting in China” by Prof. Jun Yang, and was included in the teaching materials of postgraduate study of blasting. Finally, this text was also introduced in “Theory and Technology of Engineering Blasting,” chiefly edited by Prof. Yalun Yu, as well as the “Blasting Handbook,” chiefly edited by me.

The author has a solid theoretical foundation and rich experience accumulated during long-term on-site engineering practice. More recently, in particular, more than twenty years of working in Hong Kong has conferred upon Zou better access to the most advanced theories and technologies on rock excavation in the world. All of these are fully delineated in this book.

The author has worked tirelessly in dedicating himself to the field of rock excavation, a fact that is reflected in his completion of this book in both English and Chinese within the three years since his retirement. I agreed with pleasure and enthusiasm to write this foreword at his request, and I hereby recommend this book to all my colleagues in the field and those learning the nuances of rock excavation. I also sincerely hope that the younger generation would uphold this same scientific spirit in the future and contribute to the theory and technology of rock excavation in the service of humanity.

Professor Xuguang Wang
Academician
China Academy of Engineering

Foreword II

Rock excavation changes the profile of rock mass of the earth's surface, in which blasting is the most effective method. In layman's terms, blasting is a matter of "changing the ground with a big bang," through varying amounts of magic and bewilderment. The blasting expert, however, feels a rush of excitement and joy after every successful blast as he reshapes the earth by his own hand. It is with this sense of wonder that the author writes this book and applies his over 50 years of professional experience.

After retiring and reviewing every blast in his career, Zou felt compelled to pass on his singular knowledge to the younger generation of earth scientists and engineers. After accumulating all of his practical experience and knowledge, Zou wrote "Theory and Technology of Rock Excavation for Civil Engineering" within three years. Though the book adds only a drop to the sea of knowledge, it reflects his passion and expertise on a deep and palpable level, while providing an invaluable account to the next generation of rock excavation experts.

This book covers a wide range of topics pertaining to rock excavation, from geological structure to engineering properties, from the equipment and tools of excavation to explosives and blasting accessories, from smooth blasting to pre-reinforcement of rock mass to tunnel lining, from technical data of construction technology to the expert system of excavation design, from safety and environmental protection to project management systems, all illustrated in detail. In many sections of the book the author imparts unique insights, and the abundance of reliable data and reference materials can be used as a handbook. As such, it is an invaluable text in the field of rock engineering.

The author had completed three years of postgraduate studies and cultivated the theoretical thinking. In this book, especially in the field of blasting theory, Zou breaks through the traditional theory of "blasting craters" and instead asserts that the rock mass of an object is a structural mass which had been cut by various weakened planes to form a structure of "natural blocks". Blasting, then, is the second fragmentation of the "natural blocks" and results in a new distribution. Under this foundation, Zou applies the theory of strain wave transportation to develop the BMMC computer program, which more profoundly reflects the process of rock blasting.

Having worked extensively in Hong Kong, Zou has encountered and embraced many opportunities to work with advanced equipment, materials, as well as new theoretical ideas and methods, all of which constitute the contents of this book. Hong Kong is a small city with a massive population, and with strict requirements on safety, environmental protection and sound construction management, this case study is introduced in the book as a valuable reference for similar conditions in mainland China.

Despite differences in regions and industries, individual words and terms not in line with the mainland can still be understood. Some terms like “chief shotfier” evoke vivid imagery, and may be accepted and popularized someday in the future.

China contains the largest quantity and scale of rock excavation in the world. I sincerely hope the young people who are engaging in rock excavation are enlightened by this book, contribute to the great rejuvenation of China and construct a more beautiful world.

China
May 2016

Professor Xiaohe Xu
Northeastern University

Preface

In order to develop and improve their living space and to get resources from the earth, ever since human entered a civilized society, surface and deep excavations have become an important part of civil engineering and mining, from the early simple manual excavations to modern sophisticated blasting and mechanical excavation mankind has experienced a long history of thousands of years, has accumulated a lot of knowledge. A variety of theories have been developed to explain the diversified issues appeared during the process of excavation with the development of the science and technology.

This book summarizes the technical progress and various theories on excavation in recent decades, developed by scholars of various countries, including the author of this book, major focus on rock excavation, which is more difficult and challenging compared to excavation of other material. Taking into account that mining engineering has more complex features, this book covers essentially the scope of civil engineering, while majority of the content is also applicable to mining engineering. As rock blasting is still the most important means of rock excavation, theories and technologies of rock blasting are particularly expounded in more detail than other excavation methods.

The author used to work in an underground mine for 11 years on various technical and management roles, worked in the field of research for another 11 years as well. During this period he got a chance to have a short term of study on mining and visit Sweden in 1984. Later he came to work in Hong Kong for over 25 years. A wealth of practical experience and knowledge has been accumulated during his long-term on-site technical work. Dedication to the research further developed and raised the author's theoretical level. His professional career in Hong Kong enables the author to have more accesses to the worldwide advanced technologies and theories in the field of rock excavation. The 50 years professional experience and accumulated knowledge has laid a solid foundation for this book.

In reading various publications in the field of rock excavation it is noted that the authors of the Europe and United States, including India, Japan and South Africa, rarely introduce the works and valuable theoretical contributions, which are still in application, made by Chinese and Russia (former Soviet Union) scholars.

By striking a balance, while introducing Europe and United States scholars' works, Chinese and Russia scholars' are mentioned as well.

The book is divided into three parts.

The first part is the basis. It includes basic knowledge and relevant theories on the rock and rock mass—the objectives of the excavation engineering, the basic knowledge, techniques and basic theories on the drilling and blasting and explosive materials.

The second part is on surface rock excavation. After introducing various methods of non-blast techniques, highlights of surface blasting, the application of computer simulation and CAD technique are illustrated in detail.

The third part is on underground rock excavation. Before comprehensively illustrating technology and theories on the drilling and blasting, ventilation, loading and haulage, ground reinforcement and support, computer application for underground excavation, varieties on mechanical excavation methods including tunneling, shaft and TBM technique are also described in detail.

Due to the complex working environment for rock excavation, especially when explosives are used for rock blasting, the comprehensive description of varieties on safety issues during excavation and necessary safety precautions and security measures are provided as much as possible in the book.

To be more practical, a variety of technical methods and data from various sources are provided in the book, making it a reference book covering both theoretical and practical applications.

The targeted audience of this book are engineers, researchers and academics engaged in rock excavation, but is equally applicable as a teaching reference for teachers and students in civil and mining engineering. The author sincerely hopes that this book could be of some help to readers.

At the age of seventy, the author spent three years to finish this book. It could be the last and one of the most important achievements in his whole life. It is hoped that the younger generation would share his experience and knowledge from this book.

The author received strong support and encouragement from the friends and colleagues during the writing process.

Special thanks are given to:

The younger schoolmate, Dr. Ruilin Yang (Principal Research Fellow in Orica), Mr. Archie Cheung (AMS Hong Kong), and Dr. Li Liang (Professor of Northeastern University, China). They helped the author in reviewing some chapters.

Mr. Guy Bridges, the Director of Aecom Asia, who reviewed the manuscript completely, gave valuable comments and helped me to improve the English of this book.

The staff of Atlas Copco and Sandvik who offered a lot of information and reviewed and improved the relevant contents in the book, and

My classmate and best friend, Prof. Baozhi Chen (former Dean of the School of Resources and Civil Engineering, Northeastern University, China) gave me active encouragement and strong support starting from the outset of the planning period of this book.

In the book the author shows his affection to his alma mater—Northeastern University in China. Some important achievements of the author's predecessors, which are yet glorious so far, are introduced in this book. Professor Xiaohe Xu, the supervisor of the author and a respected scholar, wrote the Second Foreword of the book, even at the age of over 80 years, which also greatly encouraged and inspired author.

The author is especially honored that the respectable academician, China Academy of Engineering, President of China Society of Explosives and Blasting, Prof. Xuguang Wang wrote the First Foreword for the book. The author would like to take this opportunity to express his sincere appreciation.

Hong Kong
July 2016

Dingxiang Zou

Acknowledgments

The author wishes to express his sincere gratitude to the following government departments, professional societies, manufacturing companies, publishers, and other professionals, as listed below, for their permission, facilities and release of technical data, release of literature, information, material and encouragement. I sincerely apologize if I missed anyone who also contributed for the publishing of this book.

The Head of the Geotechnical Engineering Office and the Director of the Civil Engineering and Development Department, the Government of the Hong Kong Special Administrative Region for the permission to publish the figures, tables and text concerned and give me a chance to introduce the successful experience to the worldwide on the geotechnical engineering and management of explosives and blasting works.

A.A. Balkema/Rotterdam
AECOM Asia Company Ltd.
Archie Mining Services Limited and Mr. Archie Cheung
Association of Geotechnical & Geoenvironmental Specialists (Hong Kong)
American Rock Mechanics Association
Atlas Copco
Australian Centre for Geomechanics (ACG)
Australasian Tunnelling Society
BARTZ
Beijing University of Science and Technology, Prof. Qu, Shijie
Balavto Ltd. Ajdovščina
BASF Construction Chemicals Europe AG
Canadian Institute of Mining, Metallurgy and Petroleum
Cardox International Limited
CFI Technologies LLC
China University of Geosciences
China Engineering Consultants, Inc., Taiwan

City University of Hong Kong and Prof. Raymond Wong
Controlled Blasting Solutions Ltd.
Durham Geo Slope Indicator (DGSI)
Dyno Nobel Asia Pacific PTY LTD
Editorial Office of Journal of Rock Mechanics and Geotechnical Engineering (JRMGE)
Electronic Journal of Geotechnical Engineering
ELSEVIER Inc.
EPC-FRANCE
European Federation of Explosives Engineers (EFEE)
Herrenknecht GmbH
ILF Consulting Engineers, Austria
Institute of Rock and Soil Mechanics Chinese Academy of Sciences and JRMGE
Institution of Mechanical Engineers
International Tunnelling Association (AITES/ITA)
International Society of Explosives Engineers
John Wiley and Sons Inc.
LLC Blast Maker
Luleå University of Technology
Mr. Partha Das Sharma and the Weblog www.miningandblasting.wordpress.com
Northeastern University, China
Orica and Dr. Yang, Ruilin
Oslo Norwegian Group for Rock Mechanics
Product Technology Ltd.
Rock Drill
Sandvik and Tamrock
Scientific Research Publishing Inc. and Mr. Bijan Ehsanzadeh
Sinosteel Maanshan Institute of Mining Research Co., Ltd.
Sing Tao Publishing Group Ltd.
SISGEO S.r.l.
Southwest Jiaotong University, China and Prof. J.C. Zhang
Swedish Detonic Research Foundation (SveDeFo)
Taylor & Francis Group
The Australasian Institute of Mining and Metallurgy (The AusIMM)
The Institute of Materials, Minerals and Mining, UK
The Norwegian Tunnelling Society
Tunnelseis
Tunnel Talk direct by design
Metallurgy Industry Press 冶金工业出版社
Science Press 科学出版社
National Defense Industry Press 国防工业出版社

Ordnance Industry Press 兵器工业出版社
China Railway Press 中国铁道出版社
«Engineering Blasting» «工程爆破»
«Blasting» «爆破»
Tsinghua University Press 清华大学出版社

Contents

Part I
Basics

Chapter 1
Geology

1.1 Categories of Rock

The word "category" instead of "classification" which is mostly used in some books and articles is used for distinguishing different types of rocks in the Earth in this section. In Sect. 1.5, the word "classification" will be used to classify the rocks into different classes qualitatively and quantitatively according to their different characters and performances during the excavating process.

It is necessary and very useful for an engineer working in the field of rock excavation to understand some basic knowledge about the main categories of different rock types. It has more than academic interest. It is assist in understanding field conditions and dealing with them in the most effective manner.

The three basic rock types are given as:

- Igneous.
- Sedimentary.
- Metamorphic.

1.1.1 Igneous Rocks

Igneous are the parents of all other rocks. They are formed from the cooling and hardening of molten rock deep inside the crust of the Earth. Magma (rock in a molten form) at high temperature is usually charged with varying amounts of liquids and gases which are mostly dissipated when the rock cools and solidifies. That molten rock is called *magma rock or magma* (Fig. 1.1).

When magma is forced upward out of the original magma pool, but does not pour out onto the ground surface, it will slowly cool in the new location. It is called intrusive rock because it has been intruded into the new location. Sometimes the

D. Zou, *Theory and Technology of Rock Excavation for Civil Engineering*,
DOI 10.1007/978-981-10-1989-0_1

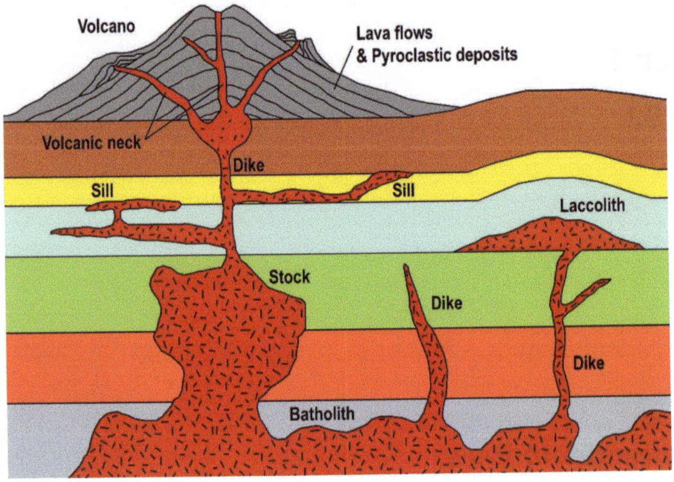

Fig. 1.1 Schematic presentation of igneous rocks in the Earth's crust

intrusive rock may be extremely large, extending for hundreds of miles. Such a rock mass may be the core of a mountain range. A huge mass of intrusive rock masses is called *batholiths*.

Intrusive igneous rocks cool slowly, producing a coarse texture with mineral grains visible to the naked eye. The minerals that form are determined by the chemistry of the magma and the way that it cools (relatively slowly or quickly, steadily, or variably). The grains are typically interlocking and of more-or-less the same size.

Dikes are tabular igneous bodies formed vertically or across sedimentary bedding. Those formed horizontally or parallel to beddings are called *sills*.

When magma pours out onto the Earth's surface, it cools, solidifies, and forms an *extrusive rock* or *lava* (sometimes also called *volcanic*). Extrusive igneous rocks cool quickly, which causes very small crystals to form, if any at all. This produces fine-grained rocks, which, without a microscope, can be identified only by color. Like the intrusive rocks, the minerals formed reflect the chemistry of the magma. Colors vary from white to black, with pink, tan, and gray being common intermediate colors. The texture of these rocks can also be influenced by the amount of gas trapped in the lava when it cools.

If the presence of SiO_2 content exceeds 62 %, the rock is considered as acidic (Felsic); from 62 to 52 % intermediate; from 52 to 45 % basic (Mafic); and <45 % ultra-basic (Ultramafic). The acid rocks, such as granite, are more abrasive and harder than the basic rocks, for example basalt, but they are more resistant to the impact.

Fig. 1.2 Photographs of igneous rocks (Courtesy of China University of Geosciences)

The following are some very popular igneous rocks (Fig. 1.2):
Intrusive Igneous Rocks:

- Granite. Granite is a common type of intrusive, acidic, igneous rock which is granular and phaneritic in texture. This rock consists mainly of quartz, mica, and feldspar. It is typically hard and strong and is often used for aggregates or building stone.
- Diorite. Diorite is a gray-to-dark gray intermediate intrusive igneous rock composed principally of plagioclase feldspar (typically andesine), biotite, hornblende, and/or pyroxene. It may contain small amounts of quartz, microcline, and olivine.
- Gabbro. Gabbro is also similar in some respects to granite and diorite, but is the next step toward a higher content of the ferromagnesian minerals. In gabbro, these dark minerals are dominant; in diorite, the feldspar is dominant.

Extrusive Igneous Rocks:

- Rhyolite. Rhyolite is a light-colored, fine-grained, extrusive igneous rock that typically contains quartz and feldspar minerals.
- Andesite. Andesite is an extrusive igneous and fine-grained volcanic rock like rhyolite except that is darker in color, and the chief feldspar component is plagioclase rather than orthoclase.
- Basalt. Basalt is a common extrusive igneous (volcanic) rock formed from the rapid cooling of basaltic lava exposed at or very near the surface of a Planet or Moon. Basalt is a fine-grained rock with a composition like that of gabbro, getting its dark color from the ferromagnesian minerals iron and magnesium.
- Tuff and Volcanic Breccia. These are formed from materials which have been ejected from volcanoes and solidified into rock. Tuff is composed of ash and the finer particles. Breccia is formed of the larger particles.

1.1.2 Sedimentary Rocks

Sedimentary rocks are formed by the deposition of material at the Earth's surface and within bodies of water. Sedimentation is the collective name for processes that cause mineral and/or organic particles (detritus) to settle and accumulate or minerals to precipitate from a solution. Particles that form a sedimentary rock by accumulating are called sediments.

Before being deposited, sediment was formed by weathering and erosion in a source area and then transported to the place of deposition by water, wind, ice, mass movement, or glaciers which are called agents of denudation. Sediments are transformed into rock by cementing them, usually by calcite, silica, or iron oxides that glue the fragments together. Compaction is achieved when fragments are squashed together (Fig. 1.3).

Fig. 1.3 Forming procedures of sedimentary rocks

Fig. 1.4 Sedimentary rocks (Courtesy of China University of Geosciences)

Sedimentary rocks are deposited in layers as strata, forming a structure called bedding. Their texture ranges from very fine grained to very coarse. Colors include red, brown, gray, yellow, pink, black, green, and purple.

The following are some very common sedimentary rocks (Fig. 1.4):

- Conglomerate. Conglomerate is a clastic sedimentary rock that contains large (greater than 2 mm in diameter) rounded clasts. The space between the clasts is generally filled with smaller particles and/or chemical cement that bind(s) the rock together.
- Breccia. Breccia is a term most often used for clastic sedimentary rocks that are composed of large angular fragments (over 2 mm in diameter). The spaces between the large angular fragments can be filled with a matrix of smaller particles or mineral cement that binds the rock together.
- Sandstone. Sandstone is a sedimentary rock composed of sand-sized grains of mineral, rock, or organic material. It also contains a cementing material that binds the sand grains together and may contain a matrix of silt- or clay-sized particles that occupy the spaces between the sand grains.

- Shale. Shale is a fine-grained sedimentary rock that is formed from the compaction of silt and clay-size mineral particles that we commonly call "mud." This composition places shale in a category of sedimentary rocks known as "mudstones." Shale is distinguished from other mudstones because it is fissile and laminated. "Laminated" means that the rock is made up of many thin layers. "Fissile" means that the rock readily splits into thin pieces along the laminations.
- Limestone. Limestone is a sedimentary rock composed primarily of calcium carbonate ($CaCO_3$) in the form of the mineral calcite. It most commonly forms in clear, warm, shallow marine waters. It is usually an organic sedimentary rock that forms from the accumulation of shell, coral, algal, and fecal debris. It can also be a chemical sedimentary rock formed by the precipitation of calcium carbonate from lake or ocean water.
- Dolomite. Dolomite, also known as dolostone and dolomite rock, is a sedimentary rock composed primarily of the mineral dolomite, $CaMg(CO_2)_3$. Dolomite is found in sedimentary basins worldwide. It is thought to form by the postdepositional alteration of lime mud and limestone by magnesium-rich groundwater.
- Iron ore. Earth's most important iron ore deposits are found in sedimentary rocks. They formed from chemical reactions that combined iron and oxygen in marine and freshwaters. The two most important minerals in these deposits are hematite (Fe_2O_3) and magnetite (Fe_3O_4) iron oxides.
- Coal. Coal is an organic sedimentary rock that forms from the accumulation and preservation of plant materials, usually in a swamp environment. Coal is a combustible rock and along with oil and natural gas it is one of the three most important fossil fuels.
- Mudstone. Mudstone (also called mudrock) is a fine-grained sedimentary rock whose original constituents were clays or muds. Grain size is up to 0.0625 mm (0.0025 in.) with individual grains too small to be distinguished without a microscope.

1.1.3 Metamorphic Rocks

Metamorphic rocks arise from the transformation of existing rock types, in a process called metamorphism, which means "change in form." The original rock (protolith) is subjected to heat (temperatures greater than 150–200 °C) and pressure (1500 bars), causing profound physical and/or chemical change. The protolith may be sedimentary rock, igneous rock, or another older metamorphic rock (Fig. 1.5).

The following are some very common metamorphic rocks (Fig. 1.6):

- Gneiss. Gneiss is a common and widely distributed type of rock formed by high-grade regional metamorphic processes from pre-existing formations that were originally either igneous or sedimentary rocks. It is foliated (composed of

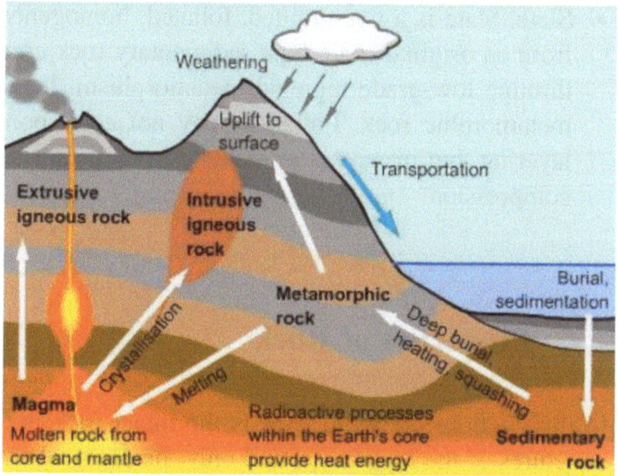

Fig. 1.5 Rock cycle and metamorphic rocks

Fig. 1.6 Metamorphic rocks (Courtesy of China University of Geosciences)

layers of sheet-like planar structures). The foliations are characterized by alternating darker and lighter colored bands, called "gneissic banding."

- Quartzite. Quartzite is a hard, non-foliated metamorphic rock which was originally pure quartz sandstone. Sandstone is converted into quartzite through heating and pressure usually related to tectonic compression within orogenic belts. Pure quartzite is usually white to gray, though quartzites often occur in various shades of pink and red due to varying amounts of iron oxide (Fe_2O_3). Other colors, such as yellow and orange, are due to other mineral impurities.
- Marble. Marble is a metamorphic rock that forms when limestone is subjected to the heat and pressure of metamorphism. It is composed primarily of the mineral calcite ($CaCO_3$) and usually contains other minerals such as clay minerals, micas, quartz, pyrite, iron oxides, and graphite. Under the conditions of metamorphism, the calcite in the limestone recrystallizes to form a rock that is a mass of interlocking calcite crystals. A related rock, dolomitic marble, is produced when dolostone is subjected to heat and pressure.

- Slate. Slate is a fine-grained, foliated, homogeneous metamorphic rock derived from an original shale-type sedimentary rock composed of clay or volcanic ash through low-grade regional metamorphism. It is the most fine-grained foliated metamorphic rock. Foliation may not correspond to the original sedimentary layering, but instead is in planes perpendicular to the direction of metamorphic compression.

1.2 Properties of Rock

In this section, the word rock is the intact rock portion excluding any geological structures of the rock mass, and only the physical and mechanical properties of the rock are discussed.

The main physical properties of rock include density, porosity, hardness, abrasivity, permeability, and wave velocity.

- Density. Density is a measure of mass per unit of volume. Density of rock material varies and often related to the porosity of the rock. It is sometimes defined by the unit weight and specific gravity. Most rocks have density between 2500 and 2800 kg/m^3.
- Porosity. Porosity describes how densely the material is packed. It is the ratio of the non-solid volume to the total volume of material. The value is typically ranging from less than 0.01 for solid granite to up to 0.5 for porous sandstones. It may also be represented in percent terms by multiplying the fraction by 100 %.
- Hardness. Hardness is the characteristic of a solid material expressing its resistance to permanent deformation. Hardness of a rock material depends on several factors, including mineral composition and density. A typical measure is the Schmidt rebound hardness number (Fig. 1.7).

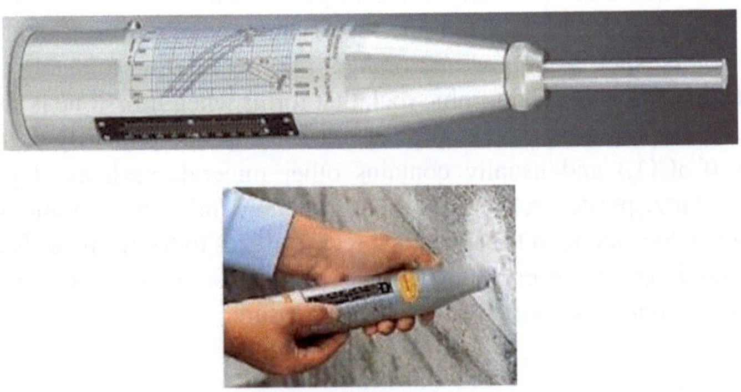

Fig. 1.7 Schmidt Hammer rebound hardness test (*Source* [10])

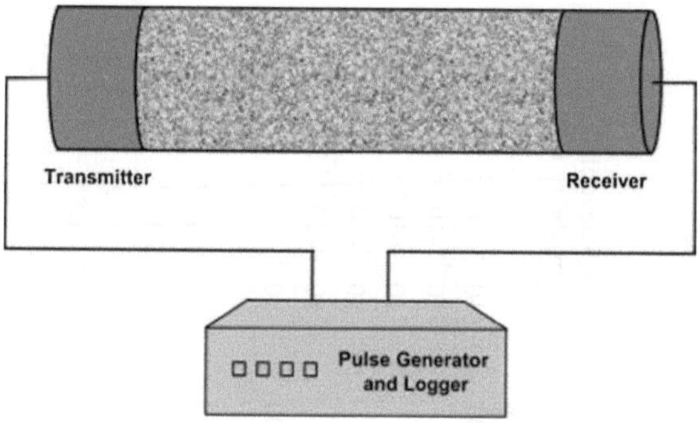

Fig. 1.8 Wave velocity measurement

- Abrasivity. Abrasivity measures the abrasiveness of a rock material against other materials, e.g., steel. It is an important measure for estimate wear of rock drilling and boring equipment. Abrasivity is highly influenced by the amount of quartz mineral in the rock material. The higher quartz content gives higher abrasivity. Abrasivity measures are given by several tests. Cerchar and other abrasivity tests are described later.
- Permeability. Permeability is a measure of the ability of a material to transmit fluids. Most rocks generally have very low permeability. Permeability of rock material is governed by porosity.
- Wave velocity. Measurement of wave velocity is often done by using *P*-waves and sometimes, *S*-waves (Fig. 1.8). The velocity measurements provide correlation to physical properties in terms of compaction degree of the material. Wave velocities are also commonly used to assess the degree of rock mass fracturing at large scale, using the same principle, and it will be discussed in a later section of the book.

The following table gives some typical values of the physical properties of some popular rocks (Table 1.1):

The main mechanical properties of rock material include compressive strength, tensile strength, shear strength, Young's Modulus, Poisson's ratio, and other engineering properties of rock materials.

- Compressive strength. Compressive strength is the capacity of a material to withstand compressive forces. The most common measure of compressive strength is the uniaxial compressive strength or unconfined compressive strength. Usually compressive strength of rock is defined by the ultimate stress. It is one of the most important mechanical properties of rock material, used in design, analysis, and modeling.
- Young's modulus and Poisson's ratio. Young's Modulus is the modulus of elasticity measuring of the stiffness of a rock material. It is defined as the ratio, for small strains of the rate of change of stress with strain. Similar to strength, Young's Modulus of rock materials varies widely with rock type.

Table 1.1 Physical properties of some fresh rock materials

Rock	Dry density (g/cm^3)	Porosity (%)	Schmidt Hardness Index	Cerchar Abrasivity Index	*P*-wave velocity (m/s)	*S*-wave velocity (m/s)	Coefficient of permeability (m/s)
Igneous							
Granite	2.53–2.62	1.02–2.87	54–69	4.5–5.3	4500–6500	3500–3800	10^{-14}–10^{-12}
Diorite	2.80–3.00	0.10–0.50		4.2–5.0	4500–6700		10^{-14}–10^{-12}
Gabbro	2.72–3.00	1.00–3.57		3.7–4.6	4500–7000		10^{-14}–10^{-12}
Rhyolite	2.40–2.60	0.40–4.00					10^{-14}–10^{-12}
Andesite	2.50–2.80	0.20–8.00	67	2.7–3.8	4500–6500		10^{-14}–10^{-12}
Basalt	2.21–2.77	0.22–22.1	61	2.0–3.5	5000–7000	3660–3700	10^{-14}–10^{-12}
Sedimentary							
Conglomerate	2.47–2.76			1.5–3.8			10^{-10}–10^{-8}
Sandstone	1.91–2.58	1.62–26.4	10–37	1.5–4.2	1500–4600		10^{-10}–10^{-8}
Shale	2.00–2.40	20.0–50.0		0.6–1.8	2000–4600		
Mudstone	1.82–2.72		27				10^{-11}–10^{-9}
Dolomite	2.20–2.70	0.20–4.00			5500		10^{-12}–10^{-11}
Limestone	2.67–2.72	0.27–4.10	35–51	1.0–2.5	3500–6500		10^{-13}–10^{-10}
Metamorphic							
Gneiss	2.61–3.12	0.32–1.16	49	3.5–5.3	5000–75000		10^{-14}–10^{-12}
Schist	2.60–2.85	10.0–30.0	31	2.2–4.5	6100–6700	3460–4000	10^{-11}–10^{-8}
Phyllite	2.18–3.30						
Slate	2.71–2.78	1.84–3.64		2.3–4.2			10^{-14}–10^{-12}
Marble	2.51–2.86	0.65–0.81			5000–6000		10^{-14}–10^{-11}
Quartzite	2.61–2.67	0.40–0.65		4.3–5.9			10^{-14}–10^{-13}

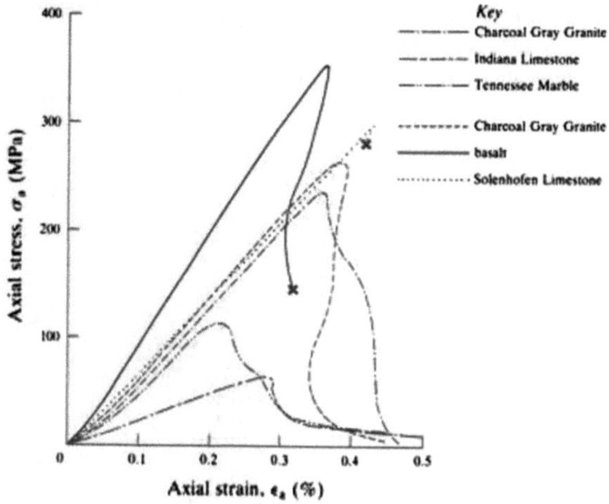

Fig. 1.9 Stress–strain curves of several rocks (*Source* [10])

Poisson's ratio measures the ratio of lateral strain to axial strain, at linearly elastic region. For most rocks, the Poisson's ratio is between 0.15 and 0.4.

- Strain at failure. Strain at failure is the strain measured at ultimate stress. Rock generally fails at a small strain, typically around 0.2–0.4 % under uniaxial compression.

  Brittle rocks, typically crystalline rocks, have low strain at failure, while soft rock, such as shale and mudstone, could have relatively high strain at failure. Strain at failure sometimes is used as a measure of brittleness of the rock. Figure 1.9 shows the complete stress–strain curves of several rocks [10].
- Tensile strength. Tensile strength of rock material is normally defined by the ultimate strength in tension; i.e., the maximum the tensile stress, the rock material can withstand. Rock material generally has a low tensile strength due to the existence of microcracks in the rock. The existence of microcracks may also be the cause of rock failure suddenly in tension with a small stress. The most common tensile strength determination is by the Brazilian test (Fig. 1.10). Tensile strength is calculated from failure load (P), specimen diameter (D), and the specimen thickness (t) by the following formula:

$$\sigma_t = \frac{0.636P}{Dt}$$

- Shear Strength. Shear strength is used to describe the strength of rock materials, to resist deformation due to shear stress. Rock resists shear stress by two internal mechanisms, cohesion and internal friction. Cohesion is a measure of internal bonding of rock material. Internal friction is caused by contact between particles and is defined by the internal friction angle. Shear strength of rock material can be determined by direct shear test and by triaxial compression tests (Fig. 1.11).

Fig. 1.10 Brazilian *tensile test* (*Source* www.ibf.kit.edu/img/brazilian.jpg)

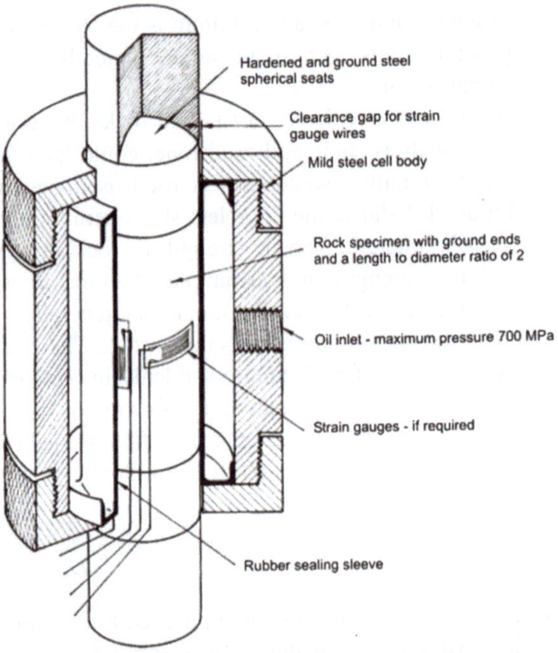

Fig. 1.11 Triaxial compression test (*Source* http://inside.mines.edu/EMI-Physical-Property-Measurements)

Tensile and shear strengths are important as rock fails mostly in tension and in shearing, even when the loading may appears to be compression. Rock generally has high compressive strength so failure in pure compression is not common unless in areas of high ground stress.

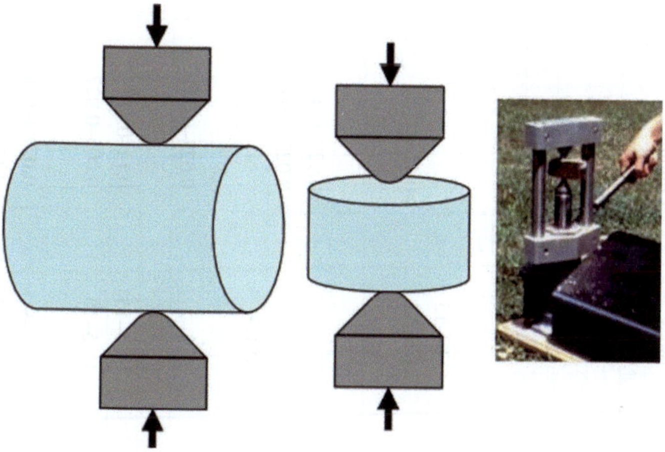

Fig. 1.12 Point loading index test (*Source* [10])

- Other Engineering Properties of Rock Materials
  - Point load strength index—a simple index test for rock material. It given the standard point load index, $I_{s(50)}$, calculated from the point load at failure and the size of the specimen, with size correction to an equivalent core diameter of 50 (Fig. 1.12).
  - Fracture toughness—it measures the effectiveness of rock fracturing.
  - Brittleness.
  - Indentation.
  - Swelling—some rocks swell when they are situated with water.

The following table gives some typical values of the mechanical and engineering properties of some common rocks (Table 1.2):

1.3 Geological Structures of Rock Mass

The practices of rock excavation show us the effects of the geological structures of rock mass are more important than the rock properties themselves to the excavation process and the stability of excavation space during excavation and afterward.

There are three types of geological structures:

(1) Primary structures: those which develop at the time of formation of rock (e.g., sedimentary structures and some volcanic structures).

(2) Secondary structures: which are those that develop in rocks after their formation as a result of their subjection to external forces within the Earth's crust. There may be a single force, or combination of forces resulting in ground stresses, tectonic forces, hydrostatic forces, pore pressures, and temperatures

Table 1.2 Mechanical and engineering properties of rock materials

Rock	UC strength (MPa)	Tensile strength (MPa)	Elastic modulus (GPa)	Poisson's ratio	Strain at failure (%)	Point Load Index $I_{S(50)}$(MPa)	Fracture mode *I* toughness
Igneous							
Granite	100–300	7–25	30–70	0.17	0.25	5–15	0.11–0.41
Diorite	100–350	7–30	30–100	0.10–0.20	0.30		>0.41
Gabbro	150–250	7–30	40–100	0.2–0.35	0.30	6–15	>0.41
Rhyolite	80–160	5–10	10–50	0.2–0.4			
Andesite	100–300	5–15	10–70	0.2		10–15	
Basalt	100–350	10–30	40–80	0.1–0.2	0.35	9–15	>0.41
Sedimentary							
Conglomerate	30–230	3–10	10–90	0.1–0.15	0.16		
Sandstone	20–170	4–25	15–50	0.14	0.20	1–8	0.027–0.041
Shale	5–100	2–10	5–30	0.1			0.027–0.041
Mudstone	10–100	5–30	5–70	0.15	0.15	0.1–6	
Dolomite	20–120	6–15	30–70	0.15	0.17		
Limestone	30–250	6–25	20–70	0.30		3–7	0.027–0.041
Metamorphic							
Gneiss	100–250	7–20	30–80	0.24	0.12	5–15	0.11–0.41
Schist	70–150	4–10	5–60	0.15–0.25		5–10	0.005–0.027
Phyllite	5–150	6–20	10–85	0.26			
Slate	50–180	7–20	20–90	0.20–0.30	0.35	1–9	0.027–0.041
Marble	50–200	7–20	30–70	0.15–0.30	0.40	4–12	0.11–0.41
Quartzite	150–300	5–20	50–90	0.17	0.20	5–15	>0.41

stresses. As a result of these forces and their magnitude, rocks are continuously undergoing varying degrees of deformation, resulting in the formation of different kinds of structural features.

(3) Compound structures: form by a combination of events some of which are contemporaneous with the formation of a group of rocks taking part in these "structures." Usually, these kinds of structures present a kind of unconformity which is a contact between two rock units in which the upper unit is usually much younger than the lower unit.

In this section, we will discuss the following four major geological structures:

- Folds,
- Faults,
- Discontinuities: bedding planes, joints and fractures, and
- The compound structures—unconformities as well.

1.3.1 Folds

Folds are bends or flexures in the Earth's crust (see Photographs 1.5 and 1.6) and can therefore be identified by a change in the amount and/or direction of dip of rock units.

Most folds result from the ductile deformation of rocks when subject to compression or shear stress. A fold convex upward is anticline, and the one convex downward is syncline (Fig. 1.9). The extent of folding and its ultimate shape depends on the intensity and duration of internal forces, as well as the properties of the rock material. Rock folds may develop in any rock type; however, they are more common in sedimentary and igneous rocks. Following Fig. 1.13 shows various types of folds (Fig. 1.14):

Fig. 1.13 Folds (image from Wikipedia)

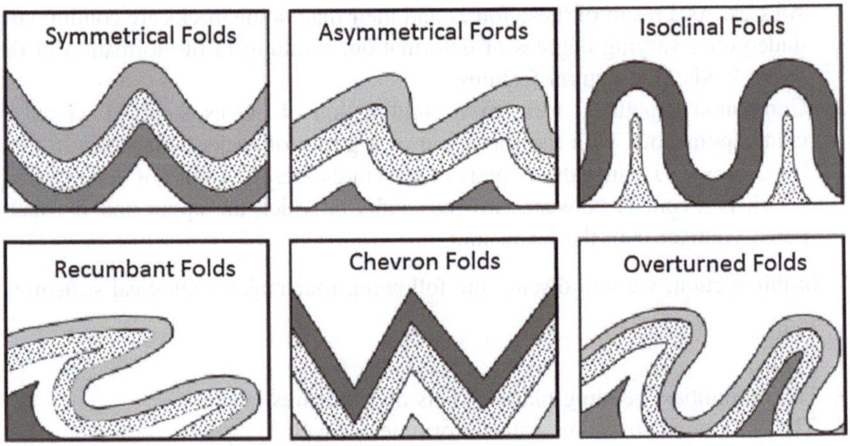

Fig. 1.14 Various types of folds

1.3.2 Faults

Faults are fractures in crustal strata along which rocks have been displaced (Fig. 1.15). The amount of displacement may vary from only a few tens of millimeters to several 100 km. In many faults, the fracture is a clean break; in others,

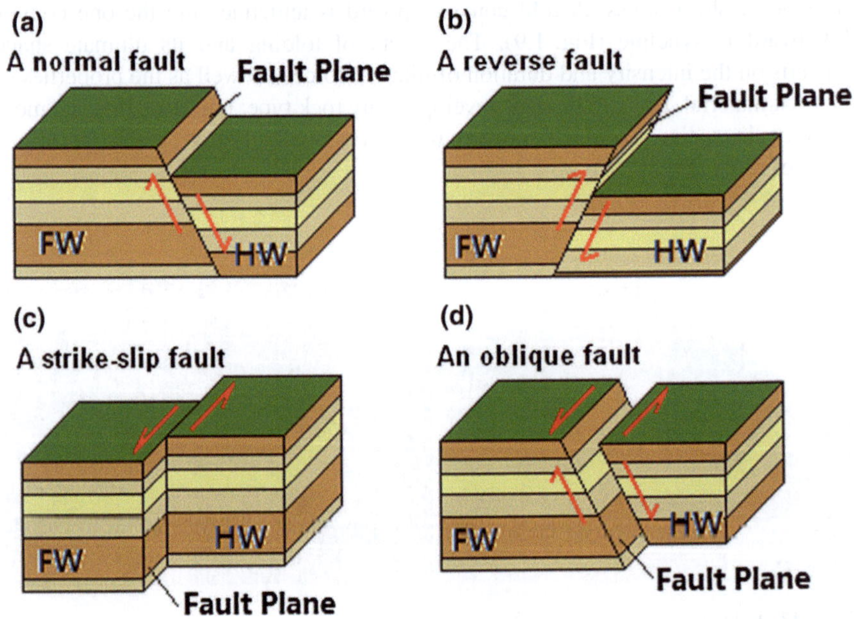

Fig. 1.15 Types of faults: **a** normal fault, **b** reverse fault, **c** wrench or strike-slip fault, **d** oblique-slip fault. *FW* footwall; *HW* hanging wall

Fig. 1.16 Fault and fault zone

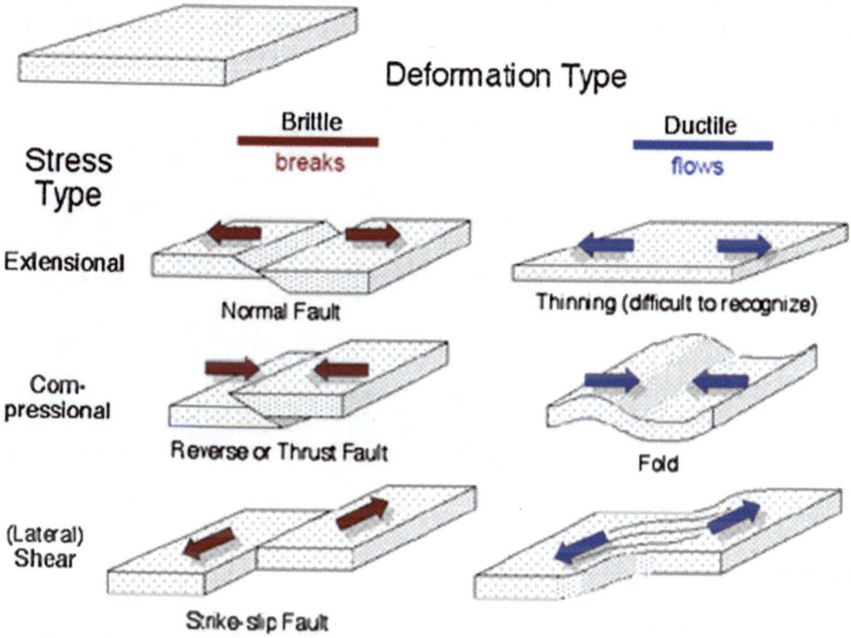

Fig. 1.17 Faults and folds and how do they form

the displacement is not restricted to a simple fracture, but is developed throughout a fault zone (Fig. 1.16).

Unlike folds which form predominantly by compression stress, faults result from either tension, compression, or shear (Fig. 1.17).

1.3.3 *Discontinuities: Bedding Planes, Joints, and Fractures*

A discontinuity in geotechnical engineering is a plane or surface that marks a change in physical or chemical characteristics in a soil or rock mass. A discontinuity can be, for example, a bedding, schistosity, foliation, joint, cleavage, fracture, fissure, crack, or fault plane. Discontinuities vary in size from small fissures to huge faults.

- Bedding Planes

Bedding in rock forms from the spatial position of rocks in the Earth's crust. Sedimentary and metamorphic rocks usually occur in the form of layers of strata bounded by roughly parallel surfaces. In their original, undisturbed bedding sedimentary rocks are arranged almost horizontally (see Figs. 1.18, 1.19); less

Fig. 1.18 Bedding in rock

Fig. 1.19 Eroded and cut bedding rock by water—Zhangjiajie World Geopark

frequently they have an initial dip in one direction or flexures caused by the relief of the surface on which they were deposited. Disturbances of the initial bedding of rocks or their dislocation result from two types of causes: endogenic, including tectonic movements, and exogenic—the action of surface and, especially, groundwaters that causes landslides, avalanches, and dissolving of rocks.

- Joints and fractures

Joints are fractures along which little or no displacement has occurred and present within all types of rocks. A group of joints that run parallel to each other are termed a joint set, whereas two or more joint sets that intersect at a more or less constant angle are referred to as a joint system. If one set of joints is dominant, then the joints are known as primary joints, and the other set or sets of joints are termed secondary.

Types with respect to formation

- Tectonic joints (Fig. 1.20)

Tectonic joints are formed during deformation episodes whenever the differential stress is high enough to induce tensile failure of the rock, irrespective of the *tectonic* regime. They will often form at the same time as *faults*. Measurement of tectonic joint patterns can be useful in analyzing the tectonic history of an area because they give information on stress orientations at the time of formation.

Fig. 1.20 Tectonic joints

Fig. 1.21 Sheet joints in granite

Fig. 1.22 Column joints of basalt in Hong Kong National Geopark

- Unloading joints (Release joints)—Sheet joints (Fig. 1.21)

Joints are most commonly formed when uplift and erosion remove the overlying rocks, thereby reducing the compressive load and allowing the rock to expand laterally. Joints related to uplift and erosional unloading have orientations reflecting the principal stresses during the uplift. Care needs to be taken when attempting to understand past tectonic stresses to discriminate, if possible, between tectonic and unloading joints.

- Cooling joints—Columnar Joints (Fig. 1.22)

Joints can also form via cooling of hot rock masses, particularly lava, forming *cooling joints*, most commonly expressed as vertical *columnar jointing*. The joint systems associated with cooling typically are polygonal because the cooling introduces stresses that are isotropic in the plane of the layer.

1.3.4 *The Compound Structures—Unconformities*

An unconformity is a surface (or contact) along which there was no fracturing (i.e., not a fault or joint) and which represents a break in the geologic record. An unconformity therefore indicates a lack of continuity of sedimentary deposition in an area, resulting in rocks of widely different ages occurring in contact with each other. In many cases (Fig. 1.23), unconformities represent a buried erosional surface. In such cases, erosion of the older units results in their fragmentation into smaller pieces. As soon as the deposition resumes, these fragments may consolidate to form a rock known as breccia (if the fragments are angular) or conglomerate (if the fragments are rounded). Because the breccia or conglomerate occurs at the base of the younger units lying on top of the unconformity surface, and because their fragments are derived from the units below this surface, the conglomerates or breccias are known as basal conglomerates or basal breccias.

1.3.5 *Geometric Representation of Structural Elements*

1.3.5.1 Definition of Geological Terms

There are two ways to define a geological plane (refer to Fig. 1.24): *Strike/Dip* and *Dip Direction/Dip.*

Strike of a plane is the trace of the intersection of that plane and a horizontal surface.

Fig. 1.23 Compound structure—unconformities

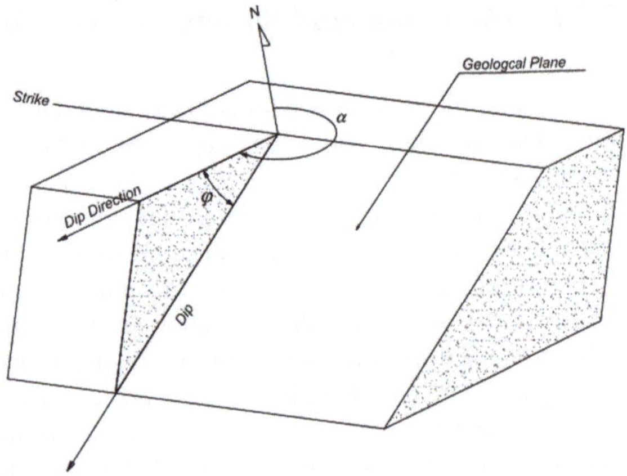

Fig. 1.24 Definition of geological terms

Dip of a plane is the angle (φ) between that plane and a horizontal surface.

Dip Direction is defined by the direction of *dip* with respect to north (α).

Geotechnical engineers, particularly those who make extensive use of computers in their analyses, have trend to use dip direction in preference to strike as means of defining the orientation of planes. If the dip direction and dip of a plane are recorded as 240/20, this is more concise than that for strike and dip, N30W/20SW, especially when processing large number of geological data by computer.

1.3.5.2 Field Survey of a Geological Plane

The device usually used for measuring geological planes is a compass clinometer. Figures 1.25, 1.26, and 1.27 show the mostly used two types of them, and how to use it to measure the attitude of a geological plane.

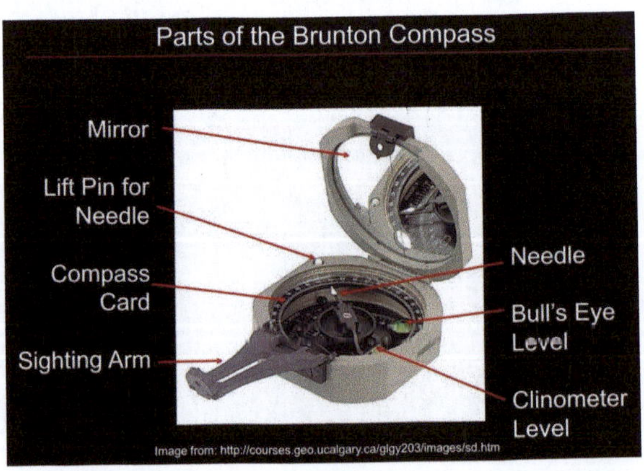

Fig. 1.25 Structure of the Brunton compass

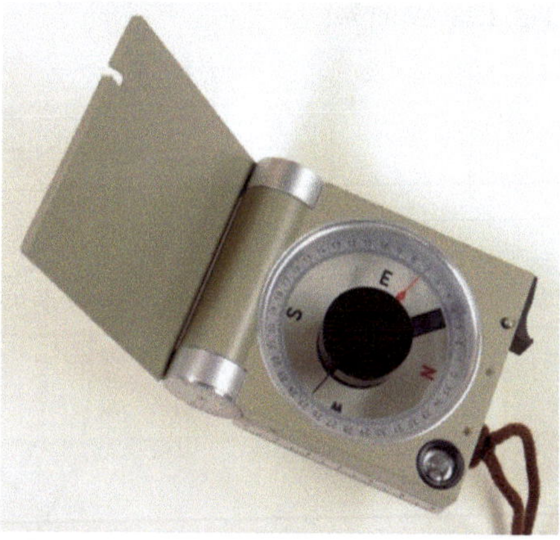

Fig. 1.26 A compass clinometer

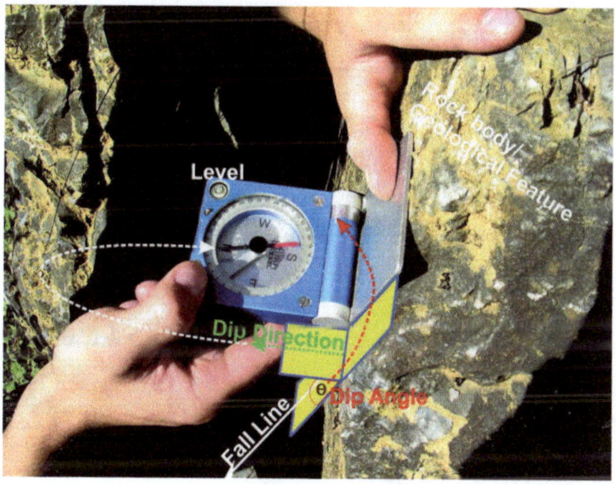

Fig. 1.27 Measure the attitude of a geological plane

1.3.5.3 Stereographic Projection

Data from a discontinuity survey are usually plotted on a stereographic projection. The use of Spherical Projections, Commonly the Schmidt or Wulff net, an equal angle projection (see Fig. 1.28), means that traces of the planes on the surface of the "reference sphere" can be used to define the dips and dip directions of discontinuity planes. In other words, the inclination and orientation of a particular plane is represented by a great circle or a pole (see Figs. 1.29, 1.30), normal to the plane, which are traced on an overlay placed over the stereonet (see Fig. 1.31). A detailed description for the method whereby great circles or poles are plotted on a stereogram has been explained in some books [1–3]. When recording field observations of the directions and amount of dip of discontinuities, it is convenient to plot the poles rather than the great circles. The poles then can be contoured in order to provide an expression of orientation concentration. This affords a quantitative

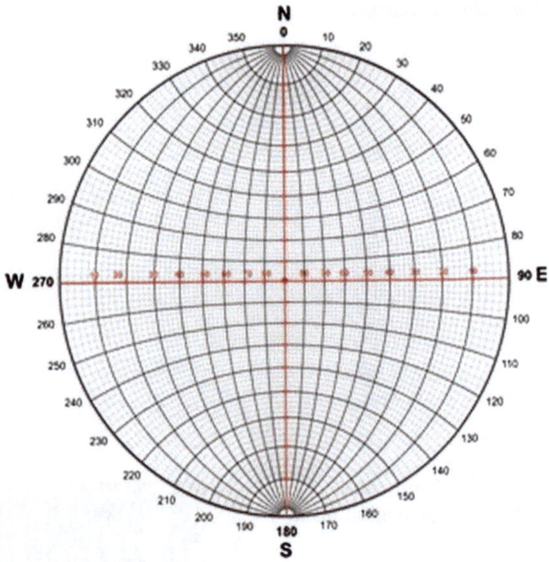

Fig. 1.28 Wulff net

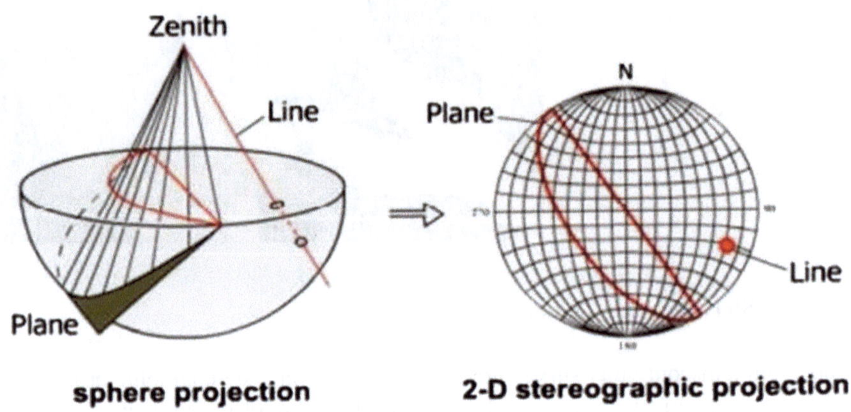

Fig. 1.29 Stereographic projection

appraisal of the influence of the discontinuities on the engineering behavior of the rock mass concerned (Fig. 1.32).

Stereographic projection is made fast and easy through computer software; analyzing data is done much faster than by hand. Several different stereonet softwares are available from the Web, some for a fee and others for free. Some commonly used softwares are WinWulff (JCrystalSoft), OSXstereo (by Nestor Cardozo), OpenStereo (by Jorge), Stereonet (by Rick Alimendinger), dips (by RocSȧence), etc.

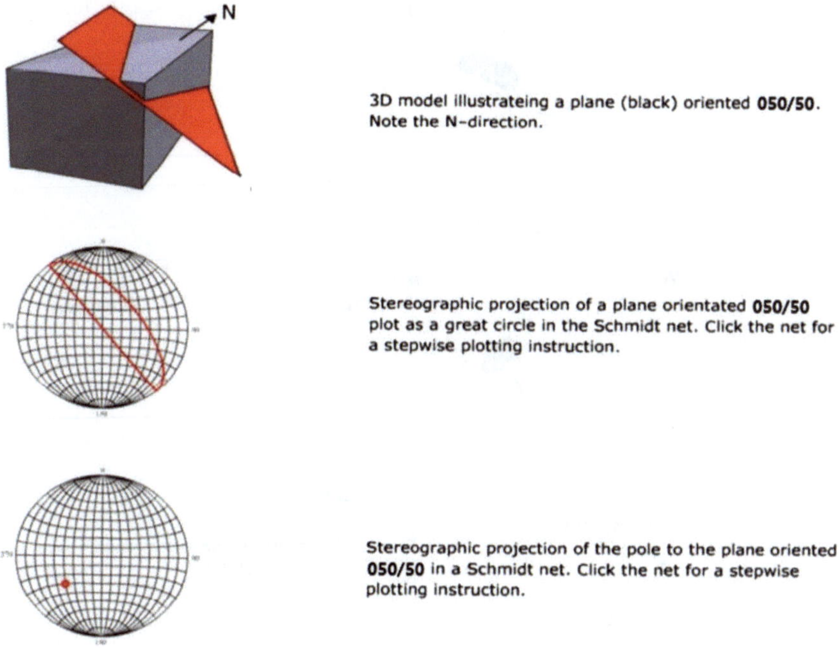

Fig. 1.30 A particular plane is represented by a great *circle* or a *pole*

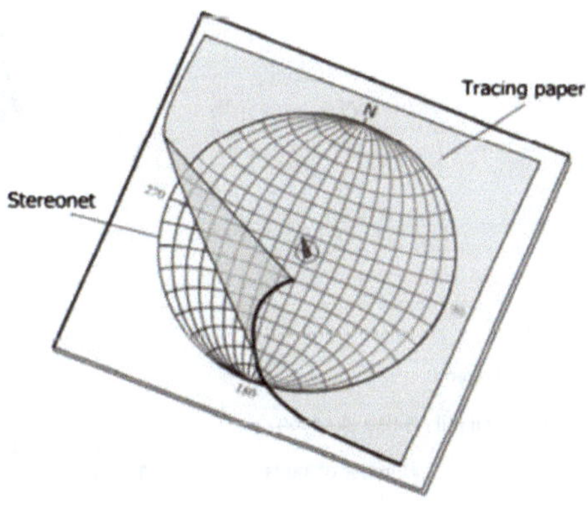

Fig. 1.31 Traced on an overlay placed over the stereonet

1.4 Properties of Rock Mass and Their Effects to Rock Excavation

Rock mass property is governed by the properties of intact rock materials and of the discontinuities in the rock mass. The behavior of rock mass is also influenced by the conditions that the rock mass is subjected to, primarily, the in situ stress and groundwater.

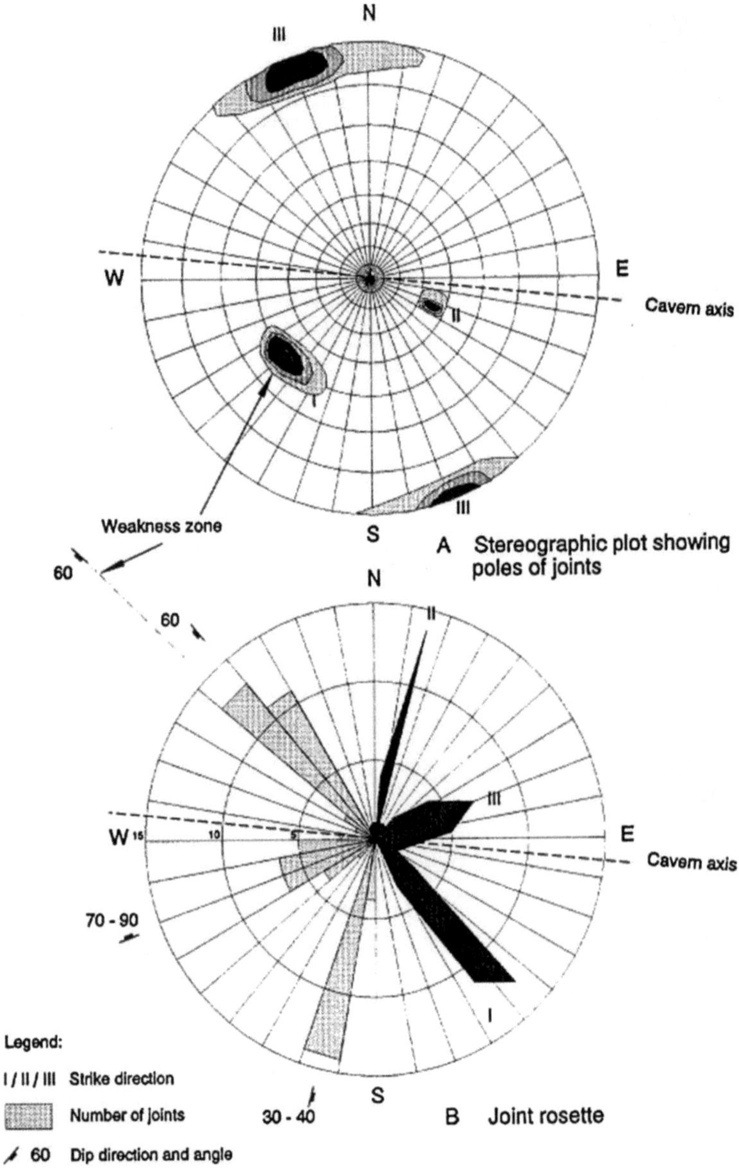

Fig. 1.32 Polar stereographic projection and joint rosette (Reprinted from ref. [4] with the permission of GEO, CEDD, HK)

Rock mass is a matrix consisting of rock material and discontinuities. As discussed earlier, rock discontinuity that distributed extensively in a rock mass is predominantly joints. Faults, bedding planes, and dyke intrusions are the localized features and therefore, are dealt individually. Properties of rock mass therefore are governed by the factors of rock joints and rock material, as well as boundary conditions, as listed below:

Prime Factors Governing Rock Mass Property

Joint parameter:
- Number of Joint Set
- Orientation
- Spacing
- Aperture
- Surface Roughness
- Weathering and alteration

Material Parameters:
- Rock Density
- Modulus of Elasticity
- Rock Strength

Boundary Conditions:
- Groundwater Pressure and Flow
- In situ Stress

The behavior of rock changes considerably from a continuous of intact rock to a discontinuity of highly fractured rock mass. The existence of rock joints and other discontinuities play an important role in governing the behavior and properties of the rock mass.

1.4.1 Characterization of Discontinuities in a Rock Mass

The characterization of discontinuities in a rock mass is a complex operation that requires field identification of geologic factors the affect their shear strength and, as a result, the stability of the rock mass. These factors are given below:

- Number of joint (discontinuities) sets and their orientation
  Several sets of discontinuities are often developed in a rock mass, three to four sets being most common. The geologist uses statistical measurements to define different sets of joints and shows them the pole plots. Figure 1.33 is an example.
- Spacing (interval)
  The statistical distribution of distances between discontinuities is measured for each set; this makes it possible to calculate the average size of blocks in the rock mass and the fracture density. Figure 1.34 shows an example of the spacing distribution of a joint set.
- Persistence (fracture length)
  The persistence or length of a discontinuity is a measure of the extent of development of discontinuity surface. For the different sets, the geologist wants to know whether the discontinuities have a significant spatial extent (such as stratification joints and fault) or a minor extent (jointing). Fracture trace length is also related to fracture surface area. As some of the discontinuities are more

(a)

Sr. No.	Surface	Strike	Dip direction	Dip amount
1	S_0	N - S	E	45°
2	T_1	E - W	Vertical	–
3	T_2	N - S	W	45°
4	S_1	N 68°- N 248°	N 338°	80°
5	S_2	N 112°- N 292°	N 202°	80°

Fracture orientation data

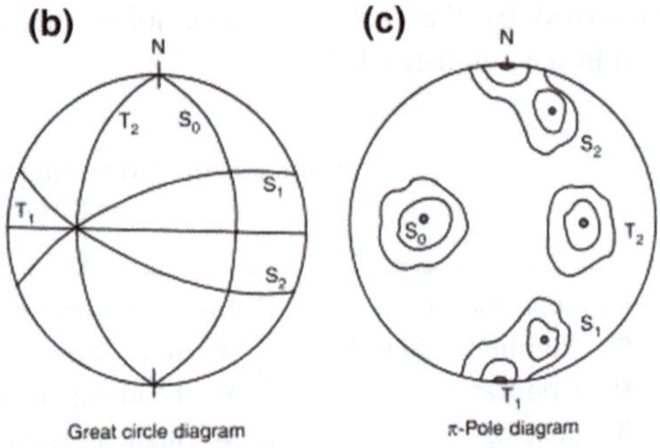

Fig. 1.33 An orientation data of discontinuities (**a**) and their great circle (**b**) and pole diagrams (**c**)

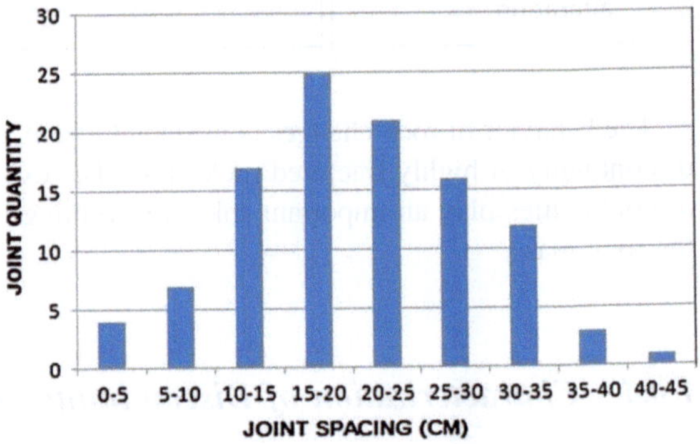

Fig. 1.34 Joint spacing distribution of a set of joints

persistent and continuous than others, it becomes a very important parameter in controlling groundwater flow. Persistence is rather difficult to quantify, as it would differ in the dip and strike direction. It can be measured by observing the discontinuity trace length in an exposure, in both dip and strike directions.

- Aperture
  Aperture is the perpendicular distance separating the adjacent rock walls of an open discontinuity, in which the intervening space is air or water filled. Aperture may vary from very tight to wide. Commonly, subsurface rock masses have small aperture. Tensile stress may lead to larger aperture or open fractures. Often shear fractures have much lower aperture values than the tensile fractures.
- Roughness (asperity)
  Fracture walls are not flat parallel smooth surface but contain irregularities, called roughness or asperity. In the field, the geologist characterizes the roughness of the joints according to the size of the asperities and their shapes. In the case of roughness that is elongated lengthwise, such as ripple marks, the relative orientation of the feature and the shear movement must be determined because it has a great influence on the share strength.

- Filling
  Open joints can be filled with air, water, or loose materials, e.g., clay, fault gouge, breccias, chert, and calcite. In the case of solid materials, the material must be carefully described because it will determine the shear strength of the joint if the rocky asperities of the walls are not in contact (clay filling should be screened as a priority).
- Hydrogeological behavior of joints.
  The hydrogeological behavior of joints is essential to determine whether groundwater occupies the discontinuities permanently or temporarily.
- Weathering condition of the rock body.
  In the case of contact between the walls, the strength of rock is to some extent influenced by the weathering state of rocks in general and the rock near the joints in particular. It should be emphasized that discontinuities play a significant role in the propagation of weathering throughout a rock mass.

1.4.2 Field Investigations on Rock Mass Properties

The characterization of the above properties of rock mass in the field is done at the outcrop, in pre-existing underground cavities, and in boreholes. One of the major difficulties is estimating the bias that affects measurements taken at the outcrop (e.g., the opening is exaggerated by the decomposition of the rock mass, surface movement, weathering, and other factors) when we want to estimate those characteristics within the rock mass. Observation in underground cavities is also subjected to these biases: plastic deformation near the walls, fracture due to excavation, etc. All these phenomena must be taken into account in the study of joints in rock mass.

1.4.3 Groundwater

Groundwater is the water located beneath the Earth's surface in soil pore spaces and in the fractures of rock formations. The principal source of groundwater is meteoric water, that is, precipitation. The amount of water that infiltrates into the ground depends on how precipitation is dispersed. The depth at which soil pore spaces or fractures and voids in rock become completely saturated with water is called the water table. Groundwater is in constant motion, although the rate at which it moves is generally slower than it would move in a stream because it must pass through the intricate passageways within the rock. The rate of groundwater flow is controlled by two properties of the rock: porosity and permeability.

Porosity is the percentage of the volume of the rock that is open space (pore space). This determines the amount of water that a rock can contain.

Permeability is a measure of the degree to which the pore spaces are interconnected, and the size of the interconnections.

1.4.4 *The Effects of Rock Mass Properties to the Excavation*

1.4.4.1 Rock Strength and Joint Distribution to Blasting Fragmentation

Rock with high strength and widely spaced joints has a higher risk of bad fragmentation in blasting. Weak rock with intensive joints will commonly give good fragmentation during blasting (Figs. 1.35, 1.36).

Fig. 1.35 Bad fragmentation by blasting hard and less jointed rock mass

Fig. 1.36 Good fragmentation by blasting weathered and highly jointed rock

Fig. 1.37 An open joint in rock mass caused high airblast in project DAR, HK, 2009

1.4.4.2 Open Joints Cause Explosive Gas Venting

Open joints in a rock mass may cause a disruption in explosive energy distribution. This may lead to a number of adverse consequences, such as the venting of explosive gases, airblast, flyrock, and the creation of large boulders in the muckpile. The following Photographs show an open joint in the rock mass blasted which caused a strong airblast of 135 dBL at 200 m distance in the project of DAR in Hong Kong (Fig. 1.37).

1.4.4.3 Wedge Failure on the Free Face Cause Flyrock

Two sets of joints intersect at an angle to form a wedge failure on the free face of the blasting bench may cause flyrock due to the reduction of burden. Figure 1.38 shows a flyrock accident in the project of Jordan Valley, Hong Kong, in 2003.

Wedge failure is also caused by the damage of the slope wall to cause danger to public.

Fig. 1.38 Wedge failure by two sets of joints on the free face caused flyrock in blasting

Fig. 1.39 Wedge failure on slope wall

1.4.4.4 Fracture Zone in Rock Mass May Cause Bench Top Flyrock

When a fracture zone exists in the rock mass which is blasted, a top flyrock accident may be caused if there is a lack of sufficient stemming in the blast holes (see Figs. 1.39, 1.40)

1.4.4.5 Slope Failure by Discontinuities

Figure 1.41 shows the unfavorable discontinuity may cause the slope failure even though the slope is formed by control blasting technique.

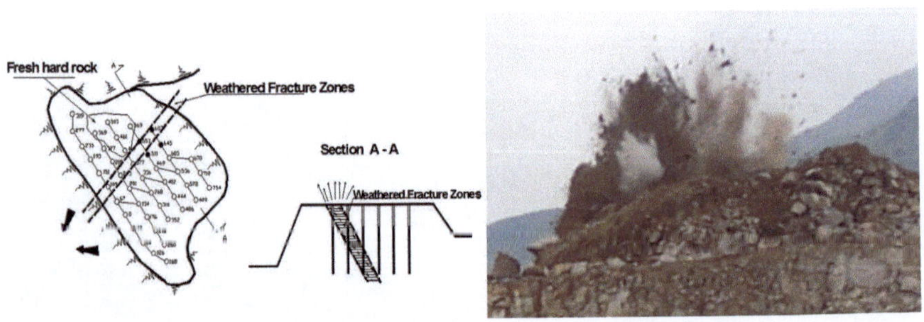

Fig. 1.40 Flyrock accident from the bench top caused by a weathered fracture zone in rock mass, 2003, Hong Kong

Fig. 1.41 Slope failure by discontinuities

1.4.4.6 Rock Failure During Tunnel Excavation Caused by Discontinuities

Stability problem in blocky, jointed rock mass is generally associated with gravity falls of blocks from roof and sidewalls. Following Figs. 1.42 and 1.43 show the common situation.

1.4.4.7 Groundwater Affects Underground Excavation

Groundwater can be a serious problem for some underground excavation projccts (Fig. 1.44). Effective control of the groundwater must be carried out before and during the excavation process.

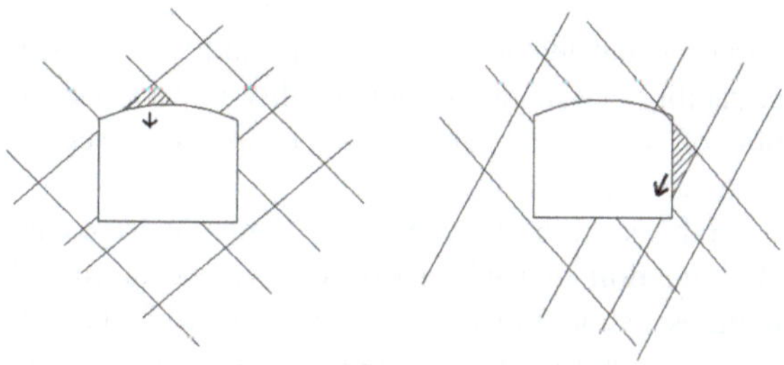

Fig. 1.42 Failure models occurred in tunnel due to block jointed rock mass

Fig. 1.43 Rock sliding failure from sidewall. 2001, East Side Access project in New York (www.dailymail.co.uk/news/)

Fig. 1.44 Groundwater affects tunnel construction

1.5 Classification of Rock Sturdiness

Rock mass classification systems can be used for engineering design, project planning, stability analysis, equipment selection, and cost estimation. These are based on empirical relations between rock mass parameters and engineering applications, such as tunnels, slopes, foundations, and excavatability. There are various standards and schemes of the rock classifications based on the engineering purposes and features, such as drillability classification, blastability classification, cuttability classification, abradability (wear) classification, and stability classification of rock mass. They will be discussed separately in different chapters and sections of this book.

In this section, the general or comprehensive classification that we define as the **classification of rock sturdiness** will be discussed. The word of **sturdiness** (or **firmness)** should involve theoretically all the properties and characters of rock and rock mass, but in fact only few indexes or rates of rock or rock mass are adopted as the principles of division (rules) to govern the assignment into the classes in practice.

Summaries of some important classification systems are presented in this section and although every attempt has been made to present all of the pertinent data from the original texts, there are numerous notes and comments which cannot be included. Interested reader should make every effort to read the cited references for a full appreciation of the use, applicability, and limitations of each system.

Particularly, the classification system developed by Prof. M. M. Protodyakonov and his son, M. M. Protodyakonov (the son, almost same name with his father), will be introduced in this section, as it has been in use for more than 100 years until today in Russia (former Soviet Union), eastern Europe, and China but seldom in Western countries, including Hong Kong. So, it will be of benefit to know this system for the engineers who have a project in these countries and work together with the local engineers.

1.5.1 *Protodyakonov's Rock Classification*

In 1909, Russia's Prof. M. M. Protodyakonov first raised the concept of "Rock Sturdiness" and carried out a systematical study on it. He addressed that: "generally, the word of sturdiness involves various respects of rock behaviors. It is a comprehensive concept essentially." It involves its anti-break ability and the ability of remaining stability as well. He further addressed that "rock sturdiness in all aspects of performance is insistent." This means that if two kinds of rock have a sturdiness ratio, f_1:f_2, e.g., drilling, then they also have similar sturdiness ratios in other aspects, such as blasting and excavation stability. He adapted a coefficient to describe sturdiness of the rock. The coefficient is called **Protodyakonov's Coefficient, "f" of a rock,** commonly in Russia and China.

In 1926, he stated that the value of "f" for a kind of rock can be determined by seven working indexes: uniaxial compressive strength, the work to break 1 cm^3 of rock by man power, productivity of drill hole by man power during one shift, productivity of surface excavation, explosive (black powder) consumption for blast 1 m^3 of rock, productivity of tunneling workers, and excavation velocity of drift. Then seven "f_i" values are made from the seven indexes. The average value of the seven "f_i" is regarded as the rock's "f" coefficient. Because the tedious process and most working conditions had been greatly changed, almost all the working indexes were died out except the uniaxial compressive strength was remained to be the critical index to determine the value of "f" coefficient:

$$f = \frac{R}{100} \tag{1.1}$$

where R is the uniaxial compressive strength of the rock and the unit is kg/cm^2.

There are totally 10 classes (add 4 subclasses) in Protodyakonov's rock classification system and the maximum value of "f" is 20 (see Table 1.3). That means the maximum uniaxial compressive strength of rock is only 2000 kg/cm^3. But in practice there are some very strong rocks in which uniaxial compressive strength is over 3000–4000 kg/cm^3. So, in 1955, Professor L. I. Baron revised the formula of (1.1) as:

$$f = \frac{R}{300} + \sqrt{\frac{R}{30}} \tag{1.2}$$

In 1950, Dr. M. M. Protodyakonov (the son) raised a simple method to determine the value of "f" of a rock—"Smash Method." Figure 1.45 shows the devices of the smash method.

Table 1.3 Protodyakonov's classification of rock sturdiness

Category	Sturdiness level	Description of rock	*f*
I	Sturdiest	The hardest, toughest, and most dense quartzites and basalts	20
II	Very sturdy	Very sturdy granitic rock, quartz porphyry, siliceous schist, weaker quartzites Sturdiest sandstone and limestone	15
III	Sturdy	Granite (dense) and granitic rock. Very sturdy sandstone and limestones. Quartz veins Sturdy conglomerate. Very sturdy iron ore	10
IIIa	Sturdy	Limestones (sturdy), weaker granites Sturdy sandstones, marble, dolomites and pyrites	8
IV	Rather sturdy	Ordinary sandstones. Iron ore	6
IV	Rather sturdy	Sandy schists. Schistose sandstones	5
V	Moderate	Sturdy shale. Non-sturdy sandstones and limestones Soft conglomerates	4
Va	Moderate	Various schists (non-sturdy). Dense marl	3
VI	Rather soft	Soft schists. Very soft limestones, chalk, rock salt, gypsum. Frozen soil, anthracite Ordinary marl. Weathered sandstones, cemented shingle and gravel, rocky soil	2
VIa	Rather soft	Detritus soil. Weathered schists, compressed shingle and detritus, sturdy bituminous coal, hardened clay	1.5
VII	Soft	Clay (dense). Soft bituminous coal, sturdy alluvium, clayey soil	1.0
VIIa	Soft	Soft sandy clay, loess gravel	0.8
VIII	Earthy	Vegetable earth, peat, soft loam, damp sand	0.6
IX	Dry loose Rock	Sand, talus, soft gravel, piled-up earth, substances extracted coal	0.5
X	Flowing rock	Shifting sands, swampy soil, rare-fractioned loess, and other rare-fractioned soils	0.3

In this method, the tested rock is broken into small pieces with diameters of 20–40 mm and divided into 5 portions, each weighed 25–75 g.

Smash each portion of rock pieces in the smash tube (left in the figure) with the hammer (dropping from the tube top) 5 times ($n = 5$) and then pour out of the smashed rock pieces. After all 5 portions of rock pieces are smashed one by one, all smashed rock pieces are mixed together and screened with a screen of 0.5-mm passing holes. All passed powder is poured into the graduated cylinder (right in the figure) to measure the height of the powder (l). Using the following formula, the sturdiness coefficient f can be calculated:

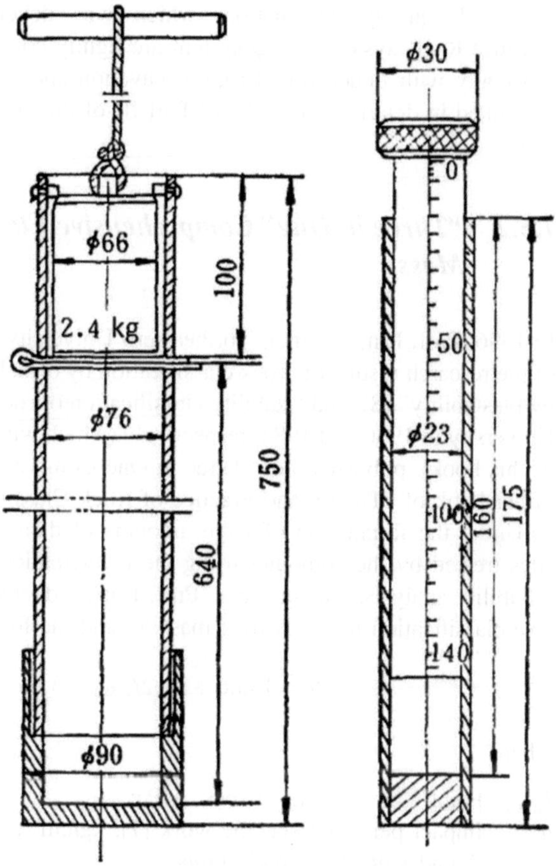

Fig. 1.45 Smash tube (*left*) and graduated cylinder (*right*)

$$f = 20\frac{n}{l} \tag{1.3}$$

In 1975, the smash method was published as the National Standard of Soviet Union although it was amended slightly afterward.

Protodyakonov's classification system has been widely used in China for a long time and even till today but more scientific and practical classification system of rock sturdiness and its subsystems, i.e., rock drillability system, rock blastability system, and rock mass stability system, have being studied and got very important achievements. These achievements will be introduced later in this book.

The most common classification systems used in Western countries and Hong Kong today are the RMR system published by Bieniawski in 1973 and the Q-system first described in 1974 by Barton et al. which is based on the RQD index (the rock quality designation, which was developed by Deere in 1967). Because RMR system and Q-system are tightly connected with the rock stability, especially with underground rock excavation and its reinforcement, they will be illustrated in detail in Chap. 21 of Part III of this book.

1.5.2 "Three in One" Comprehensive Classification of Rock Mass

In 1996, Prof. Lin, Yunmei, Northeastern University (NEU), PRC, under the basis of the research results of "rock classification by drillability" [7], "rock classification by blastability" [8], and "stability classification of rock mass" [9], developed by the University in 1980 and 1984, respectively (which will be discussed in later chapters of this book), published her "Three in One Comprehensive Classification System" in her book of "Theory and Practice of Rock Classification" [6].

Under the foundation of large amounts of data collected in different working sites treated by the computer using the mathematical methods of cluster analysis, reliability analysis, and statistics, Prof. Lin conducted a "three in one comprehensive classification index of rock mass S" and the following formula:

$$S = 135.6 + 10.2 I_{s(50)} + 3.3a + 21.5 v_r \tag{1.4}$$

where

$I_{s(50)}$ Point load strength of rock, MPa;
a Impact penetrant specific work [7], kg/cm^3 (1 kg/cm^3 = 0.1 J/cm^3);
v_r Sound velocity in rock, km/s

The three in one comprehensive classification of rock mass is given below (Table 1.4):

Table 1.4 Three in one comprehensive classification of rock mass

Class	Index S	Stability	Drillability	Blastability
I	>550	Most stability	Most difficult	Most difficult
II	550–450	Stability	Difficult	Difficult
III	450–350	Medium stability	Moderate	Moderate
IV	350–250	Unstable	Easy	Easy
V	<250	Most unstable	Very easy	Very easy

1.5.3 China's "Standard for Engineering Classification of Rock Masses" GB50218-2014

In 1994, Chinese government organized the experts and scholars from relevant units (academies, universities, institutions of design and research) of China to study and work out a standard of engineering classification of rock masses and further updated in 2014 as the national standard, GB50218-2014 and forced to be carried out in China from May 1, 2015.

1.5.3.1 Factors of Classification for Basic Quality of Rock Masses

The standard addressed that the basic quality of a rock mass should be determined by two factors: rock's hardness and intactness of the rock mass and these two factors should be determined by both the way of qualitative gradating and the way of quantitative indexes.

1.5.3.2 The Qualitative Division of the Hardness Degrees of Rock

Thc qualitative grades of rock hardness and degrees of weathering are shown in Tables 1.5 and 1.6.

1.5.3.3 The Qualitative Division of the Intactness Degree of Rock Masses

The qualitative division of the intactness degree of rock masses should be in accordance with Table 1.7, and the binding degree of structural planes should be determined in accordance with the features of the structural planes given in Table 1.8.

1.5.3.4 The Quantitative Indexes Determination and Division for Rock Masses

a. The quantitative index of rock should adapt **uniaxial compressive strength of saturated rock (R_c).** R_c should be the **measured value**. If R_c cannot be measured directly, it can be calculated using the following formula:

$$R_c = 22.82 \mathrm{I}_{s(50)}^{0.75} \tag{1.5}$$

Table 1.5 Qualitative grades of rock hardness

Description		Qualitative determination	Representative rocks
Hard rock	Hard	When be hammered, there is a crisp sound, rebounding, handshaking, and hardly to be crushed Immersing in water, no water absorption action almost	Unweathered or lightly weathered granite, syenite, diorite, diabase, basalt, andesite, gneiss, quartz schist, siliceous slate, quartzite, siliceous cemented conglomerate, quartz sandstone, siliceous limestone, etc.
	Fair hard	When be hammered, there is a crisp sound, lightly rebounding, handshaking lightly, and rather hardly to be crushed Immersing in water, there is a slight water absorption action	1. Lightly weathered hard rock; 2. unweathered to slightly weathered ignimbrite, marble, slate, limestone, dolomite, calcareous cemented sandstone, etc.
Soft rock	Fair soft	When be hammered, there is no crisp sound, no rebound and easily to be crashed After soaking, the nails can carve prints	1. Slightly* weathered hard rock; 2. lightly weathered fair hard rock; 3. unweathered to slightly weathered Tuff, Phyllite, argillaceous sandstone, sandy mudstone, marl, siltstone, shale, etc.
	Soft	When be hammered with a hoarse sound and a dent, no rebound and easily be crashed Hand can break apart after soaking	1. Strongly** weathered hard rock; 2. slightly to strongly weathered fair hard rock; 3. slightly weathered fair soft rock; 4. unweathered mudstone and etc.
	Very soft	When be hammered with a hoarse sound and a deep dent, no rebound and easily be crashed by hand After soaking can shape into groups	1. Completely weathered all kinds of rocks; 2. A variety of semi-diagenetic rocks

* and ** seem to be turned around in the original Chinese document

Table 1.6 Qualitative degree of rock weathering

Description	Degree of weathering
Unweathering	No change in texture and structure, rock looks fresh
Slightly weathering	Essentially unchanged in texture, structure. and mineral color, part of fracture surfaces is rendered by ferromanganese materials
Weakly weathering	Texture and structure were damaged partly. Minerals' color was visibly changed. Some weathered minerals or weathering sandwich appears in the fracture surfaces
Strongly weathering	Most texture and structure were damaged. Minerals' color was obviously changed. Feldspar, mica, and other minerals were weathered to be secondary minerals
Completely weathering	Texture and structures were completely damaged. All minerals except quartz were weathered to Earth

Table 1.7 Qualitative division of the intactness degree of rock mass

Description	Developed degree of structural planes		Binding degree of main structural planes	Categories of main structural planes	Corresponding types of structures
	Set number	Average spacing (m)			
Intact	1–2	>1.0	Binding well or fair	Joints, fractures, and beddings	Massive or very thick layered structure
Fair intact	1–2	>1.0	Binding badly	Joints, fractures and beddings	Blocky or thick layered structure
	2–3	1.0–0.4	Binding well or fair		Blocky structure
Fair broken	2–3	1.0–0.4	Binding badly	Joints, fractures, beddings, and faults	Fractural blocky or medium thick layered structure
	≥3	0.4–0.2	Binding well		Inlaid **cataclastic** structure
			Fair binding		Medium to thin layered structure
Broken	≥3	0.4–0.2	Binding fair or badly	All categories of structural planes	Fractured blocky structure
		<2			Cataclastic structure
Highly broken	Disorder		Binding very badly		Granular structure

Table 1.8 Division of the extent of binding of structural planes of rock masses

Description	Features of structural planes
Good binding	Open gap less than 1 mm, no fillings.
Good binding	Open gap 1–3 mm, cemented by siliceous or iron; open gap >3 mm, structural plane rough and cemented by siliceous
Fair binding	Open gap 1–3 mm, cemented by calcareous or mud; open gap >3 mm, cemented by iron or calcareous
Bad binding	Open gap 1–3 mm, flat structural planes, cemented by mud or mud and calcareous; open gap >3 mm, mostly filled with mud or debris
Very bad biding	Filled by mud or mud-adding debris, filling thickness is greater than the undulating differs

where $\boldsymbol{I_{s(50}}$ is the measured **point load strength index** of the rock.

b. The correspondence between the uniaxial compressive strength of saturated rock ($\boldsymbol{R_c}$) and the qualitative grades of rock hardness can be determined in accordance with Table 1.9:

Table 1.9 Correspondence between R_c and the qualitative grades of rock hardness

R_c (MPa)	>60	60–30	30–15	15–5	<5
Hard grade	hard	Fair hard	Fair soft	soft	Very soft

c. The quantitative index of intactness degree of rock masses should adapt the **Intactness Index** of rock masses (**K_v**). K_v should be the **measured value**. If K_v cannot be measured directly, it can be obtained in accordance with Table 1.10 using the **Volumetric Joint Count of Rock Mass (J_v):**

$$K_v = (V_{\mathrm{pm}}/V_{\mathrm{pr}}) \tag{1.6}$$

where

V_{pm} Velocity of elastic longitudinal wave in rock mass;
V_{pr} Velocity of elastic longitudinal wave in rock

and

$$J_v = S_1 + S_2 + \cdots + S_n + S_k. \tag{1.7}$$

where

S_n The number of the n set of joints along the measuring line per meter;
S_k The number of joints out of any joint set per cubic meter of rock mass

(Table 1.10)

d. The correspondence between the **Intactness Index of rock masses (K_v)** and the qualitative division of the intactness degree of rock masses can be determined in accordance with Table 1.11.

Table 1.10 Correspondence between J_v and K_v

J_v (No/m^3)	<3	3–10	10–20	20–35	>35
K_v	>0.75	0.75–0.55	0.55–0.35	0.35–0.15	<0.15

Table 1.11 Correspondence between K_v and qualitative division of the intactness degree of rock masses

K_v	>0.75	0.75 0.55	0.55–0.35	0.35–0.15	<0.15
intactness degree	Intact	Fair intact	Fair broken	Broken	Highly broken

1.5.3.5 Classification of the Basic Quality of Rock Masses

The classification of the basic quality of rock mass should be determined by combination of the qualitative features (given in Tables 1.5 and 1.6) and the **Index of Rock Basic Quality (*BQ*)** in accordance with Table 1.12:

The Index of Rock Basic Quality (BQ) is calculated by the following formula:

$$\mathrm{BG} = 100 + 3\mathrm{R_c} + 250K_\mathrm{v} \tag{1.8}$$

where

R_c Uniaxial compressive strength of saturated rock, (MPa);
K_v Intactness index of rock masses

Note: If $R_c > 90K_c + 30$, taking $R_c = 90K_c + 30$ (MPa) and K_c into the formula (1.8);

If $K_v > 0.04R_c + 0.4$, taking $K_v = 0.04R_c + 0.4$ and R_c into the formula (1.8).

The Index of Rock Basic Quality (BQ) should be amended according to some factors like the characteristics of different types of projects to consider groundwater status, the initial stress state, the relationship between the orientation of the engineering axis or the direction of the trend line and the main weak structural planes, and the effect of the surface water for the slop rock mass.

The amended Index of Rock Basic Quality [BG] can be calculated using the following formula:

$$[\mathrm{BG}] = \mathrm{BQ} - 100(K_1 + K_2 + K_3) \tag{1.9}$$

where

Table 1.12 Classification of the basic quality of rock masses

Basic quality class	Qualitative features of basic quality of rock mass	Index of basic quality of rock mass (BQ)
I	Hard rock, intact rock mass	>550
II	Hard rock, fair intact rock mass; fair hard rock intact rock mass	550–451
III	Hard rock, fair broken rock mass; fair hard rock or interbedded soft and hard rocks, fair intact rock mass; fair soft rock, intact rock mass	450–351
IV	Hard rock, broken rock mass; fair hard rock, fair broken to broken rock mass; fair soft rock or interbedded soft and hard, mainly soft rock, fair intact to fair broken rock mass; soft rock, intact to fair intact rock mass	350–251
V	Fair soft rock, broken rock mass; soft rock, fair broken to broken rock mass; all soft rock and all broken rock mass	≤250

K_1 Correction factor for groundwater effect;
K_2 Correction factor for the effect of the attitude of main weak structural planes;
K_3 Correction factor for the effect of the initial stress state

The correction factors of K_1, K_2, and K_3 can be determined in accordance with the following tables: (Tables 1.13, 1.14, and 1.15)

For the details of this National Standard, please refer to the literature [5] cited.

Table 1.13 Correction factors for groundwater effect, K_1

Groundwater status	BQ				
	>550	550–451	450–351	350–251	≤250
Wet or dripping, $P \le 0.1$ or $Q \le 25$	0	0	0–0.1	0.2–0.3	0.4–0.6
Like rain or inrush, $0.1 < P \le 0.5$ or $25 < Q \le 125$	0–0.1	0.1–0.2	0.2–0.3	0.4–0.6	0.7–0.9
Inrush, $P > 0.5$ or $Q > 125$	0.1–0.2	0.2–0.3	0.4–0.6	0.7–0.9	1.0

Note 1. P is the water pressure of underground rock mass around the engineering project (MPa)
2. Q is the water quantity out from each 10 m cavern (L/min·10 m)

Table 1.14 Correction factor for effect of main weak structural planes

Relationship between structural planes and engineering axis	The angle between structural planes and engineering axis <30° and the dip of structural planes are 30°–75°	The angle between structural planes and engineering axis >60° and the dip of structural planes >75°	Other combination
K_2	0.4–0.6	0–0.2	0.2–0.4

Table 1.15 Correction factor for the effect of the initial stress state, K_3

Radio of strength of rock mass and stress $\left(\frac{R_c}{\sigma_{max}}\right)$	BQ				
	>550	550–451	450–351	350–251	≤250
<4	1.0	1.0	1.0–1.5	1.0–1.5	1.0
4–7	0.5	0.5	0.5	0.5–1.0	0.5–1.0

References

1. Bell FG (2007) Engineering geology, 2nd edn. B. H. of Elsevier. ISBN: 9780750680776
2. Haakon Fossen (2010) Structural geology. Cambridge University Press. ISBN: 9780521516648-8
3. Aurele Parriaus (2009) Geology: basics for engineers. CRC Press. ISBN: 9780415461658
4. GEOGUIDE 4, (1992) Guide to cavern engineering. Geotechnical Engineering Office, Civil Engineering Department, Hong Kong
5. Standard for Engineering Classification of Rock Mass, GB50218-2014 (in Chinese) on 27 Aug. 2014 and to be carried out from 1 May 2015. 工程岩體分級標準,中華人民共和國國家標準 GB50218-2014, 2014-8-27 由中華人民共和國住房和城鄉建設部 & 中華人民共和國國家品質監督檢驗 檢疫總局聯合發佈. 2015-5-1 開始執行。
6. 林韻梅等 (1996) 岩石分級的理論與實踐. 冶金工業出版社, Lin Y et al (1996) Theory and practice of rock classification (in Chinese). Metallurgical Industry Press
7. 費壽林等 (1980) 鑿碎法岩石可鑽性分級. 冶金工業出版社. 北京. Fei S et al (1980) Rock drillabilty classification with the method of impact penetrant (in Chinese). Metallurgical Industry Press, Beijing
8. 鈕強 (1984) 我國礦山岩石爆破性分級的研究. 爆破 1984(1). Niu Q (1984) The study of rock blastability classification for mines in China (in Chinese). Blasting 1
9. 林韻梅. (1985). 圍岩穩定性的動態分級法. < 金屬礦山 > 1985(8) pp 2–6. Lin, Yunmei, (1985), *The Dynamic Classification Method for the Stability of Rock Mass.* (in Chinese). Blasting, No.8
10. Rao UKM (2013) Rock mechanics. National Mission Project on Pedagogy (Main Phase), Last updated on June 07 2013, Current Stage: Draft. http://www.ide.iitkgp.ernet.in/Pedagogy3/fullcourse.jsp?COURSE_ID=118. Accessed 11 July 2016

Chapter 2
Rock Drilling

Drilling, in the field of rock excavation by drilling and blasting, is the first and essential operation carried out, and its purpose is to drill holes, with the adequate geometry and distribution within the rock masses, where the explosive charges will be placed along with their initiating devices. Even in the rock excavation with non-blasting method, drilling sometimes is also needed for creating some spaces to place the chemical expanding agent or insert a breaking tool or only creating a free face for rock breaking by mechanical tools.

The systems of rock drilling have been developed with various means such as mechanical methods (percussion, rotary, rotary-percussion), thermal method (flame, plasma, hot fluid, freezing), hydraulic method, sonic method, and laser ray method. Nevertheless, mechanical drilling system is always the most economic and convenient method and widely used in mining and civil engineering field at present. Therefore, in this book, only the mechanical means will be discussed.

2.1 Mechanism of Rock Breakage by Drilling and Drillability of Rock

2.1.1 Mechanism of Rock Breakage During Drilling

The general types of rock breakage during drilling by mechanical method, including percussion drilling, rotary drilling, and rotary-percussion drilling, are three kinds of basic mechanism: percussion-penetration, pressured roller, and cut (see Fig. 2.1).

During the process of drilling the tool (percussive drilling bit, roller-disk and studded roller-disk cutter, rotary tricone bit, or drag tools), the first action is push

D. Zou, *Theory and Technology of Rock Excavation for Civil Engineering*,
DOI 10.1007/978-981-10-1989-0_2

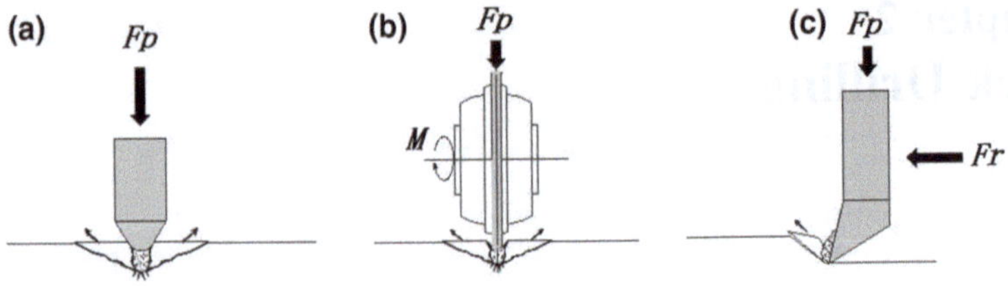

Fig. 2.1 General types of rock breakage during drilling: **a** percussive-penetration; **b** pressured roller; **c** cut

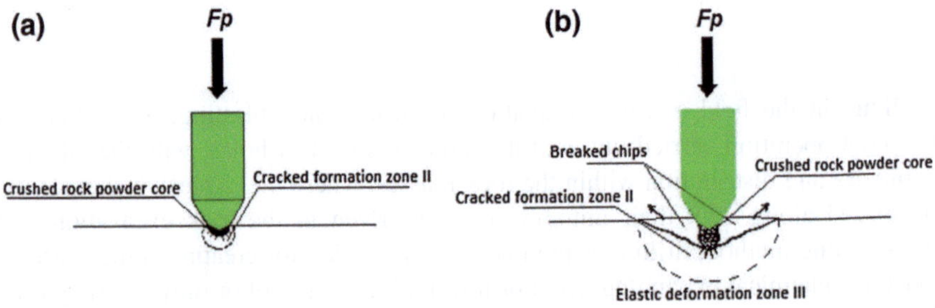

Fig. 2.2 Rock breakage by tool penetration

(or percussion), the tool penetrates into (indentation) and breaks (by Fp) the rock surface, then expands the breakage by continual percussion together with rotation of the bit, or pressured-rolling by thrust force (Fp) and torque (*M*) or continual cut by push force (Fr) under the thrust force (Fp). The tool penetrates and breaks the rock surface by a static (thrust) force or impact (percussion) force; this is the basic process of the rock breakage by mechanical method.

The process of tool penetrating the rock surface can be divided into four phases as follows [1] (Fig. 2.2):

- Crushed zone

As the tool tip begins to dent the rock surface, stress grows with the increasing load and the material is elastically deformed, zone III in Fig. 2.2. At the contact surface, irregularities are immediately formed and a zone of crushed rock powder core develops beneath the indenter (the bottom or insert of the tool). The crushed core comprises numerous microcracks that pulverize the rock into powder of extremely small particles. About 70–85 % of the indenter's work is consumed by the formation of the crushed zone. The crushed core transmits the main force component into the rock.

- Crack formation

As the process continues, dominant cracks begin to form in the rock, phase (a) in Fig. 2.2. This initial stage of restricted growth is described as an energy barrier to full propagation. The placement of major cracks depends on the indenter shape. Generally, the dominant placement of major cracks with blunt indenters, such as a sphere, is located just outside the contact area, pointing down and away from the surface.

- Crack propagation

After the energy barrier has been overcome, spontaneous and rapid propagation follows. At a lower depth than the contact dimension, the tensile driving force falls below that necessary to maintain growth, thus the crack again becomes stable. The crack is then said to be "well developed."

- Chipping

When the load reaches a sufficient level, the rock breaks and one or more large chips are formed by lateral cracks propagating from beneath the tip of the indenter to the surface. This process is called surface chipping, phase (b) in Fig. 2.2. Each time a chip is formed, the force temporarily drops and must be built up to a new, higher level to achieve chipping. Figure 2.3 describes the "leapfrogging" progress of the indenter as it penetrates the rock surface [2].

During the process of loading–penetration, there are two facts mentioned by the researchers [2]:

- From Fig. 2.3, it shows that the load–penetration curves for each subrising sections have substantially the same slope. That means the increase in penetration depth is nearly a constant when unit load is increased. The dropping sections of the curves are in relation to the stiffness of the loading mechanics; it is not fully dependent on the rocks being dented;
- Secondly, the bottom angle of the crater (called "natural breaking angle") formed by crushing and chipping are almost always within a range about 120°–150° (see Fig. 2.4). Table 2.1 gives the values of the natural breaking angle of some rocks.

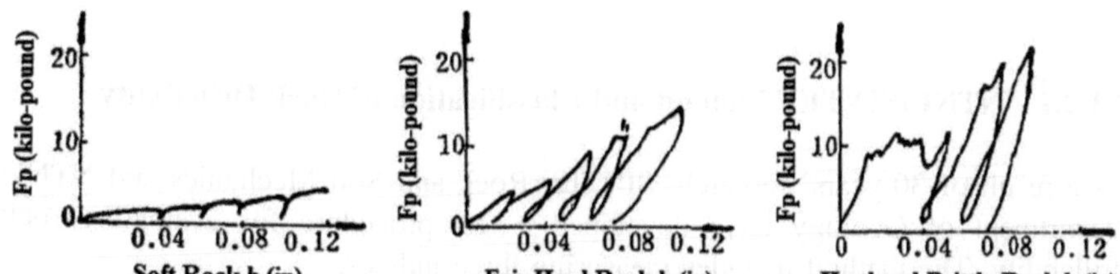

Fig. 2.3 Load (Fp)—Penetration (*h*) profiles of various rocks (Courtesy of Coal Industry Press, China, Ref. [2])

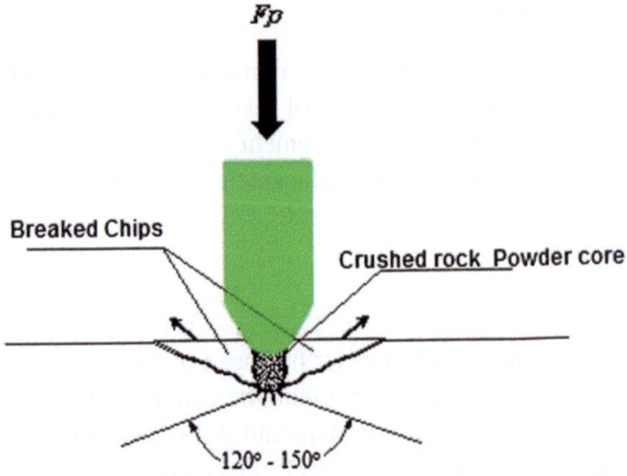

Fig. 2.4 Natural breakage angle

2.1.2 *Drillability of Rock and Its Classification*

Drillability is the resistance of rock to penetration by a drilling technique, and it is a term used to describe the influence of numbers of parameters on the drilling rate (drilling velocity) and the tools wear of the drilling machine. Penetration of rocks is influenced by rock properties as well as machine parameters.

The purpose of studying the drillability of rock is for:

- Choosing a suitable drilling method, equipment, and technology to achieve best results on project progress and economy;
- Estimation of the drilling rate and working life of the drilling tools to offer the basic data of project planning;
- Offering reliable data of rock performance for design and improvement of drilling machines.

So, studying rock drillability and its classification is a basic technical work for rock excavation, mining, and geological and petroleum exploration.

Since 1927, B.F. Tillson introduced the concept of "rock drillability," researchers in many countries carried out lots of work on the rock drillability and its classification. In this book, the work of NTNU/SINTEF (Norway) and Northeast University (China) will be introduced.

2.1.2.1 NTNU/SINTEF Method and Classification of Rock Drillability

As a result of 30 years' research, SINTEF Rock and Soil Mechanics and NTNU Department of Geology have developed a test procedure for evaluating rock drillability. The method includes measuring three indices:

- Drilling rate index (DRI);
- Bit wear index (BWI); and

Table 2.1 Natural breakage angles (φ) of some rocks courtesy of coal industry press, Ref. [2]

Rock	Soft shale	Clay shale	Dense limestone	Soft sandstone	Hard sandstone	Coarse-grained marble	Basalt	Diabase	Fine-grained granite	Hard quartzite
φ	116	128	116	130	144	130	146	126	140	150

- Cutter life index (CLI).

Testing on more than 3400 rock samples from all over the world provides unique and sound basis for correlation, updating, and further development. These data make it possible for us to give cost and time estimates for:

- Tunnel driving with TBM;
- Tunnel driving with conventional methods; and
- Rock quarrying.

a. Drilling Rate Index (DRI)

The DRI is assessed on the basis of two laboratory tests: the brittleness value (S_{20}) test and The Sievers' J value (Sj) miniature drill test.

The brittleness value S_{20} is an indirect measure of rock resistance to crack growth and crush. S_{20} is determined by the Swedish Stamp Test (Fig. 2.5).

The crushed and sieved aggregate, sizes ranging 16.0–11.2 mm, is placed in a mortar and then struck 20 times with a 14-kg hammer. The mortar aggregate volume corresponds to that of a 0.5-kg aggregate with a density of 2.65 tons/m^3.

S_{20} equals the percentage of undersized material that passes through 11.2-mm mesh after droptest. S_{20} is presented as a mean value of three or four parallel tests.

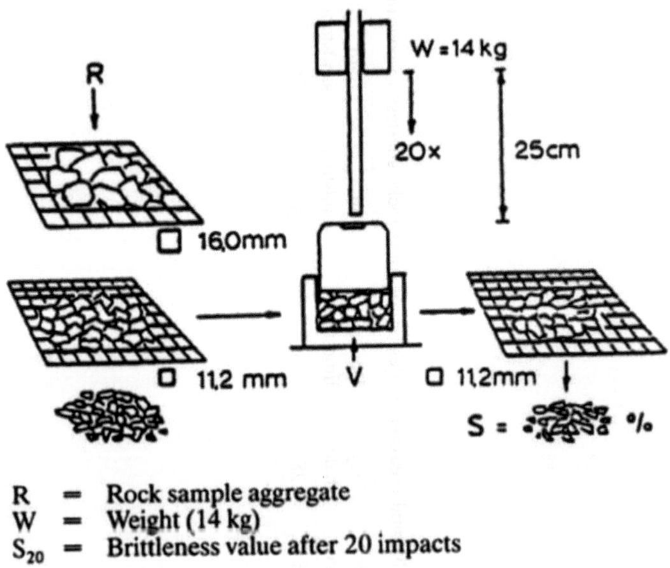

Fig. 2.5 Outline of the brittleness value by stamp test (Reproduced from Ref. [4] by permission of Sandvik)

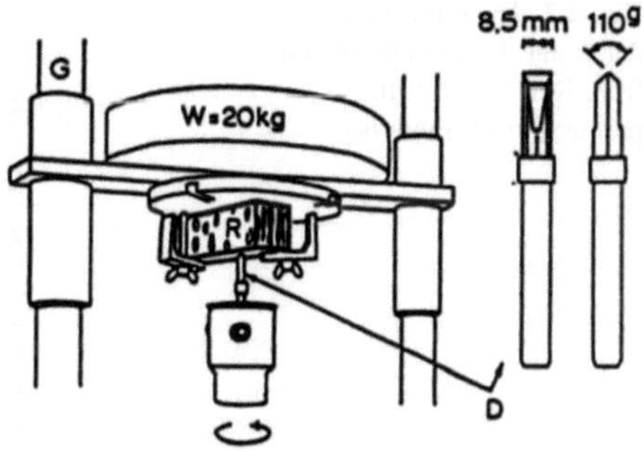

Fig. 2.6 Sievers' miniature drill test (Reproduced from Ref. [4] with the permission of Sandvik)

The *Sj* miniature drill test is also an indirect measure of rock resistance to tool indentation (surface hardness). The apparatus of miniature drill is shown in Fig. 2.6.

The hole depth in the rock sample is measured after 200 revolutions in 1/10 mm. A mean value of four to eight test holes is used.

The orientation of the rock specimen can affect test results.

Therefore, the *Sj* value is always measured for holes parallel to rock foliation. In coarse-grained rocks, care must be taken to ensure that a representative number of holes are drilled in the different mineral grain types.

The DRI is determined by the diagram shown in Fig. 2.7 [4]. The DRI can also be seen as the brittleness value corrected for its *Sj* value.

A qualitative DRT drillability rating is shown in Table 2.2 (Tamrock, [4]).

b. Bit Wear Index (BWI)

The BWI is assessed on the basis of two laboratory tests, the abrasion value (AV) test and abrasion value cutter steel (AVS) test [3].

The AV test constitutes a measure of the rock abrasion or ability to induce wear on tungsten carbide. The development of the AVS test was based on the AV test method. The same test equipment as for the AV measures the AVS, but the latter uses a test piece of steel taking from a TBM cutter ring. The AVS constitutes a measure of rock abrasion or ability to induce wear on cutter ring steel. The abrasion powder used for both the AV and AVS is normally prepared by the use of test material from the extractions used to determine S_{20} and should hence be regarded as representative and homogenized sample material. An outline of the AV and the AVS tests is shown in Fig. 2.8.

AV is defined as the weight loss of the test piece in milligrams after 5-min testing. AVS is defined as the weight loss of the test piece in milligrams after 1 min of testing. The AV and AVS tests are normally performed on 2–4 test pieces.

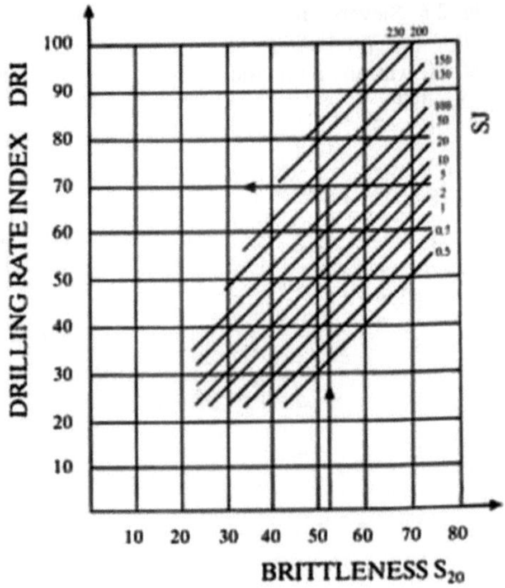

Fig. 2.7 Diagram used to determine DRI (Reproduced from Ref. [4] with the permission of Sandvik)

Table 2.2 Qualitative DRI drillablity

Rating	DRI
Extremely low	21
Very low	28
Low	37
Medium	49
High	65
Very high	86
Extremely high	114

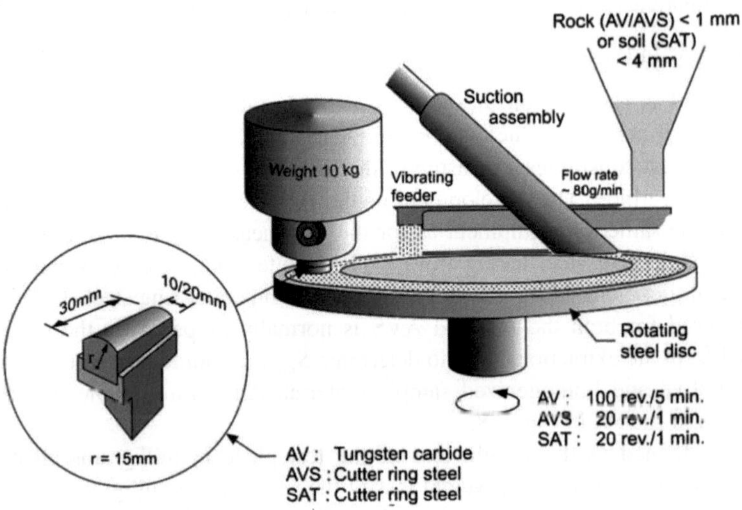

Fig. 2.8 Outline of the AV and AVS test (*source* [3])

Table 2.3 Classification or rock drillability by NTNU/SINTEF (quoted from [3])

Class	S_{20} value (%)	*Sj* value (mm/10)	AV (mg)	AVS (mg)
Extremely high	≥66.0	≤2.0	≥58.0	≥44.0
Very high	60.0–65.9	2.1–3.9	42.0–57.9	36.0–44.0
High	51.0–59.9	4.0–6.9	28.0–41.9	26.0–35.9
Medium	41.0–50.9	7.0–18.9	11.0–27.9	13.0–25.9
Low	35.0–40.9	19.0–55.9	4.0–10.9	4.0–12.9
Very low	29.1–34.9	56.0–85.9	1.1–3.9	1.1–3.9
Extremely low	≤29.0	≥86.0	≤1.0	≤1.0

c. Classification of drillability

The NTNU/SINTEF database does presently contain recorded test results for nearly 3200 samples from various rock excavation projects. NTNU/SINTEF tests show very good reproducibility and consistency. The following table shows the classification or rock drillability of NTNU/SINTEF [3] based on statistical analysis of the test values recorded in the database so far (Table 2.3).

2.1.2.2 Rock Drillability Classification Using the Method of Impact Penetrate

In 1980, Northeastern University, China (NEU), published their research result of rock drillability classification using the method of impact penetrate with two indexes of "specific impact penetrate work" and "abrasion width of bit" and developed two sets of measurement apparatus.

a. Concept of impact penetrate-specific work (IPSW)

The work consumed for impact penetrate on a unit volume of rock is called "impact penetrate-specific work (IPSW)." It is the basic physical quantity for the percussion (rotary-percussion) drilling of rock. During the process of impact penetrate, the relationship between the impact work applied (*A*) and the IPSW (*a*) is shown in the following Fig. 2.9 for some rocks. In the figure, it shows that there is a critical value of impact work Ac for the tested rock. When the applied impact work *A* is less than a certain value Ac, the value of IPSW is not stable and varies greatly as the small impact force only produces a scar and small powder cannot produce any chipping. When impact work *A* is greater than Ac, IPSW reach a plateau. The phenomenon tells us that the impact work as a main parameter of the test apparatus must be greater than the critical value, Ac, of any rock to be tested.

b. Test apparatus—impact penetrate apparatus (IPA)

The apparatus of the IPA is shown in Fig. 2.10. The weight of hammer (5) is 4.0 kg. The hammer free fall height along the guide rod (4) is 1.0 m. The hammer impacts the body (2) with an I-type bit connected in the bottom and the bit chisels

Fig. 2.9 Impact penetrate apparatus (Reproduced from Ref. [5] with the permission from Metallurgy Industry Press)

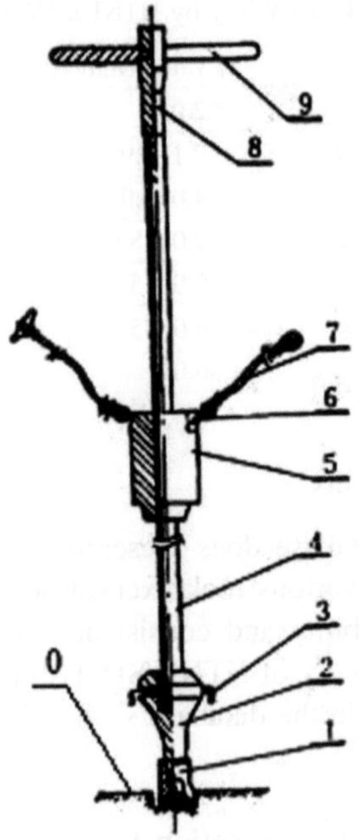

the rock. After every impact, the bit is turned 15° by the top handle of the rod. The diameter of the bit is 40 ± 0 and made with Type YG-11G tungsten carbide insert. Insert angle is 110°. For measuring the abrasion of the bit, a new bit (or newly grinded bit) must be used for every test. The rock face to be tested is placed horizontally, and a shallow nest is previously prepared manually for locating the tested bit. The net depth of the drilled hole, H, is measured and recorded after total 480 impacts for each rock specimen. The IPSW, a, can be calculated using the following formula:

$$a = \frac{A}{V} = \frac{nA_0}{\frac{\pi}{4}d^2 H} = \frac{480 \times 39.2}{\frac{\pi}{4} \times 4.1^2 \times \frac{H}{10}} = \frac{14252}{H}\ \mathrm{J/cm^2} \tag{2.1}$$

where:

- a impact penetrate-specific work (IPSW), J/cm^3;
- A total impact work of 480 freely falling of the hammer, J;
- V rock volume to be broken after 480 impacts, cm^3;
- N total impact times, n = 480;
- A_0 work of single impact, A_0 = 39.2 J;
- D actual hole diameter after drilling, d = 41 mm (bit diameter = 40 mm); and
- H net depth, mm.

After the test of IPSW, the abrasion of the bit is measured as well. The measurement is carried out using a reading microscope, expressed as "b" in mm, shown

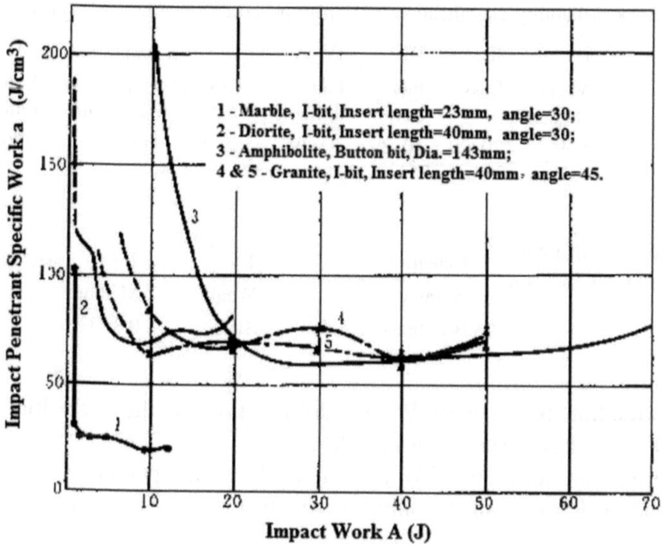

Fig. 2.10 Relationship between impact work and IPSW (Reproduced from Ref. [5] with the permission from Metallurgy Industry Press)

Fig. 2.11 Measurement of bit abrasion (Reproduced from Ref. [5] with the permission from Metallurgy Industry Press)

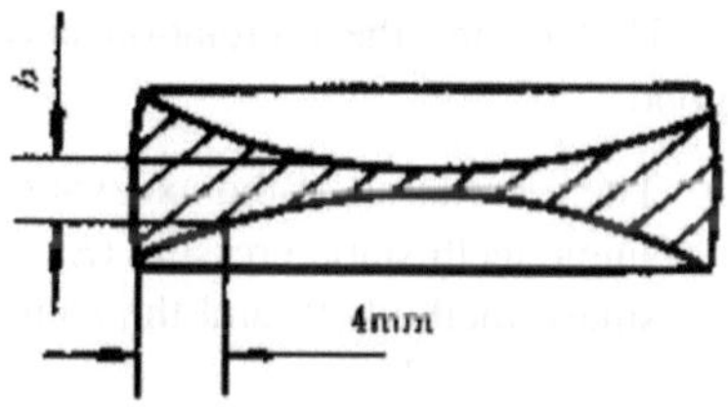

in Fig. 2.11. The abrasion value is the average value measured at the two ends of the bit edge after 480 impacts.

c. Classification of rock drillability

In the system of rock drillability using the method of impact penctrant, rock drillability is divided into seven classes and three categories according to both the index of impact penetrant-specific work (IPSW) "a" and the index of bit abrasion "b." The classification is shown in Tables 2.4 and 2.5 (Courtesy of Metallurgy Industry Press, Ref. [5]).

For verifying the classification system, total 2532 samples of 96 kinds of representative rocks from more than one hundred mines and working sites were tested. These tests show very good reproducibility and consistency. The correlation study between the drillability classification and the practical drilling effects of the production equipment, such as percussive rock drill, down-the-hole drill, rotary drill with rolling tricone bit, and TBM, also was carried out.

Table 2.4 Rock drillability classification by impact penetrant-specific work index "a"

Class	I	II	III	IV	V	VI	VII
Drillability	Very easy	Easy	Fair easy	Fair difficult	Difficult	Very Difficult	Extremely difficult
IPSW "a" (kg m/cm^3)	≤19	20–29	30–39	40–49	50–59	60–69	≥70

Table 2.5 Rock drillability classification by bit abrasion index "b"

Category	1	2	3
Abrasive	Weak	Medium	Strong
Bit abrasion "b"	≤0.2	0.3–0.6	≥0.7

- The correlation between IPSW and drilling time of the rock drill 7655 and 73-200 DTH drill is shown in Fig. 2.12.
- The correlation between IPSW and the drilling time per meter borehole, t, of 45-R and 60-R rotary drill (rolling tricone bit) is expressed with the regressive equation:

  45-R rotary drill: $t \approx 0.11a$ (the correlation coefficient of the equation is: 0.94);
  60-R rotary drill: $t \approx 1 + 0.1a$ (the correlation coefficient of the equation is: 0.98).

That means the correlation between IPSW and the tested rotary drill is very good.

- The correlation of the excavation speed (v) of SJG-53-12 TBM with IPSW (a), single teeth static pressure (K), rock compressive strength (R), Protodyakonov's smash method (f), and the sound velocity (V_L):

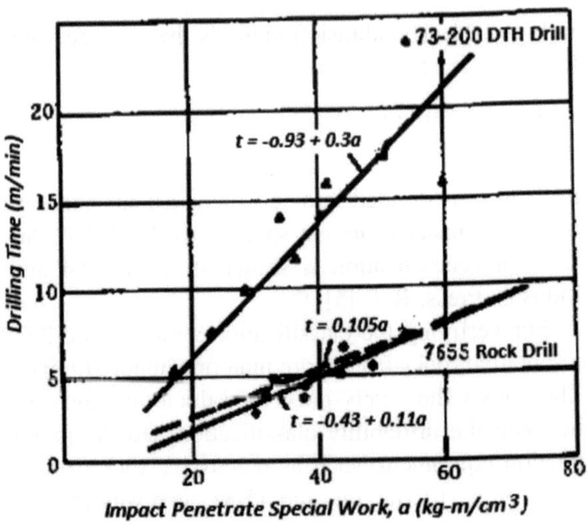

Fig. 2.12 Correlation between IPSW and drilling time of handheld rock drill and DTH drill (Reproduced from Ref. [5] with the permission from Metallurgy Industry Press)

Table 2.6 Correlation between TBM boring speed, v, and different indexes of drillability

Test method	Relationship	Regressive equation	Correlation coefficient	Sample quantity
IPSW	v–a	$v = 41.67a^{-1.04}$	−0.885 (>0.561)	20
Single teeth static pressure	v–K	$v = 0.35 + 0.014K$	0.717 (>0.684)	13
Rock compressive strength	v–R	$v = 1.61 - 0.41R$	0.63 (∼0.632)	10
Smash method	v–f	No correlation		19
Sound velocity	v–V_L	No correlation		23

SJG-53-12 TBM ($\phi = 5.2$ m, thrust = 396 t, cutterhead rotation speed: 5.79 r/min, cutter disk number = 64) bored 346-m tunnel in biotite gneiss.

The correlation of the TBM boring speed v with the above-said indexes is shown in Table 2.6 (Courtesy of Metallurgy Industry Press, Ref. [5]).

It is self-evident that the index of IPSW (a) has the best correlation with the TBM boring speed.

The sample tests also show that the index of bit abrasion (b) has a very close relationship with the consumption of drilling tools. As an example, Table 2.7 shows the test results from a large open iron mine.

Table 2.7 Rock drillability versus bit life (courtesy of Metallurgy Industry Press, Ref. [5])

Rock	Average IPSW a (J/cm^3)	Average bit abrasion b (mm)	Bit life (m)		
			DTH drill	45-R rotary drill	60-R rotary drill
Biotite quartz schist	240–280	0.1–0.2	120–150	350–450	500–600
Green clay amphibolite	200–250	0.1–0.2	120–150	350–400	400–450
Black green amphibolite	≈300	0.1–0.2	60–100	300–350	350–400
Dark green amphibolite	500–550	0.2–0.3	25–40	150–200	
Mixed granitic rocks	300–400	0.2–0.4	40–60	250–350	250–350
Third layer red iron ore	300–350	0.6–0.8	25–40	200–300	250–300
Third layer Iron asbestos ore	450–500	0.6–0.9	25–40	≈200	150–200
Third layer gray iron ore	450–550	0.5–1.0	12–16	80–100	100–140
1st and 2nd layers iron ore	550–600	0.8–1.3	8–12		≈100

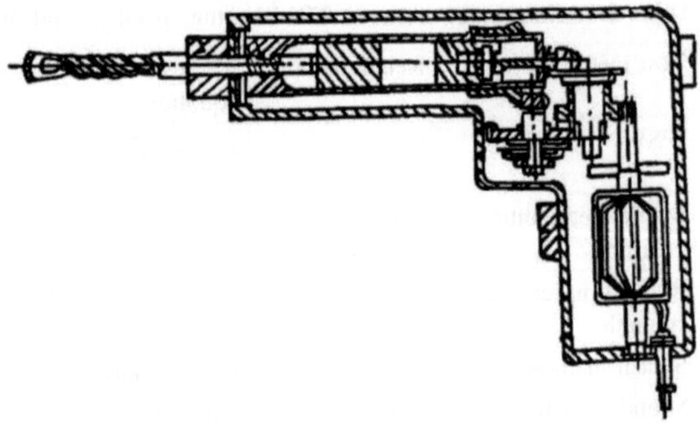

Fig. 2.13 WZ-1 mini-drillgauge (courtesy of "Metal Mines," China, Ref. [6])

d. Mini-drillgauge (MDG)

For convenience of test the rock drillability, the WZ-1 type of mini-drillgauge (MDG) was designed by the same research group of Northeast University in 1991 [6]. In fact the mini-drillgauge is a reformed electric mini-impact drill and a special designed drill bit. Its structure is shown in Fig. 2.13.

For testing the rock drillability, the working parameters of WK-1 MDG are determined as follows:

Working voltage: 220 V,
Motor speed: 1000 r/min,
Single impact work: 2.6 J, Impact frequency: 3800/min
I-type bit diameter: 14 ± 0.5 mm, Insert angle: 110°.

Drilling time for each test is set to 1 min. The impact penetrant-specific work measured by the WZ-1 MDG is calculated with the following formula:

$$a_w = \frac{A}{V} = \frac{n \times A_w}{\frac{\pi}{4} \times D^2 \times \frac{H}{10}} = \frac{360 \times 2.6 \times 10}{\frac{\pi}{4} \times 1.5^2 \times H} = \frac{55909}{H} \tag{2.2}$$

where

A, V same as formula (2.1);
n impact frequency, $n = 3800$/min;
A_w single impact work, $A_w = 2.6$ J
D drillhole diameter, $D = 1.5$ cm (bit diameter: 1.4 ± 05 cm);
H net depth of drilled hole

The measurement of bit abrasion is shown in Fig. 2.14.

The stability and reliability of the performance of WZ-1 MDG and the reproducibility and consistency of the data measured by MDG had been tested using ten apparatuses of WZ-1 MDG and different rocks. The test results are satisfied. The dispersion coefficient of the drilled depth in a same rock sample by ten apparatuses is 4.55 %. The maximum relative error of the average drilled hole depth on different rock types using 3 MDG, which have different used time, is less than 4.0 %.

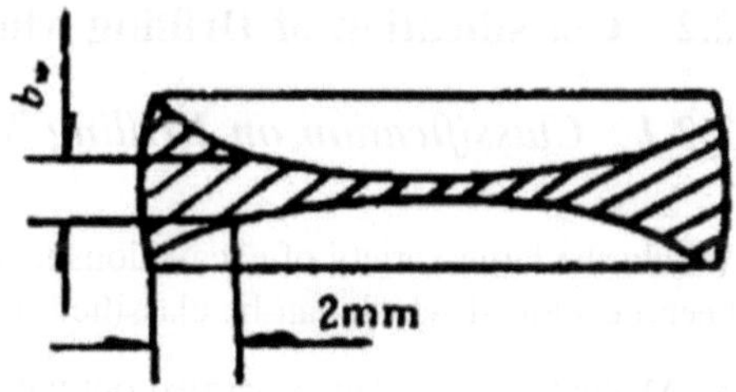

Fig. 2.14 Measurement of bit abrasion for WZ-1 MDG (courtesy of "Metal Mines," China, Ref. [6])

Table 2.8 Rock drillability classification by impact penetrant-specific work (IPSW) index

Class	IPSW (J/cm^3)		Drillability	Representative rock
	IPA "a"	WZ-1 MDG "a_w"		
I	≤186	≤400	Very easy	Shale, coal, tuff
II	187–284	401–650	Easy	Limestone, sand shale, peridotite, green clay amphibolites, mica quartz schist, dolomite
III	285–382	651–900	Fair easy	Granite, limestone, olive schist, bauxite, migmatite, amphibolite
IV	383–480	901–1150	Fair Difficult	Granite, siliceous limestone, gabbro, porphyrite, pyrite, magnetite quartzite, gneiss, skarn, marble
V	481–578	1151–1350	Difficult	Martite, magnetite quartzite (Nan Fen), cangshan gneiss, in fine-grained granite, dark green amphibolitc
VI	579–676	1351–1550	Very difficult	Martite (Gu Shan), lamprophyre, magnetite quartzite (Nan Fen's 1st and 2nd layer iron ore), dense skarn
VII	≥677	≥1551	Extremely difficult	Martite (baiyunebo), magnetite quartzite (Nan Fen)

Table 2.9 Rock drillability classification by bit abrasion index

Category	Bit abrasion (mm)		Abrasion	Representative rock
	IPA "b"	WZ-1 MDG "b_w"		
1	≤0.2	<0.2	Weak	Shale, coal, tuff, marble, amphibolite, peridotite, diabase, dolomite, bauxite, phyllite, skarn
2	0.3-0.6	0.2-0.4	Medium	Granite, diorite, gabbro, sandstone, sandshale, siliceous limestone, siliceous marble, migmatite, leptynite, gneiss, skarn
3	≥0.7	≥0.5	Strong	Pyrite, martite, magnetite quartzite, quartzite, hard gneiss

The rock drillability classification for some representative rocks using the method of IPSW and two kinds of apparatuses are shown in Tables 2.8 and 2.9 below (courtesy of "Metal Mines," China, Ref. [6]).

2.2 Classification of Drilling Machines

2.2.1 Classification on Drilling Manner

Within the large variety of excavations using explosives, numerous machines have been developed which can be classified into two types of drilling Manners:

- Manual drilling. This is carried out with light equipment that is handheld by the drillers. It is used in small operations where, due to the size, other machinery cannot be used or its cost is not justified.

The modern handheld rock drills are developed trending to be lighter, more convenient, and more efficient. Except the widely used pneumatic handheld drill, some new energy sources, like hydraulic, electricity and internal combustion engine, are also developed (Figs. 2.15, 2.16, 2.17, 2.18).

- Mechanized drilling. The drilling equipment is mounted upon rigs with which the operator can control all drilling parameters from a comfortable position. These structures or chassis can themselves be mounted on the wheels or tracks and either be self-propelled or towable (Figs. 2.19, 2.20).

Fig. 2.15 Handheld pneumatic rock drill (Reproduced with the permission from Atlas Copco)

Fig. 2.16 Handheld hydraulic rock drill (Reproduced with the permission from Atlas Copco)

Fig. 2.17 Internal combustion rock drill (courtesy of Luoyang Bytain Trading Co., Ltd.)

Fig. 2.18 Handheld electric rock drill (courtesy of HILTI power tools)

Fig. 2.19 Crawler rig for surface drilling (Reproduced with the permission from Atlas Copco)

Fig. 2.20 Wheel-mounted rig for underground drilling (Reproduced with permission from Sandvik)

2.3 Classification on Drilling Methods

The two most used mechanical drilling methods are rotary-percussion and rotary.

- Rotary-percussive methods. These are the most frequently used in all type of rocks, the top hammer, as well as the down-the-hole hammer.
- Rotary methods. These are subdivided into two groups, depending upon if the penetration is carried out by crushing with tri-cones or by cutting with drag bits. The first system is used in medium-to-hard rocks, and the second in soft rocks, see Fig. 2.21.

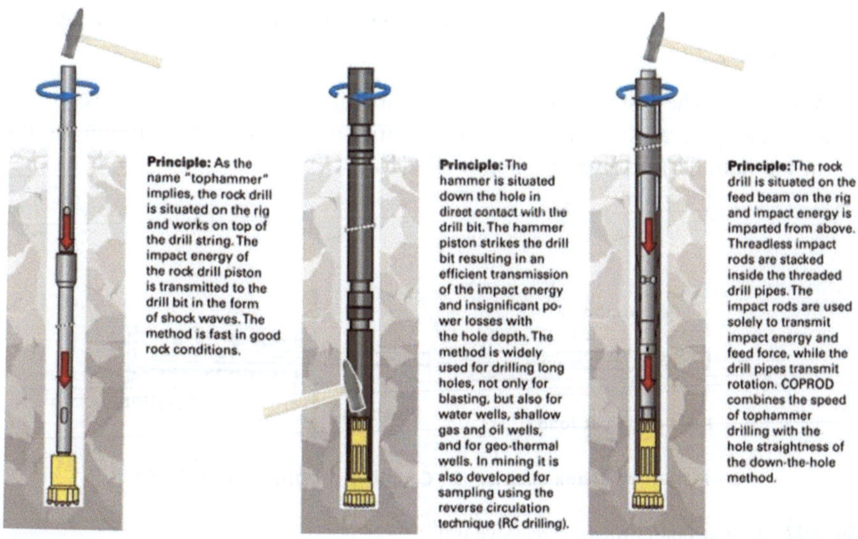

Fig. 2.21 Working methods of rotary-percussive drilling (Reproduced with the permission from Atlas Copco)

For engineering blasting, rotary-percussive drills have being used for more than 200 years from the manual drilling (excluding manual hammer drilling) to the modern hydraulic rigs. The mostly used rotary-percussive drills are the types of top hammer within the range of drillhole diameter from 38 to 127 mm. The down-the-hole drills are used for the larger holes with a diameter from 75 to 200 mm. The down-the-hole drills with a hole diameter of 150 mm were used in the blasting works for the construction of Hong Kong New Airport at Chek Lap Kok Island.

Rotary drills mostly are used in large open pit mines for drilling large diameter blastholes (127–440 mm) with the rolling cone rock bits. Other rotary drills with cutting action using the drag bits are used for soil, e.g., for installing soil nails on a slope, soft rock, like coal, or overburden drilling.

2.4 Rotary-Percussive Drilling

Rotary-percussive drilling is the most classic system for drilling blastholes and widely used in mining and civil engineering since the middle of nineteenth century starting with steam power then by compressed air. The appearance of hydraulic power in the sixties of last century has given a new boost to this method, complementing and widening its field of application.

According to the difference in the working modes of the major performances, hammer impact and rod/bit rotary, the rotary-percussive drilling are classified into two groups:

- Top hammer method including the newly developed COPOROD;
- Down-the-hole hammer method (DTH), also known as ITH (in-the-hole).

Rotary-percussive drilling is based upon the combination of the following four actions (Fig. 2.22):

- Percussion: The piston inside the rock drill strikes the tail end of the rod or bit itself and generates shock waves that are transmitted to the bit through the rod (in top hammer) or directly upon it (DTH).

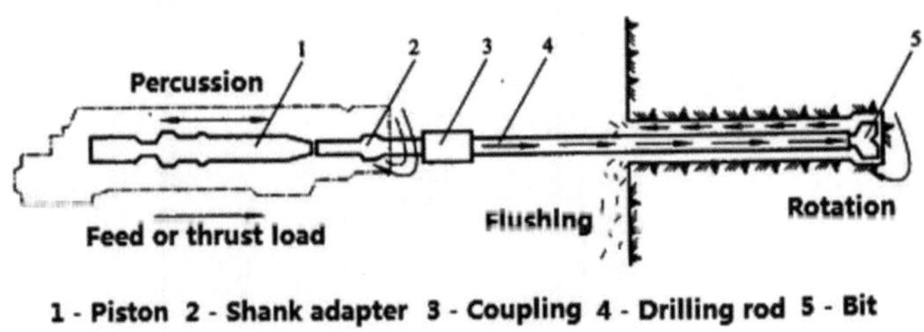

Fig. 2.22 Basic action in rotary-percussive drill

- Rotation: The rotary mechanism rotates the rod (in top hammer) or tube (in COPROD) or DTH hammer. With this movement, the bit is turned so that the impacts are produced on the rock in different positions.
- Feed or thrust load: Feed force is required to keep the shank in contact with the drill and the drill bit in contact with the rock. This ensures maximum impact energy is transferred from the piston to the rock.
- Flushing: Flushing is used to remove the rock cutting from the drillhole and to cool bit. The flushing medium—air, water, mist, or foam—is forced to the bottom of the drillhole through the rod's flushing hole and the hole in the drill bit.

2.4.1 Top Hammer Drilling

Top hammer drilling is the most widely used mode of rotary-percussive methods from handheld to drilling rigs. In percussive top hammer drilling, the impact energy is generated when the piston is striking the adapter (or tail end of the rod in handheld drill). This energy is transmitted from the rock drill via the shank adapter, drill rod, and drill bit to the rock, where it is used for crushing. The top hammer method is primarily used for drilling in hard rock for hole diameters up to 5 in (127 mm), and the main advantage is the high penetration rate in good solid rock conditions. Handheld pneumatic rock drill is used for small hole diameters while rig mounted hydraulic rock drill is commonly used for hole diameters above 1 5/8 in (41 mm). Heavy hydraulic rock drill with an impact power of up to 40 kW is used for large hole diameters up to 5 in.

2.4.1.1 Pneumatic Rock Drills

Pneumatic rock drill is equipped with valves (or piston itself in some handheld drills) to change the direction of compressed air into the cylinder, so that the compressed air pushes the piston with reciprocating strikes on the adapter or the tail of the drill rod through which the shock wave is transmitted to the bit where the chisel crushes rock. Along with each strike of the piston, the drill rod was rotated a certain angle (5°–15°) by a spirally fluted rifle bar or by independent rotary mechanism. The flushing system consists of a tube that allows the passage of air or water to the inside of drill steel. Figure 2.23 is a handheld rock drill, and Fig. 2.24 shows the structure of a typical handheld rock drill.

The pneumatic top hammer drill with independent rotary mechanism usually has more power even when the piston has the same size because the rifle bar is eliminated and the working surface of the piston on which the compressed air acts is increased. Another advantage of independent rotary mechanism is the percussion, and the rotation speed can be adjusted independently to suit the rock type to be

Fig. 2.23 Handheld rock drill (Reproduced with the permission from Atlas Copco)

drilled. But the independent rotary mechanism increased the weight of the drill so that it can only be used for the drill mounted in a rig.

Despite the quick development of the hydraulic hammer drill, pneumatic top hammer drill mounted on the rig has only gradually been replaced by hydraulic drills; it still occupies some market in mining and construction works due to its simplicity, reliability, easy repair, and low capital cost.

Table 2.10 gives the technical specification of some typical handheld pneumatic rock drills.

2.4.1.2 Hydraulic Rock Drill

At the end of the sixties and beginning of the seventies, a great technological advance took place in rock drilling with the development of hydraulic hammers. These new, high-power rock drills not only doubled drilling capacities but also improved the drilling environment. The introduction of hydraulics to rock drilling also led to improvements in drilling accuracy, mechanization, and automation.

The general working principle of a hydraulic percussive rock drill is presented in Fig. 2.25.

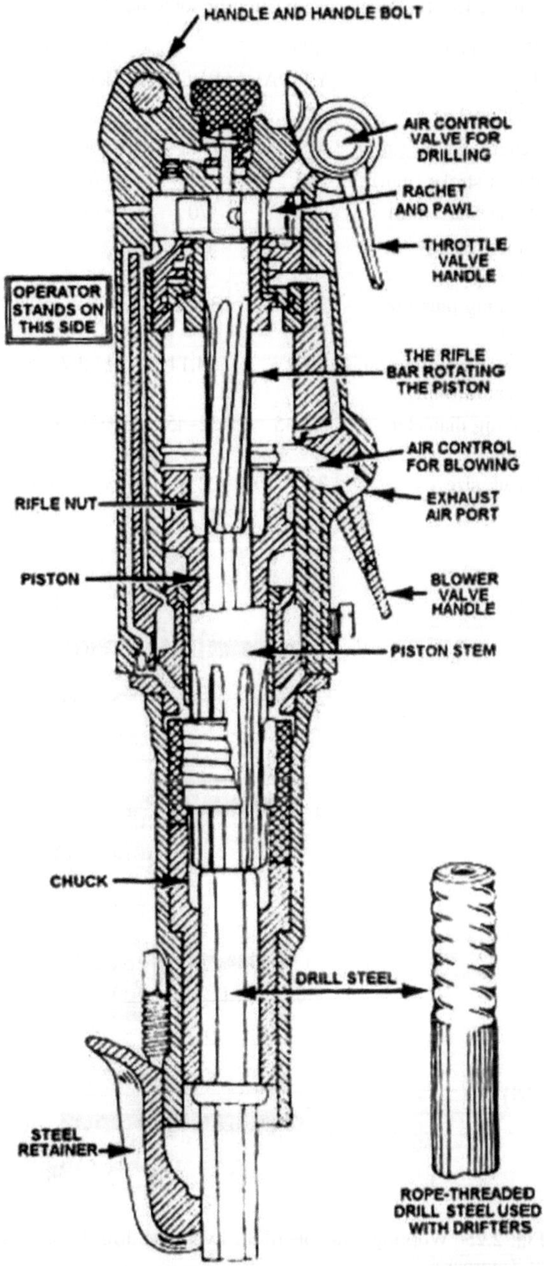

Fig. 2.24 Operational parts of handheld drill

Table 2.10 Technical specification of some handheld pneumatic rock drills

Manufacture	Atlas		Sandvik		Tianshui, China	
Type	YT29A	BBC16W	RD245	RD855	YT24	YT28
Weight (kg)	27	28.5	26	27	24	26
Length overall (mm)	659	705	660	615		
Piston stroke (mm)	60	55	68	60	70	60
Piston Dia. (mm)	82	70	66.7	85	70	80
Impact energy (at 5 bar) (J)	≥70	≤70			62	63
Working pressure (bar)	3.5–5	3.5–5			4–5	4–5
Air consumption (at 5 bar) m^3/min	≤3.9	≤4.14	2.7	3.5	3.36	3.48
Drilling diameter (mm)	32–45	32–45	32–45	32–45	34–45	34–45
Shank size	22 × 108	22 × 108	22 × 108	22 × 108	22 × 108	22 × 108

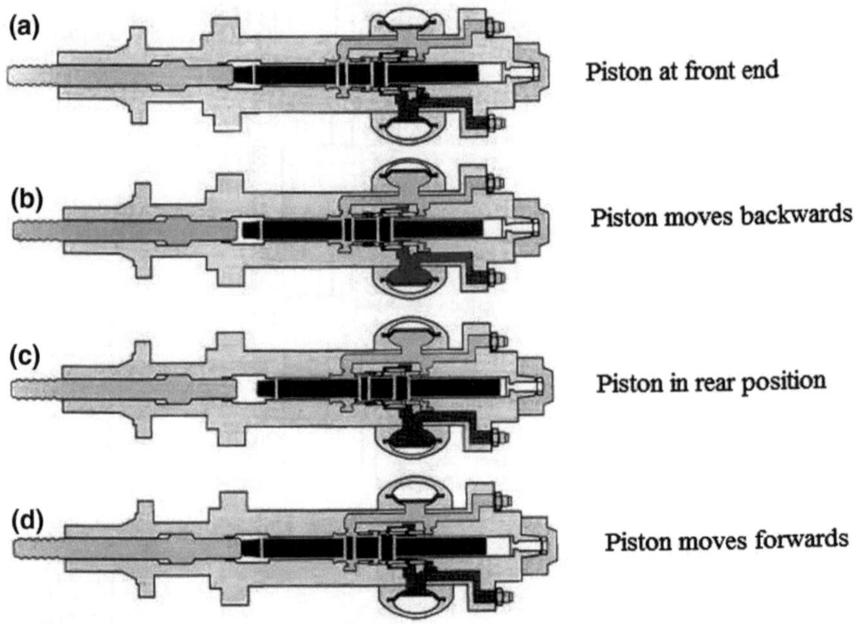

Fig. 2.25 Working principle of the hydraulic drill (Reproduced from Ref. [1] with the permission of Tamrock)

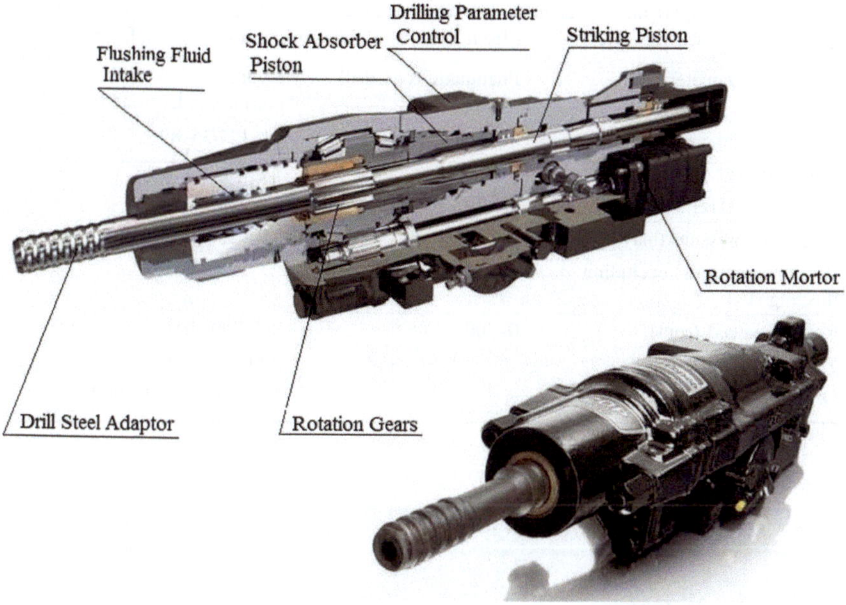

Fig. 2.26 Hydraulic rock drill and its cross section (Reproduced from Ref. [9] with the permission from Atlas Copco)

A hydraulic drill is composed basically of the same elements which are shown in Fig. 2.26.

Although in the beginning hydraulic drill rigs were mostly used in underground operation, but at present it has been widely used in both underground and surface drilling except some very small projects and places where hydraulic drill rigs cannot be used. But on the other hand, they also have some disadvantages: high initial investment, more complex, and costly repairs than those for pneumatic drills, requiring better organization and preparation of maintenance personnel. Figures 2.19 and 2.20 in Sect. 2.1 show the hydraulic drill rigs that are used in surface and underground excavation. A comparison of general working parameter range of top hammer drill rigs between pneumatic and hydraulic drills are shown in Table 2.11.

The COPROD drilling system, newly developed by Atlas Copco, combined the advantages of top hammer and DTH drilling. In this system, the inner drill rods transmit strike power to the drill bit and out tubes transfer rotation, adding stiffness to the string and improved flushing efficiency. Practices show that the COPROD system offers unique features for drilling holes straight and fast, especially suitable for the fractured rock conditions, in spite of its higher initial cost. Figure 2.27 shows the COPROD drill string, and Fig. 2.28 shows an Atlas Copco ROC F7CR equipped COPROD system in the Jackomini Quarry, Austria, drilling 89-mm diameter holes in the fractured rock.

Table 2.11 Comparison of general working parameters ranging between pneumatic and hydraulic rock drill rigs (Reproduced from Ref. [9] with permission from Atlas Copco)

Working parameter	Pneumatic rock drill (for Rig)	Hydraulic rock drill (surface/underground)
Drillhole diameter (mm)	48–102	64–127/35–89
Impact power (kW)	7.2–12	12–40/12–22
Impact rate (Hz)	33–39	36–75/40–73
Hydraulic pressure (bar)		210–230/190–250
Air consumption—excluding flushing (l/s)	175–354	
Rotation speed (rpm)	0–300	0–220/0–380
Rotation pressure (bar)		200–210/210
Drill steel torque (Nm)		1000–3500/520–1000
Flushing Air Pressure (bar)		8–12/10

Fig. 2.27 COPROD drill string (Reproduced with the permission from Atlas Copco)

2.4.1.3 Hydraulic Drilling Rigs

Since the seventies, rock drilling technique has seen rapid development following the development of the hydraulic technique. These new, high-power rock drills not only doubled drilling capacities but also improved drilling environment. The

Fig. 2.28 ROC F7CR with COPROD system (Reproduced with the permission from Atlas Copco)

hydraulics to rock drilling also led to improvement in drilling accuracy, mechanization, and automation.

The following photographs (Figs. 2.29 and 2.30) show the basic components of the hydraulic rock drilling rigs working in surface and underground excavation.

For the surface drilling rigs, the power pack, including the hydraulic power supply and air compressor, is driven by a diesel engine. But for the underground drilling jumbo, it usually is driven by electric motors for reducing the air pollution of the underground working space. When a surface drilling rig works, the main source of noise is the top hammer drifter. The recently introduced silenced drill rig is for use especially in urban areas where noise levels are restricted. A soundproofing enclosure kit is designed for the drilling components, resulting in a 10 dB (*A*) external noise reduction. Figure 2.31 is a silenced Smart Rig, ROC D7C, manufactured by Atlas Copco, which was firstly used in 2006 in Finland and 2008 in Hong Kong (Fig. 2.31).

2.4.2 *Down-the-Hole (DTH) Drilling*

Down-the-hole (DTH) drilling, also known as in-the-hole (ITH) drilling, is a method in which the percussive hammer works in the hole during drilling. In this system, the hammer (piston) strikes directly on the bit, and no energy is lost through joins in the drill string (Figs. 2.21, 2.32). The piston strikes the drill bit directly,

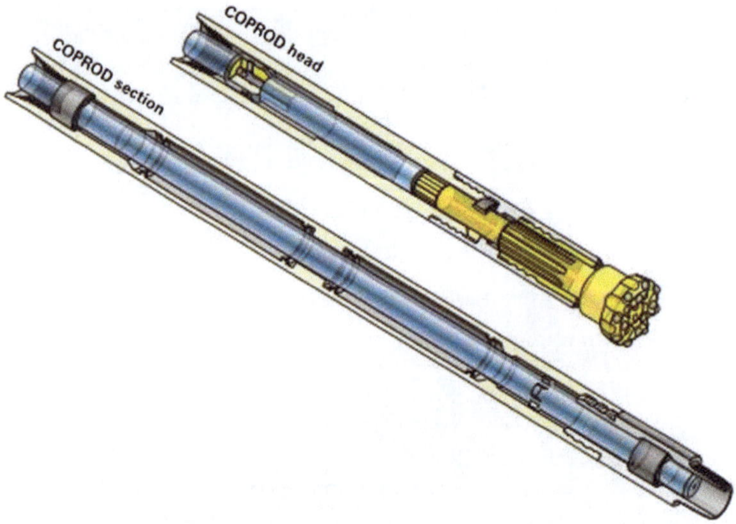

Fig. 2.29 Basic components of the surface drilling rig (Reproduced with the permission from Atlas Copco)

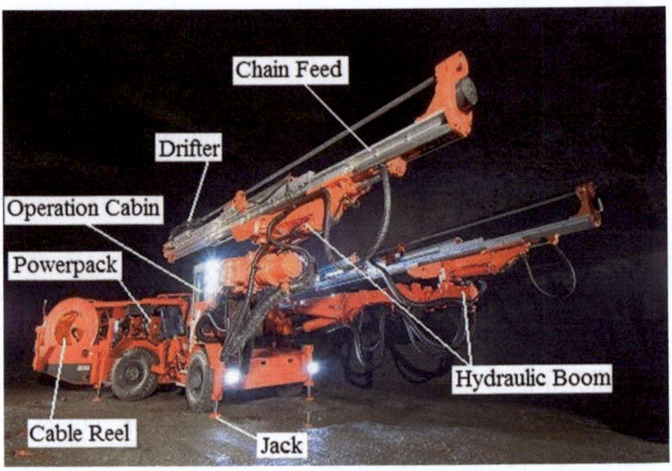

Fig. 2.30 Basic components of the underground drilling jumbo (Reproduced with the permission of Sandvik)

while the hammer casing gives straight and stable guidance of the drill bit. This results in minimal deviation and great hole wall stability, even in fissured or otherwise demanding rock. The driving fluid of the hammer is compressed air that is supplied through a tube which serves as support and makes the hammer turn. The rotation and thrust force are carried out by two separate hydraulic motors (or

Fig. 2.31 Silenced drilling rig (Reproduced with the permission from Atlas Copco)

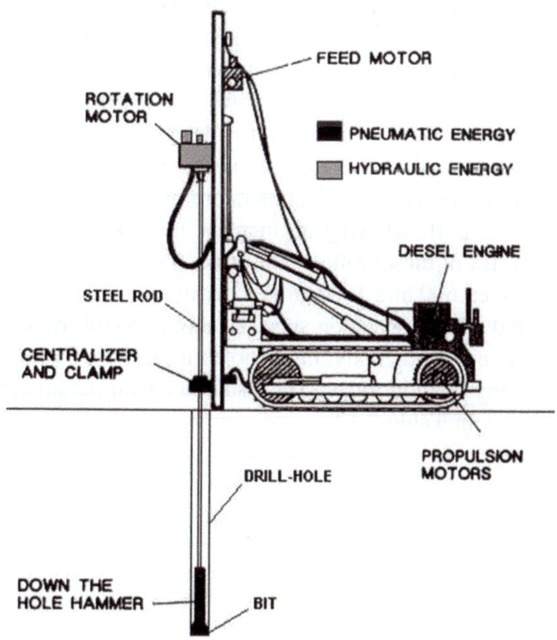

Fig. 2.32 Down-the-hole drilling rig

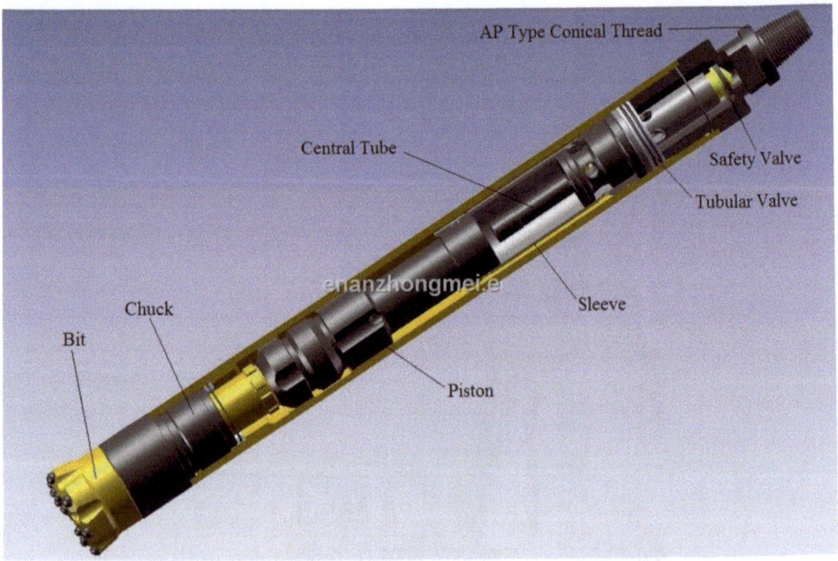

Fig. 2.33 General structure of DTH drill hammer (courtesy of Henan Zhongmei Drill Tool Co. Ltd., China)

pneumatic motors) mounted on the surface rig. Flushing is carried out with the exhaust air of the hammer through the holes in the drill bit. Since the annulus between the drill pipes and the hole wall is comparative small, a high flushing velocity is maintained, which contributes further to hole quality. DTH method is widely used to drill long holes, not only for blasting, but also for water wells, shallow gas, oil wells, and for geo-thermal wells. From an environmental point of view, as the hammer is working in the hole, the noise emission from DTH drilling is comparatively low. This is of particular advantage when drilling in densely populated areas. Figure 2.33 shows the general structure of an original DTH drill hammer. Today's DTH hammer design is much simpler than the original one which had a butterfly valve incorporated to direct the air alternately to the top part of the piston. The valveless hammers are operated through ribbings or projections of the piston itself, allowing an increase in strike frequency, lowering air consumption, and risk of dieselization. The strike frequency for down-the-hole hammer is usually between 600 and 1600 strikes per minute. The air pressure used is usually between 6 and 24 bar. Rotation speed is about 25–100 rpm, and the feeding force is varied between 6 and 20 kN. In the hole range 100–254 mm, DTH drilling is the dominant drilling method today. The main technical parameters of some DTH hammers are shown in Table 2.12.

Table 2.12 Technical parameters of some DTH drill hammers

Hammer	Drilling dia. (mm)	Hammer dia. (mm)	Hammer weight (kg)	Working pressure (bar)	Air consumption (m^3/min)
Atlas Copco					
COP 34	92–105	83.5	27	12–24	6–16.8
COP 54	134–152	120	57	12–24	10.8–24
COP 64 Gold	156–178	142	96	12–28	12–36
SANDVIK					
EH550r, 4″	110–133	98	38	10–24	5.1–15.9
EH550r, 5″	130–152	126	73	10–24	8.64–24.21
EH550r, 6″	152–203	150	82	10–24	9.8–27.5
Ingersoll Rand					
DHD340A	110–125	98	32.1	10.4–24.1	
QL5/QL50	140–152	125	60.3	10.4–24.1	
China					
ZX-115	110–140	98	36	12–20	3.5–15
ZX-150	152–165	136	98	7–21	8.5–25

2.5 Rotary Drilling

The rotary drills include two kinds of drilling method: rotary crushing with tricone and fixed-type bits. The fixed-type bits, such as claw or drag bits, have no moving parts and cut through rock by shearing it. Thus, these bits are limited to the softest materials. The primary difference between rotary drilling and other methods is the absence of percussion.

2.5.1 *Rotary Drilling with Rolling Tricone Bits*

Rotary crushing is a method, which was originally used for drilling oil wells, but it is nowadays also employed for the blasthole drilling in large open pits and hard species of rocks. In most rotary applications, the preferred bit is the tricone bit. Tricone bits rely on crushing and spalling the rock. This is accomplished through transferring downforce, known as pulldown, to the bit while rotating in order to drive the carbides into the rock as the three cones rotate around their respective axis. All rotary drilling requires high feed pressure and slow rotation. The relationship between these two parameters varies with the type of rock. In soft formations, low pressure and higher rotation rate and vice versa are the logic usually followed.

Fig. 2.34 Crawler-mounted rotary drilling rig (Reproduced with the permission from Atlas Copco)

Rotation of the tricone bit is provided by a hydraulic or electric motor-driven gearbox (called a rotary head) that moves up and down the tower via a feed system and through a pipe transferring to the bit. Feed system utilizes cables, chains, or rack-and-pinion mechanisms driven by hydraulic cylinders, hydraulic motors, or electric motors. A typical rotary drilling rig with tricone bits is shown in Figs. 2.34 and 2.35 shows the typical structure of a tricone bit (Table 2.13).

2.5.2 *Rotary Drilling with Drag Bits*

A drag bit with no moving parts that cut by a combination of shearing (cutting action) and gouging, predominantly used in softer sedimentary rock types.

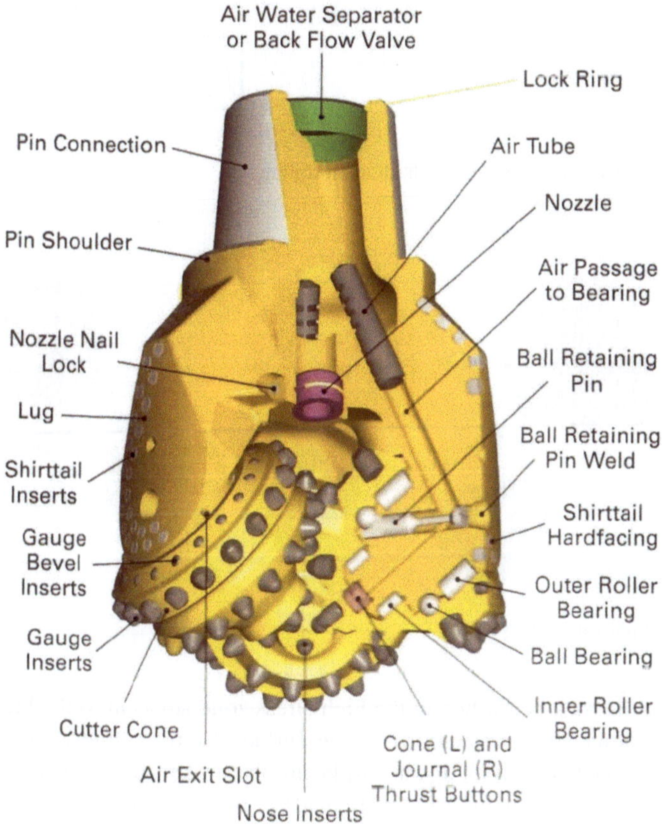

Fig. 2.35 Structure of a typical tricone bit (Reproduced with permission from Atlas Copco)

Table 2.13 The range of working parameters of rotary drilling rig with tricone bits

Nominal hole size (mm)	102–406
Hole depth (m)	12–85
Feed system	
Pulldown force (kN)	111–534
Weight on bit (kg)	11,300–56,700
Rotation torque (k Nm)	4.7–25.7
Diesel engine/electric motor (kW)	336–1230

The cutting action of a drag-type rotary drill bit is performed by a variety of tools, including blade and diamond drills as well as rope, chain, and rotary saws. Regardless of the geometry of the device, drag action at the cutting surface by two forces: thrust, a static load acting normally; and torque, the tangential force component of a rotational moment acting on the rock surface.

The mechanism of penetration in drag bit drilling is as follows (Fig. 2.36): (a) As the cutting edge of the bit comes in contact with rock, elastic deformation

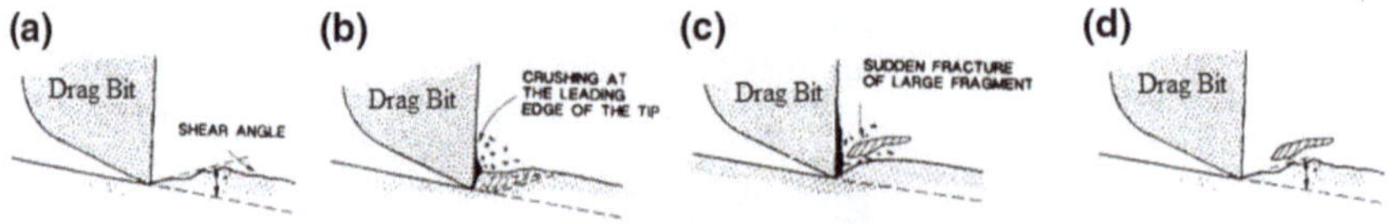

Fig. 2.36 Mechanism of penetration in drag bit drilling

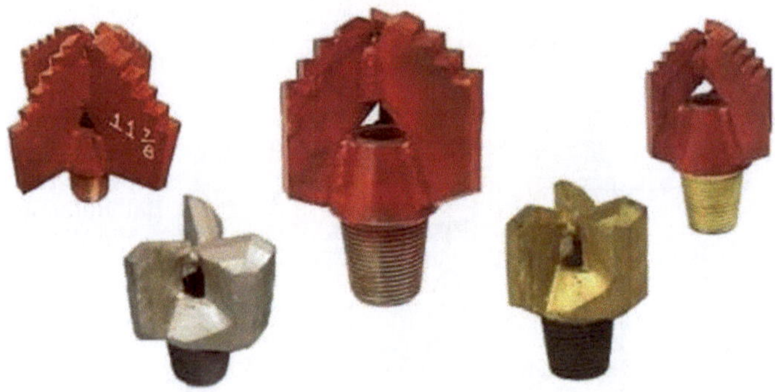

Fig. 2.37 Wing rotary drag bits

occurs; (b) the rock is crushed in the high-stress zone adjacent to the bit; (c) cracks propagate along shear trajectories to the surface forming chips; and (d) the bit moves to contact solid rock again, displacing the broken fragments.

The edges of the drag bit usually are made of tungsten carbide or other materials such as synthetic diamonds or polycrystal, which vary in shape and angle. Figure 2.37 shows some drag bits.

The cuttings of drilling are eliminated with a flushing fluid that can be air, in surface operation, water, or humid air in underground operations.

The auger drag drill, in which a hollow-stem augur is rotated into the ground, needs no mud or flushing. Figure 2.38 is a two-wing auger drag drill for coal and soil drilling. The continuous-flight augurs convey the cuttings continuously to the surface.

2.6 Rotary-Percussive Drilling Accessories

To drill a hole, apart from the rock drill, some drilling accessories are required. Except the integral drills, the extension drill steel is usually made up of the following elements: shank adaptor, coupling sleeves, extension rods, and drill bits (Fig. 2.39). We will discuss them in this section one by one.

Fig. 2.38 Auger drill with two-wing drag bit

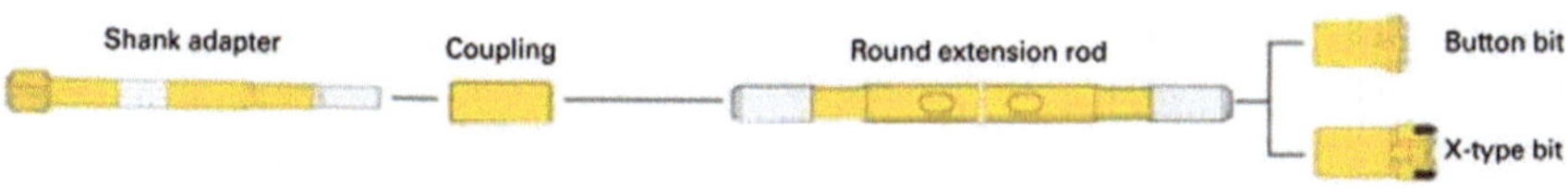

Fig. 2.39 Drilling accessories of rotary-percussive drills (Reproduced with the permission from Atlas Copco)

2.6.1 Integral Drill Steels

An integral drill steel as shown in Fig. 2.40 consists of a rod with a forged shank at one end and a forged bit with cemented carbide inserts at the other end. Thus, each

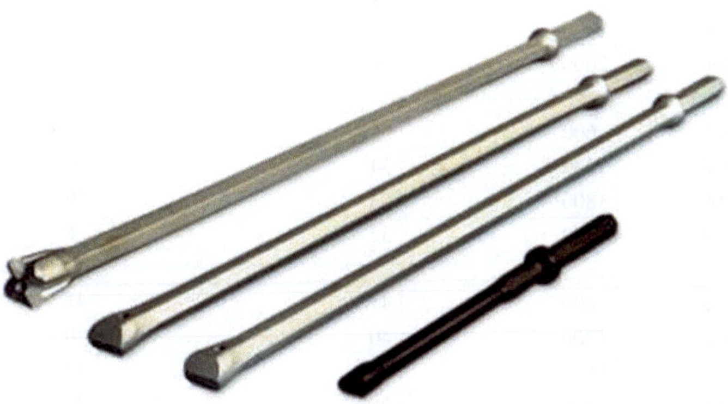

Fig. 2.40 Integral drill steels

drill steel is of a specific length and cannot be extended. Once the first drill steel has drilled all the way into the rock, it is withdrawn and replaced by the longer one to drill further into the rock. The integral drill steels are available in increasing lengths with reduced diameters as shown in Table 2.14. The diameter of the longest steel should be selected as per the size of explosive cartridge. The most common integral drill steel is chisel-type (I-type) and other types include multiple-insert steel, button

Table 2.14 Integral drill steel series

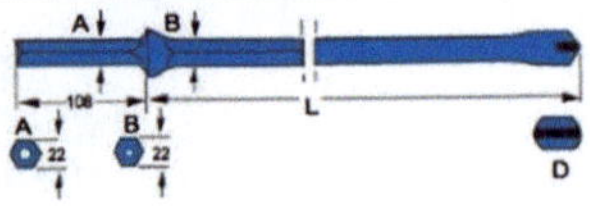

Shank (mm)	Length (mm)	Bits diameter, *D* (mm)	Weight (kg)
11 Series			
Hex 22 × 108	800	34	3
Hex 22 × 108	1600	33	5.5
Hex 22 × 108	2400	32	8
Hex 22 × 108	3200	31	10.5
Hex 22 × 108	4000	30	12.9
Hex 22 × 108	4800	29	15.4
12 Series			
Hex 22 × 108	800	40	3
Hex 22 × 108	1600	39	5.5
Hex 22 × 108	2400	38	8
Hex 22 × 108	3200	37	10.5
Hex 22 × 108	4000	36	12.9
Hex 22 × 108	4800	35	15.4
13 Series			
Hex 22 × 108	400	34	1.8
Hex 22 × 108	800	33	3.1
Hex 22 × 108	1200	32	4.3
Hex 22 × 108	1600	31	5.5
Hex 22 × 108	2000	30	6.7
16 Series			
Hex 22 × 108	600	35	2.4
Hex 22 × 108	1200	34	4.3
Hex 22 × 108	1800	33	6.1
Hex 22 × 108	2400	32	8
17 Series			
Hex 22 × 108	600	41	41
Hex 22 × 108	1200	40	40
Hex 22 × 108	1800	39	39
Hex 22 × 108	2400	38	38

steels, double chisel steels, and cross edge bit steels. The integral drill steels are usually used in handheld drills or light rigs, and the steels are manually changed.

2.6.2 Type of Threads

As shown in Fig. 2.39, the drill string, that is the shank, the drill rod, and the bit, is jointed together by threads. The string must be firmly connected to give efficient energy transmission. At the same time, the thread needs to be fairly easy to open when coupling and uncoupling rods. These features are required even when the threads are part of a long drill string.

The ease with which the thread can be tightened and loosened depends on multiple factors but mainly the thread design which includes pitch, angle between flanks, material, and surface properties.

The most common threads are the *R*, *T*, and *S* (Fig. 2.41).

The *R* thread has a small pitch and a large angle between the flanks. *R* thread is used in small blastholes with drill rods of 22–38 mm and powerful independent rotation rock drills

The *R* thread has a small pitch and a large angle between the flanks. *R* thread is used in small blastholes with drill rods of 22–38 mm and powerful independent rotation rock drills with air flushing.

The *T* thread is used in most drilling conditions with drill rod diameters of 38–51 mm. It has a greater pitch and smaller flank angle than the *R* thread.

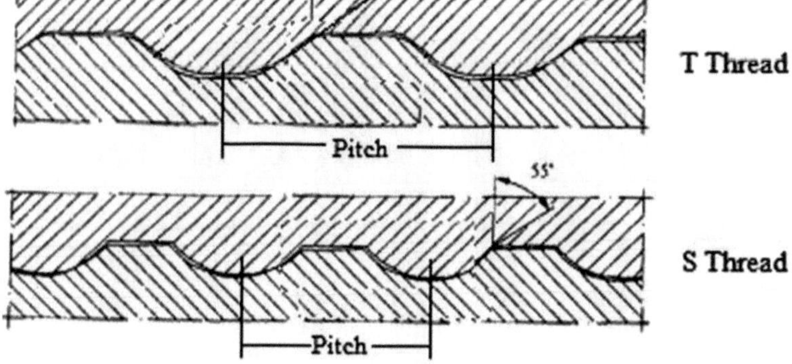

Fig. 2.41 Profiles of the *R*, *T*, and *S* threads

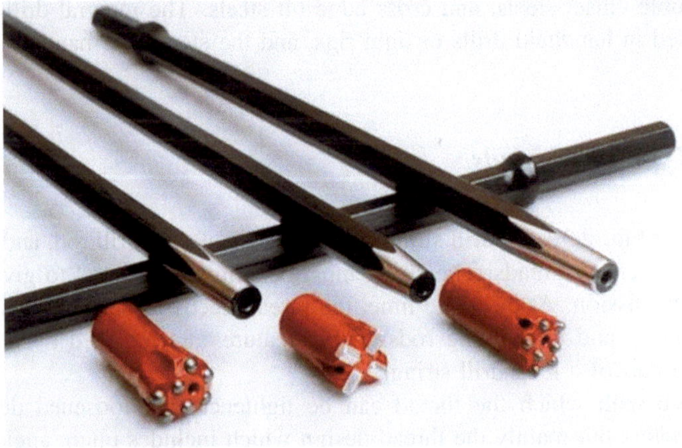

Fig. 2.42 Tapered drilling rods and bits (Reproduced with the permission of Sandvik)

The *S* thread has the same angle between the flanks as the *T* thread but a smaller pitch and used in large extension rods of 51 mm.

The tapered drill steels and bits without thread are still used at present especially in handheld drilling (Fig. 2.42).

2.6.3 Shank Adapters

The shank adapter is the component that enables percussive impact and rotation to be transmitted from the rock drill to the drill string. Shank adapter is made of high-quality special steel with excellent wear resistance. The hard surface is obtained by carburization. Shank adapters are specially designed for particular rock drills (Fig. 2.43). Where there is a separate flushing system, the flushing medium enters the shank through a side hole between the splines.

Fig. 2.43 Shank adapters (Reproduced with the permission of Sandvik)

2.6.4 *Drill Rods*

Drill rods can be divided roughly into five categories (Fig. 2.44):

- Shank rods
- Drifter rods
- Extension rods
- Drill tubes
- Guide rods.

The length and diameter of the rods depend on the hole/bit size and hole depth. Percussive rods have a hole inside to provide a means of flushing for the bit end.

Shank rods have an integral shank at one end and thread for the bit at the other end, similar to integral steels.

2.6.4.1 Drifter Rods

Drift rods have threads at both ends, with the shank end thread usually bigger than the bit end thread. The shank end of the rod may have a female or male thread (Figs. 2.44, 2.45). The drifter rod is designed for fast drilling of short holes, especially for the underground drilling Jumbos.

Typical Drifter rod lengths are 10, 12, 14, 16, and 20 ft. Today with modern drilling and blasting knowhow, rods shorter than 10 ft have become quite rare. There are two types of rods: hexagonal and round. The diameter of rods varies: 25, 28, 32, 35, and 39 mm.

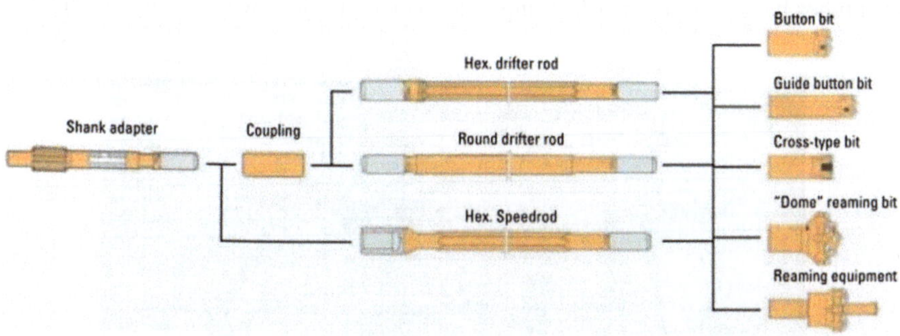

Fig. 2.44 Drilling accessories for underground drills (Reproduced with the permission from Atlas Copco)

Fig. 2.45 Drifter rods (Reproduced with the permission of Sandvik)

2.6.4.2 Extension Rods

Extension rods are made for either light or heavy drilling, although the heavy extension is more commonly used. The threads at both ends of the rod have the same dimensions. Both threads can be either male or female (Fig. 2.46). The lengths of rods vary from 10 to 20 ft for surface drilling and 3 to 6 ft for underground drilling. The diameters vary from 32 to 52 and 60 to 87 mm for heavy drilling.

2.6.4.3 Drill Tubes

With the application of top hammer hydraulic rock drills to the drilling of large diameter and long blastholes, some drill tubes have been recently used in both surface and underground drilling, which are similar to those used in down-the-hole drilling, especially when drilling in difficult rock formation with soft, broken rocks. Drill tubes have a large inner flushing hole, which provides water or air for effective

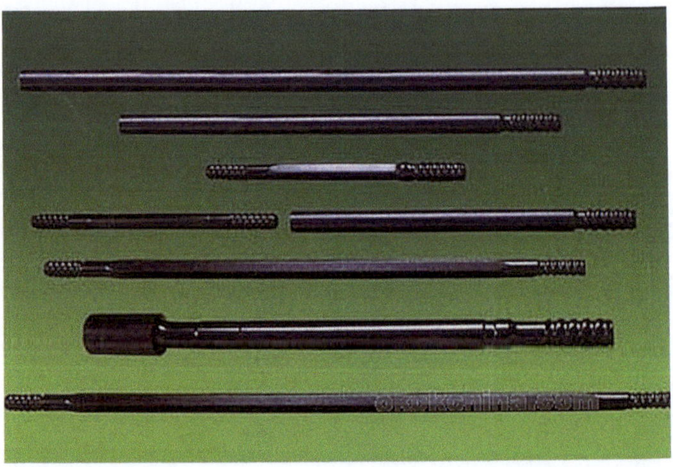

Fig. 2.46 Extension rods (Reproduced with the permission of Sandvik)

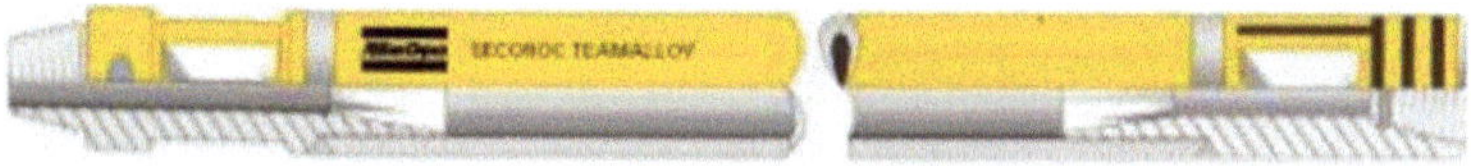

Fig. 2.47 Drill tube (Reproduced with the permission from Atlas Copco)

flushing. They have a female–male connection and do not require any couplings. The tight joints enable maximum energy transfer. As they have larger diameters and more rigidity, the deviations and irregular blasthole walls are reduced (Fig. 2.47). The diameter of the drilling tube is only slightly smaller than the recommended optimal hole size. The tube diameters are varied between 76 and 127 mm, and lengths varied between 1.5 and 6.1 m.

Drill tubes for DTH drilling usually have a diameter of 76–140 mm and lengths of 4.0–6.0 m.

There are also guide tubes, used immediately after the guide or retrace bit, which have two guiding sections to improve accuracy and hole straightness. The guide tube has a female thread at one end and a male thread at the other.

2.6.5 Couplings

Couplings are required to extend the drill string to the desired length and to join the rods together in such a way that the energy is transferred from one rod to the next all the way to the bit. The correct type of coupling promotes accurate drill steel connection and minimizes energy loss in the joint. To prevent overthreading, couplings have a stopping point in the middle. The type of coupling depends on the drilling conditions and the selected rod type (Fig. 2.48). Figure 2.49 shows the cross sections of three types of couplings.

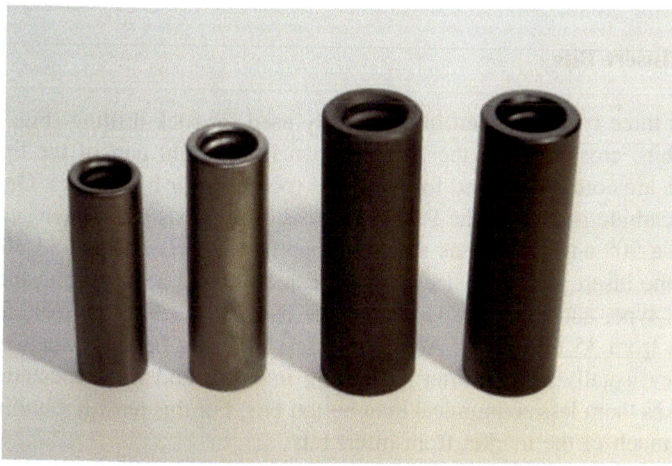

Fig. 2.48 Couplings (Reproduced with the permission of Sandvik)

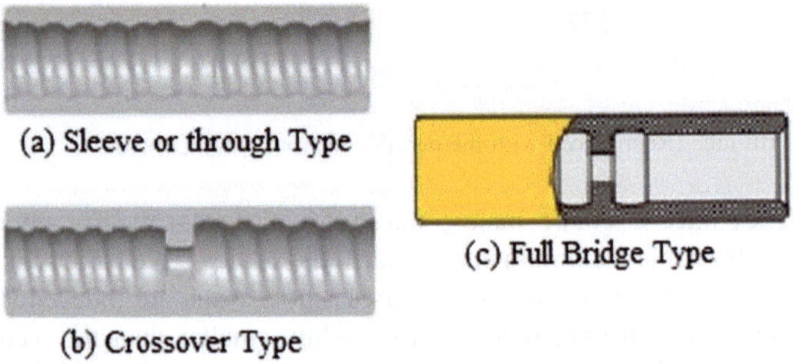

Fig. 2.49 Three types of couplings

2.6.6 *Drill Bits*

There are two types of drill bits for rotary-percussive drilling:

- Insert Bits, and
- Button bits.

For the two types of bits, there are some design characters in common:

a. The rods are threaded to the end of the bit thread so that the transmission of impact energy is as direct as possible to the rock.
b. The bits have a series of central and lateral openings through which the flushing fluid is injected and they have channels through which the rock particles produced pass upwards.
c. The bits are designed to be slightly conic, with the widest part in contact with the rock so as to counteract the wear and avoid an excessive adaptation to the blasthole wall.

2.6.6.1 Insert Bits

There are three types of insert bits presently used for rock drilling (Fig. 2.50): the chisel (I-bit), cross bit, and the X-bit (shown in the right part of the figure). The chisel bits are commonly used for handheld rock drill for hard rocks. One piece of tungsten carbide is fixed in the I-bit. The cross bit consists of four tungsten carbide inserts at a 90° angle, whereas the X-bit has four inserts at 75° and 105° angles between the insert pair. The size of insert can be varied according to the drillhole size, rock type, and the abrasiveness of the rock. Insert bits are manufactured in diameters from 35 to 64 mm. Although insert bits may be less expensive to purchase, they usually have shorter regrinding intervals and life expectation, which often makes them less economical than button bits. For this reason, button bits have captured much of the market from insert bits.

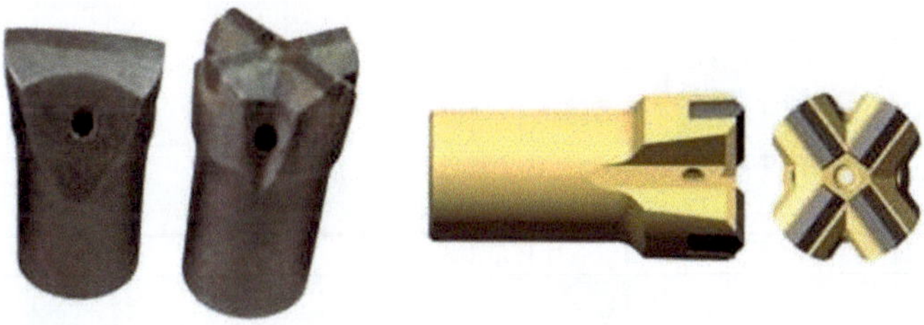

Fig. 2.50 Insert bits

2.6.6.2 Button Bits

The button bit is the most popular type of bit in use today for big hole, high production, and blasthole drilling. These bits have buttons or cylindrical inserts of tungsten carbide distributed in various patterns on the face. They are manufactured in diameters that go from 50 to 251 mm. See Fig. 2.51.

The bit face is so designed that it can achieve the following important tasks:

- Allow for rock chips to clear and avoid recutting;
- Hold gauge and retain cleaning flutes;
- Present the most effective impact alignment of carbides to break and chip the rock, and
- Drill straight.

Fig. 2.51 Button bits

Fig. 2.52 Tungsten carbide

The carbide buttons have several basic shapes and are made of various materials. The carbide material normally contains 6–12 % cobalt, and it is usually classified as soft, standard, and hard, see Fig. 2.52.

- Soft material is generally used in soft, abrasive rock to allow carbide wear to move approximate bit body wear and avoid excess carbide extension.
- Standard material is used for general drilling conditions.
- Hard materials are used for very hard, abrasive formations.

Button size generally increases with bit diameter, which allows for higher rotation speeds. Bit bodies are generally of steel composition, and the various grades and styles of carbide inserts are press-fitted into the body. For some applications, the body steel may be hardened all the way through and carburized.

2.6.6.3 Special Bits

There are some specially designed bits for the particular application:

1. Retrac bits: When collaring or other problems cause tight steels, the "retrac" bit body help to ream the bit out of the hole. A typical retrac bit has a long, large diameter with edges. The large body helps it to drill straight holes, and the edges enable the drill string to be withdrawn when spalling has occurred.
2. Reaming bits: The reaming button bits are used underground to drill the large parallel cut holes. These bits usually are used with pilot rods or extension rods and reaming bit adaptors.

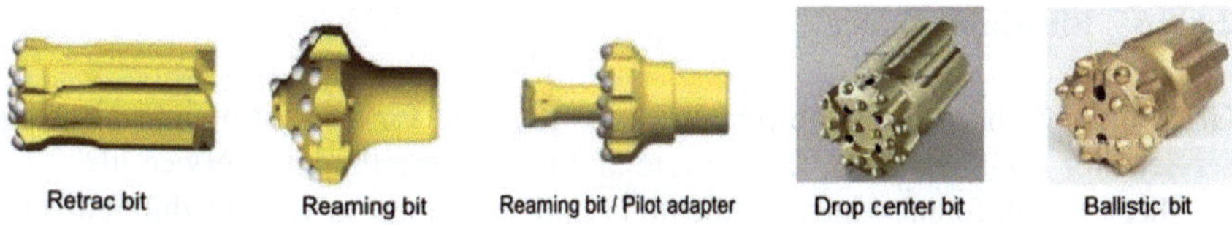

Fig. 2.53 Special drilling bits

3. Drop center bits: The drop center bits have excellent flushing characteristics, as the flushing hole of the bit is in the center of the face. They are used in soft rocks that are easy to drill.
4. Ballistic bits: The ballistic bits have bullet-shaped buttons which are longer than the standard and give high penetration rates and a more efficient flushing for soft rock formation (Fig. 2.53).

2.6.6.4 Down-the-Hole Hammer Bits

The bits for DTH hammers have shanks incorporated upon which the piston strikes directly. The most common diameters of these bits go from 85 to 250 mm, although larger ones exist.

Both insert (cross and X inserts) and button bits are used for DTH hammers, but button bits are the most commonly used and good for any type of rock. Figure 2.54 shows the common DTH bits designed for different rock formations. The manufacturers, like Atlas Copco, Sandvik, and Ingersoll-Rand, have similar series of bit design.

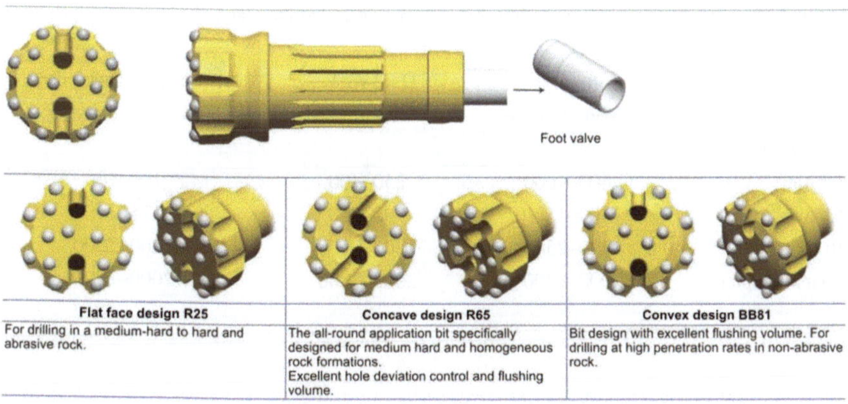

Fig. 2.54 Basic button bit face design used for DTH (courtesy of Mitsubishi, Ref. [7])

2.6.6.5 Bit Maintenance

Grinding and sharpening equipment is also required for drill bit maintenance and services for optimizing the drilling rate and to increase the bit's service life.

There are different kinds of methods and machines to regrind different bits: button bits, insert bits, and integral drill steels. The drill equipment suppliers usually supply the services of grinding and sharpening of bits together with the maintenance service of the equipment.

2.6.7 *The Service Lifetime of Drilling Accessories*

The drill bit re-grinding and service lifetime depends on the geology conditions (especially rock abrasiveness), bit configuration and quality, drilling conditions (flushing, rotation, percussion, and feed), and handling. The bit lifetime can be as much as 800 drilled meters or as little as 30 drilled meters. Acceptable lifetime estimates can be based on various laboratory tests and drilling information from similar conditions.

The drill rod lifetime is measured in rod meters, a term expressing the total number of meters drilled by all the rods in a drill string for a given number of hole meters. Table 2.15 (from [4], Tamrock) can be referred to estimate the service lifetime of extension rods and tubes.

Shank adapters transmit impact, rotation torque, and feed force. Constant exposure to high impact energy levels is directly related to the shank lifetime. Abnormal drilling conditions such as too high or too low percussion pressure and too low fed pressure reduce the service life of the shank.

Table 2.15 Service lifetime for drilling bits, extension rods/tubes, and shank adapters (courtesy of Tamrock)

Component	Soft rock	Hard rock	Hard rock
	Non-abrasive	Medium-abrasive	Very-abrasive
Button bits			
45 mm	600–1000	250–400	100–200
51 mm	1000–1500	500–600	175–250
Rods	1200–2500	1200–2500	1200–2500
Shank adapters	2000–3000	2000–3000	2000–3000

2.7 Selection of Rock Drill and Accessories

2.7.1 *Fields of Application for Different Drilling Methods*

By taking into account the rock drillability and the drilling diameter, the following Fig. 2.55, as a reference, gives the fields of application for different drilling methods as the function of rock drillability and drillhole diameters.

2.7.2 *Principles of Selection of Drilling Equipment for Surface Excavation*

The following principles should be considered firstly for selection of drilling equipment for surface excavation:

- Scale and complexity of the project, the total amount of materials to be excavated, and the project schedule for excavation.
- Geological conditions of the project, especially the rock drillability.
- Environment conditions, including the distance from the residential area, slopes, and other sensitive receivers (structure, building, and utilities), relative laws and regulations.
- The conditions and costs of maintenance and services for the equipment.
- Initial investment for purchase of drilling equipment.

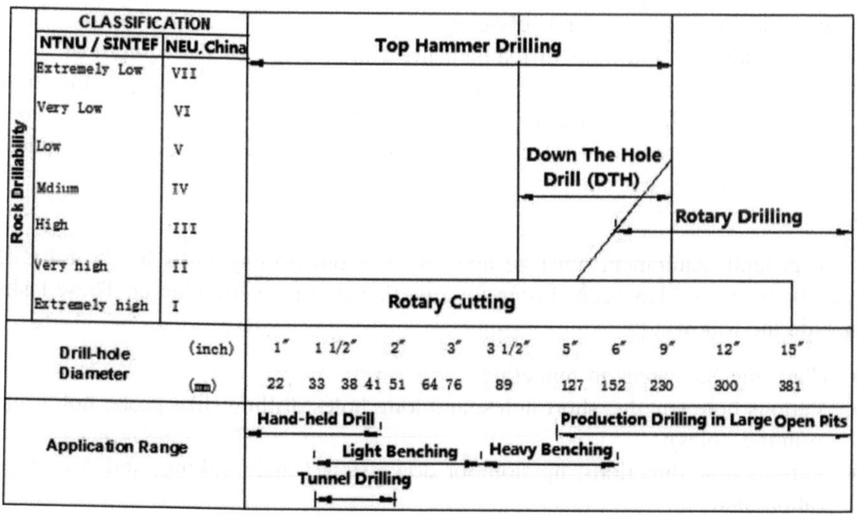

Fig. 2.55 Field of application for drilling methods as function of the rock drillability and the drillhole diameters

Under these conditions, the selection of drilling equipment can be determined step by step:

- The scale of each blasting, blasthole diameters, maximum drillhole depth;
- Total amount of drillhole length, including blastholes and other holes, such as soil nails/rock bolts, and drilling productivity are required;
- Drilling method, capacity of drilling equipment, type and modes of drilling equipment, number of drills to be used for the project; and
- Estimating the quantity of drill accessories to be used for the project.

2.7.3 *Selection of Drilling Equipment for Underground Excavation*

Similar to the surface excavation but more complicated, the following factors must be taken into account:

- Scale and complexity of the project. The total amount of materials to be excavated and the project schedule for excavation.
- Geological conditions of the project, especially the rock drillability.
- Environment conditions, including the distance from the residential area, slopes, other sensitive receivers (structure, building, and utilities), relative laws and regulations.
- Compatibility with other excavation equipment, like loading and hauling, for the job. The equipment must be technically advanced but compatible with existing machines and anything else being purchased. This compatibility must also extend to maintenance and servicing.
- The conditions and costs of maintenance and services for the equipment.
- Detailed calculations are necessary to determine which equipment is the most economical, efficient, practical, and technically suitable.

  During the procedure of selection of drilling equipment, the following technical aspects must be considered:
- Versatility.

In general, equipment must be able to carry out drilling tasks in a variety of conditions, even it has been chosen for a particular construction target. These tasks include the following:

- Changing face areas in tunneling;
- Various hole lengths, short holes, and long holes drilling (like probe holes and grouting holes);
- Various hole directions, up-slope or down-slope, shaft sinking, and rise excavation; and
- Bolting: amount and frequency of bolting, different bolt types, length, and size.

The best possible solution in each drilling group will be able to perform efficiently while covering all major drilling tasks.

- Selection of carrier.
  Three types of carrier are available: rail-mounted, crawler-mounted, and wheel-mounted, and deciding which type is suitable for a particular job depends on several factors. Table 2.16 gives the basic criteria of selection. Rail-mounted drill jumbo is very rarely used for construction projects.
- Selection of the boom
  The following aspects should be taken into account:

  1. Coverage. Booms have a slightly different coverage when they are mounted on a drilling rig. The effective boom coverage must be sufficient for the whole tunnel face, including the upper and lower corners of the tunnel. The coverage changes with the number of booms and when the jumbo carrier is changed, since it affects the boom mounting distance and height. Usually a drilling jumbo can be equipped with one to three booms and the coverage changes from smallest of 12 m^2 to largest of 230 m^2. Figure 2.56 shows the coverage of two manufactures' jumbos.
  2. Selection of feed. The length of the feed, which decides the length of the hole and the round, is determined by the excavation timetable and any rock mechanical and geological restrictions. The drifter dimension and selected rod length also should be considered when choosing the feed.
  3. Selecting a boom and rod changer for long hole drilling.
  4. Selecting a boom for bolt drilling.

Table 2.17 gives the main specifications of some drilling jumbos for tunneling manufactured by Atlas Copco and Sandvik.

2.7.4 *Selection of Drilling Accessories*

2.7.4.1 The Features of Button Bit Design

Most percussive button bit is designed with a basic face profile and button layout similar to one of those shown in Fig. 2.57 below.

The typical shapes to be used of carbide and their characters are shown in Fig. 2.58.

With different combinations of bit face profiles, carbide shapes, and carbide materials, there are various models of drill bits to suit different rock conditions and penetration rate. The typical bit models include the following:

* CV: convex/ballistic front bit;
* FB: full-ballistic bit;
* FF: flat front bit;
* FF HD: flat front heavy duty bit;

Table 2.16 Carrier selection criteria

Tunnel length/Face area	Road/floor conditions	Curves	Speed/moving frequency	Angle of slope	Carrier
Long/wide-narrow	Good	Wide	Moderate-high/moderate-often	Horizontal	Rail-mounted
Short-long/wide-narrow	Good-moderate	Tight-moderate	High/moderate-often	Moderate-horizontal	Wheel-mounted
Fairly short/wide	Rough	Wide	Slow/rare	Very steep	Crawler-mounted

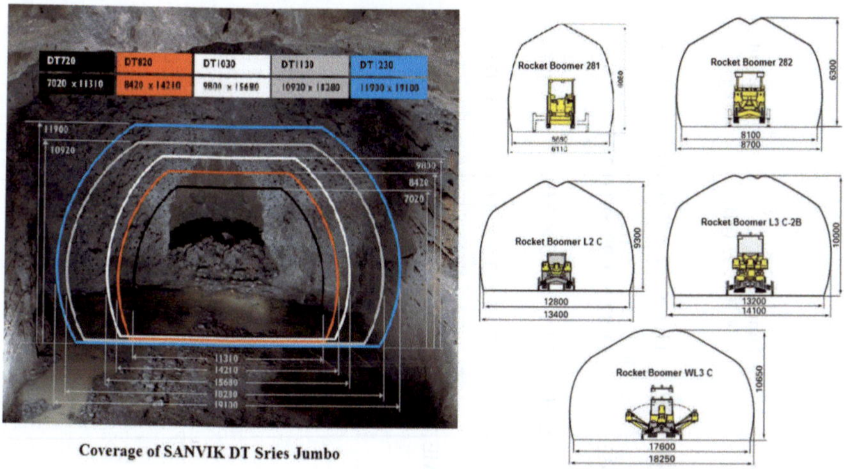

Coverage of SANVIK DT Sries Jumbo

Coverage of Atlas Copco's Jumbo

Fig. 2.56 Coverages of Sandvik's DT series and Atlas Copco's jumbos (no scale)

* DC: drop center bit;
* DC XHD: drop center extra heavy duty

Refer to Fig. 2.59.

2.7.4.2 Selection of Drilling Bits

The button bit is the most popular type of bit in use today for big hole, high production, and blasthole drilling. This is mainly because the button bit has a higher penetration rate and better wear resistance than the insert bit under the intensive hammering and tough rock conditions and requires less maintenance. However, when very straight holes are required to be drilled, the cross or X-type insert bit has its advantage. Figure 2.60 shows how to select the button bit under different rock conditions.

2.7.4.3 Drilling Rod Selection

Top hammer Rods Selection for Surface Drilling

For bench drilling, three types of drill rods can be chosen:

- Surface hardened rods, in which only the thread parts are hardened, are the toughest but have the lowest fatigue strength. They are the good choice when drilling in faults or folded formations, when driller is a green hand as they are the cheapest rods;

Table 2.17 Main Specifications for Some Drilling Jumbos for Tunneling

Company/model	Coverage (m^2)	Hole Dia. (mm)	Boom	Feed	Rock drill	Dimension $L \times W \times H$ (m)	Weight (kg)
ATLAS COPCO							
Rocket Boomer 281	Up to 31	43–76/64–89	1 × BUT 28	1 × BMH 2831-BMH 2849	1 × COP1838ME/COP 1838HF	11.7 × 1.7 × 2.8(2.1)	9300
Rocket Boomer 282	Up to 45	64–89	2 × BUT 28	2 × BMH 2831-BMH 2849	2 × COP1838ME	11.82 × 1.98 × 3.0(2.3)	17,500
Rocket Boomer L2 C	Up to 104	43–76/64–89	2 × BUT 35G	2 × BMH 6814-BMH 6820	2 × COP1838ME/COP 1838HF	14.17 × 2.5 × 3.01	23,600
Rocket Boomer L3 C-2B	Up to 114	43–76	3 × BUT 35G	3 × BMH6800-series	3xCOP1838ME	17.17 × 2.5 × 3.66	50,000
Rocket Boomer WL3 C	Up to 163	43–76	3 × BUT 35G	3 × BMH6800-series	3 × COP1838ME	17.22 × 3.01 × 3.66	43,000
SANDVIK							
DT921i	12–125	43–64	2 × SB100i	2 × TF5i 12–21 ft	2 × RD525	15.4 × 3.25 × 4.13	29,000
DT1121i	20–183	43–64	2 × SB150i	2 × TF5i 12–21 ft	2 × RD525	17.78 × 3.86 × 4.69	37,700
DT1131i	20–183	43–64	3 × SB150i	3 × TF5i 12–21 ft	3 × RD525	17.78 × 3.86 × 4.69	44,000
DT1231i	20–211	43–64	3 × SB150i	3 × TF5i 12–21 ft	3 × RD525	17.78 × 3.86 × 4.78	45,500
DT1331i	20–232	43–64	3 × SB150i	3 × TF5i 12–21 ft	3 × RD525	17.78 × 3.86 ×.965	50,000

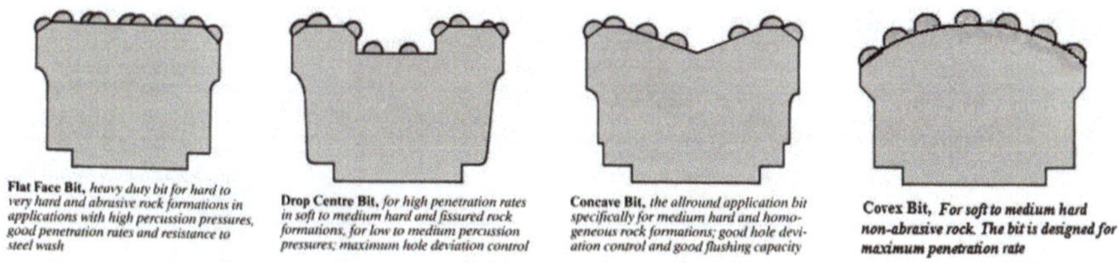

Fig. 2.57 Typical button bit face profiles (Reproduced with the permission of Sandvik)

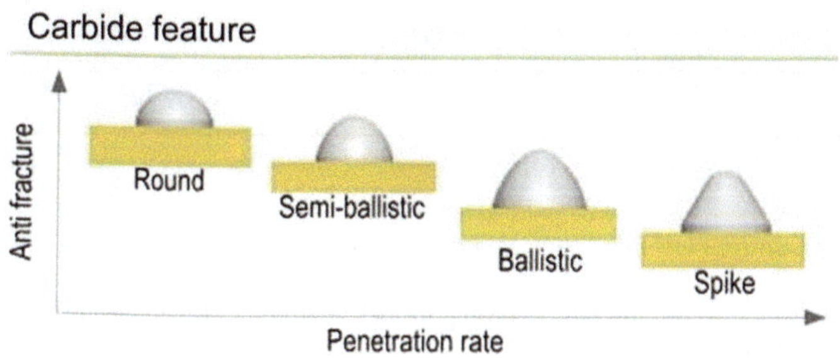

Fig. 2.58 Carbide shapes and characters (courtesy of Mitsubishi, Ref. [7])

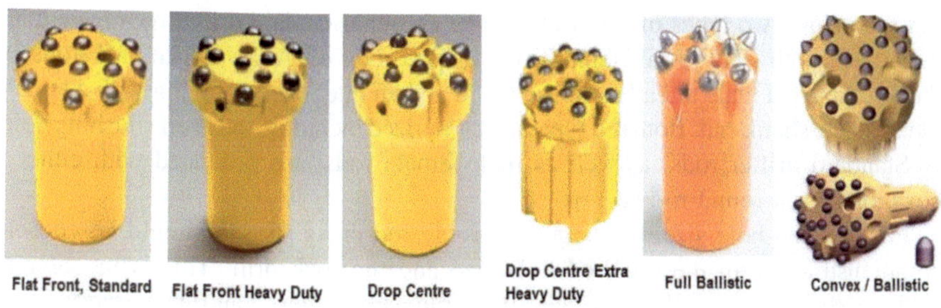

Fig. 2.59 Typical models of button bits (Reproduced with the permission from Atlas Copco)

- Carburized rods, where all surfaces, including the inside of the flushing hole, are hardened. Carburized rod has better wear resistance and a higher fatigue life compared to surface hardened rod; and
- Carburized male/female rods (Speedrods of Atlas Copco, MF-rods of Sandvik), which have male thread in one end and a integrated female coupling in opposite end. As the integrate coupling is used, the energy loss in the joints of carburized male/female rods are about 50 % less than normal carburized rods which use the standard couplings.

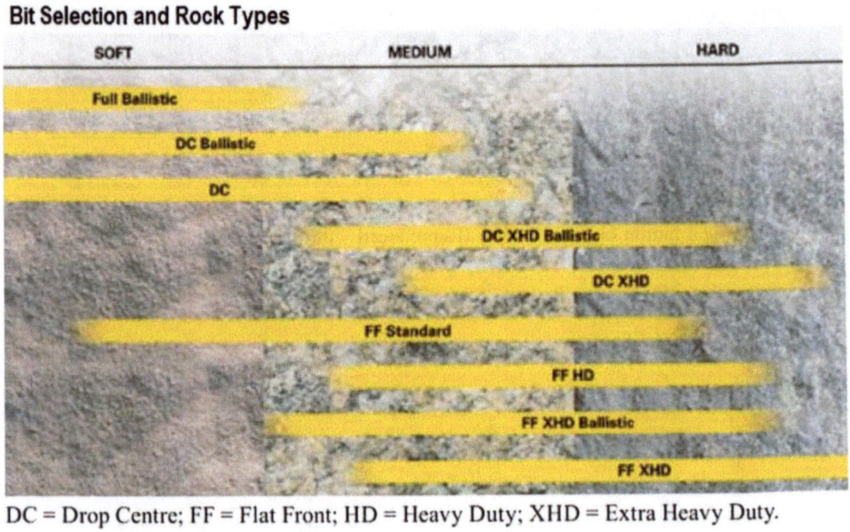

Fig. 2.60 Selection of drilling bits according to rock conditions (Reproduced from ref. [8] with the permission from Atlas Copco)

Rods Selection for Underground Jumbos

For tunneling, two types of drill rods can be chosen. Standard drifter rods have male threads at both ends. Male/female rods (Speedrods of Atlas Copco, MF-rods of Sandvik) have a male thread at the front end and an integrated coupling with a female thread at the shank end. Both rod types are carburized including the inside flushing hole.

Standard drifter rods, as well as male/female rods, are produced with either a hexagonal or a round rod section.

For a given hole size, the largest possible rod cross section should be chosen, commensurate with the required hole size and the rock drill. This is in order to achieve the best possible service life, hole straightness, and penetration rate. Normally, a rod with a T38 or R38 thread in the shank end will be chosen. The long middle section of the drifter rod is generally hexagonal, with a 32- or 35-mm cross section. Round 39-mm rods are getting more and more common, especially if the hole is 4 m and longer. The bit end of the rod is similar and has a smaller thread in order to fit the small bits and hole size used. Hexagonal rods are today's standard, while round rods, diameter 39 mm, have started to become of more and more common. The round rod is a stiffer rod, because of more material in the cross section. Round rods give straighter holes and are therefore recommended when hole deviation is a problem. But using the round rod the flushing properties for cleaning the cuttings out of the hole are not good as the hexagonal rod. This can result in higher risk of jamming round rods when drilling in the fractured rock formation, mainly when drilling 45-mm holes or smaller. In homogeneous rock, this is normally not a problem.

References

1. Sandvik/Tamrock (1999) Rock excavation handbook—for civil engineering.
2. 徐小荷, 余靜 (1984) 岩石破碎學. 煤炭工業出版社. Xiaohe X, Jing Y (1984) Theory of rock breakage. Coal Industry Press (**in Chinese**)
3. Dahl F et al (2011) Classification of properties influencing the drillability of rock, based on the NTNU/SINTEF test method. Tunn Undergr Space Technol 28(2012):150–158
4. Puhakka T (1997) Underground drilling and loading handbook. Tamrock Editor
5. 費壽林等 (1980) 鑿碎法岩石可鑽性分級. 冶金工業出版社. Fei S et al (1980) Rock drillability classification with the method of impact penetrant. Metallurgical Industry Press (**in Chinese**)
6. 單守智 (1994) 用微型鑿測器測定岩石的可鑽性. 金屬礦山 1994(1):4–10. Chan S (1994) Measuring rock drillability by mini-drillgauge. Metal Mine 1994(1):4–10
7. Mitsubish (2012) Rock tools—RT01B, Mitsubish Materials Corporation. http://mrt.mitsubishicarbide.com. Accessed 14 Apr 2016
8. Atlas Copco (2009) Surface drilling, 4th edn. (Reference Booklets (CD) from Atlas Copco Rock Drills AB, First edition 2010)
9. Atlas Copco (2006) Face drilling, 3rd edn. (Reference Booklets (CD) from Atlas Copco Rock Drills AB, First edition 2010)

Chapter 3
Explosives

3.1 History of Explosives Development

* ***Black powder (gunpowder)*** is the first explosive substance mankind learnt to use and also one of the four great inventions of ancient China. As early as 492, Chinese alchemists had noted that saltpeter burns with a purple flame, allowing for practical efforts at purifying the substance, one of the most important ingredients in black powder. The first reference to black powder in China occurs in a Daoist work from the mid-ninth century, the *Zhengzhou miaodao yaolüe*. However, the earliest surviving chemical formula dated to 1044 in the Wujing Zongyao (武經總要), a Chinese collection of military information (see Fig. 3.1).

"Many western history books over the years have stated that the Chinese used this discovery only for fireworks, but that is not true. Song Dynasty military forces as early as 904 A.D. used gunpowder devices against their primary enemy, the Mongols" [1].

"By the mid- to late-eleventh century, the Song government had become concerned about gunpowder technology spreading to other countries. The sale of saltpeter to foreigners was banned in 1076. Nonetheless, knowledge of the miraculous substance was carried along the Silk Road to India, the Middle East, and Europe. In 1267, a European writer made reference to gunpowder, and by 1280 the first recipes for the explosive mixture were published in the west" [1]. The another theory of how gunpowder came to Europe is that it was brought to Europe during the Mongol invasion in the first half of the thirteenth century, or during the diplomatic and military contact.

* Black powder was used in civil engineering and mining as early as the fifteenth century. The earliest surviving record for the use of black powder in mines comes from Hungary in 1627. It was introduced to Britain in 1638 by German miners, after which records are numerous.

D. Zou, *Theory and Technology of Rock Excavation for Civil Engineering*,
DOI 10.1007/978-981-10-1989-0_3

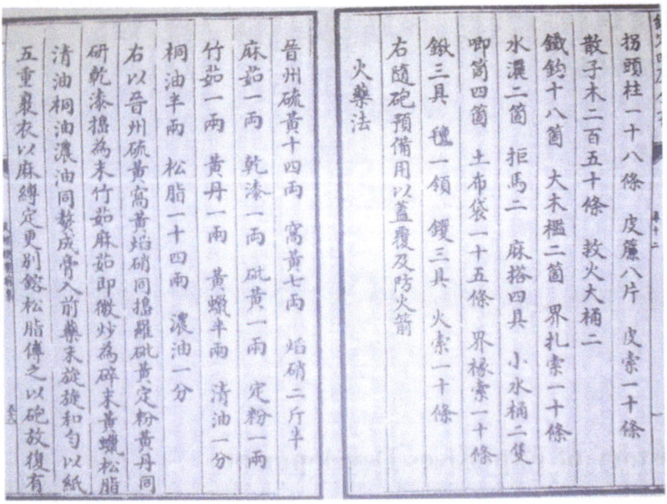
拐頭柱一十八條 皮簾八片 皮索一十條
散子木二百五十條 救火大桶二
鐵鉤十八箇 大木檑二箇 界扎索一十條
水濃二箇 拒馬二 麻搭四具 小水桶二隻
唧筒四箇 土布袋一十五條 界樣索一十條
鍬三具 钁一領 鏝三具 火索一十條
右隨砲預備用以蓋覆及防火箭
火藥法
晉州硫黃十四兩 窩黃七兩 焰硝二斤半
麻茹一兩 乾漆一兩 砒黃一兩 定粉一兩
竹茹一兩 黃丹一兩 黃蠟半兩 清油一分
桐油半兩 松脂一十四兩 濃油一分
右以晉州硫黃窩黃焰硝同搗羅砒黃定粉黃丹同
研乾漆搗為末竹茹麻茹即微炒為碎末黃蠟松脂
清油桐油濃油同熬成膏入前藥末旋旋和勻以紙
五重裹衣以麻縛定更別鎔松脂傅之以砲放復有

Fig. 3.1 Earliest known written formula for black powder, from Chinese 武经总要 in 1044, Ref. [1]

Black powder used in the mining industry is considered to mark the end of the Middle Ages and the beginning of the Industrial Revolution.

* In 1846, Italian chemist Ascanio Sobrero (1812–1888) invented the first modern explosive, ***nitroglycerin***, by treating glycerin with nitric and sulfuric acids. Sobrero's discovery was, unfortunately for many early users, too unstable to be used safely. Nitroglycerin readily explodes if bumped or shocked. This inspired Swedish inventor Alfred Nobel (1833–1896) in 1862 to seek a safe way to package nitroglycerin. In the mid-1860s, he succeeded in mixing it with an inert absorbent material, kieselguhr. His invention was called ***dynamite***.

* Dynamite replaced gunpowder as the most widely used explosive. But Nobel continued experimenting with explosives and in 1875 invented a ***gelatinous dynamite***, an explosive jelly. It was more powerful and even a little safer than the dynamite he had invented 9 years earlier. The addition of ammonium nitrate to dynamite further decreased the chances of accidental explosions. It also made it cheaper to manufacture.

* In 1867, Ohlsson C. J. and Norrbin J. H. proposed mixed explosive patent which is made of ***ammonium nitrate*** and various fuels. It laid the foundation of competitive development between the ***ammonium nitrate explosives*** and dynamite. After that, various types of ammonium nitrate (AN types) were widely used in the world.

* In 1943, Consolidated Mining and Smelting Co. of Canada produced the ***porous ammonium nitrate prills.*** About 94 % of porous prilled ammonite nitrate mixed with about 6 % fuel oil is simply called ***ANFO.*** Due to its very low cost and ease to use, ANFO has been quickly used worldwide and is called a ***blasting agent***

instead of explosive. It accounts for an estimated 80 % of the 6,000,000,000 lb (2.7×10^9 kg) of explosives used annually in North America in 2012.

* Dr. Melvin A. Cook was awarded US patent number 2,930,685 on March 29, 1960, for his invention of a practical explosive composition containing water as an essential ingredient. This was the birth of the water-based explosives, ***slurry/water gel***. As his invention, Dr. Cook awarded the Nobel Prize in 1968.

* In 1969, Blubm H. F. of Atlas Chemical Industry Co. Ltd invented the ***emulsion explosive***. Emulsion explosive is a water-in-oil emulsion and a solid oxidizing agent such as ammonium nitrate. Emulsion explosive was the new development of the water-based explosives. Today, emulsion explosives have almost replaced slurry/water gel and share domination of the worldwide explosive market with ANFO.

3.2 Characters of Explosion of Explosives

There are three kinds of explosion: physical explosion, chemical explosion, and nucleus explosion.

Physical explosion is also called mechanical explosion. For example, the water is heated in a steam boiler or a pressure cooker. Steam is generated, and the steam pressure gradually builds up and eventually reaches a point when it overcomes the structural or material resistance of its container and an explosion occurs. Such a mechanical explosion should be accompanied by high temperature, rapid escape of gases or steam, and a loud noise. But in physical or mechanical explosion, there is no new substance produced. Steam is still vaporous water in the above example. No new substance produced during the procedure of explosion is the principal character of physical explosion.

Chemical explosion is caused by the extremely rapid conversion of a solid or liquid explosive compound into gases having a much greater volume than the substances for which they are generated. During chemical explosion, the original substance(s) is (are) transferred to some gaseous new substances.

Nucleus explosion may be induced by fission, the splitting of the nucleus of atoms, or fusion, the joining together under greater force of nuclei of atoms. Nuclear fission or fusion occurs only in extremely heavy elements which are atomically unstable or radioactive. When fission or fusion occurs, a tremendous release of energy, heat gas, and shock take place.

Explosion of the explosive is the chemical explosion.

Explosion of the explosive is characterized by the following features:

- ***High rate*** of reaction;
- ***High temperature*** exothermic; and
- ***High pressure*** to the surrounding materials by the large volume of gases evolved.

The thermite reaction represented by the following equation is not an explosion, since no gases are evolved although there is a very high temperature (3000 °C) generated during the reaction.

$$Fe_2O_3 + 2Al \rightarrow Al_2O_3 + 2Fe + 205\,\mathrm{k\,cal/mol} \quad (3.1)$$

Some chemical reactions also produce a great volume of gases and may have a high rate of reaction, but no heat is released, even needing to absorb heat from surrounded substances, or only very low heat released, they are not explosion. The following two decomposing equations are examples:

$$(NH_4)_2C_2O_4 \rightarrow 2NH_3 + H_2O + CO + CO_2 - 263.3\,\mathrm{KJ} \quad (3.2)$$

$$C_uC_2O_4 \rightarrow C_u + 2CO_2 + 23.6\,\mathrm{KJ} \quad (3.3)$$

1 kg of coal burning in the air can generate about 32,660 kJ of heat which is much more than the heat released by 1 kg TNT explosive (4187 kJ but in 0.00001 s), but the reaction of coal burning is very slow and may take more than 1 h time and is not an explosion.

When a block of explosive is made to detonate, the gases produced may have 10,000–15,000 times greater volume than the original volume of the explosive. The expansion of gases generated is very rapid, reaching velocities of 7000–8000 m/s. The temperature generated may reach 3000–4000 °C. The entire conversion process takes only a fraction of a second and is accompanied by shock and loud noise.

Detonation of explosives can be initiated by (a) mechanical action, i.e., impact or shock wave from another detonation, (b) high temperature, and (c) direct flame.

3.3 Types of Chemical Decomposition of Explosives and the Detonation Process

3.3.1 Types of Chemical Decomposition of Explosives

Explosives, depending upon the conditions to which they are exposed, can offer a different behavior than would be expected from their explosive nature. The chemical decomposition processes of an explosive compound are thermal decomposition, combustion, deflagration, and detonation.

- Thermal decomposition. The thermal decomposition of an explosive may take place under the room condition, but the processes of decomposition are very slow. It may take years, days, hours, or a fraction of a second. The speed of decomposition will increase along with the rising temperature of the condition that the explosive is exposed to. The behavior of the thermal decomposition of explosives is very important for the storage of explosive materials. Controlling the storage volume and stacked manner of explosives, to maintaining a good ventilation of the warehouse are the necessary measures to control the temperature of the warehouse and keep the stability of explosives during their storage.

- Combustion. Under certain conditions, most explosives can be burned without explosion. Combustion reaction starts from a local part of the explosive and then propagates to other parts along the surface or axial of the explosive with a slow speed. The usual reaction speed is several millimeters to tens of centimeters per second. If the surrounding pressure and temperature are not changed, the combustion will continue till all explosive burns out.
- Deflagration. Deflagration is a rapid high-energy release combustion event that propagates through the explosive material at subsonic speeds, driven by the transfer of heat. Deflagration is a characteristic of low explosive materials, such as gunpowder, flares, and fireworks.
- Detonation. Detonation is a chemical reaction process accompanied by the release of a large number of energy. Detonation involves a supersonic exothermic front accelerating through the explosive material that eventually drives a shock front propagating directly in front of it. The shock front is called detonation wave and capable of passing through the explosive material at great speeds, typically thousands of meters per second. After the detonation wave sweeps through the media, it results in detonation products of high temperature and pressure.

There is no essential difference between deflagration and detonation, but different propagation speeds only. Detonation is propagated with a constant speed which is much greater than the speed of sound, and deflagration propagation velocity is generally lower than the speed of sound and unstable (See Fig. 3.2; [3]). Detonation is the fullest form of the chemical reaction of the explosive, and the reaction releases the most energy.

3.3.2 *Detonation Process of an Explosive*

As described above, the detonation consists of the propagation of a chemical reaction that moves through the explosive at supersonic speed, transforming it into new chemical compounds. The basic characteristic of these reactions is that it is initiated and sustained by a supersonic shock wave–detonation wave (Fig. 3.3).

The first theory of detonations was proposed by D. L. Charpman in 1899 (England) and E. Jouguet in 1905 (France). In this theory, it is assumed that chemical reactions take place instantaneously inside the shock wave, and the reaction zone length would shrink to zero. In both sides of the reaction zone, each parameter of the initial state and the final state can be linked with the three conditions of mass, momentum, and energy conservation laws refer to [4]:

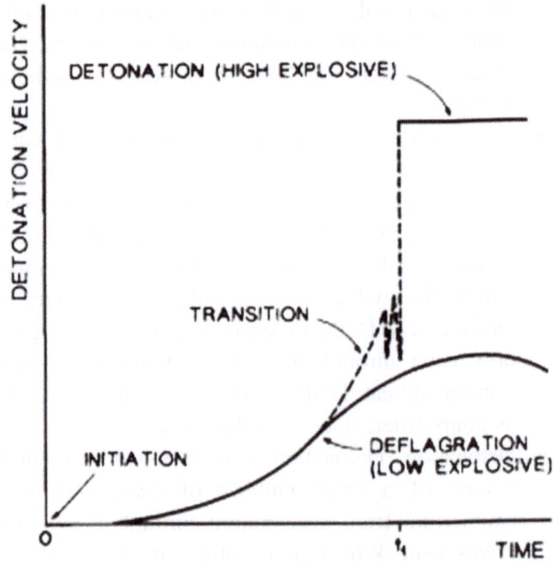

Fig. 3.2 Deflagration and detonation (Reproduced from Ref. [3] with the permission from Taylor & Francis Book, UK)

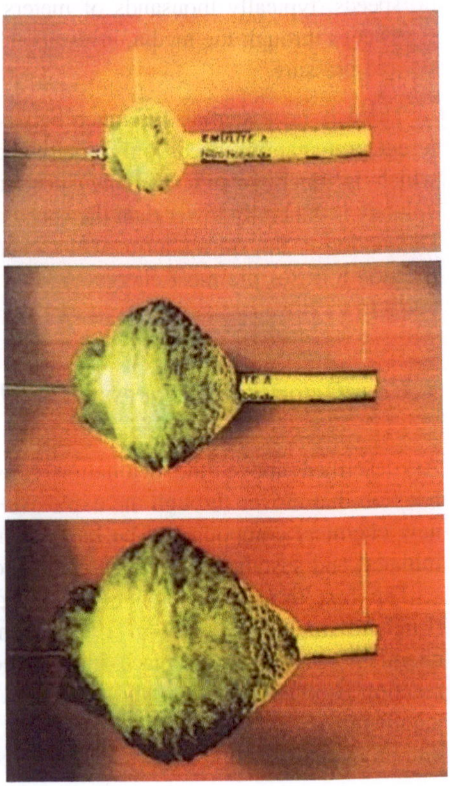

Fig. 3.3 Detonation of an unconfined explosive (Nitro Nobel)

$$\text{Mass conservation law:} \quad \rho_1 u_1 = \rho_2 u_2 \tag{3.4}$$

$$\text{Momentum Conservation Law:} \quad P_1 + \rho_1 u_1^2 = P_2 + \rho_2 u_2^2 \tag{3.5}$$

$$\text{Energy conservation Law:} \quad h_1 + \tfrac{1}{2}u_1^2 = h_2 + \tfrac{1}{2}u_2^2 \tag{3.6}$$

where

ρ_1 Explosive density of the initial state;
ρ_2 Density of the substances within the reaction zone;
u_1 Detonation wave speed;
u_2 Speed of the substance of the flow of reaction products;
P_2 Pressure on the plane of reaction zone (called CJ plane), i.e., detonation pressure;
P_1 Initial pressure;
h_2, h_1 Energy during detonation and before detonation.

Equation (3.6) is called Rankine–Hugoniot condition or Rankine–Hugoniot relation.

To solve the mass and momentum conservation Eqs. (3.4) and (3.5), the following equation can be obtained:

$$(\rho_1 u_1)^2 = (\rho_2 u_2)^2 = \frac{(P_2 - P_1)}{\left(\frac{1}{\rho_1} - \frac{1}{\rho_2}\right)} = \frac{P_2 - P_1}{V_1 - V_2} \tag{3.7}$$

where

V_1 Initial specific volume of explosive, $V_1 = 1/\rho_1$;
V_2 Specific volume of reaction products on the plane of reaction zone, $V_2 = 1/\rho_2$.

The equation of (3.7) is an equation of a line which passes through the point of (P_1, V_1). This line is called the **Rayleigh line.**

A solution of all equations of motion (Eqs. 3.4–3.6), and eliminating both u_1 and u_2 gives the **Hugoniot** relationship (Fig. 3.4), which may be written in terms of total enthalpy:

$$h_2 - h_1 = \frac{1}{2}(P_2 + P_1)(V_1 - V_2) \tag{3.8}$$

On the plane of (P, V), Eq. (3.8) is a curve, called **Hugoniot curve.**

As the detonation process is an exothermic reaction, Eq. (3.8) should be:

$$h_2 - h_1 = \frac{1}{2}(P_2 + P_1)(V_1 - V_2) + Q_e \tag{3.9}$$

It can be seen that due to the release of detonation reaction heat Q_e during the detonation wave propagation process, so that the specific internal energy of the

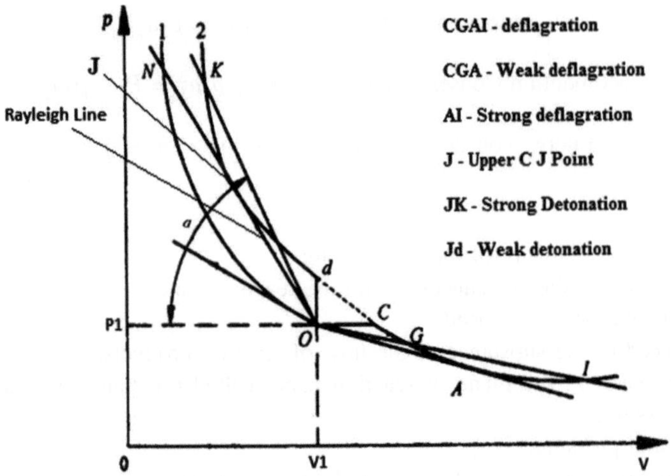

Fig. 3.4 Hugoniot curve on *P* versus *V* plane for one-dimensional steady process (Courtesy of Ordnance Industry Press, Ref. [5])

detonation products is further increased, the formula (3.9) is also known as exothermic Hugoniot equation.

The state equation of detonation products can be written as:

$$h = h(P, V) \tag{3.10}$$

As there are four equations in five unknowns, to determine the value of all detonation parameters must identify the fifth equation. To this end, Chapman and Jouguet made the famous hypothesis (called CJ assumptions or conditions): A stable state of detonation products promises Hugoniot curve corresponding to the Rayleigh line and the tangent point of *J*, namely CJ point (Fig. 3.4); the point of detonation velocity u_J is the minimum that can prove that the *J* point has the following relationship:

$$u_{1,j} = u_{2,j} + c_j \tag{3.11}$$

where

$u_{1,j}$ Detonation wave speed at *J* point;
$u_{2,j}$ Speed of the substance of the flow of reaction products at *J* point;
c_j Sonic speed of the substance of the flow of reaction products at *J* point;

Based on the exothermic Hugoniot equation, the detonation parameters of gas can be approximately calculated as follows (Fig. 3.5):

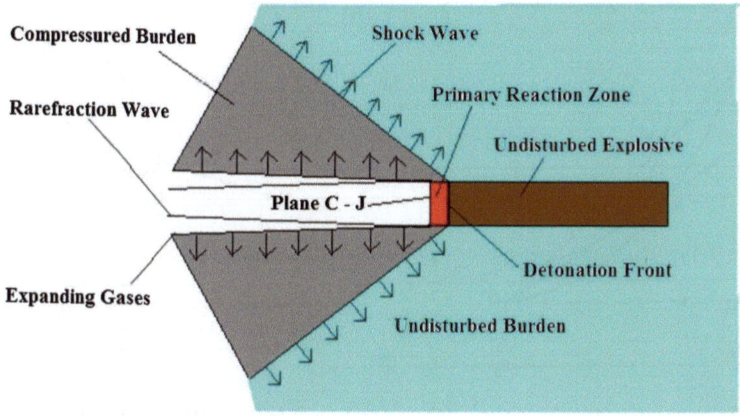

Fig. 3.5 Detonation process of an explosive charge

a. Particle velocity at the C - J plane: $u_{2,j} = \frac{1}{k+1} u_{1,j}$ (3.12)

b. Pressure at the C - J Plane: $P_j = \frac{1}{k+1} \rho_1 u_{2,j}^2$ (3.13)

c. Detonation velocity: $D = u_1 = \sqrt{2(k^2 - 1)Q_v}$ (3.14)

d. Density of substance at C - J plane: $\rho_j = \frac{k+1}{k} \rho_1$ (3.15)

e. Temperature of substance at C - J plane: $T_j = \frac{1}{n_j R} \frac{kD^2}{(k+1)^2}$ (3.16)

where n_j is the amount of moles of gas and R is the ideal gas constant.

Generally, the calculated results of detonation velocity using CJ theory are approximately close to the experimental values of practical detonation velocity. That indicates that the CJ theory is basically correct. However, the conducted values of detonation pressure and densities of detonation products through precision measurement during the gaseous detonation are about 10–15 % lower than the theoretical values obtained by CJ theory, and the measured Mach of the detonation products is about 10–15 % higher than the Mach calculated with CJ theory. That means that CJ theory is an approximate theory. Furthermore, the reaction zone of explosive detonation actually exists within a certain width, and the width of some reaction zones is quite large; therefore, the assumption that chemical reactions take place instantaneously inside the shock wave and the reaction zone length would shrink to zero is not appropriate. This shows that the internal structure of the detonation wave requires in-depth research.

Significant amendments to the CJ theory defects are raised by Zel'dovich (Я.Б.Зельдович, former Soviet Union 1940), Von Neumnn (USA 1942), and W. Doring (Germany 1943) independently, called ZND theory or ZND model.

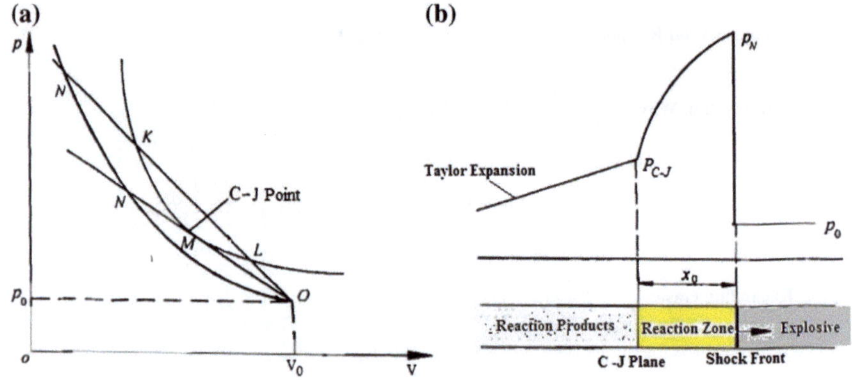

Fig. 3.6 ZND detonation model (Courtesy of Ordnance Industry Press, Ref. [5])

In ZND model, the detonation wavefront is composed by a cutting-edge shock and followed by the chemical reaction zone, they propagate along the explosives at the same speed, and the end of the reaction zone corresponds to the plane CJ state, called CJ surface. According to this model, the course occurred within the detonation wavefront is as follows (see Fig. 3.6a): First, the original explosive is compressed strongly by the shock wave and immediately jumps from the initial state O (v_O, p_O) to the compressed state N (v_N, p_N) point. Temperature and pressure suddenly increase, and the detonation speed of the chemical reaction is excited. Following the continuous reaction, the state gradually changes along the Rayleigh line from point N to the final state point M, the reaction zone reaches the final state, and the chemical reaction heat is released completely. ZND detonation model can also be represented by Fig. 3.6b. After the pressure behind the shock wave suddenly jumping to P_N (called Von Neumann peak), the pressure sharply decreases following the chemical reaction proceeding till the point $P_{c\text{-}j}$ (CJ Pressure) at the end of the reaction. After the CJ plane, it is the isentropic expansion flow area of detonation products, called Taylor expansion wave. The pressure with this area fells gently with the expansion.

Although the ZND model modified and developed CJ model, it is still not a perfect model. In fact, the chemical reaction of the reaction zone cannot be so in order. The factors, such as uneven density of the reaction zone, viscous media, thermal conduction, and diffusion dissipative effects, are likely to cause distortion to the structure of detonation reaction zone, such as the phenomena of spiral detonation and cellular structure observed during gas detonation.

3.4 Oxygen Balance of Explosives

Currently, explosives that are extensively used mainly compose the four elements C, H, O, and N. Some also contain Cl, F, and S and some metal elements such as Al and Mg. The explosives composed of C, H, O, and N can be written as $C_aH_bO_cN_d$.

Among them, C and H are combustible agents and O is oxidant. During the explosion, explosive's molecules rupture and a redox reaction takes place, and these elements reassemble into new stable products, mainly CO_2, H_2O, CO, N_2 and O_2, H_2, C, NO, NO_2, CH_4, $C_2N_2NH_3$, HCN, and the like. The type and quantity of the products are subjected to the influence of pressure and temperature during explosion of the explosion, but also related to the amount of combustible agents and oxidizers contained in the explosive [9]. Generally, oxygen balance and oxygen coefficient indicate the relative oxygen content of explosives and combustible elements.

- Oxygen balance is an expression that is used to indicate the degree to which an explosive can be oxidized. If an explosive molecule contains just enough oxygen to convert all of its carbon to carbon dioxide, all of its hydrogen to water, and all of its metal to metal oxide with no excess, the molecule is said to have a zero oxygen balance. The molecule is said to have a positive oxygen balance if it contains more oxygen than that is needed and a negative oxygen balance if it contains less oxygen than that is needed.

For the type of $C_aH_bO_cN_d$ explosives, the oxygen balance can be formulated as:

$$\text{O.B.} = \frac{\left[c - \left(2a + \frac{b}{2}\right)\right] \times 16}{M} \tag{3.17}$$

where

16 atomic weight of oxygen; and
M Molecular weight of explosives.

When $c - \left(2a + \frac{b}{2}\right) > 0$, there is a surplus of oxygen in the explosive after the completion of oxidation of the combustible elements. This situation is called a positive oxygen balance. Such explosives are called positive oxygen balance explosives.

When $c - \left(2a + \frac{b}{2}\right) = 0$, the explosive's oxygen just completes oxidation of combustible elements. This situation is called a zero oxygen balance. Such explosives are called zero oxygen balance explosives.

When $c - \left(2a + \frac{b}{2}\right) < 0$, the explosive elements of oxygen are insufficient to completely oxidize the combustible. This situation is called negative oxygen balance. Such explosives are called negative oxygen balance explosives.

Example 1 For nitroglycerin, $C_3H_5(ONO_2)_3$, or $C_3H_5O_9N_3$, its oxygen balance is:

$$\text{O.B.} = \frac{\left[9 - \left(2 \times 3 + \frac{5}{2}\right)\right] \times 16}{227} = +0.035, \quad \text{or} +0.035\ \text{g/g of explosive.}$$

Example 2 For PETN, $C(CH_2ONO_2)_4$, or $C_5H_8O_{12}N_4$, its oxygen balance is:

$$\text{O.B.} = \frac{\left[12 - \left(2 \times 5 + \frac{8}{2}\right)\right] \times 16}{315} = -0.101, \quad \text{or} \ -0.101 \ \text{g/g of explosive.}$$

For a mixed explosive, the oxygen balance is the algebraic sum of the products of the oxygen balance of each component multiplied by their weight percent in the explosive.

Example 3 #2 rock explosive consists of ammonium nitrate (85 %), TNT (11 %), and wood powder (4 %). The oxygen balances of these three components are +0.2 g/g, −0.74 g/g, and −1.38 g/g, respectively. The oxygen balance of #2 rock explosive is:

$$\begin{aligned} \text{O.B.} &= \frac{+0.2 \times 85 + (-0.74) \times 11 + (-1.38) \times 4}{100} \\ &= 0.0334 \ \text{g/g of explosive.} \end{aligned}$$

Table 3.1 gives the values of oxygen balance of some explosives and substances.

- The oxygen coefficient indicates the degree of saturation of the explosive molecule with oxygen. For the type of $C_aH_bO_cN_d$ explosives, the formula of oxygen coefficient is:

$$A = \frac{2c}{4a + b} \times 100\,\% \tag{3.18}$$

This means that the oxygen coefficient is the percentage of oxygen contained in the explosive and the oxygen that is required to completely oxidize carbon and hydrogen contained in the explosive.

Take example 1 above, Nitroglycerin, $C_3H_5(ONO_2)_3$ or $C_3H_5O_9N_3$, its oxygen coefficient is:

$$A = \frac{2 \times 9}{4 \times 3 + 5} \times 100\,\% = 105.9\,\%, \text{and}$$

Table 3.1 Oxygen balance values of some explosives and substances

Name and symbol	Molecular formula	Weight of atomic or molecular	O.B. (g/g)
Trinitrotoluene (TNT)	$C_7H_5(NO_2)_3$	227	−0.740
Hexogen (RDX)	$C_3H_6N_3(NO_2)_3$	222.1	−0.216
Octogen (HMX)	$C_4H_8N_4(NO_2)_4$	296.2	−0.216
Tetryl (Te)	$C_7H_5\ N(NO_2)_4$	287.2	−0.474
Nitroglycerin (NG)	$C_3H_5(NO_3)_3$	227	+0.035
Pentaerythritol tetranitrate (PETN)	$C_5H_8(NO_3)_4$	316.2	−0.101
Dinitrotoluene (DNT)	$C_7H_6(NO_2)_2$	182.1	−1.142
Mercury fulminate	$HgC_2N_2O_2$	284.7	−0.113
Ammonium nitrate	NH_4NO_3	80	+0.200
Potassium nitrate	KNO_3	101	+0.396
Sodium nitrate	$NaNO_3$	85	+0.471
Al (powder)	Al	27	−0.889
Light diesel oil	$C_{18}H_{32}$	224	−3.420
Wood powder	$C_{15}H_{22}O_{10}$	362	−1.370
Paraffin	$C_{18}H_{38}$	254.5	−3.460
Magnesium	Mg	24.31	−0.658
Sodium nitrite	$NaNO_2$	69	+0.348
Sulfur	S	32.0	−1.00
Span-80	$C_{22}H_{42}O_6$	428	−2.39

Example 4 For trinitrotoluene (TNT), $C_7H_5O_6N_3$, its oxygen coefficient is:

$$A = \frac{2 \times 6}{4 \times 7 + 5} \times 100\,\% = 35.36\,\%.$$

3.5 Thermochemistry of Explosives

3.5.1 Detonation Heat of Explosives

Explosive contains a huge amount of chemical energy. Through the form of explosion reaction, it releases its full potential energy and transforms into mechanical work on the surrounding medium. The energy released as heat is one of the important performance parameters of the explosive.

Detonation heat is the heat released by one unit mass of explosives (usually refers to 1 mol or 1 kg explosives) during detonation.

As the detonation reaction of explosives is extremely rapid, the detonation process can be considered in a constant volume condition, so the detonation heat is generally expressed in Qv.

3.5.1.1 Calculation of Detonation Heat of Explosives

In 1840, a Swiss-born Russian chemist and physician, Germain Henri Hess, published the famous ***Hess's law of constant heat summation***, also known as ***Hess's law.*** The law states that the total enthalpy change during the complete course of a reaction is the same whether the reaction is carried out in one step or in several steps.

In other words, if a chemical change takes place by several different routes, the overall enthalpy or heat change is the same, regardless of the route by which the chemical change occurs (provided that the initial and final conditions are the same).

According to the Hess's law, while calculating the detonation heat, the thermal effects of explosives can be illustrated with Hess's triangle, as shown in Fig. 3.7.

We envision there are three states: The state 1 is the stable components of the explosive, state 2 is the explosive, and state 3 is the detonation products. According to the Hess's law, it is clear that the thermal effects ($Q_{1\text{-}3}$) of the process from stable components of the explosive converting to detonation products are equal to the sum of the thermal effects ($Q_{1\text{-}2}$) of these components to form the explosive by adding the thermal effects ($Q_{2\text{-}3}$) of explosive converted to the detonation products. It is:

$$Q_{1-3} = Q_{1-2} + Q_{2-3} \quad \text{or} \quad Q_{2-3} = Q_{1-3} - Q_{1-2} \tag{3.19}$$

where

Q_{1-2} Formation heat of explosive;
Q_{1-3} Formation heat of detonation products; and
Q_{2-3} Detonation heat of explosive.

Equation (3.19) shows that the detonation heat of an explosive is equal to the formation heat of the detonation products subtracted from the formation heat of the explosives. So, by knowing the explosion explosive chemical reaction equation, formation of the explosive, and the formation heat of detonation products, the

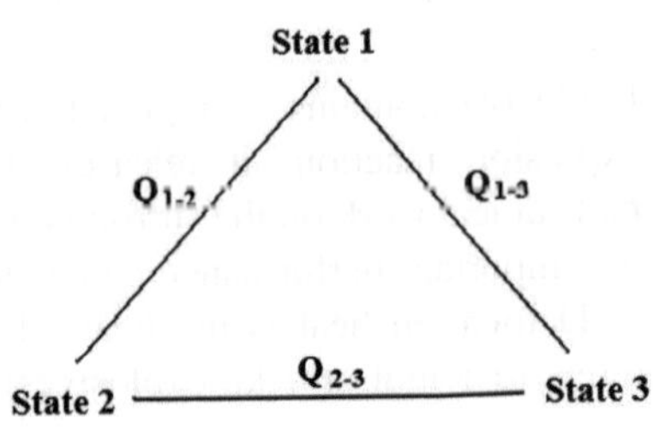

Fig. 3.7 Hess' Triangle for calculation of detonation heat

detonation heat of the explosive can be calculated. Formation heat of explosives and detonation products can be obtained from the relevant chemical manuals.

Example: The chemical reaction equation of ANFO is given as:

$$3NH_4O_3 + 1CH_2 \rightarrow CO_2 + 7H_2O + 3N_2$$

We want to calculate the detonation heat of ANFO.

In some manual, we can get that the formation heat of NH_4O_3 is 87.3 kcal/mol and the formation heat of CH_2(fuel oil) is 7.0 kcal/mol. The formation heats of CO_2 and H_2O are 94.1 kcal/mol and 57.8 kcal/mol, respectively. Then, according to formula (3.19), we can get the detonation heat of ANFO as follows:

$$\begin{aligned} Q_{\mathrm{mp}} &= Q_{\mathrm{mp}(1-3)} - Q_{\mathrm{mp}(1-2)} = (94.1 + 7 \times 57.8) - (3 \times 87.3 + 7) \\ &= 229.8\,kcal \end{aligned}$$

The molecular weight of the ANFO is:

$$P_m = 3 \times 80.1 + 14 = 254.3g$$

The detonation heat of explosive's (ANFO) resulting kilogram is:

$$Q_{\mathrm{kp}} = \frac{229.8\,\mathrm{kcal}}{254.3\,\mathrm{g}} \times \frac{1000\,\mathrm{g}}{\mathrm{kg}} = 903.7\,\mathrm{kcal/kg}$$

As the formation heat which were taken above for all explosive components and detonation products are derived at a condition of constant pressure, the calculated detonation heat should be converted to the condition of constant volume:

$$Q_{\mathrm{mv}} = Q_{\mathrm{mp}} + 0.577 \times n_{\mathrm{pg}}$$

where

n_{pg} Number of moles of the gaseous products, in this case:

$$n_{\mathrm{pg}} = 1 + 7 + 3 = 11$$

So,

$$\begin{aligned} Q_{\mathrm{mv}} &= 229.8 + 0.577 \times 11 = 236.15\ \mathrm{kcal/mol.} \\ Q_{\mathrm{kv}} &= \frac{236.15\,\mathrm{kcal}}{254.3\,\mathrm{g}} \times \frac{1000\,\mathrm{g}}{\mathrm{kg}} = 928.6\ \mathrm{kcal/kg.} \end{aligned}$$

Table 3.2 gives the formation heat of some substances and explosives under the conditions of constant pressure and 18 °C or 291 °K.

Table 3.2 Heats of formation and molecular weight of some substances and explosives under the conditions of constant pressure and temperature of 18 °C or 291 °K

Substance/explosive	Molecular formula	Mole weight	Heat of formation (kcal/mol)
Corundum	Al_2O_3	102.0	399.1
Fuel oil	CH_2	14.0	7.0
Nitromethane	CH_3O_2N	61.0	21.3
Nitroglycerin	$C_3H_5O_9N_3$	227.1	82.7
PETN	$C_3H_8O_{12}N_4$	316.1	129.4
Trinitrotoluene (TNT)	$C_7H_5O_6N_3$	227.1	17.5
Carbon monoxide	CO	28.0	26.9
Carbon dioxide	CO_2	44.0	94.5
Water	H_2O	18.0	57.8
Ammonium nitrate	NH_4NO_3	80.1	87.4
Aluminum	Al	27.0	0.0
Carbon	C	12.0	0.0
Nitrogen	N	14.0	0.0
Nitrogen oxide	NO	30.0	−21.6
Nitrogen dioxide	NO_2	46.0	−12.2
Mercury fulminate	$HgC_2N_2O_2$	284.7	−64.1
Hexogen (RDX)	$C_3H_6N_3(NO_2)_3$	222.1	−15.64
Sodium nitrate	$NaNO_3$	85	−111.72
Octogen (HMX)	$C_4H_8N_4(NO_2)_4$	296.2	−17.9
Lead azide	PbN_6	291.0	−115.5

3.5.1.2 Calorimetric Determination of Detonation Heat of Explosives

Detonation heat is an important parameter to measure the explosive energy, experimentally determining that the detonation temperature is important. The measurement usually is taken in the bomb calorimeter. Figure 3.8 is the device designed by the Lawrence Livermore National Laboratory, USA [6]. Figure 3.9 is a commonly used detonation calorimeter in China [8].

The device in Fig. 3.9 is divided into three layers. The outer layer is a wooden cask, the middle layer is a heat reservation jacket with which the inner wall of the jacket is polished and chromed, and the innermost layer is the calorimeter barrel made of stainless steel for which the inner and outer surfaces are polished. The calorimeter barrel is filled with distilled water of quantitative and constant temperature for measurement of heat. The space between the wooden cask and the heat conservation jacket is filled with heat reservation material to insulate heat exchange with the outside. Bomb calorimeter is placed in the center of the calorimeter barrel The tested explosive is suspended inside the bomb. Since the high temperature and pressure during the detonation process produce a strong destruction, the strength of the bomb must be very high and the bomb wall should be thick enough.

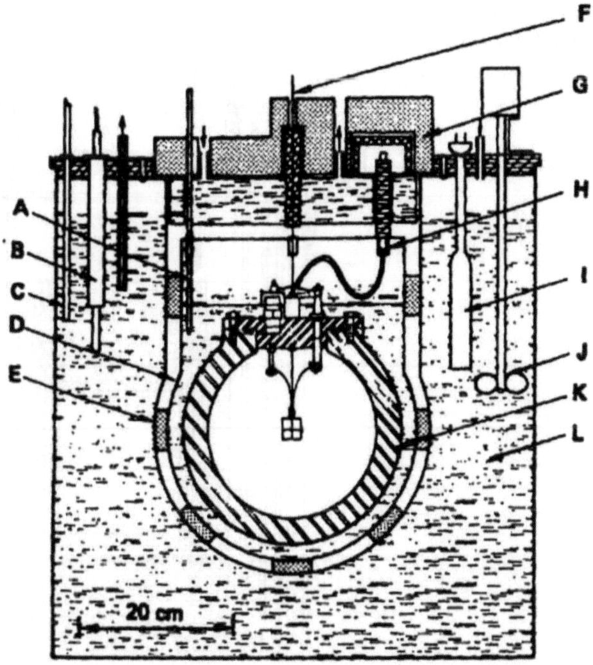

A, quartz thermometer; B, nickel resistance thermometer; C, mercury-in-glass thermometer; D, calorimeter bucket with lid; E, Styrofoam support blocks; F, support cable; G, Styrofoam insulation; H, firing-lead connector; I, knifeblade heater; J, stirrer; K, bomb; and L, constant-temperature jacket.

Fig. 3.8 The bomb calorimeter (Approved public release distribution unlimited, Ref. [6])

In the experiment, the sample of explosives together with a detonator is suspended at the lid of the bomb and then the bomb lid is closed. Air is withdrawn from the bomb and replaced with nitrogen, and then it is evacuated to create a vacuum. The bomb is placed inside the calorimeter barrel. The barrel is filled with accurately weighed distilled water until the bomb is fully submerged. The bomb is maintained at a stable temperature keeping thermostatic for about an hour, temperature of the water in the barrel is recorded, then explosives sample is detonated. The temperature of the water will continue to rise. By recording the highest temperature measured of the water, the detonation heat of the explosive sample can be calculated with the following formula:

$$Q_{\mathrm{kv}} = \frac{(C_w + C_d) \times (T_1 - T_0) - q}{m} \tag{3.20}$$

Fig. 3.9 Calorimeter of detonation heat, refer to [7]

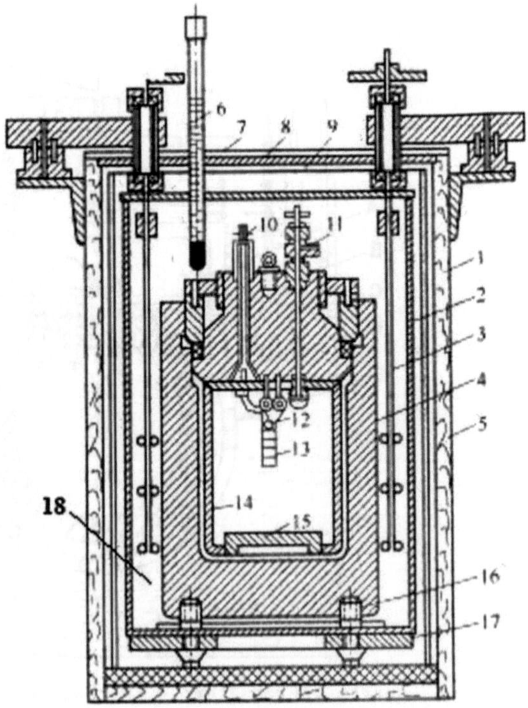

1- Wooden Cask: 2 Calorimeter Barrel; 3-Stirrer; 4-Bomb; 5-Heat Reservation Jacket; 6-Thermometer; 7, 8, 9-Lid; 10-Firing-Lead Connector; 11-Exhaust Pot; 12-Elec. Detonator; 13-Explosive Charg ; 14-Lined Barrel; 15-Pad: 16-Support Bolts; 17-Bottom Bracket; 18-Water.

where

Q_{kv} Detonation heat of the tested explosive under the condition of constant volume, kj/kg;
C_w Total heat capacity of the distilled water to be used, kj/°C;
C_d Heat capacity of the experiment device, expressed with the heat capacity of equivalent water, kj/°C;
q Detonation heat of the detonator, kj;
T_1 The highest temperature of the water after detonation, °C;
T_0 Water temperature before detonation, °C and
m Weight of the sample explosive, kg.

It has to be indicated that the measured value of the detonation heat of the tested explosive is only an approximation as the effects of various experiment conditions.

3.5.2 Detonation Temperature of Explosives

Explosive's detonation temperature is the maximum temperature of the detonation products heated by the released detonation heat. Detonation temperature is also one of the important parameters of the explosive.

So far, the direct measurement of the detonation temperature is still very difficult. This is because the explosion produces a very high temperature (typically up to several thousand degrees) in a very short time (10^{-6} s), and then the maximum temperature rapidly declines along with the expansion of the detonation products. Current method for measuring the detonation temperature is determined using the color temperature of the instant detonation products. The measured temperature is generally higher than the actual temperature.

As the experimental measurement of detonation temperature cannot yet get an exact result and the measurement method is quite complex, the theoretical method is usually used to calculate the explosives approximate detonation temperature. To simplify the theoretical calculation of detonation temperature, three assumptions should be made first:

(1) Explosion process can be regarded as a constant volume process;
(2) The detonation process is adiabatic, and all detonation heat is used to heat the detonation products; and
(3) Heat capacity of detonation products is just a function of the temperature, and pressure has nothing to do with the products.

It is obvious that the third assumption is not the actual situation during detonation process, so that some small deviation may be caused but still within the accepted range.

There are several methods to calculate the detonation temperature of an explosive. Only, the method of calculation of detonation temperature starting from the average heat capacity of the detonation products is introduced in this book.

According to the above assumptions, the detonation heat can be expressed as:

$$Q_v = \sum \overline{C}_{\mathrm{vi}} \times t \tag{3.21}$$

where

Q_v Explosive's detonation heat under constant volume;
$\overline{C}_{\mathrm{vi}}$ The average heat capacity of detonation product i within the temperature from 0 to t °C; and
t Detonation temperature.

Generally, the relationship between heat capacity and temperature can be expressed as:

$$\overline{C}_{vi} = a_i + b_i t + c_i t^2 + d_i t^3 + \cdots \quad (3.22)$$

For an approximate calculation, only the first and second items are taken in the calculation. That means that the heat capacity is considered to be a linear relationship, like:

$$\overline{C}_{vi} = a_i + b_i t \quad (3.23)$$

$$\sum \overline{C}_{vi} = a + bt \quad (3.24)$$

$$Q_v = \sum \overline{C}_{vi} \times t = a + bt \quad (3.25)$$

So, the detonation temperature can be obtained as:

$$t = \frac{-a + \sqrt{a^2 + 4bQ_v}}{2b} \quad (3.26)$$

Using this method to calculate the detonation temperature, the components of the detonation products and their heat capacities must be known. Kast and Beyling gave the average heat capacities' expressions of molecules of some substances:.

For 1 mol of two-atom gas	$\overline{C}_v = 4.8 + 4.5 \times 10^{-4}t$
For 1 mol of water vapor	$\overline{C}_v = 4.0 + 21.5 \times 10^{-4}t$
For 1 mol of three-atom gas	$\overline{C}_v = 9.0 + 5.8 \times 10^{-4}t$
For 1 mol of four-atom gas	$\overline{C}_v = 10 + 4.5 \times 10^{-4}t$
For 1 mol of five-atom gas	$\overline{C}_v = 12 + 4.5 \times 10^{-4}t$
For carbon, C	$\overline{C}_v = 6.0$
For salt, NaCl	$\overline{C}_v = 28.3$
For AlO_3	$\overline{C}_v = 23.86 + 67.3 \times 10^{-4}t$

Example Calculation of the detonation temperature of TNT knowing the chemical reaction equation of the detonation of TNT is:

$$C_6H_2(NO_2)_3CH_3 = 2CO_2 + CO + 4C + H_2O + 1.2H_2 + 1.4N_2 + 0.2NH_3 + 226.08 \text{ kcal/mol}$$

Calculate the heat capacities of detonation products:

For two-atom gas: $\overline{C}_v = (1 + 1.2 + 1.4) \times (4.8 + 4.5 \times 10^{-4}t) = 17.28 + 0.00126t$

For water (H_2O): $\overline{C}_v = 4.0 + 21.5 \times 10^{-4}t$

For 3 atoms, CO_2: $\overline{C}_v = 2 \times (9.0 + 5.8 \times 10^{-4}t) = 18.0 + 0.00116t$

For 4 atoms, NH_3: $\overline{C}_v = 0.2 \times (10 + 4.5 \times 10^{-4}t) = 2.0 + 0.00009t$

For carbon, C: $\overline{C}_v = 4 \times 6.0 = 24.0$
All the heat capacities of detonation products are:
$\sum \overline{C}_{vi} = a + bt = 65.28 + 0.00502t$
So we get: $a = 65.28$, $b = 0.00502$ and $Q_v = 226.08$ kcal/mol. Substituting them to the equation of (3.26), we get:

$$t = \frac{-a + \sqrt{a^2 + 4bQ_v}}{2b}$$
$$= \frac{-65.28 + \sqrt{65.28^2 + 4 \times 0.00502 \times 226.08 \times 1000}}{2 \times 0.00502} = 3260\ ^\circ\text{C}$$

Or $T = 3260 + 273 = 3533$ K.

As the values of $\overline{C}_v$ calculated using Kast's expressions are generally a little lower, so the value of calculated detonation temperature t is a little higher as well.

3.5.3 *Detonation Volume of Explosives*

Detonation volume refers to the occupied volume of the products of a unit mass (usually 1 kg) of explosives after detonation in the standard state. The common unit is l/kg.

Explosive gases generated during detonation are the working medium of explosives to the surrounding world, explosive's power and its detonation volume of generated gases is closely related. If the detonation volume of an explosive is greater, then its ability to do work is stronger and it has greater power.

According to the Avogadro's law, "equal volumes of all gases, at the same temperature and pressure, have the same number of molecules," and the following formula can be used to calculate the detonation volume of an explosive if its chemical reaction equation is known.

$$V = \frac{\sum n_j \times 1000}{\sum m_i \text{Me}_i} \times 22.4\ (\text{l/kg}) \tag{3.27}$$

where

V Detonation volume of the explosive;
$\sum n_j$ Total amount of moles of the detonation products;
m_i The moles of component I of the explosive;
Me_i Molecular weight of component I of the explosive; and
22.4 the volume of 1 g-mol of any gas under normal conditions.

Taking AN-TNT (Amato 80/20) as an example, its chemical reaction equation is:

$$11.35NH_4NO_3 + C_7H_5O_6N_3 \rightarrow 7CO_2 + 25.2H_2O12.85N_2 + 0.425O_2 + 1135\ \text{kcal}$$

According to formula (3.27):

$$\sum n_j = 7 + 25.2 + 12.85 + 0.425 = 45.475$$

$$m_{NH_4NO_3} = 11.55, \quad Me_{NH_4NO_3} = 80, \quad m_{TNT} = 1, \quad Me_{TNT} = 227.$$

So, the detonation volume of AN-TNT (Amato 80/2 is:

$$V = \frac{\sum n_j \times 1000}{\sum m_i Me_i} \times 22.4 = \frac{45.475 \times 1000}{11.35 \times 80 + 1 \times 227} = 897.48\ (\text{l/kg})$$

3.5.4 *Detonation Pressure and Velocity of Detonation of Explosives*

Detonation pressure and velocity are the very important parameters to describe the power and energy of an explosive.

M. J. Kamlet et al. carried out a lot of research and analysis about the C-, H-, O-, N-type explosives and raised the empirical formulas to calculate the detonation velocity and pressure of such explosives (refer to [5]):

$$P_J = K\phi\rho_0^2 \tag{3.28}$$

$$D_J = A\phi^{1/2}(1 + B\rho_0) \tag{3.29}$$

$$\phi = NM^{\frac{1}{2}}Q^{\frac{1}{2}} \tag{3.30}$$

where

K, *A*, and *B*	are the experimental constants. $K = 15.58$, $A = 1.01$, and $B = 1.30$;
P_J	Detonation pressure at the CJ plane, kbar;
D_J	Detonation velocity, m/s;
ρ_0	Initial density of explosive, g/cc;
ϕ	Explosive's eigenvalue;
N	Moles of generated gases of 1 g of the explosive, mol/g;
M	Average molecular weight of generated gas components, g/mol; and
Q	Chemical reaction heat of 1 g of the explosive.

The chemical reaction equation of the $C_aH_bO_cN_d$-type explosion is determined by the principles of the maximum exothermic. The chemical reaction equation can be written as:

$$C_aH_bO_cN_d \rightarrow \frac{b}{2}H_2O + \left(\frac{c}{2} - \frac{b}{4}\right)CO_2 + \left(a - \frac{c}{2} + \frac{b}{4}\right)C + \frac{d}{2}N_2 + Q_V \quad (3.31)$$

$$N = \frac{b + 2c + 2d}{48a + 4b + 64c + 56d} \quad (3.32)$$

$$M = \frac{88c + 56d - 8b}{b + 2c + 2d} \quad (3.33)$$

$$Q = \frac{28.9b + 47 \times \left(c - \frac{b}{2}\right) - Q_{f.exp}}{12a + b + 16c + 12d} \quad (3.34)$$

where

$Q_{f.exp}$ Formation heat of the explosive.

3.6 Classification of Explosives

There are different standards or methods of the classification of explosives. In this book, the commonly used standards or methods will be introduced.

3.6.1 Classification by Composition

There are two fundamentally different kinds of explosive materials, namely *single explosive substance* and *composite explosive mixtures.*

Single explosives are chemical substances that contain one well-defined molecule all that is needed for an explosion. The molecule decomposes into mainly gaseous reaction products, such as CO_2, N_2, and H_2O. The solid explosives, trinitrotoluene ($C_7H_5(NO_2)_3$, TNT), Pentaerythritol tetranitrate ($C_5H_8(NO_3)_4$, PETN), hexogen ($C_3H_6N_3(NO_2)_3$, RDX), octogen ($C_4H_8N_4(NO_2)_4$, HMX), and the liquid explosives, nitroglycerin ($C_3H_5(NO_3)_3$, NG) and nitromethane (CH_3NO_2, NM), are examples of single explosive substances.

A composite explosive can be a mixture of two single explosive substances, a mixture of a fuel and an oxidizer, or an intermediate mixture containing one or more single explosive substances together with fuel and/or oxidizer ingredients.

Most rock blasting explosives are composites containing both single explosive substances and several fuel and oxidizers. In addition, they often contain ingredients such as ballast materials or water that do not add energy to the chemical reaction but are used to modify the explosive's consistency or flow properties. Rock blasting explosive system is described in detail under a separate heading below.

3.6.2 Classification by Sensitivity

- *Primary Explosives*. A primary explosive is an explosive that is extremely sensitive to stimuli such as impact, friction, heat, static electricity, or electromagnetic radiation. A relatively small amount of energy is required for initiation. Primary explosives are often used in detonators or are used to trigger larger charges of less sensitive secondary explosives. Primary explosives are commonly used in blasting caps and percussion caps to translate a physical shock signal. As an example, lead azide (PbN_6) is used as igniting charge in detonators.
- *Secondary Explosives*. A secondary explosive is less sensitive than a primary explosive and requires substantially more energy to be initiated. Because they are less sensitive, they are usable in a wider variety of applications and are safer to handle and store. Secondary explosives are used in larger quantities in an explosive train and are usually initiated by a smaller quantity of a primary explosive. Examples of secondary explosives include TNT, NG, NM, and RDX. Most industrial explosives, such as dynamites, slurries, water gels, and emulsion, are high explosives as well.
- *Tertiary Explosives*. Tertiary explosives, also called *blasting agents*, are so insensitive to shock that they cannot be reliably detonated by practical quantities of primary explosive and instead require an intermediate explosive booster of secondary explosive (according to this definition, ANFO and bulk emulsion are tertiary explosives). These are often used for safety and the typically lower costs of material and handling. Primary users are large-scale mining and construction operations. They have also been used for terrorist attacks, because of the sometimes ready availability of large quantities of precursors (e.g., nitrate fertilizers). Ammonium nitrate and fuel oil mixture (ANFO) is an example of a tertiary explosive.

3.6.3 Classification by Detonation Velocity

- *Low explosives*. Low explosives are compounds where the rate of decomposition proceeds through the material at less than the speed of sound. The decomposition is propagated by a flame front (deflagration) which travels much more slowly through the explosive material than a shock wave of a high explosive. Under normal conditions, low explosives undergo deflagration at rates that vary from a few centimeters per second to approximately 400 meters per second. It is possible for them to deflagrate very quickly, producing an effect similar to a detonation. This can happen under higher pressure or temperature, which usually occurs when ignited in a confined space.

A low explosive is usually a mixture of a combustible substance and an oxidant that decomposes rapidly (deflagration); however, they burn more slowly than a high explosive, which has an extremely fast burn rate.

Low explosives are normally employed as propellants. Included in this group are gunpowders and light pyrotechnics, such as flares and fireworks.

- *High Explosives*. High explosives are explosive materials that detonate, meaning that the explosive shock front passes through the material at a supersonic speed. High explosives detonate with explosive velocity ranging from 3 to 9 km/s.

They are normally employed in mining, rock excavation, demolition, and military applications. They can be divided into two explosives classes differentiated by sensitivity: primary explosive and secondary explosive. The term *high explosive* is in contrast with the term *low explosive*, which explodes (deflagrates) at a lower rate.

3.6.4 Classification by Purpose (Application)

According to the purposes of usage, explosives can be classified into *military explosives* and *commercial (industrial) explosives*. Explosives for military purpose have developed in a direction totally different from those intended for rock blasting which is the major purpose of commercial explosives.

- Usually, the costs of production and usage for commercial purposes are much lower than those of the military explosives. Most commercial explosives used for large-scale blasting are produced near the point of use, or even on site. The raw materials, processing, transportation, mixing and loading into the blast holes, and a small profit margin are all included in a low price for the commercial explosives;
- The requirement for storage life is not very long for commercial explosives due to the quick consumption in rock excavation. For example, cartridge explosives may have a shelf life of 1–2 years. Bulk blasting agents should be capable of "sleeping" loaded in the drillholes for a week or even a month, but in most cases, the blast is fired within a day of being loaded. In contrast, military explosives may have to be stored for 10–20 years. In fact, some gun ammunition for the Second, even First, World War is still in storage in many explosives magazines.
- Usually, the density and explosion energy of military explosives are the range of 1.5–1.9 g/cm^3 and 5–6.5 MJ/kg, and the commercial explosives are much lower

than them and are the range of 0.80 (ANFO) to 1.60 (gelatin dynamites) and 2.5–4.0 MJ/kg, respectively.

The most common military explosive still in vary large-scale use is TNT, which replaced the original military explosive picric acid during and after the First World War.

3.6.5 *Classification by IMDG Code [17]*

The International Maritime Dangerous Goods (IMDG) Code was developed as a uniform international code for the transport of dangerous goods by sea covering such matters as packing, container traffic, and stowage, with a particular reference to the segregation of incompatible substances. The code is updated and maintained by the DSC Sub-Committee of the International Maritime Organization every 2 years.

Substances (including mixtures and solutions) and articles subject to the provisions of this code are assigned to one of the classes 1–9 according to the hazard or the most predominant of the hazards they present. Some of these classes are subdivided into divisions. Explosives belong to Class 1 which is subdivided into six divisions:

Class 1: Explosives

Division 1.1: substances and articles which have a mass explosion hazard,
Division 1.2: substances and articles which have a projection hazard but not a mass explosion hazard,
Division 1.3: substances and articles which have a fire hazard and either a minor blast hazard or a minor projection hazard or both, but not a mass explosion hazard,
Division 1.4: substances and articles which present no significant hazard,
Division 1.5: very insensitive substances which have a mass explosion hazard,
Division 1.6: extremely insensitive articles which do not have a mass explosion hazard.

For the safety of transportation and storage, the Compatibility Group Code for Class 1 dangerous goods are defined. 13 letters are used to designate 13 compatibility groups:

A: Primary explosive substance (1.1A).

B: An article containing a primary explosive substance and not containing two or more effective protective features. Some articles, such as detonator assemblies for blasting and primers, cap type, are included (1.1B, 1.2B, 1.4B).

C: Propellant explosive substance or other deflagrating explosive substance or article containing such explosive substance (1.1C, 1.2C, 1.3C, 1.4C). These are bulk propellants, propelling charges, and devices containing propellants with or without means of ignition. Examples include single-, double-, and triple-based and composite propellants, solid propellant rocket motors, and ammunition with inert projectiles.

D: Secondary detonating explosive substance or black powder or article containing a secondary detonating explosive substance, in each case without means of initiation and without a propelling charge, or article containing a primary explosive substance and two or more effective protective features (1.1D, 1.2D, 1.4D, 1.5D).

E: Article containing a secondary detonating explosive substance without means of initiation, with a propelling charge (other than one containing flammable liquid, gel, or hypergolic liquid) (1.1E, 1.2E, 1.4E).

F: Containing a secondary detonating explosive substance with its means of initiation, with a propelling charge (other than one containing flammable liquid, gel, or hypergolic liquid) or without a propelling charge (1.1F, 1.2F, 1.3F, 1.4F).

G: Pyrotechnic substance or article containing a pyrotechnic substance, or article containing both an explosive substance and an illuminating, incendiary, tear-producing or smoke-producing substance (other than a water-activated article or one containing white phosphorus, phosphide, or flammable liquid or gel or hypergolic liquid) (1.1G, 1.2G, 1.3G, 1.4G). Examples include flares, signals, incendiary or illuminating ammunition, and other smoke- and tear-producing devices.

H: Article containing both an explosive substance and white phosphorus (1.2H, 1.3H). These articles will spontaneously combust when exposed to the atmosphere.

J: Article containing both an explosive substance and flammable liquid or gel (1.1J, 1.2J, 1.3J). This excludes liquids or gels which are spontaneously flammable when exposed to water or the atmosphere, which belong to group H. Examples include liquid- or gel-filled incendiary ammunition, fuel-air explosive (FAE) devices, and flammable liquid-fueled missiles.

K: Article containing both an explosive substance and a toxic chemical agent (1.2K, 1.3K)

L: Explosive substance or article containing an explosive substance and presenting a special risk (e.g., due to water activation or presence of hypergolic liquids, phosphides, or pyrophoric substances) needing isolation of each type (1.1L, 1.2L, 1.3L). Damaged or suspect ammunition of any group belongs to this group.

N: Articles containing only extremely insensitive detonating substances (1.6N).

S: Substance or article so packed or designed that any hazardous effects arising from accidental functioning are limited to the extent that they do not significantly hinder or prohibit firefighting or other emergency response efforts in the immediate vicinity of the package (1.4S).

Fig. 3.10 Shipping labels of Class 1 dangerous goods

The commonly used commercial explosive products are classified as follows according to the IMDG Code:

- Commercial explosives, including cartridge explosives and ANFO, cast boosters, and detonating cords, belong to 1.1D;
- All kinds of detonators belong to 1.1B;
- Bulk emulsion (sensitized) belongs to 1.5D, but the emulsion matrix (before sensitizing) belongs to Class 5.1 (oxidizing substances) and does not belong to explosives; and
- Safety fuse and fuse igniter belong to 1.4S.

Figure 3.10 is the shipping labels of Class 1 Dangerous Goods:

3.7 Properties of Explosives

Knowing the properties of explosives is critical to understand the performance, and meaningful predictions can be made in blast design. These properties are detonation velocity, density, detonation pressure, water resistance, shelf life, and fume class. For a given explosive, these properties vary with the manufacturer.

3.7.1 Density

The density of an explosive may be expressed in terms of specific gravity. The specific gravity is the ratio of the density of explosive to the density of water under standard conditions. The specific gravity of commercial explosives ranges from 0.6 to 1.7 g/cc. With few exceptions, denser explosives give higher detonation velocities and pressure, but for any explosive, there is a critical density, above which it cannot reliably detonate, for example, for TNT—1.78 g/cc and ANFO—above 1 g/cc.

Density is an important consideration when choosing an explosive. For difficult blasting conditions or where fine fragmentation is required, a dense explosive is usually necessary. In easily fragmented rock or where fine fragmentation is not needed, a low-density explosive will often suffice. Low-density explosives are particularly useful in the production of riprap or other coarse products. The density of an explosive is also important when working under wet conditions. An explosive with a specific gravity of <1.0 will not sink in water.

3.7.1.1 Detonation Velocity

Velocity of detonation (VOD) is the velocity, with which the detonation waves move through a column of explosives. The factors that affect VOD are charge density, diameter, confinement, initiation, and aging of the explosive.

In general, the larger the diameter is, the higher is the VOD until a steady velocity is reached. For every explosive, there is a minimum critical diameter at which the detonation process once initiated, will support itself in the column. The influence of hole diameter on the VOD for various types of explosives is shown in Fig. 3.11.

The detonation velocity when measured at a confined condition of any explosives is larger than that measured at an unconfined condition. The confined detonation velocity of commercial explosives varies from 1500 to 6700 m/s [3].

There are diverse methods to measure VOD [11], among which the following are emphasized:

(a) D'Autriche method.

It is an old and simple method. In this method, the detonation wave propagating from two separate points of an explosive column via a detonating cord, of which the VOD is known, bound on a lead plate collide (see Fig. 3.12). The distance of the collision mark from the midpoint is measured. The VOD of the measured explosive can be determined from the following formula:

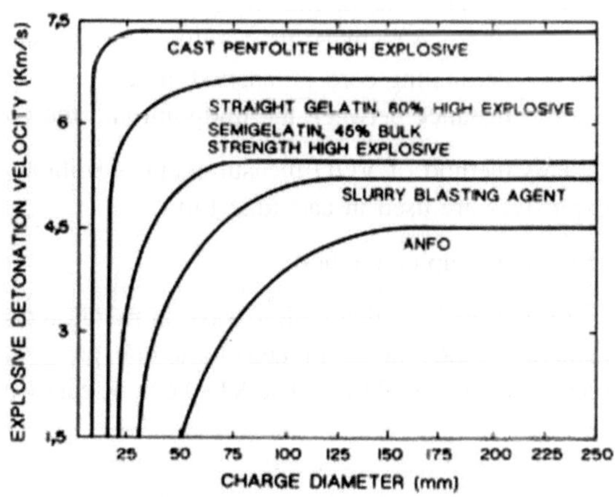

Fig. 3.11 Influence of the charge diameter upon the detonation velocity (Ash 1977) (Reproduced from Ref. [3] with the permission from Taylor & Francis Book, UK)

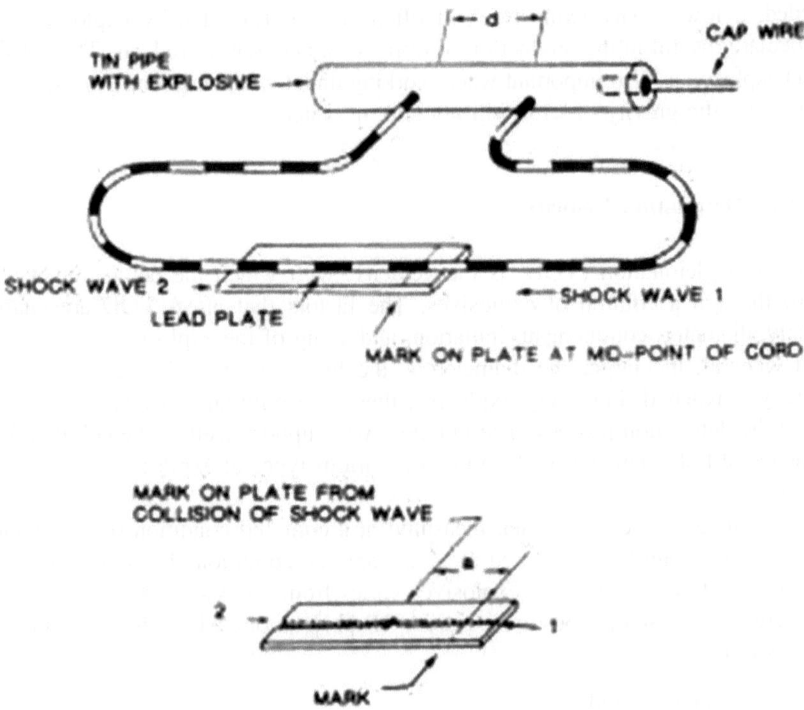

Fig. 3.12 D'Autriche method of VOD measurement

$$\mathrm{VOD}_e = \frac{\mathrm{VOD}_c \times d}{2a} \quad (3.35)$$

where

VOD_e VOD of explosive measured;
VOD_c VOD of the detonating cord;
d distance between two prefixed points on the explosive column where the detonating cord is inserted; and
a distance between the midpoint and the collision mark on the lead.

This method of VOD measurement is suitable for unconfined space where the explosives are used in cartridge form.

(b) Photographic method

In this method, detonation wave is monitored continuously using a high-speed camera. Explosion is an auto-luminous process. The light emitted is captured continuously in real time. The VOD can be easily calculated from the motion video.

(c) Discrete point (Point-to-Point) electric method

Point-to-point VOD systems are basically supported by electronics start and stop timer. Two cables have one end embedded in the explosive column, a known distance (L) apart. The other end of the sensor cables is attached to the VOD recorder, where the start and stop signals are recorded. When detonation reaches the first sensor, timing clock is started and the following sensor cable sends the stop signals when detonation reaches to it. The distance between sensor cables are known, and thus, VOD can be calculated. Figure 3.13 is one of these types of VOD instrument and its operating principle. Figure 3.14 is the same but more advanced instrument for VOD measurement of explosive in blasthole.

(d) Resistance Wire Continuous VOD Method

This method basically follows the Ohm's law ($E = RI$) where E—voltage, R—resistance, and I—current. In this method, ionization caused by explosion provides an electric short continuously and causing the voltage drop monitored by the instrument which is equivalent to the change in resistance value and the constant current. Thus, a voltage drop can be acted as single or double sensors, respectively, at the same time. Figure 3.15 is one of these types of instrument and working principle [12].

(e) Method Based on Fiber-optic

In this method, optical fiber is used which is capable of detecting and transmitting a light signal accompanying the detonation wave front. This method is point-to-point type wherein the first cable signals the start, whereas the second cable placed at a known fixed distance stops the timing clock. The fixed distance between probes divided by the timed clock directly gives the VOD value.

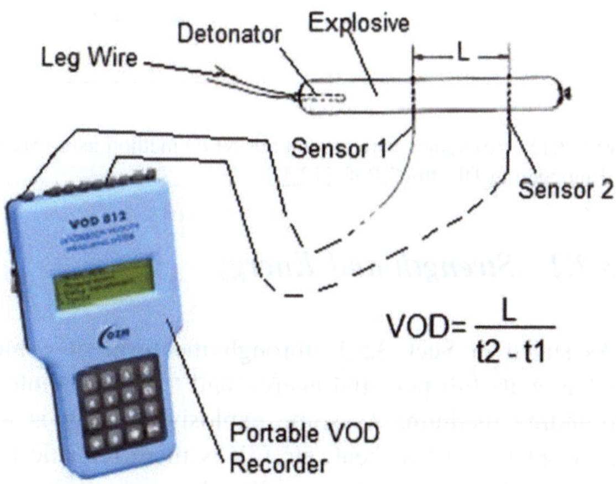

Fig. 3.13 Point-to-point electric method of VOD measurement

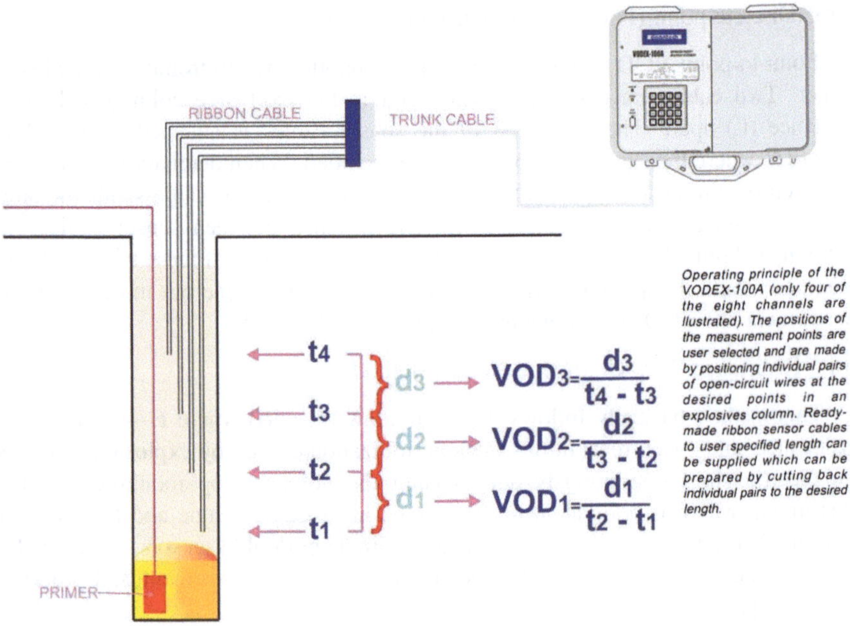

Fig. 3.14 Operation principle of the VODEX-100A for blasthole VOD measurement

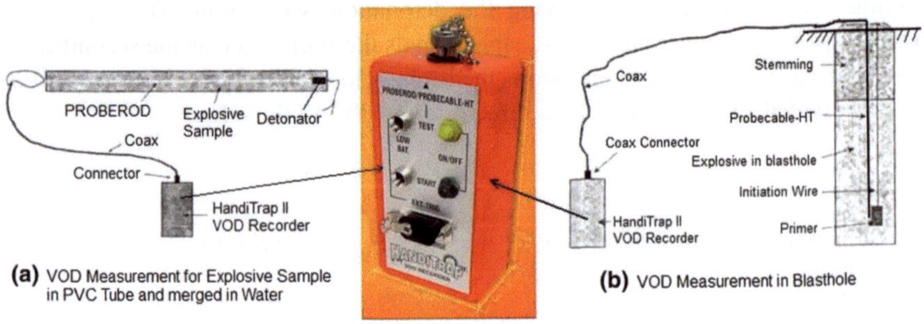

Fig. 3.15 Resistance wire continuous VOD method using Handi Tap II VOD record (Courtesy of "Engineering Blasting" Ref. [12])

3.7.2 *Strength and Energy*

As stated in Sect. 3.5.1, through the form of explosion reaction, the explosive releases its full potential energy and transforms into mechanical work on the surrounding medium. As some explosive energy is always wasted (vented to the atmosphere, lost as heat, etc.), it is more realistic to express explosive energy in terms of the amount of energy that the user can expect to have available to do useful work. This parameter is termed the *strength* of an explosive (refer to [10]).

There are different terms to express the strength of an explosive. The most useful indication of the strength of an explosive is the relative effective energy: *relative weight strength* (*RWS*) and *relative bulk strength* (*RBS*), referring to the strength of an explosive in so much percentage of another explosive that is taken as a standard, usually the standard ANFO (i.e., 94 % AN, 6 % fuel oil, density = 0.8 g/cc) which has the assigned value of 100.

There are various practical methods to measure the strength or the effective energy of an explosive. The following are the commonly used methods.

3.7.2.1 Trauzl Test

In this test, a lead cylinder (20 cm × 20 cm) is used (Fig. 3.16). Ten grams of the explosive to be tested is inserted into the 2.5-cm hole, sand stemmed, and detonated. The volume of the cavity formed is determined. This volume is a measurement of the expansion work performed by the well-coupled explosive. A correction must be carried out according to Table 3.3 given below when the test is not under the standard test temperature of the lead cylinder of 15 °C. Meyer (1977) reported the results of some explosives using this method (Table 3.4.).

3.7.2.2 Brisance—Hess Test

Brisance is the shattering capability of a high explosive, determined mainly by its detonation pressure. Brisance is of practical importance for determining the effectiveness of an explosion in fragmenting rock, shells, bomb casings, grenades, structures, and the like. Fragmentation occurs by the action of the transmitted shock wave, the strength of which depends on the detonation pressure of the explosive. Generally, the higher the pressure, the finer the fragments generated. High detonation pressure correlates with high detonation velocity. There are various methods to measure the brisance of an explosive. Figure 3.17 is the most commonly used method—Hess test. The value of the crushing height of the lead cylinder, B, is the brisance of the tested explosive.

Fig. 3.16 Trauzl lead block test. **a** Before test. **b** After test

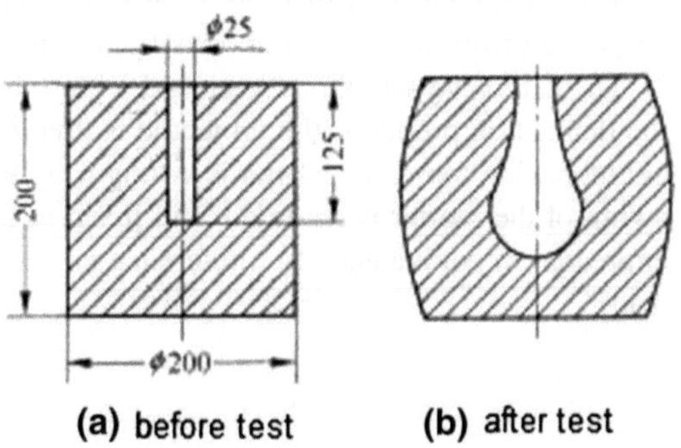

Table 3.3 Correction for the test result with Trauzl method

Temperature (°C)	−10	0	5	8	10	15	20	25	30
Correction (%)	10	5	3.5	2.5	2	0	−2	−4	−6

Table 3.4 Test results of some explosives using Trauzl test method by Meyer (1977)

Explosive	Cavity size (cm^3)
Nitroglycol	610
Nitroglycerin	530
PETN	520
TNT	300
Guhr dynamite	412
ANFO	316

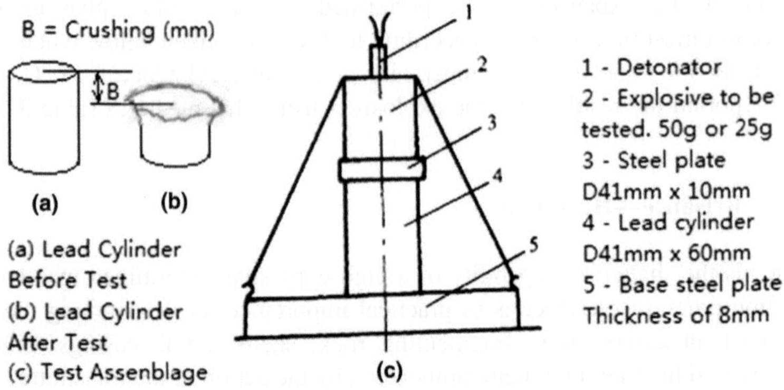

Fig. 3.17 Hess test for explosive brisance measurement

3.7.2.3 Ballistic Mortar Test

The ballistic mortar apparatus is used for testing of relative working ability (power, strength) of explosives.

The equipment is a massive steel mortar enclosed by a heavy steel projectile, suspended from the pendulum axis by a long pendulum arm (Fig. 3.18).

The explosive sample is detonated within the mortar cavity. Under the action of the detonation products, the projectile is fired out of the mortar, whereas due to the counteracting force, the mortar is swung from its position backward. The maximum swing of the mortar is recorded, and it serves as a measure of the explosive power expressed in equivalence to a reference explosive (blasting gelatin, TNT, or picric acid).

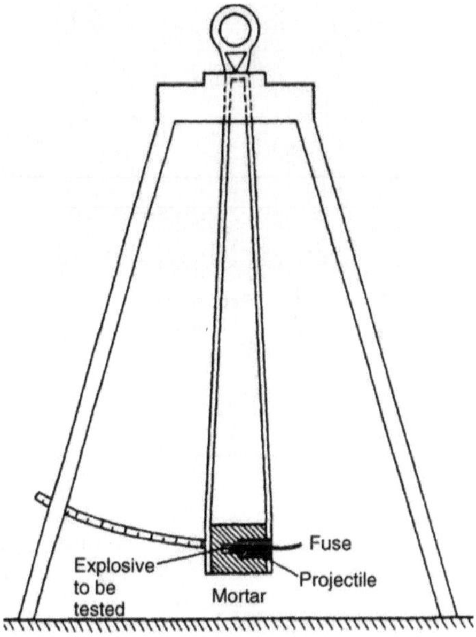

Fig. 3.18 Ballistic mortar test

3.7.2.4 Underwater Test

In this method, a small explosive charge is detonated underwater at sufficient depth to contain the gas bubble. The shock pressure pulse in the water is measured at a point distant from the charge.

This method is based on the hypothesis that "shock energy" from an explosive under water measures the explosive's shattering action in other materials, such as rock, and that "bubble energy" from the underwater explosion was a measure of the "heaving action" of the explosive. The shock energy in the test is the compressional strain energy radiated from an underwater explosive charge and is derived by measuring the area under the pressure squared time curve at a known distance from the explosion. The bubble energy is the potential energy of the displaced water at the maximum size of the bubble. It is derived by measuring the elapsed time between the shock wave and the pulse emitted by the first collapse of the gas bubble, knowing the ambient hydrostatic and atmospheric pressure action on the gas bubble. The total explosive energy is calculated to be the sum of the shock wave and bubble energies. The underwater energy test has been widely used throughout the world by both commercial explosives industry and the military.

In addition to measuring shock wave and bubble energies, underwater tests also can measure the shock wave impulse, another indicator of explosive strength. The shock wave impulse is derived by measuring the area under the pressure–time curve for a selected integration time interval at a known distance from the explosion.

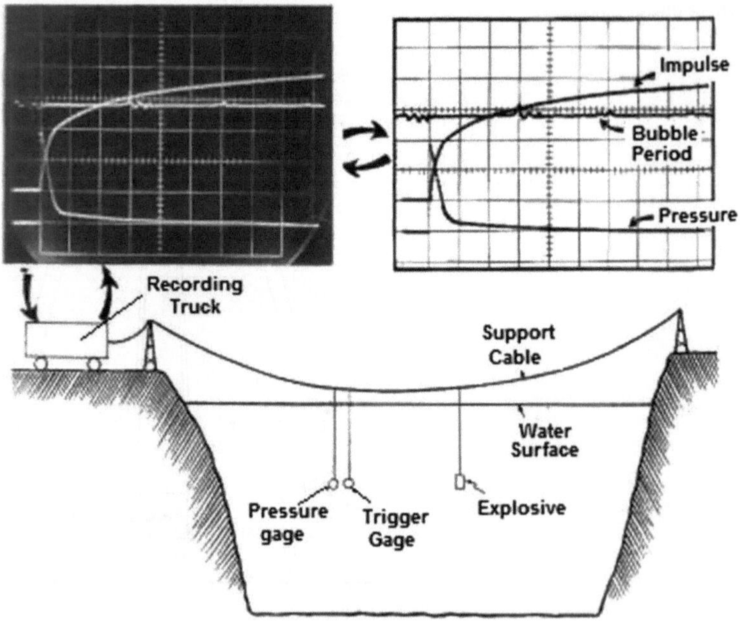

Fig. 3.19 Underwater test to assess explosive's energy (Courtesy of ISEE [13])

Figure 3.19 is the schematic diagram of a typical underwater test configuration and oscilloscope record used to determine shock wave impulse.

The experiments show that underwater test is a useful tool in evaluating the relative strengths of various explosives. The experiments also show that bubble energy values frequently overrate an explosive's ability to fragment and heave hard rock, but are more closely related to its capability of displacing weak rock.

3.7.2.5 Blasting Crater Test

The crater test method based on the theory of Livingston (1962) can be used for rating the blasting performance of various explosives in a given formation. In the crater test, a given size charge (height equal to six times the hole diameter) is buried at various depth and fired. The achieved crater volumes and critical depth where no breakage occurs are measured. The crater volume (see Fig. 3.20 [16]) can be calculated as:

$$V = \frac{1}{12} \times \pi \times d^3 h = 0.2618\, d^3 h \tag{3.36}$$

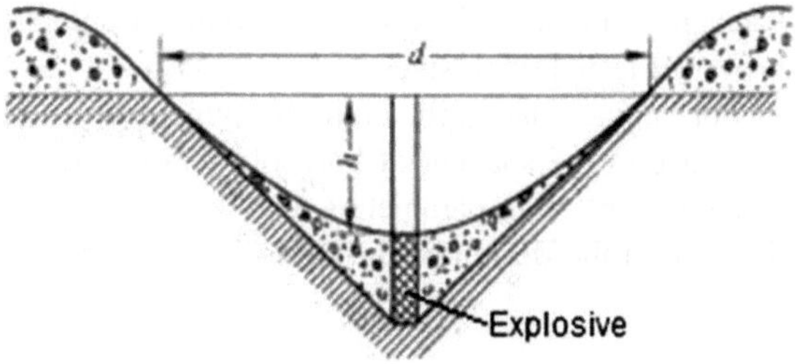

Fig. 3.20 Blasting crater test

where

V Volume of the blasted crater, m^3;
d Diameter of the formed crater, m; and
h Depth of the formed crater, m.

The main problem of this method is the difficulty in having an available test area of homogeneous rock and in the number of shots necessary.

3.7.3 Sympathetic Detonation

Sympathetic detonation is an explosion caused by the transmission of a detonation wave through any medium from another explosion. The initiating explosive is called donor explosive, and the initiated one is known as receptor explosive. In the case of a chain detonation, a receptor explosive can become a donor one.

The shock sensitivity, also called gap sensitivity, which influences the susceptibility to sympathetic detonations, can be measured by gap tests (see Fig. 3.21). The method consists in measuring the maximum gap of d (in cm), at which a primed cartridge can make another non-primed receiver cartridge explode, both being aligned with reference to their axis and well in contact with a ground or metal surface, or even inside tubes of different materials or in the open air. The test should be repeated through increasing the size of the gap until non-sympathetic detonation. The maximum gap in 3 successive tests with fully sympathetic detonation is the test result.

The gap value of sympathetic detonation is affected by some factors, such as the density of the receptor explosives, the diameter and quantity of tested explosives,

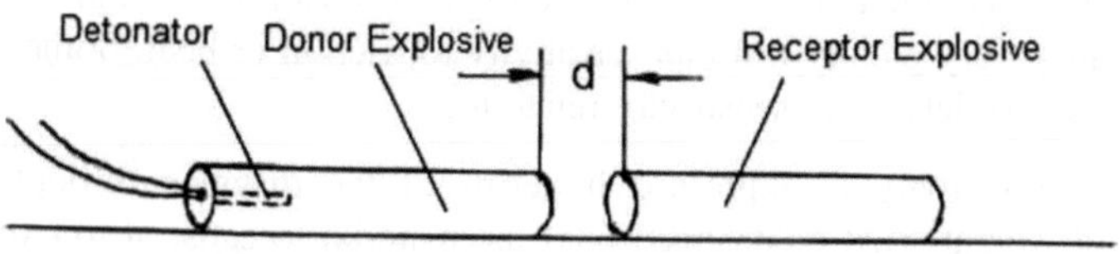

Fig. 3.21 Test of sympathetic detonation

package materials of explosives, confined situation and the firing direction and layout of tested explosives.

For engineering application, the sympathetic detonation value has an instructive reference to the spacing of borehole-segmented charges, misfire handling, and the reasonable blasting parameters. It is also an important basis to determine the safety distance in the design of the explosives factory and explosives warehouse.

3.7.4 *Sensitivity*

Sensitivity of explosives is the degree to which an explosive can be initiated by external actions, such as shock wave, impact, heat, or friction. All explosive compounds have a certain amount of energy required to initiate. If an explosive is too sensitive, it may go off accidentally. A safer explosive is less sensitive and will not explode if accidentally dropped or mishandled. However, such explosives are more difficult to initiate intentionally. Sensitivity is one of the most important indexes not only for blasting works but also for the safety of storage, transportation, and application. In fact, the sympathetic detonation that is discussed above is also an index of sensitivity to shock wave.

3.7.4.1 Detonator or Cap Sensitivity

This index is defined as the capacity of an explosive to be initiated into detonation in a sustained manner by the power of a detonator. This test, often referred to as the cap sensitivity test, not only characterizes an explosive's ease of initiation by a standard detonator, but also is used to classify products for safety in storage, transportation, and use.

A no. 8 strength detonator, where the charge corresponds to 2 g of mercury fulminate (80 %) and potassium chlorate (20 %), or a charge of equivalent pressed PETN, is used as the standard by the explosives industry.

3.7.4.2 Sensitivity to Heat, Impact, and Friction

A variety of tests are conducted to measure an explosive's susceptibility to initiation by accident. When the term sensitivity is used, care must be taken to clarify what kind of sensitivity is under discussion. The relative sensitivity of a given explosive to impact may vary greatly from its sensitivity to friction or heat. Some of the test methods used to determine sensitivity relate to:

- Impact—Sensitivity is expressed in terms of the distance through which a standard weight must be dropped onto the material to cause it to explode.

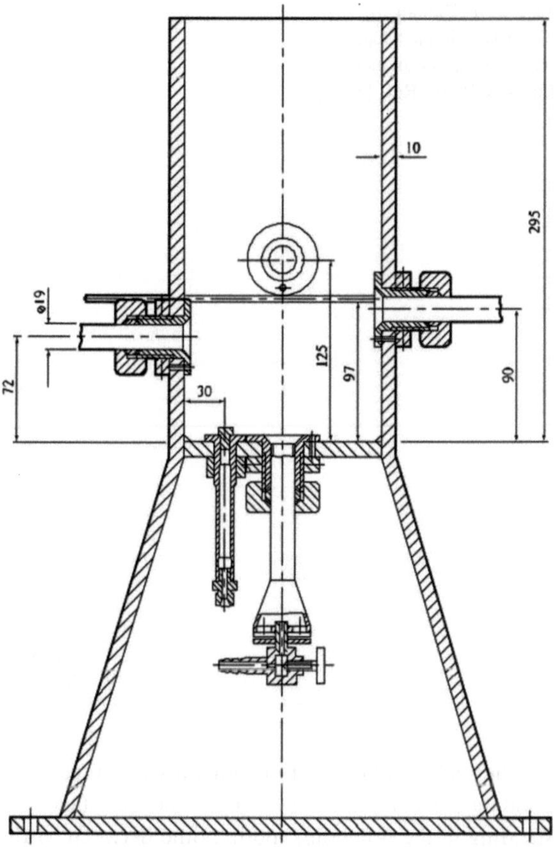

Fig. 3.22 Heating and protective device (Courtesy of Ordnance Industry Press, Ref. [18])

- Friction—Sensitivity is expressed in terms of what occurs when a weighted pendulum scrapes across the material (it may snap, crackle, ignite, and/or explode).
- Heat—Sensitivity is expressed in terms of the temperature at which flashing or explosion of the material occurs.

United Nations published the "Recommendation on the Transport of Dangerous Goods, Manual of Tests and Criteria" since 1986 (first edition). Figures 3.22, 3.23, 3.24, and 3.25 are parts of the recommended test apparatus (refer to [18]).

3.7.5 *Water Resistance*

An explosive's water resistance is a measure of its ability to withstand exposure to water without deteriorating or losing sensitivity.

Fig. 3.23 Bureau of explosives impact machine (Courtesy of Ordnance Industry Press, Ref. [18])

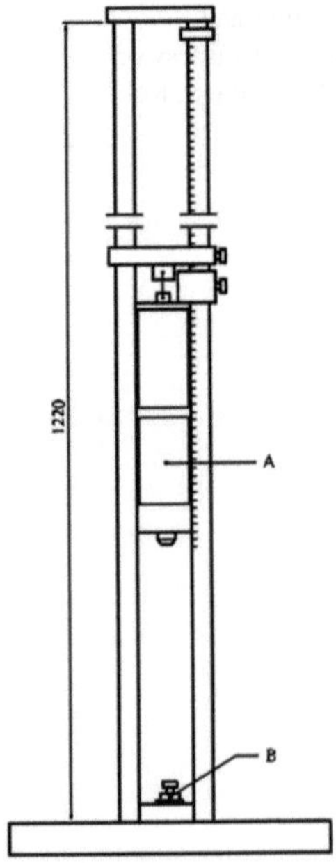

(A) Drop Weight
(B) Sample assembly

The scale of classification generally accepted goes from null, limited, good, very good, and excellent. In dry work, water resistance is of no consequence. If water is standing in the borehole, and the time between loading and firing is fairly short, an explosive with a water-resistant rating of "good" is sufficient. If the exposure is prolonged, or if the water is percolating through the borehole, "very good" to "excellent" water resistance is required.

Among the commercial explosives which are widely used in the world at present, water gels and emulsion explosives have a very good water resistance. Ammonium nitrate/fuel mixtures (ANFOs) have no inherent water resistance, as ammonium nitrate is very soluble in water. The thin film of fuel oil offers little protection. The blended emulsion/ANFO explosives have a range of water resistance, which varies from very good to a little resistance according to the emulsion percentage blended.

3.7.6 *Fumes and Fume Classification*

Ideally, detonation of a commercial explosive produces water vapor, carbon dioxide, and nitrogen. In addition, undesirable poisonous gases such as carbon monoxide and nitrogen oxides are usually formed. These gases are known as fumes,

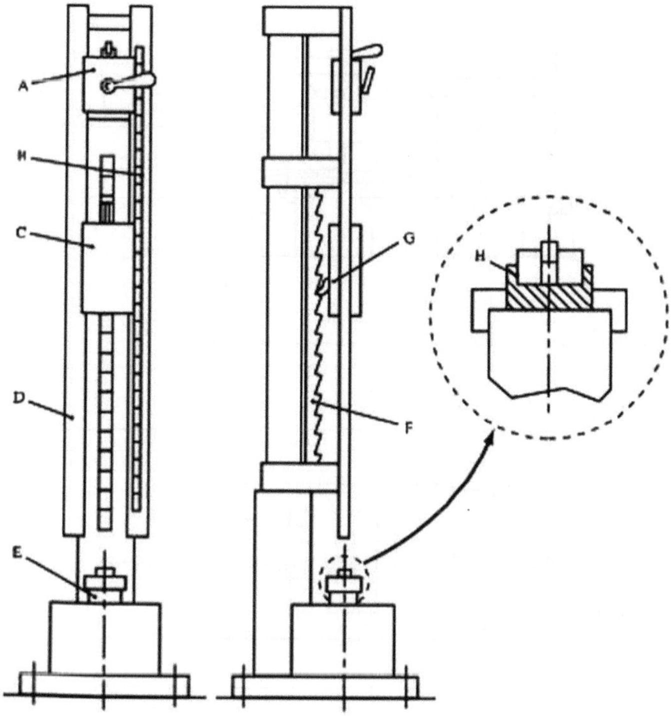

(A)	Catching and dropping device	(B)	Graduated scale
(C)	Drop weight	(D)	Guide column
(E)	Anvil	(F)	Rack
(G)	Ratchet to catch rebound	(H)	Magnified view of roller assembly

Fig. 3.24 Impact apparatus (Courtesy of Ordnance Industry Press, Ref. [18])

and the fume class of an explosive indicates the nature and quantity of the undesirable gases formed during detonation. Better ratings are given to explosives producing smaller amounts of fumes. For open work, fumes are not usually an important factor. In confined spaces, however, the fume rating of an explosive is important. In any case, the blaster should ensure that everyone stays away from fumes generated in a shot. Carbon monoxide gradually destroys the brain and the central nervous system, and nitrogen oxides immediately form nitric acid in the lungs.

Some factors that increase fumes are poor product formulations, inadequate priming, insufficient water resistance, lack of confinement, and reactivity of the product with the rock or other material being blasted.

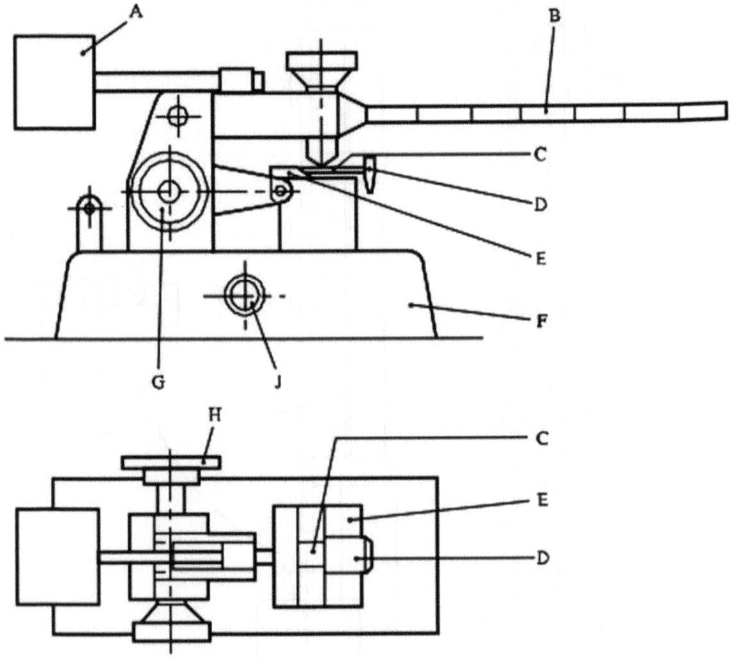

(A) Counter weight
(B) Loading arm
(C) Porcelain plate held on carriage
(D) Adjusting rod
(E) Movable carriage
(F) Steel base
(G) Handle for setting the carriage at the starting position
(H) Points the direction to electric motor drive
(J) Switch

Fig. 3.25 BAM friction apparatus (Courtesy of Ordnance Industry Press, Ref. [18])

Fume Classification.

In the USA, there are two different fume classifications for explosives. The type of classification depends whether or not the explosive is an MSHA (Mines Safety and Health Administration)-approved explosive—"permissible." (Permissible explosives are for use in underground coal mines and other mines or underground construction projects classified as gassy, where the use of regular, non-permissible explosives could ignite a gas or duct explosion in the air adjacent to the blast.) No explosive is approved as permissible if it generates more than 71 L of toxic gases per 456 g (2.5 cu. Ft of toxic gases per pound). Permissible grades of explosives do not carry fume markings.

US Bureau of Mines gives the classification for permissible as below:

- Class A: 0–53 L (0–1.87 cu.ft) noxious gases per 681 g (1.5 lb) explosive;
- Class B: 53–106 L (1.87–3.74 cu.ft) noxious gases per 681 g (1.5 lb) explosive.

All non-permissible grades explosives are classified to the Institute of Makers of Explosives (IME) classification: Fume classification standard based upon the volume of poisonous gases emitted by cartridge of 200 g is below:

- Fume Class 1: 0.00–0.453 dm^3 (0.16 cu.ft) noxious gases;
- Fume Class 2: 4.53–9.34 dm^3 (0.33 cu.ft) noxious gases; and
- Fume Class 3: 9.34–18.96 dm^3 (0.67 cu.ft) noxious gases.

Depending upon that classification, the first class explosives can be used for any type of underground work, those of second class only in well ventilation areas, and those of third class only in surface blasting.

There are several ways to determine fume concentrations. The commonly used test is carried out in the so-called Bichel Bomb with 200 g of explosive sample.

3.7.7 Desensitization

As it is stated in Sect. 3.7.1 that for any explosive, there is a critical density, above which it cannot reliably detonate. The increase of an explosive's density can be caused by some external pressures and the sensitivity of the explosive can be diminished to the point that it cannot detonate or has only weak detonation. This kind of phenomenon is called "desensitization."

The desensitization of an explosive can be caused by a hydrostatic pressure or a dynamic pressure. The first one is usually only present in very deep blastholes and is not common for this reason. In dynamic desensitization, the following situations are usually observed:

- Desensitization by in-hole detonating cord.

As the detonation velocity of the detonating cords (usually above 6000 m/s) is much higher than the commercial explosive (usually water gel and emulsion: 3500–5000 m/s, ANFO: 2000 m/s), the shock wave produced by the detonation of in-hole detonating cord compresses the explosive cartridge and increases the density of explosive or breaks the structure of the composition of the explosive (e.g., compressing or breaking the bubbles of the hot points for water gels and chemically sensitized emulsion explosives) and decreases the sensitivity of the explosives.

- Desensitization by channel effect.

Channel effect is known as the phenomenon that the detonation of the explosive charge self-suppresses—energy gradually decays until the detonation extinguishes due to the existence of a crescent-shaped space between the explosive charge and the borehole inner wall.

The common explanation of this phenomenon is that a precursor air shock wave (PAS), which propagates ahead of the detonation front in an air channel, precompresses and desensitizes the unreacted explosive charges. Under some conditions, the PAS causes detonation failure.

The research by M. A. Cook and L. L. Udy (IRECO, USA) revealed that the precursor air shock wave (PAS) is a plasmasphere produced by the detonation of explosive charge which precompresses and desensitizes the unreacted explosive charges. The experimental research demonstrated that the difference between PAS velocity and detonation velocity was the primary factor for the channel effect and the decrease of the PAS velocity can prevent the channel effect. Increasing the surface roughness of either explosive charge or borehole inner wall is one of the effective methods to reduce the PAS velocity and prevent the channel effect.

- Exerted pressure by adjacent charges.
  Sometimes, desensitization may be caused by the detonation of adjacent charges fired in advance in the following situation:
  - A shock wave that passes through the charge from the adjacent ones;
  - Lateral deformation of the borehole, compressing the charge, due to the rock movement or underground water, especially blasting in weak or highly fissured rock mass;
  - Compression of charge by pushing of the intermediate stemming material; and
  - By infiltration of explosion gases through fissures or fractures opened in the rock mass.

Appropriate blasting design of blasthole pattern and delay timing to suit the actual situation of rock mass can prevent desensitization of explosives during blasting.

3.7.8 Stability and Shelf Life for Storage

Stability is the ability of an explosive to be stored without deterioration.

The following factors affect the stability of an explosive:

- Chemical constitution. In the strictest technical sense, the word "stability" is a thermodynamic term referring to the energy of a substance relative to a reference state or to some other substances. It is generally recognized that certain groups such as nitro ($-NO_2$), nitrate ($-ONO_2$), and azide ($-N_3$), are intrinsically labile. Kinetically, there exists a low activation barrier to the decomposition reaction. Consequently, these compounds exhibit high sensitivity to flame or mechanical shock. The chemical bonding in these compounds is characterized as predominantly covalent, and thus, they are not thermodynamically stabilized by a high ionic-lattice energy. Furthermore, they generally have positive enthalpies of formation, and there is little mechanistic hindrance to internal molecular rearrangement to yield the more thermodynamically stable (more strongly bonded) decomposition products.

- Temperature of storage. The rate of decomposition of explosives increases at higher temperatures. All standard military explosives may be considered to have a high degree of stability at temperatures from −10 to +35 °C, but each has a high temperature at which its rate of decomposition rapidly accelerates and stability is reduced. As a rule of thumb, most explosives become dangerously unstable at temperatures above 70 °C.
- Exposure to sunlight. When exposed to the ultraviolet rays of sunlight, many explosive compounds containing nitrogen groups rapidly decompose, affecting their stability.
- Electrical discharge. Electrostatic or spark sensitivity to initiation is common in a number of explosives. Static or other electrical discharge may be sufficient to cause a reaction, even detonation, under some circumstances. As a result, safe handling of explosives and pyrotechnics usually requires proper electrical grounding of the operator.

The stability of the explosives is one of the properties that are related to the maximum storage time of these substances, so that their effects in a blast will not be reduced. The manufacturers of any explosive product will specify the shelf life for storage of the product under the normal storage conditions to ensure the quality of the product during the specified storage time.

3.8 Commercial Explosives (Industrial Explosives)

All commercial explosives that are extensively used in industry should meet the following basic requirements [2]:

- Have a low mechanical sensitivity and proper initiation sensitivity, which should ensure not only the safety during the process of production, storage, transport, and use, but also a convenient and reliable initiation during the blasting operation;
- Have a good blasting performance with enough strength for meeting various rock characters;
- Its chemical composition should have a zero oxygen balance or near-to-zero oxygen balance to ensure less toxic fumes during blasting and no or very few poisonous components in its composition;
- Have a proper and stable shelf life for storage and does not deteriorate or lose efficiency within the period of its shelf life;
- Have a wide range and cheap sources of raw materials; and
- Have a simple manufacturing process and safe operation.

The commercial explosives for industrial use are divided into two large groups:

- Cartridge (conventional) explosives. These explosives usually have a cap sensitivity and are made up in various sizes of diameter and length for convenient use. They include:
  - Nitroglycerin (NG)-based explosives. They are usually called as dynamites. There are three basic types of dynamite: gelatin, semigelatin, and granular.
  - AN-TNT-based explosives—ammonite. It is a civilian explosive, generally comprising a mixture of ammonium nitrate, TNT, and other flammable materials. Typically, it is used for quarrying or mining purposes. It is a popular civil engineering explosive in Russia, Eastern Europe, and China.
  - Water-based explosives. They include two different forms: (1) water gel/slurry and (2) emulsion.
  - Permissible explosives. They are used for coal mines.
- Blasting agents. These mixtures, with few exceptions, do not contain ingredients classified as explosive. The following are the most common:
  - ANFO,
  - ALANFO,
  - Bulk slurries or water gels,
  - Bulk emulsions,
  - Heavy ANFO.

3.8.1 Nitroglycerin (NG)-Based Explosives

Nitroglycerin (NG) has been in use for more than 140 years as the mainstay of the commercial explosives industry, and usually, it was called dynamite. Nitroglycerin explosive was discovered by Sobrero in 1846, and Alfred Nobel (Fig. 3.26.)

Fig. 3.26 Alfred Nobel (1833–1896)

developed it into a commercial-scale explosive in 1875. Alfred Nobel discovered that a large quantity of nitroglycerin could be absorbed into kieselguhr and made safe to transport and use. Over the years, various percentages of nitroglycerin and diverse materials have been mixed to produce different types and grades of dynamites. They are packed in cylindrical cartridges, 20 mm diameter and larger, with length ranging from 200 to 1000 mm. Various paper shells or wrappers are used to package and protect it from moisture. The overall weight percentage quantity and type of wrapper have an important influence on explosives fume production, water resistance, tampability, and loadability.

There are three basic types of dynamite: granular, semigelatin, and gelatin (Fig. 3.27).

- Granular dynamites. Granular dynamite is also called as straight dynamites. It contains 15–60 % explosive oil (nitroglycerin plus ethylene glycol). In addition, antacid, carbonaceous material, and sodium nitrate is used. Though they have high detonation velocities and good water resistance, they have high flammability and highly sensitive to shock and friction and produce large noxious fumes.
- Gelatin dynamites. There are two types of gelatin dynamites: straight gelatin dynamite and ammonia gelatin dynamite.

Straight gelatin dynamites are similar to straight dynamites, except that the explosive oil has to be collided with nitrocellulose to form a gel.

Ammonia gelatin dynamites are actually ammonium nitrate explosive sensitized with NG. Generally, they have the almost same blasting strength as most straight gelatin dynamites and are less costly.

- Semigelatin dynamites. Semigelatin dynamite is a hybrid formulation with properties between those of gelatin dynamite and granular dynamite. They combine the economy of ammonia dynamites with the water resistance and cohesiveness of ammonia gelatin explosives and contain less explosive oil, sodium nitrate, and nitrocellulose, but more ammonium nitrate.

Fig. 3.27 Dynamite explosive

3.8.2 AN-TNT-Based Explosives—Ammonite

Ammonite was formerly known as amatol which was a highly explosive material made from a mixture of TNT and ammonium nitrate and was used extensively during World War I and World War II, typically as an explosive in military weapons such as aircraft bombs, shells, depth charges, and naval mines. Amatol is rare today. A form of amatol exists under a different name—**ammonite**. Ammonite is a civilian explosive, generally comprising a mixture of ammonium nitrate, TNT, and other flammable materials. Typically, it is used for quarrying or mining purposes. It is a popular civil engineering explosive in Russia, Eastern Europe, and China.

3.8.2.1 Main Ingredients

(a) Ammonium Nitrate

The main ingredient of ammonite is ammonium nitrate (NH_4NO_3). As an oxidizing agent, its content is about 75–90 % in the ammonite.

Ammonium nitrate (AN) is a white crystalline solid at room temperature and standard pressure. It is commonly used in agriculture as a high-nitrogen fertilizer, and it has also been used as an oxidizing agent in explosives.

In room temperature, AN appears in a white crystalline form and it is also colorless. Its melting point is 169.6 °C or 337.3 °F. These crystals are rhombohedral in shape, but when they are subjected to temperatures above 32 °C, they change from α-rhombic to β-rhombic crystals and the volume increases about 3.6 %. As the one of the main raw materials of explosives, there are two forms of products from the manufacturers: crystalline and porous spherical prills (see Fig. 3.38).

AN has a strong moisture absorption, caking sclerosis. Its identifiers and properties are listed in Table 3.5.

Dry AN has a very slow chemical reaction with metal, and water can accelerate the action. But AN has no action with aluminum and tin, so that aluminum tools usually are used in the manufacturing process of AN-based explosives. Table 3.5 shows the identifiers and properties of ammonium nitrate (Fig. 3.28)

(b) TNT (Trinitrotoluene)

TNT (trinitrotoluene) is a yellow-colored solid chemical compound with the formula $C_6H_2(NO_2)_3CH_3$ and works as a sensitizer in AN-TNT-based explosives (Fig. 3.29). TNT is one of the most commonly used explosives for military and industrial applications. It is valued partly because of its insensitivity to shock and friction, which reduces the risk of accidental detonation, compared to other more sensitive high explosives such as nitroglycerin. TNT melts at 80 °C (176 °F), far below the temperature at which it will spontaneously detonate, allowing it to be poured as well as safely combined with other explosives. TNT neither absorbs nor

Table 3.5 Ammonium nitrate identifiers and properties

Identifiers	
CAS number	6484-52-2
ChemSpider	21511
UN II	T8YA51M7Y6
UN number	0222—with >0.2 % Combustible Substances 1942—with ≤0.2 % Combustible substances 2067—fertilizer 2426—liquid
RTECS number	**BR90500 00**
Properties	
Molecular formula	NH_4NO_2
Molar mass	80.052 g/mol
Appearance	White/gray solid
Density	1.725 g/cm^3(20 °C)
Melting point	169.6°
Boiling point	Approx. 201 °C decomp.
Solubility in water	118 g/100 ml (0 °C) 150 g/100 ml (20 °C) 297 g/100 ml (40 °C) 410 g/100 ml (60 °C) 576 g/100 ml (80 °C) 1024 g/100 ml (100 °C)
Structure	
Crystal structure	Trigonal
Explosive data	
Shock sensitivity	Very low
Friction sensitivity	Very low
Explosive velocity	5270 m/s
Hazards	
MSDS	ICSC 0216
EU Index	Not listed
Main hazards	Explosive
NFPA 704	
LD_{50}	2085–5300 mg/kg (oral in rate, mice)

dissolves in water, which allows it to be used effectively in wet environments. Additionally, it is stable compared to other high explosives.

TNT is poisonous, and skin contact can cause skin irritation, causing the skin to turn a bright yellow-orange color. People exposed to TNT over a prolonged period tend to experience anemia and abnormal liver functions. Blood and liver effects, spleen enlargement, and other harmful effects on the immune system have also been

(a) Crystalline (b) Porous Prills

Fig. 3.28 Ammonium nitrate

Fig. 3.29 Trinitrotoluene (TNT)

found in animals that ingested or breathed trinitrotoluene. There is evidence that TNT adversely affects male fertility. TNT is listed as a possible human carcinogen, so no developed countries have used AN-TNT-based explosives after World War II and they also have been prevented from use in China several years ago and replaced by emulsion explosives, although they were used in China, Russia, and East Europe for more than 60 years.

(3) Wood Flour

Wood flour works as a flammable agent in AN-TNT-based explosives, and it also acts as an anti-caking effect.

Table 3.6 shows the ingredients and properties of two types of ammonite.

Table 3.6 Ingredients and properties of two types of ammonite

Explosive name	Country	Ingredients (%)			Properties				
		AN	TNT	Wood flour	Density (g/cm^3)	Brisance (mm)	Working capacity(ml)	Sympathetic distance(cm)	Detonation velocity(m/s)
2# Rock	China	85 ∓ 1.5	11 ∓ 1.0	4 ∓ 0.5	0.95–1.10	≥12	≥298	≥5	≥3200
Ammonite 6	Bulgaria				0.98–1.05	≥14	≥340	≥5	3600

3.8.3 *Water-Based Explosives*

Dr. Melvin A. Cook was awarded US patent number 2,930,685 on March 29, 1960, for his invention of a practical explosive composition containing water as an essential ingredient. This was the birth of the water-based explosives. Currently, there are two different forms of water-based explosives: (1) water gel/slurry and (2) emulsion. Although water-based explosives contain water as an essential ingredient, they all have a very good water resistance.

3.8.3.1 Water Gels/Slurries

Water gel/slurry explosive is essentially based on saturated aqueous solutions of AN, often with other oxidizers such as sodium nitrate and/or calcium nitrate, in which the fuels, sensitizers, gellants, and cross-linking agents are dispersed to avoid the segregation of the solids.

Generally, there is no strict distinction between water gel explosives and slurry explosives. The main difference is that the two kinds of explosives use different sensitizers. The main sensitizer of slurry explosives is non-water-soluble explosive (e.g., TNT), metal powders (e.g., powder aluminum), and solid combustibles, and the water gel explosives use water-soluble monomethyl amine nitrate (MMAN) as a sensitizer. The sensitivity of water gel explosives is higher than that of normal slurry explosives. Usually, water gel explosives have cap sensitivity and are frequently used as cartridge explosives. Slurry explosives usually are used in open pits as the bulk, pumpable blasting agent (Fig. 3.30).

Water gel/slurry explosives have a good blasting performance. Their densities generally range from about 0.8 to about 1.60 g/cc, while most are formulated in the range of 1.00–1.35 g/cc. The detonation velocities of most water gel/slurry explosives range from 3500 to 5000 m/s. Because they contain substantial amount of water and separate oxidizer and fuel components, they are intrinsically less sensitive than the water-free, NG dynamite explosives. Water gel/slurry explosives

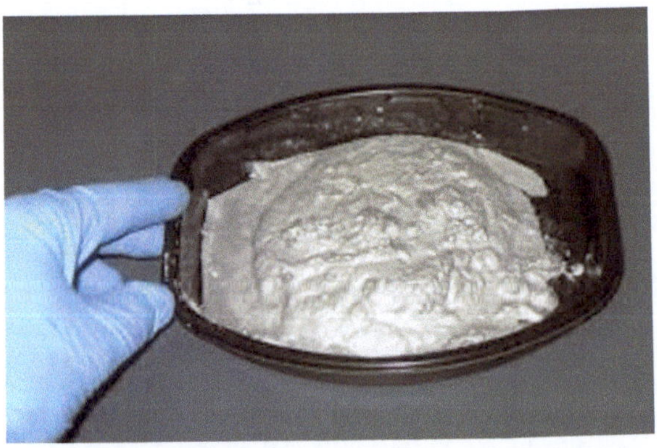

Fig. 3.30 Slurry explosives

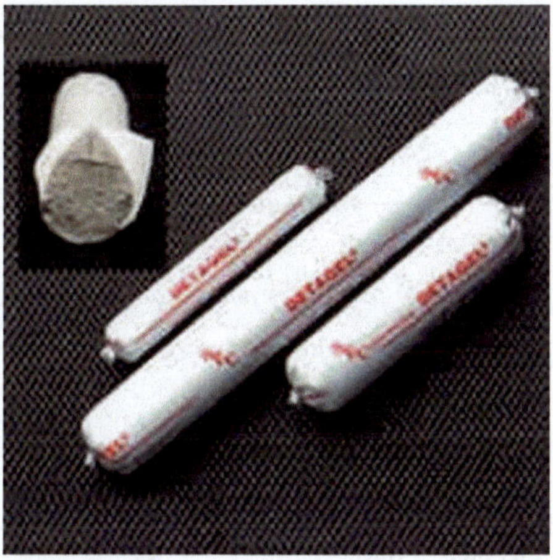

Fig. 3.31 Water gel explosives

have a very good to excellent water resistance. They can sustain a normal detonation after 24 h under 10-m-deep water. Water gel is generally chemically stable and has a long shelf life of about 1–2 years under the normal storage conditions (Fig. 3.31).

The most important disadvantage of water gel/slurry explosives is that there is an explosive ingredient, e.g., TNT or MMAN, as the sensitizer in their composition. These ingredients increase the danger and complexity during the manufacture and raw material storage. Additionally, the cost of water gel is relatively higher due to the expensive MMAN- compared to AN-TNT-based explosives. In particular, emulsion explosives have been developed and have quickly replaced water gel/slurry explosives in the world explosive market.

3.8.3.2 Emulsion Explosives

Emulsion explosives generally refer to a category of water-in-oil (W/O), emulsoid-type, water-resistant industrial explosives which are made by emulsification technique.

In the process of developing emulsion explosives, their earliest initial form, which was formed by mixing water-in-oil emulsion with ordinary water-bearing slurry explosives, was made in 1961 by R. S. Egly and others of commercial Solvents Corporation, USA, and in 1963, N.E. Gehrig of Atlas Chemical Industrial Limited, USA, further developed emulsions without slurries. Although these persons individually obtained patents, it was H. F. Bluhm of Atlas Chemical Industrial Limited, USA, who first described the techniques of emulsion explosives. Therefore, it is generally believed that water-in-oil emulsion explosives were first revealed by Bluhm on June 3, 1969.

AN emulsion is a two-phase system in which an inner or dispersed phase is distributed in an outer or continuous phase as shown in Figs. 3.32 and 3.33. In

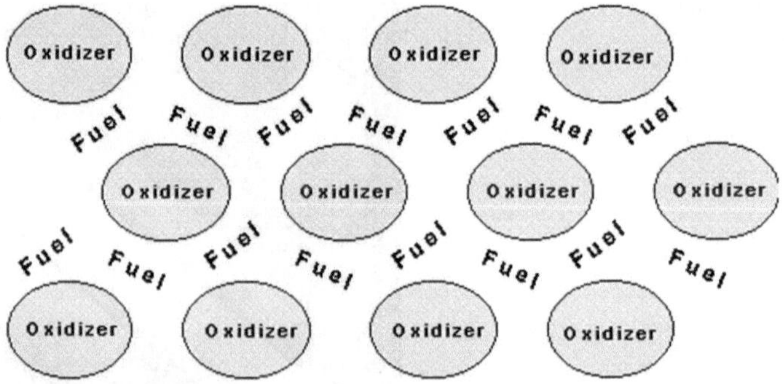

Fig. 3.32 Oxidizer surrounded by fuel

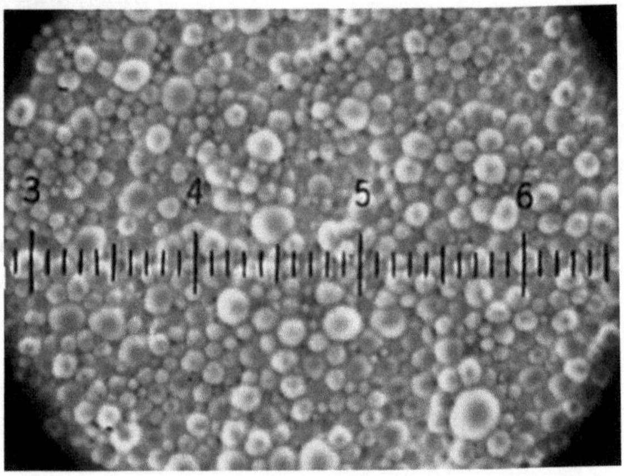

Fig. 3.33 Typical water-in-oil emulsion as seen under a microscope. Each number is 50 μm. Each small division is 5 μm

simpler terms, an emulsion is a mixture of two liquids that do not dissolve in one another. This unique feature coupled with the fact that the minute sized nitrate solution droplets are tightly compacted within the continuous fuel phase results in good intimacy between the oxidizer and fuel giving increased reaction efficiency compared to other systems.

In order to obtain an adequate sensitization of the explosives, when these do not contain liquid or solid chemical sensitizers, a physical mechanism such as gas bubbles is required, so that when the explosive is adiabatically compressed, it produces a "hot spot" phenomenon, favoring initiation as well as detonation propagation. The gasifying agents are made up of some chemical (such as sodium nitrite) or some solid particles with entrapped gases (such as hollow glass microballoons, expanded pearlite microparticles, and hollow resin microballoons).

In the late 1980s and 1990s, a new type of emulsion explosives has been developed, the powdery emulsion explosives (PEEs). It was made using an emulsification–spray drying technique. PEE is composed of 91–92.5 wt% ammonium nitrate (AN), 4.5–6 wt% organic fuels, and 1.5–1.8 wt% water. Due to its microstructure as a water-in-oil (W/O) emulsion and low water content, it has excellent detonation performance, outstanding water resistance, reliable safety, and good application.

The practice of production and application for decades shows that emulsion explosive has the following characteristics:

- Good explosive performance. Detonation velocity of small-diameter cartridge is up to 4000–5200 m/s, brisance up to 15–19 mm, sympathetic distance of 7.0–12.0 cm, and the critical diameter of 12–16 mm and can be reliably initiated with a #8 detonator.
- Excellent water resistance. A cartridge, which is exposed out of its wrapper and immersed in water for over 96 h, shows very little change in performance. Because of its high density, it is easy to sink into the water, suitable for water-filled holes in surface blasting and underwater blasting operations.
- Good safety performance. Because they contain a substantial amount of water and separate oxidizer and fuel components, they have intrinsically less sensitivity than the water-free, NG dynamite explosives. As they contain no ingredient that is an explosive in itself, and also because of the desensitizing effect of water content, one of the outstanding characteristics of emulsion explosive is having a high degree of inherent safety. The international practice has shown that emulsion explosive's mechanical sensitivity (shock sensitivity and friction sensitivity), combustion sensitivity, ignition point, rifle bullet impact sensitivity, etc., are all lower than other commercial explosives. On the other hand, the minute size of the nitrate solution droplets is tightly compacted within the continuous fuel phase resulting in good intimacy between the oxidizer and fuel and increased reaction efficiency compared to other systems, so that they have good detonation sensitivity (Fig. 3.34)
- Less environmental pollution. There is no toxic substance such as TNT in the composition of emulsion explosives, so that it solves the problems of the

Fig. 3.34 Rifle bullet impact test

environmental pollution and worker poisoning during the production process. The toxic gases produced in the explosion are relatively small.

- A wide range of sources of raw materials and production process is relatively simple. The main raw materials of emulsion explosives are ammonium nitrate, sodium nitrate, water, diesel oil, emulsifier, and a few of additive agents and they all are readily available in the market. The production equipment required and production process are relatively simple.
- Low production cost. In all commercial explosives, in addition to ANFO, the production costs of emulsion explosives are the lowest.

In general, emulsion explosives can be classified into two groups of cap sensitive and non-cap sensitive according to their detonation sensitivity. It can also be classified according to the packing into three groups: cartridge emulsion, bag packed emulsion, and pumpable bulk emulsion blasting agents (usually blending with ANFO for surface blasting application). The first group has the cap sensitivity, and the others have non-cap sensitivity and need boosters to initiate them in the boreholes. Table 3.7 shows the formulations of three types of cartridge emulsion explosives, and Fig. 3.35 shows the manufacturing flowchart of emulsion explosives.

Table 3.8 gives the properties of some cartridge emulsion explosives.

Table 3.7 Examples of formulation of emulsion explosives

Materials	Examples of series of emulsion explosives		
	EL-102-2	BME-2-2	Rock powdery emulsion
Ammonium nitrate	73.0	63.0	87–93
Sodium nitrate	10.0	9.0	
Ammonium perchlorate	–	6.2	
Water	12.0	11.0	0–8
Emulsifier	1.0	1.0–1.5	1.5–2.5
Wax	2.5 (composite)	1.5	
Vaseline	–	2.0	
Microcrystalled wax	–	0.8	
Oil	1.5	–	3.6–5.5
Density adjustor	–	2.4	
Appearance status	Cartridge with paper/plastic wrapper	Cartridge with paper/plastic wrapper	Cartridge with paper/plastic wrapper

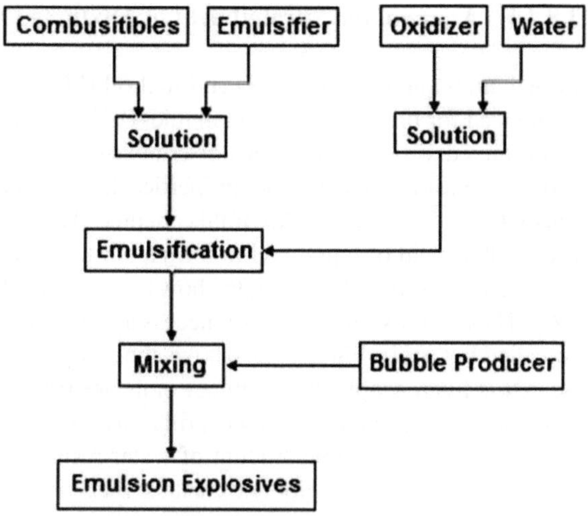

Fig. 3.35 Flowchart of emulsion production (Courtesy of Metallurgy Industry Press, China, Ref. [14])

Table 3.8 Properties of some cartridge emulsion explosives

Brand	Senatel Pulsar	Emulex 150	Powermite Max	Rock No. 1	Rock PPE
Manufacturer	Orica	Tenaga Kimia	Dyno Nobel	Nanling, China	Liming, China
Density (g/cc)	1.22	1.12–1.24	1.15	1.0–1.3	0.9–1.05
VOD (m/s)	4400	4500	4500	4500	4198
Relative weight strength* (%)	116		86		
Relative bulk strength* (%)	175	143	125		
Energy (MJ/kg)		4.8	3.2		
Brisance (mm)				16	18
Work capacity (ml)				320	352
Sympathetic distance (cm)				4	16

*Relative to ANFO = 100 % @ 0.8 g/cc

3.8.4 Bulk Blasting Agent

As defined in Sect. 3.6.2 of this chapter, explosives that are so insensitive to shock that they cannot be reliably initiated by practical quantities of primary explosive (usually #8 detonator) and instead require an intermediate explosive booster (primer) of secondary explosive are called *blasting agents* or tertiary explosives. ANFO, bulk (pumpable) emulsion, and heavy ANFO (blends of emulsion/ANFO) are blasting agents.

3.8.4.1 Ammonium Nitrate/Fuel Oil Mixtures (ANFO)

Prilled ammonium nitrate (AN) and fuel oil (FO) mixtures, known as ANFO, were introduced for blasting operation in mid-1950s. Ammonium nitrate in a proper form when mixed with carbonaceous or combustible material in appropriate proportion forms a blasting agent. The properties and features of ammonium nitrate are illustrated in Sect. 3.8.2.1 (a) of this chapter. Although many forms of AN could be used with a solid or liquid fuel to form a blasting agent, the porous prilled forms are preferred for ANFO. Figure 3.36 shows the manufacturing tower and porous prilled AN. These voids in prills are necessary to sensitize ANFO: They create the so-called hot spots. Figure 3.37 shows the AN prills' manufacturing schematic [15]. As AN is hygroscopic, it can attract moisture from the atmosphere and deteriorate. For this reason, explosive-grade prills have some protective coating, anti-caking agent, which offers some amount of water resistance to avoid caking.

The most common product used is fuel oil which, when compared to other liquids such as gasoline and kerosene, has the advantage of a higher point of volatility and, as a consequence, less risk of steam explosions. Owing to their high viscosity, they tend to coat the surface of the AN prills, filling the macropores.

The chemistry of ANFO detonation is the reaction of ammonium nitrate (NH_4NO_3) with a long-chain alkane (C_nH_{2n+2}) to form nitrogen, carbon dioxide, and water. In an ideal oxygen balanced reaction:

$$3NH_4NO_3 + CH_2 \rightarrow 3N_2 + 7H_2O + CO_2$$

ANFO is composed of approximately 94.3 % AN and 5.7 % FO by weight, producing around 920 kcal/kg and a gas volume of 970 L. In practice, a slight excess of fuel oil is added, i.e., 6.0 %. When detonation conditions are optimal, the aforementioned gases are the only products. In practical use, such conditions are impossible to attain, and blasts produce moderate amounts of toxic gases such as carbon monoxide and nitrogen oxides (NO_x). The effects of the content of fuel oil in

Fig. 3.36 Porous prilled ammonium nitrate, the 62-m-high prill manufacture tower, and loading to truck

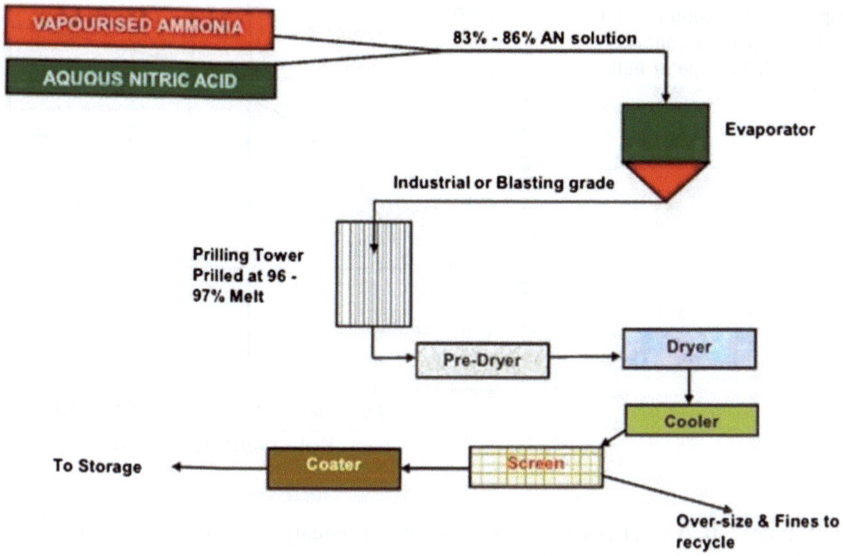

Fig. 3.37 AN prill manufacture schematic (Reproduce from Ref. [15] with the permission from the author, Mr. Partha Das Sharma)

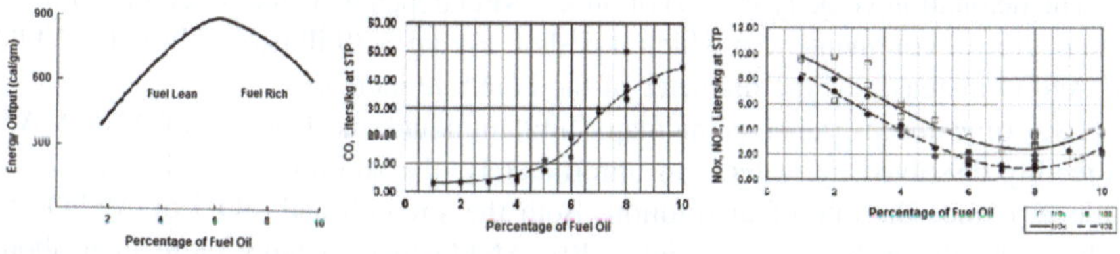

Fig. 3.38 Effect of fuel oil content to the energy output and produced toxic fumes of ANFO (Reproduced from Ref. [15] with the permission from the author of Mr. Partha Das Sharma)

ANFO to the energy output and the amount of toxic fumes are shown in Fig. 3.38 [15].

ANFO under most conditions is blasting cap insensitive, and so it is classified as a blasting agent and not a high explosive. Because it is cap insensitive, it generally requires a primer, also known as a booster (e.g., one or two sticks of cap-sensitive cartridge high explosives, or in more recent times, cast boosters of TNT/PETN or similar compositions to ensure continuation of the detonation wave train). The detonation sensitivity of ANFO is reduced as the blasthole diameter increases. In practice, the primers of 150 g are effective in charge diameter under 150 mm; above this diameter, cast primers of 400–500 g are recommended.

The loosed-poured density of ANFO is usually about 0.80 g/cm^3. To obtain densities above the loosed-poured value, some pneumatic loading machine or augering equipment can increase the density of ANFO to a value as high as about

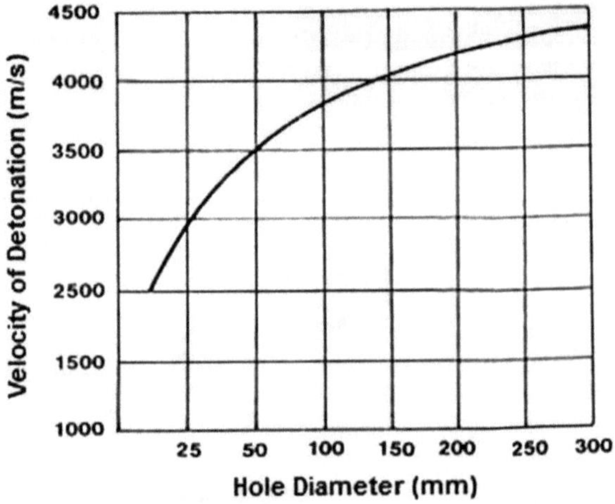

Fig. 3.39 Variation of VOD with hole dia. for confined full-coupled charge of bulk ANFO

0.95 g/cm^3. If ANFO is compressed beyond a density of about 1.20 g/cm^3, it will not detonate as it becomes dead-pressed.

The critical diameter of ANFO is influenced by its confinement and charging density. Well-mixed loosed-poured ANFO can be used successfully in blasthole diameters down to about 25 mm.

The detonation velocity (VOD) of bulk ANFO depends on the borehole diameter and degree of confinement in which it is shot. Figure 3.39 illustrates how the VOD of ANFO increases as the diameter of the borehole increases.

Lack of water resistance is the major limitation and disadvantage of ANFO. AN is readily dissolved by water, and unfortunately, the addition of 5.64 % FO does little to reduce the rate of dissolution. Both the strength and VOD of ANFO are reduced by the water content (Fig. 3.40). ANFO that contains more than about 10 % water usually fails to detonate. Wherever there is a chance of blasthole water-desensitizing ANFO and makes it less effective, efforts should be made to minimize the period between charging and firing the blast.

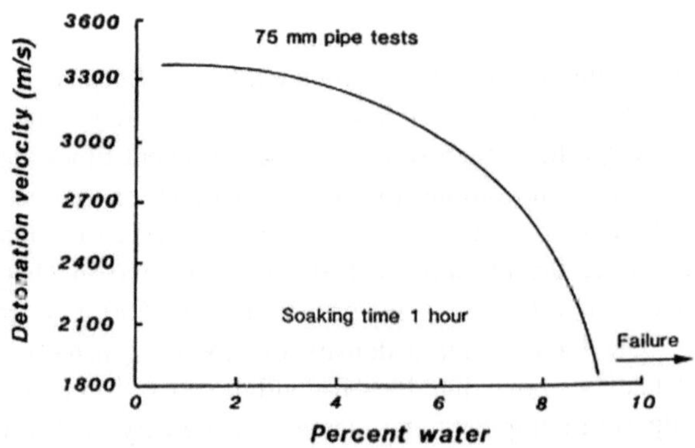

Fig. 3.40 Influence of water on ANFO performance

Fig. 3.41 On-site mixing and loading ANFO truck

Fig. 3.42 ANFO charging truck for underground excavation (Atlas Copco)

Most ANFO products can be supplied in four forms:

- Site mixing as bulk product and charging to the blastholes directly using a loading truck (Figs. 3.41 and 3.42).
- Premixed and bagged in bulk form for on-site storage or direct blasthole delivery (Fig. 3.43).
- In paper, polyethylene or burlap package (Fig. 3.44).
- In cylindrical cartridges. AN powder or crushed AN prills mixed with fuel oil, packaged in textile or cardboard tube with tough plastic liners, are used in wet blastholes over 75 mm in diameter but eliminate the advantage of direct borehole coupling.

Fig. 3.43 Charging blastholes with premixed ANFO

Fig. 3.44 Package ANFO

3.8.4.2 Bulk Emulsion and Heavy ANFO

Bulk emulsion explosives are mostly non-cap sensitive. They are usually mixed and prepared on-site and directly loaded into borehole by mix-loading truck or pumping truck. Because the type of loading trucks used is different, the mixing and loading method also vary in two forms. The first method, which is called mixing–loading truck method, is to transport the raw materials to a place near blasting area followed by emulsifying and mixing all materials in the same truck and then to pump the mixed materials into boreholes. This kind of method was used in the 1980s and early 1990s and rarely used at present as the truck needs to carry the hot solution of AN and emulsifier in the truck. The second method is called pumping–loading truck method. The emulsion matrix, which is manufactured (emulsified) in a stationary factory, is loaded to the truck, which is also called as mobile manufacture unit (MMU). This mixture is then mixed with sensitizing and other solid materials (mainly ANFO) and finally pumped into the boreholes. The structure of the truck for the second method is simpler than the first method and more reliable for operation. Figure 3.45 shows the emulsion matrix which is transported to site in drums or tank from the factory.

In general, a modern MMU has multi-functions which can mix and load ANFO, emulsion, and heavy ANFO. A modern MMU is shown in Fig. 3.46. The working procedure of the multi-function MMU is also shown in Fig. 3.47.

Heavy ANFO, which is a mixture of emulsion matrix with ANFO, gives a new perspective to the field of explosives.

ANFO leaves interstitial voids which can be occupied by a liquid explosive such as an emulsion that acts as an energizing matrix, as in Fig. 3.48.

Although the properties of the explosive depend upon the percentage of the mixture, its main advantages are obvious:

- More energy,
- Better sensitivity,
- High water resistance, and
- Possibility of charging with variation in energy along the length of the blasthole.

Fig. 3.45 Emulsion matrix was transported to site in drums from factory

Fig. 3.46 Multi-function mobile manufacture unit (MMU)

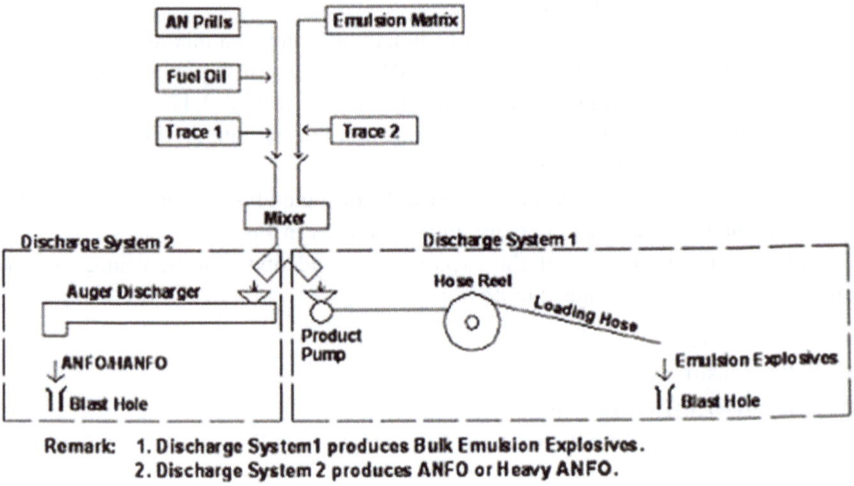

Fig. 3.47 Working procedure of the multi-function MMU

Table 3.9 gives typical explosive properties of emulsion/ANFO mixture in different proportions.

When the proportion of ANFO is greater than 30 %, the diameter of boreholes to be loaded for blasting should be larger than 127 mm because of the high viscosity of the mixture.

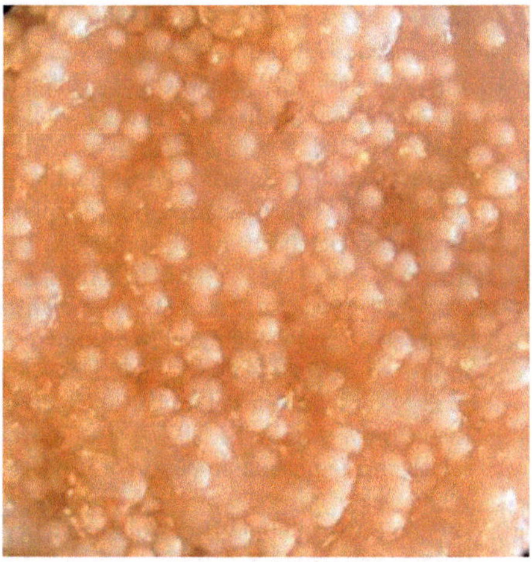

Fig. 3.48 Mix of emulsion and ANFO

Table 3.9 Typical explosive properties of emulsion/ANFO mixture in various proportions

Emulsion: ANFO	Density (g/cc)	Water resistance	Relative weight strength (%)	Relative bulk strength (%)	Detonation velocity (m/s)	Loading method
0:100	0.82	No	100	100	4450	Auger
20:80	1.05	No	108	138	5000	Auger
25:75	1.13	Bad	111	152	5150	Auger
30:70	1.20	Not good	113	166	5300	Auger
35:65	1.25	Moderate	114	174	5400	Auger
40:60	1.30	Good	116	185	5500	Auger
45:55	1.35	Excellent	117	193	5600	Auger
50:50	1.30	Excellent	113	179	5460	Auger
55:45	1.30	Excellent	112	177	5400	Auger
60:40	1.30	Excellent	110	175	5300	Pump
65:35	1.30	Excellent	1.09	173	5270	Pump
70:30	1.30	Excellent	1.07	169	5200	Pump
75:25	1.30	Excellent	1.05	167	5150	Pump
80:20	1.30	Excellent	1.04	165	5090	Pump

References

1. Szczepauski K (2015) Invention of Gunpowder. About.com, Asia History. http://asianhistory.about.com/od/asianinventions/a/InventGunpowder.htm, Updated October 15, 2015. Accessed 14 June 2016
2. 吕春緒等，工業炸藥. 兵器工業出版社，1994，北京。 Lu C et al (1994) Industrial explosives. Ordnance Industry Press, Beijing

3. Jimeno CL et al (1995) Drilling and blasting of rock. A.A.Balkema/Rotterdam/Brookfield/1995. ISBN: 90 5410 199 7. Copyright ©1995 A. A. Balkema
4. Slide A, Liudholm I (2000) On detonation dynamics in hydrogen-air-steam mixtures: theory and application to Olkiluoto Reaction building, NKS-9. ISBN 87-7893-058-8
5. 張寶平等 (2001), 爆轟物理學. 兵器工業出版社,8. 2001. Zhang B et al (2001) Detonation physics. Ordnance Industry Press, Beijing
6. Ornellas DL (1982) Calorimetric determinations of the heat and products of detonation for explosives: October 1961 to April 1982. UCRL-52821
7. 俞統昌等 (1994),絕熱型爆轟熱量計的建立和高能炸藥的爆熱測定。 < 含能材料 >, 第2卷第2期, 1994. Yu T et al (1994) An adiabatic detonation calorimeter and determination of detonation heat for high explosives. Energ Mater 2(2)
8. 金韶華等(2010),炸藥理論。西北工業大學出版社, 2010. Jin S et al (2010) Explosives theory. Northwestern Polytechnical University Press. ISBN: 978756/1227268
9. Kamlet MJ, Jacobs SJ (1967) The chemistry of detonations. I. A simple method for calculating detonation properties of C–H–N–O explosives. US Naval Ordinance Laboratory, White Oak
10. Persson PA et al (1993) Rock blasting and explosives engineering. CRC Press, Boca Raton. ISBN 0-8493-8978-X
11. Tete AD et al (2013) Velocity of detonation (VOD) measurement techniques practical approach. Ind J Eng Technol 2(3):259–265
12. 趙根等 (2008) 孔內炸藥連續爆速測試新技术。 < 工程爆破 >, 第14卷第3期, 9. 2008. Zhao G et al (2008) New measurement technology of continuous detonation velocity of explosive in blasthole. Eng Blast 14(3)
13. ISEE (1988) Blaster's handbook. 17th ed. ISBN: 1-892396-00-9
14. Wang X (1994) Emulsion explosives. Metallurgy Industry Press, Beijing. ISBN 7502415742, 9787502415747
15. Sharma PD (2016) ANFO, emulsion and heavy ANFO blends. LinkedIn Corporation © 2016. http://www.slideshare.net/sharmapd1/. And http://www.miningandblasting.wordpress.com Accessed 21 June 2016
16. 中國工程爆破協會于亞倫主編(2004),工程爆破理論与技术。冶金工業出版社,2004年,北京。 Yu Y et al (2004) CSEB, theory and technology of engineering blasting. Metallurgical Industry Press, Beijing. ISBN: 7-5024-3390-2
17. Recommendation on the Transport of Dangerous Goods (2013) Model regulations, vol 1, Eighteenth Revised Edition, United Nations, New York and Geneva. ISBN: 978-92-1-139146-6
18. 华东工学院民用爆破器材研究所译 (1986) 危险货物的运输(试验和标准)联合国编,国防工业出版社, ISBN:7-118-00056-6/U6. Recommendation on the Transport of Dangerous Goods (1986) Manual of tests and criteria, 1st ed. United Nations, New York and Geneva

Chapter 4
Initiation System

According to the functions, the initiation system includes three categories:

1. The initiating devices for generating a detonation. The most commonly used initiating devices are the detonators;
2. The articles for transferring a detonation. Detonating cord can transfer detonation wave which is generated by the initiated device to any distance as required;
3. The enhancing articles to strengthen the detonation generated by the initiating devices. The duty of this kind of articles is to strengthen the detonation which is generated by a detonator for initiating some very insensitive blasting agents, such as ANFO or bulk emulsion or heavy ANFO. These kinds of articles are usually called "boosters" or "primers."

In this chapter, all the three categories of the initiation system will be discussed, especially the detonators.

4.1 Detonators

4.1.1 Brief History on Detonators

Before the invention of modern explosives, when rock blasting was done with black powder, initiation was started by the flame slowly propagating through a black powder fuse. Since 1831, when William Bickford invented the safety fuse, the use of black powder became safer and reliable. But the black powder fuse was not suitable to initiate the more powerful nitroglycerin, which was invented in 1846 by Italian chemist Ascanio Sobrero. In 1864, Alfred Nibel applied the patents of the igniter caps for nitrogycerin and improved them in 1865. In the early stage of his invention, the igniter cap consisted of an enclosed wooden capsule containing finely powdered black powder, ignited by a black powder fuse. With further development,

D. Zou, *Theory and Technology of Rock Excavation for Civil Engineering*,
DOI 10.1007/978-981-10-1989-0_4

as described in later (1865) Nobel's Patent, the detonator was made use of a small quantity of primary explosive, mercury fulminate, pressed into a copper capsule which was crimped to the end of the black powder fuse. Nobel's invention opened the door of the development of modern initiation devices.

About 1900, the electrically igniting fuse head was used to connect to a short length of black powder fuse, in its turn connected to an ordinary detonator. This constituted the first electric delay detonator.

Later, in about 1920, came the instantaneous electric detonator in which a thin electrically heated bridge wire set on fire a minute charge of a flame-producing compound. This in turn initiated a small quantity of primary explosive such as lead styphnate or lead azide. The primary explosive detonated, initiating detonation in the main charge—usually about half gram (No. 8 detonator) of a relatively sensitive secondary explosives such as PETN, tetryl, or RDX/TNT. (Based on the quantity of base charge and primary charge quantity, the detonators are designed as detonator No. 1 to No. 8 or more. No. 6 and No. 8 are usually seen in the market, but No. 8 is much more widely used for rock blasting as it produces much stronger detonation power than No. 6 cap.)

In 1927, second delay electric detonators came under the foundation of instantaneous electric detonator and then the millisecond delay electric detonator was followed in 1946.

In 1973, the Nonel non-electric initiation system (it is also called shock tube initiating system) was invented by Nitro Nobel AB of Sweden. This system transmits the initiating signal in the form of a reaction-supported air shock wave in the air inside a narrow-gauge plastic tube which ends in a detonator. The inner wall of the tube is covered with a fine dust of explosive which reacts chemically to support the air shock wave by heating and by the expansion of the gaseous reaction products. As it is a safe, multipurpose system which combines the simplicity of fuse and detonating cord with the precision achieved with electric firing, Nonel system was applied worldwide quickly.

Due to the safety problems during manufacture and usage caused by the very high sensitivity of the primary explosives such as lead styphnate or lead azide and the environment pollution during the production of the primary explosives, the non-primary explosive detonators (NPED) came to the world in 1971 as the US patent applied by J.R. Strond. After that, more types of NPED were invented. In 1984, Nitro Nobel AB procured a China's patent, Patent No.CN 85101936, and then patented in US patent, Patent No. 4727808. In 1993, Nitro Nobel AB introduced the new detonators into the marketplace [2]. The NPED has successively replaced the conventional detonators in the electric and Nonel systems.

Since the last years of the twentieth century, more advanced initiation devices, such as laser detonators, electronic (or digital) detonators, and sound control detonators, came to the world and promoted the development of blasting technology. Especially, the digital detonators can offer great advantages of accurate timing, safety and reliability although they are more expensive than the normal electrical and Nonel systems but the cost difference has been and will be further reduced along with the development of electronic technology.

4.1.2 *Plain Detonators and Safety Fuses*

The use of plain detonator (also called ordinary detonator) and safety fuse is the oldest explosive initiation system. After the detonating explosives were introduced, and until electric methods became widespread, the plain detonator and fuse system was the dominant initiation method for small-diameter holes. Due to economics, this method remains in wide use in many areas of the world. But at present, this system has fallen into disfavor due to its high accident potential and the fact that better breakage and higher productivity are possible with modern electric and non-electric methods.

The safety fuse, or black powder fuse, consists of a black powder core wrapped in textile and covered with waterproof materials, traditionally bitumen, wax and plastics (see Figs. 4.1 and 4.2).

The black powder core used in safety fuses is the mixture of potassium nitrate (63–75 %), sulfur (15–27 %), and charcoal (10–13 %) properly balanced to get a uniform and reliable rate of burning.

Normal unconfined burning rates for fuses are in the range 100–120 s/m, corresponding to a propagation velocity of 8–10 cm/s. The tolerance allowed generally is about ±10 %. It should also show no side sparking and gases evolved during the process of burning and should vent through the textile layers uniformly.

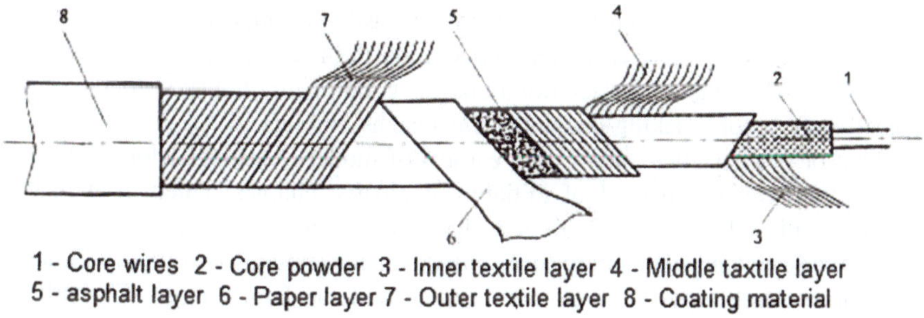

Fig. 4.1 Structure of the safety fuse. *1* Core wires. *2* Core powder. *3* Inner textile layer. *4* Middle textile layer. *5* Asphalt layer. *6* Paper layer. *7* Outer textile layer. *8* Coating material

Fig. 4.2 **a** Safety fuse and **b** a burning safety fuse

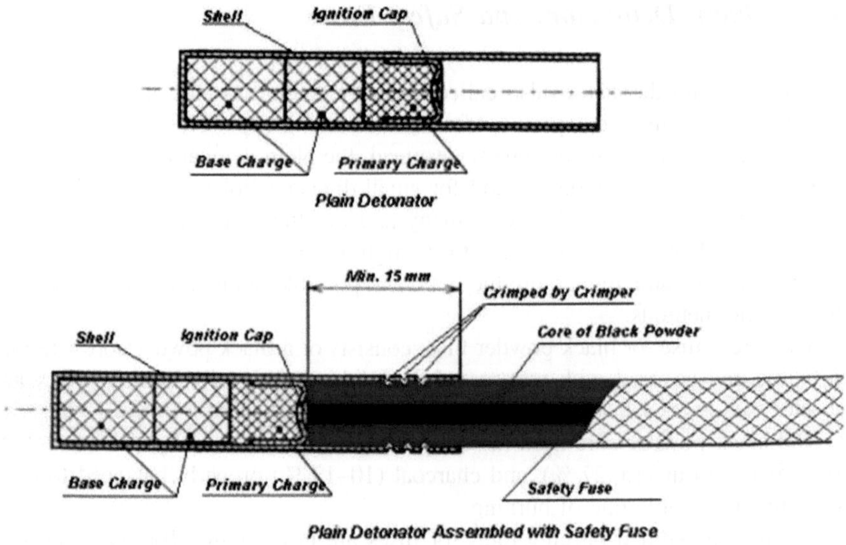

Fig. 4.3 Plain detonator and assembled with safety fuse

Both dampness and high altitude will cause the fuse to burn more slowly. Fuse should be test-burned periodically so that the blaster can keep a record of its actual burning rate. "Fast fuse" has been blamed for blasting accidents, but the fact is that this rarely if ever occurs. However, pressure on the fuse may increase its burning rate. To guard against water deterioration, it is good to cut off a short length of fuse immediately before making detonator and fuse assemblies.

Plain or ordinary detonator is the earliest of modern blasting detonators which provide non-electric method of initiating explosive charges, when used in conjunction with safety fuse. Figure 4.3 shows the general construction of such a detonator assembled with safety fuse. The detonator contains two types of charges (sometimes three types): the primary charge and the base charge. The primary charge (dinitrodiazophenol or mercury fulminate or lead azide) ensures flame pickup from the safety fuse, which in turn detonates the base charge (PETN, tetryl, or RDX/TNT) and thus detonates the explosive charge being primed with the detonator. Plain detonators store well for long periods (Fig. 4.4).

To assemble a plain detonator and fuse, the fuse is carefully cut squarely and inserted into the cap until it abuts against the explosive charge in the cap.

The cap is attached to the fuse with an approved bench or hand crimper. When crimping the cap, care should be taken so as not to crimp the zone containing the charge. Figure 4.5 shows the crimpers and the crimp on a cap.

Matches, cigarette lighters, carbide lamps, or other open flames are not suitable for igniting fuse. The following devices or articles are used for igniting safety fuse:

- Hot wire fuse lighters. It is a device similar in appearance to a fireworks sparkler. It consists of a wire covered with an ignition composition that burns

Fig. 4.4 Plain detonators

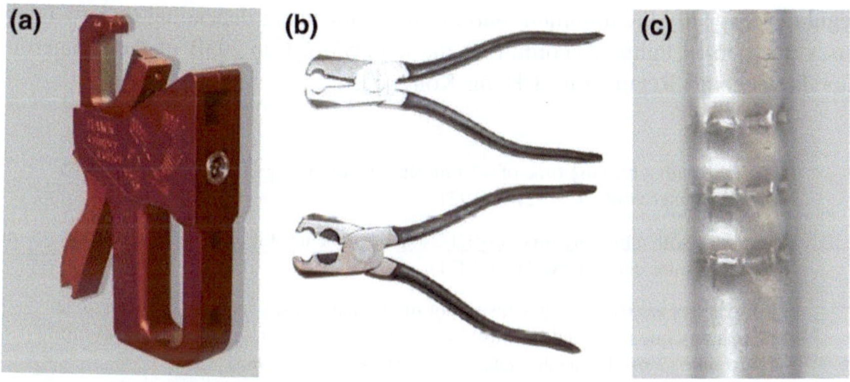

Fig. 4.5 Two kinds of hand crimpers (**a**, **b**) and the crimps on the cap (**c**)

slowly and as a fairly steady rate with an intense heat. The hot wire fuse lighter is lighted by a match and can be used to ignite fuse merely by holding the burning portion of lighter against the fresh cut end of the fuse. These lighters are supplied in several lengths;

- Fuse igniting bar. The length of the bar is about 100–150 mm and made of a paper tube with a diameter about 3–4 mm and charged with burning material such as black powder. The burning time of the fuse igniting bars are supplied in 1, 2, 3 min or longer. Same as the hot wire fuse lighter, the fuse igniting bar can be lighted by a match.
- Igniter cord and igniter cord connectors. Igniting cord is an incendiary cord developed for lighting a number of safety fuses in a series via beanhole connectors. When cord is ignited, an intense flame (which will ignite the black powder core of safety fuse) passes at a uniform rate along the cord. Beanhole connecters for igniter cord are made of aluminum tubes and assembled on the cord. The safety fuses are inserted into the open end of the tubes.

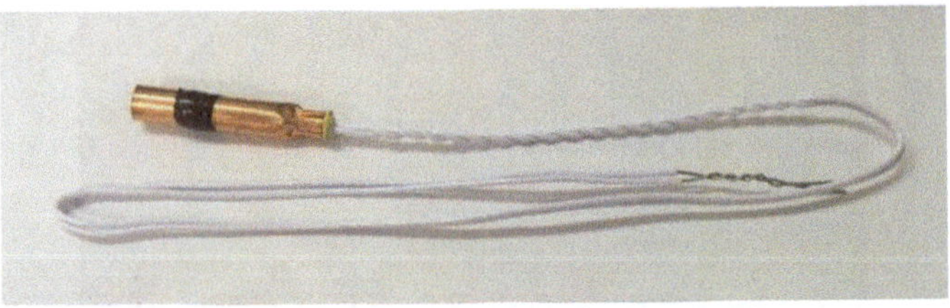

Fig. 4.6 Electric igniter for safety fuse

- Electric igniter. As shown in Fig. 4.6, this device is designed to provide a means of ignition from an electrically ignited fusehead to a pyrotechnic safety fuse.

The primary hazard of using safety fuse is the tendency of blasters to linger too long on the face, making sure that all the fuses are lit. To guard against this, regulations specify the minimum burning times for fuses, depending on how many fuses one person lights. According to the item 55 of Cap.295B of the Dangerous Goods (General) Regulation of Hong Kong [7],

(f) no fuse with a burning time of <2 min shall be used if ignition is not effected by means of igniter cord; (L.N. 21 of 1971)

(fa) no fuse with a burning time of <1 1/2 min shall be used if ignition is effected by means of igniter cord; (L.N. 21 of 1971)

(g) if matches are used to light fuses, not more than 2 fuses shall be lit by any one person at any one time, and if 2 fuses are lit, the first fuse lit shall be of not <3 min burning time (2 m of standard safety fuse); (L.N. 119 of 1983)

(h) no fuse shall be lit by means of an acetylene lamp; (L.N. 104 of 1967)

(i) if fuses are to be lit in a series by means of an igniting squib, then, except with the prior permission in writing of the Authority –

(i) the series shall consist of not more than 10 fuses,
(ii) each fuse of the series shall be longer than the preceding fuse by not <15 s burning time (150 mm of standard safety fuse), (L.N. 119 of 1983)
(iii) the person using the igniting squib shall be accompanied by another person having a box of matches or other suitable means of immediately relighting the igniting squib if it should become extinguishes, and
(iv) the fuses shall be lit seriatim commencing with the longest fuse; (L.N. 104 of 1967)

4.1.3 Electric Detonators

4.1.3.1 Structure of Electric Detonators

Electric detonators are such kind of detonators which are ignited with an electrical igniting element. In other words, electrical current is its initial energy source. In the electric detonators, the electrical current to the detonator is supplied from the power source through the circuit wiring to the detonator by means of two leg wires that are internally connected by a small length of high-resistance bridge wire. The electrical energy is converted into heat energy on passing the firing current through bridge wire. The heat energy ignites the pyrotechnic that surrounds the bridge wire on the match head assembly (called fusehead, see Fig. 4.7). The resulting flash or flame ignites the primary charge or the delay element, and these in turn set off base charge.

Electric detonators are classified as instantaneous (Fig. 4.7a) and delay detonators (Fig. 4.7b).

In instantaneous detonators, the fusehead directly ignites the primary charge of the detonator. Instantaneous detonators fire within a few milliseconds (<5 ms) after they receive the current. Instantaneous detonators are used when all the holes are to be fired simultaneously. At present, instantaneous electric detonators are usually used as a starter to ignite some non-electrical initiation systems, detonating cord system and Nonel (shock tube) system. Figure 4.8 is an example which is supplied by Dyno Nobel—an instant electric detonator housed in a plastic bunch block, facilitating easy connection to both shock tube and detonating cord.

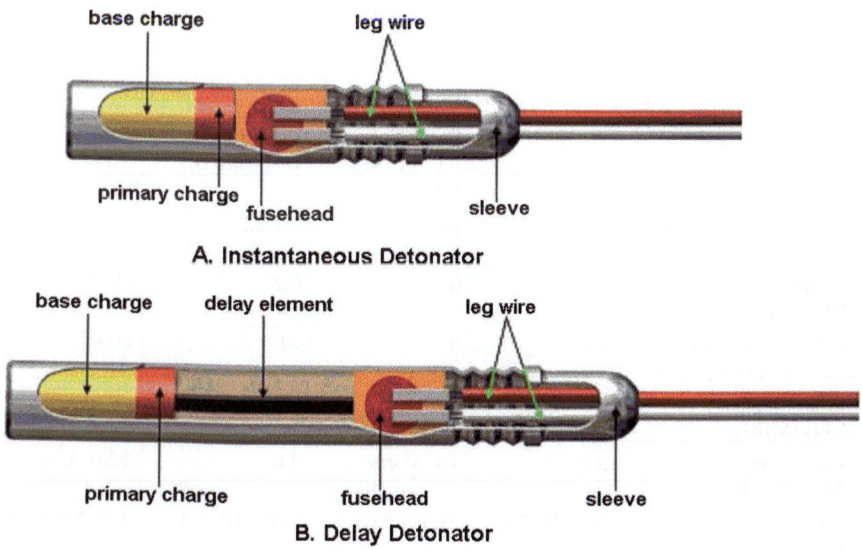

Fig. 4.7 Electric detonators

Fig. 4.8 Dyno Nobel's electric superstarter

4.1.3.2 Electrical Parameters of Electric Detonators

For safe and reliable application of electric detonators, the firing parameters, such as the safety current, all-firing current, and water pressure resistance, of the electric detonators supplied by different manufacturers must be well known by the user. Table 4.1 lists the parameters of electric detonators supplied by some manufacturers.

Figure 4.9 shows a product of ohmmeter for safely measuring the electrical resistance of electric detonators, and Fig. 4.10 shows the blasting machines that are used to fire the electric detonators.

Table 4.1 Technical parameters of electric detonators supplied by some manufacturers

Manufacturer and type	Austin Powder Rock Star I	Dyno Nobel Electric Super SP	Orica Electric Detonator SD	China GB8031 Type I
Bridge wire resistance (Ω)	1.7 ± 0.2	1.6	1.2-2.2	
No-fire current (NFC) (A)	0.18	0.3	0.2	≥0.20
All-fire current series connection (AFC) (A)	≥1.0	0.5	>1.0	≤0.45
No-fire impulse (NFI) (MJ/Ω)	0.8		0.6	
All-fire impulse (AFI) (MJ/Ω)	3.0	1.1 A^2-ms	2.0	≥2.0 A^2-ms
Water pressure resistance	0.3 MPa/24 h	250psi/20 h	0.02 MPa/6 h	0.05 MPa/4 h
Shelf life (years)	2	3	2	1.5

Fig. 4.9 Ohmmeter for measuring the resistance of electric detonators

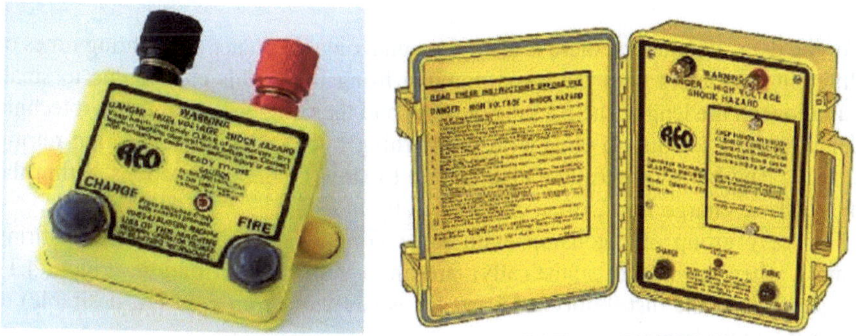

(a) Type:CD450-4J (b) Type: CD1000-9J

Fig. 4.10 Blasting machines. **a** Type:CD450-4J. **b** Type:CD1000-9J

4.1.3.3 Delay Detonators

For most blasting operations, it is an advantage to havc thc various holes fired in a predetermined sequence with specific time intervals between detonators. The most notable advantages of delay detonators are as follows:

- Reduced vibration, airblast, and flyrock;
- More predictable throw (amount and direction);
- Reduced backbreak and overbreak, with working faces left in an improved condition;
- Improved the excavation results for tunneling due to the fresh free faces for the subsequently firing holes are offered by the holes previously fired in a proper time interval.

In delay detonators, a delay element is inserted between the electrical fusehead and the primary charge (Fig. 4.7b). This delay element consists of a column of slow-burning composition contained in a thick-wall metal tube. The length and composition determine the amount of delay time introduced into the detonator.

There are three basic delay series:

1. Millisecond delay detonators;
2. Long-period delay detonators. There are also different delay series: 1/4 s delay (or 200 ms delay), half second delay, and second delay; and
3. Coal mines delay detonators which are specially manufactured for underground coal mines.

Table 4.2 gives the delay times of millisecond delay detonators of some products, and Table 4.3 gives the timing series of some long-period delay detonators for reference.

4.1.3.4 Timing Accuracy

For the delay detonators, there is always an unavoidable scatter in the firing times of different detonators with the same nominal firing time. This can be due to small variations in the length, the packing density, and the composition of the pyrotechnic delay charges, or due to change in the burning rate occurring from aging during storage. The burning rate may change due to slow oxidation reactions within the pyrotechnic charge itself, often accelerated by moisture.

For any group of detonators with same nominal firing time, the exact firing times, when evaluated statistically, are in a normal distribution. Figure 4.11 describes how the distribution of firing times can vary from tight (A, desirable) to wide (C, undesirable).

The mean or average firing time for any group of detonators typically is slightly offset from the nominal firing time. A tight distribution of the firing time and a small displacement of the mean firing time from the nominal indicate a precise and accurate delay detonator.

The effect on the blast of detonator time scatter can be of tremendous importance. Overlapping intervals can result in situations where holes are fired with a too large burden if the charge in the hole in front did not detonate properly. Coarse fragmentation, flyrock, and excessive ground vibrations result when delay precision is lost. An expected smooth blasted contour can also be spoiled if the time scatter of the detonation time of holes having detonators with the same nominal time delay becomes too large. Figure 4.12 shows firing time distributions for detonators of different precision.

In the case of millisecond delay detonators where the interval between delays usually is 25 ms, if the total deviation from nominal is <12.5 ms (half of the delay interval), no overlap will occur.

Not only must there simply be no overlap between detonation events, but a high probability must be assured that required minimum separation between detonation events is achieved. According to the research by Dick R.A. et al. of US Bureau of

Table 4.2 Nominal delay times of some products of millisecond delay electric detonators

Delay no.		0	1	2	3	4	5	6	7	8	9	10	11	12	13	14	15	16	17	18	19	20	22	24	26	28	30
Dyno Nobel	Super SP	9	25	50	75	100	125	150	175	200	225	250	275	300	325	350	375	400	425	450	475	500	550	600	650	700	750
Austin Powder	Rock Star I	0	25	50	75	100	125	150	175	200	225	250	275	300	325	350	375	400	425	450	475	500	600	700	800	900	1000
Orica	Electric MS	9	25	50	75	100	125	150	175	200	225	250	275	300	325	350	375	400	425	450	475	500					
China	Series 3	0	25	50	75	100	125	150	175	200	225	250	275	300	325	350	375	400	425	450	475	500					

Table 4.3 Nominal delay times of some products of long-period delay electric detonators

Delay no.	Austin Powder (ms)			Dyno Nobel (ms)	Orica (ms)	China GB/T 803 (s)		
	Timestar I 250	Timestar I 500	DEM-F-80	Super LP	Electric LP	1/4 second	1/2 second	Second
0			0					
1	250	500	80	25	25	0	0	0
2	500	1000	160	200	200	0.25	0.50	1.0
3	750	1500	240	400	400	0.50	1.00	2.0
4	1000	2000	320	600	600	0.75	1.50	3.0
5	1250	2500	400	800	800	1.00	2.00	4.0
6	1500	3000	480	1000	1000	1.25	2.50	5.0
7	1750	3500	560	1200	1200	1.50	3.00	6.0
8	2000	4000	640	1400	1400		3.50	7.0
9	2250	4500	720	1600	1600		4.00	8.0
10	2500	5000	800	1900	1900		4.50	9.0
11	2750	5500	880	2200	2200			10.0
12	3000	6000	960	2500	2500			
13	3250		1040	2900	2900			
14	3500		1120	3300	3300			
15	3750		1200	3800	3800			
16	4000		1280	4400	4400			
17	4250		1360	5100	5100			
18	4500		1440					
19			1520					
20	5000		1600					

(continued)

Table 4.3 (continued)

Delay no.	Austin Powder (ms)			Dyno Nobel (ms)	Orica (ms)	China GB/T 803 (s)		
	Timestar I 250	Timestar I 500	DEM-F-80	Super LP	Electric LP	1/4 second	1/2 second	Second
21			1750					
22	5500		2000					
23			2250					
24	6000		2500					
25			2750					
26			3000					
27			3250					
28			3500					
29			3750					
30			4000					

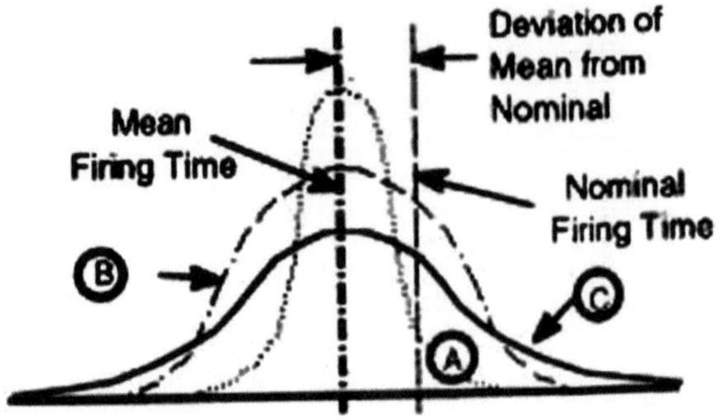

Fig. 4.11 Firing time distribution [4]

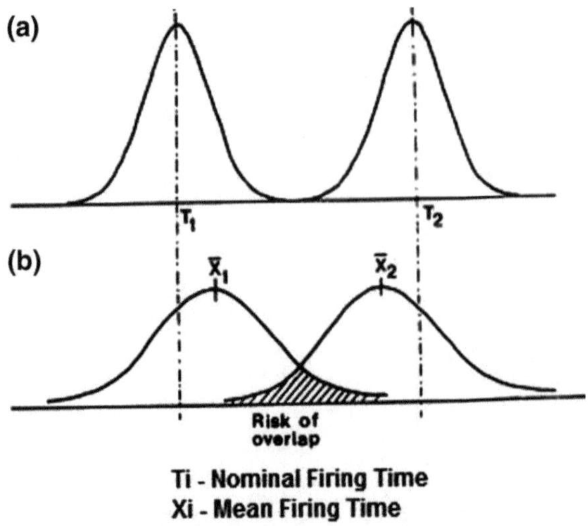

Fig. 4.12 Distribution of firing times for **a** good and **b** bad precision detonators [4]

Mines [3], it is often required that detonation events and/or explosive weight per time interval be separated by a certain specified minimum time. This is often considered to be 8 ms. Considering the rock breakage and movement caused by the detonation of explosive charge of each blasthole, long firing time intervals between blastholes are often necessary, especially in tunnel and shaft blasting.

4.1.4 Electromagnetic Detonators

The first electromagnetic detonator, Magnadet system, was developed by ICI Nobel's Explosives Ltd, UK, in 1979. It employs a high-frequency (15,000 Hz) induced electromagnetic current to initiate electric detonators without having direct electric contact with the detonator and the blasting machine. The system is shown in Fig. 4.13 (from [1]). The transformer device used is a small ferrite ring or toroid

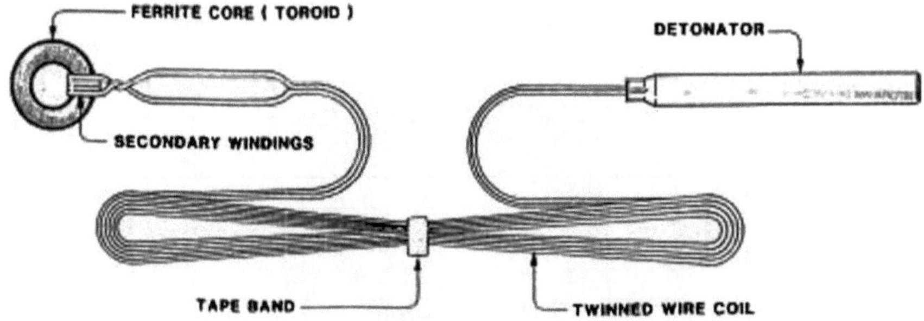

Fig. 4.13 Magnadet assembly (reproduced from Ref. [1] with the permission from Taylor & Francis Group)

which is an integral part of each specialized detonator. The detonator leg wires form the secondary side of the transformer winding. The primary side is a single wire loop common to all units within the circuit. Power is supplied from a frequency-tuned portable exploder with unique circuit-sensing features.

The Magnadet system introduces transformer coupling into electric blasting circuits. Safety features with respect to current leakage, stray currents of AC or DC source, electrostatic energy, and radio frequency radiation are comparable to "non-electric" systems, with advantages in simplicity, speed of hookup, and range of application.

4.1.5 *Shock Tube Detonators (Nonel System)*

4.1.5.1 Composition of Shock Tube Detonators

A shock tube detonator is a non-electric detonator in the form of small-diameter hollow plastic tubing used to transport an initiating signal to initiate the primary explosive or delay element in a detonator by means of a shock wave traveling the length of the tube. It was invented by Nitro Nobel AB under the leadership of Per Anders Persson, patented in 1971, and sold by them under the registered trademark *Nonel* (the contraction of *Non-el*ectric) (see Fig. 4.14).

The plastic shock tube is composed of one or more layers of plastic which are designed to enhance the physical properties (tensile strength, flexibility, and abrasion resistance) (see Fig. 4.15). The inner wall of the tube is covered with a fine dust of explosives (usually composed of HMX and aluminum powder) which reacts chemically to support the air shock wave by heating and by the explosion of the gaseous reaction products. The explosive load of the shock tube is <20 mg/m of tubing, sufficient to propagate the shock wave indefinitely through the tube, or until the conventional cap delay element is reached. The propagation velocity of the shock wave is about 2000 m/s (1980–2130 m/s), and the shock wave travels through the tube without affecting the outside surface of the tube.

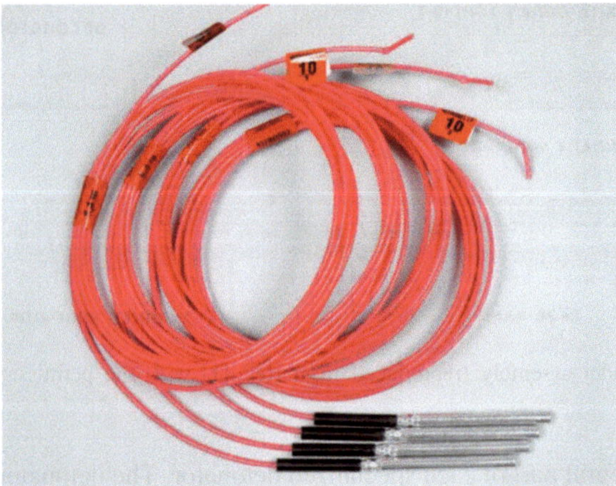

Fig. 4.14 Shock tube detonators (Nonel)

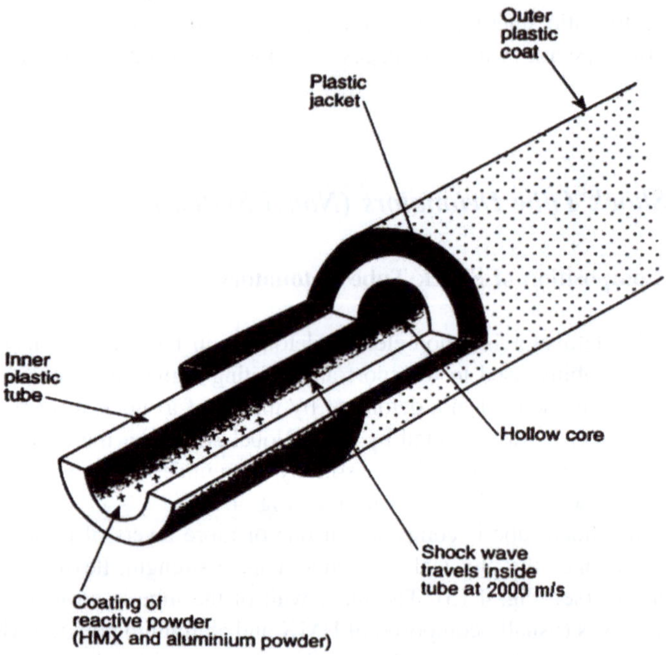

Fig. 4.15 Structure of plastic shock tube

4.1.5.2 Types of Shock Tube Detonators

A variety of shock tube system configurations are available for specific applications. Usually, there are three configurations that are widely used in rock blasting:

- Millisecond delay system;
- Long-period delay system; and
- A combination of high-precision long-period in-hole detonators and a number of short-delay surface connector units (Unidet system).

Millisecond delay system is used for bench blasting and underground blasting, long-period delay system is used for underground blasting, and shock tube downlines with shock tube trunkline delay units (Unidet system) is used for bench blasting.

- Millisecond delay system

Millisecond delay system of shock tube detonators is a conventional initiation system with a delay time of 25 ms between each interval. Table 4.4 shows the nominal delay times of millisecond delay systems provided by several manufacturers.

- Long-period delay system

Long-period delay system is an initiation system intended for underground use. The delay times between intervals in the system are generally longer in order to give enough time for blasted rock to be properly displaced in the confined space and single free face typical in tunneling. Table 4.5 shows the nominal delay times of long-period delay systems provided by several manufacturers.

- Combination of high-precision long-period in-hole detonators and a number of short-delay surface connector units (Unidet system).

This system (Unidet) is an initiation system that employs a uniform delay time in the in-hole detonators and variable delay times in the connector units on the surface. The delay time in the drillhole usually has a longer delay time which normally enables most of the in-hole detonators to be initiated on the surface before any rock displacement begins. This is then supplemented by delay times in the surface connector units, which give the desired initiation sequence.

Surface delays from 0 to 200 ms are available, which gives great flexibility in adapting the initiation sequence to suit the burden and rock characteristics.

Figure 4.16 shows the principle of initiation with this system.

To reduce the noise levels and shrapnel cutoff concerns, the base charge of the detonator in the surface connectors usually is about one-third of a normal No. 8 detonator. There is a small difference of the delay times of the surface connectors for different manufacturers. Table 4.6 gives their delay time of surface connectors.

Table 4.4 The nominal delay times of millisecond delay systems provided by several manufacturers

Austin Powder				Dyno Nobel				Orica						China [6]			
Shock Star MS				Nonel MS Series				Excel MS						Series 3			
No.	Time	No.	Time	No.	Time	No.	Time	No.	Time	No.	Time	No.	Time	No.	Time	No.	Time
0	0	14	350	1	25	15	375	1	25	15	375	40	1000	1	0	16	400
1	25	15	375	2	50	16	400	2	50	16	400	48	1200	2	25	17	450
2	50	16	400	3	75	17	425	3	75	17	425	56	1400	3	50	18	500
3	75	17	425	4	100	18	450	4	100	18	450	64	1600	4	75	19	550
4	100	18	450	5	125	19	475	5	125	19	475	72	1800	5	100	20	600
5	125	19	475	6	150	20	500	6	150	20	500	80	2000	6	125	21	650
6	150	20	500	7	175	21	550	7	175	22	550	90	2250	7	150	22	700
7	175	22	600	8	200	22	600	8	200	24	600			8	175	23	750
8	200	24	700	9	225	23	650	9	225	26	650			9	200	24	800
9	225	26	800	10	250	24	700	10	250	28	700			10	225	25	850
10	250	28	900	11	275	25	750	11	275	30	750			11	250	26	950
11	275	30	1000	12	300	26	800	12	300	32	800			12	275	27	1050
12	300	15	375	13	325	27	900	13	325	34	850			13	300	28	1150
13	325	16	400	14	350	28	1000	14	350	36	900			14	325	29	1250
														15	350	30	1350

Table 4.5 The nominal delay times (ms) of long-period delay systems provided by several manufacturers

Austin Powder				Dyno Nobel				Orica				China [6]					
Shock Star LP detonators				Nonel LP series				Excel LP				1/4 second delay series 1		1/2 second delay series 2		Second delay series 2	
No.	Time	No.	Time	No.	Time	No.	Time	No.	Time	No.	Time	No.	Time	No.	Time	No.	Time
0	0	13	3500	0	25	13	4900	1/4	100	7	2250	1	0	1	0	1	0
1	200	14	4000	1	500	14	5400	1/2	200	8	2500	2	250	2	500	2	1000
2	400	15	4500	2	800	15	5900	3/4	300	9	3000	3	500	3	1000	3	2000
3	600	16	5000	3	1100	16	6500	1	400	10	3500	4	750	4	1500	4	3000
4	800	17	5500	4	1400	17	7200	1-1/4	500	11	4000	5	1000	5	2000	5	4000
5	1000	18	6000	5	1700	18	8000	1-1/2	600	12	4500	6	1250	6	2500	6	5000
6	1200	19	6500	6	2000			2	800	13	5000	7	1500	7	3000	7	6000
7	1400	20	7000	7	2300			2-1/2	1000	14	5500	8	1750	8	3500	8	7000
8	1600	21	7500	8	2700			3	1200	15	6000	9	2000	9	4000	9	8000
9	1800	22	8000	9	3100			4	1400	16	6500	10	2250	10	4500	10	9000
10	2000	23	8500	10	3500			5	1600	17	7000						
11	2500	24	9000	11	3900			5-1/2	1800	18	8000						
12	3000	25	9600	12	4400			6	2000	19	9000						

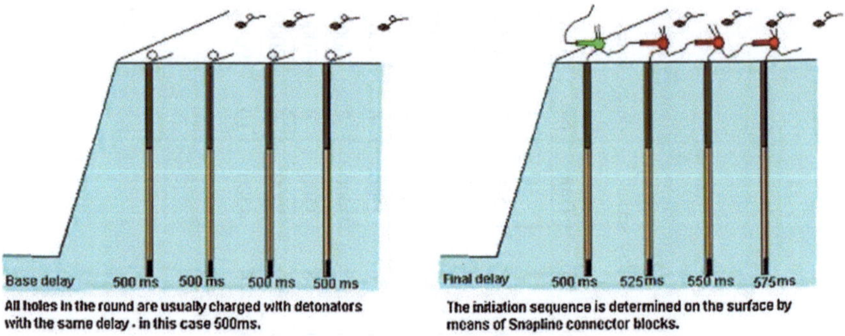

Fig. 4.16 Principle of initiation by combination of surface delay and in-hole delay (Dyno Nobel)

Figure 4.17 shows Dyno Nobel's Unidet system.

To simplify the connection work, manufacturers combined in-hole delay detonator and surface delay connector into one product. Figure 4.18 shows an example of them.

For tunneling, a bunch connector is especially designed to initiate a number of shock tubes. Using a detonating cord (5 g/m) loop attached to it (see Fig. 4.19), it can initiate up to 20 shock tubes.

4.1.5.3 Initiation Methods for Shock Tube Detonator System

There are several methods for initiating a shock tube detonator system:

1. Initiation using an electric detonator

A shock tube detonator round can be initiated with an electric detonator. The electric detonator is taped to the shock tube with the detonator bottom pointing opposite to the direction of the shock wave starting propagation in the tube (see Fig. 4.20). The electric detonator should be well covered with earth, drill cuttings, etc., as the strength of this detonator is considerably greater than that of the surface connection unit. Shrapnel from the detonator may cut off the shock tubes in the round. Note that when the electric detonator is connected to the round, the whole round is exposed to the same risks as when electric detonators are used with regard to thunderstorms, static electricity, stray currents, etc.

Shock tube detonator system can also be initiated by plain detonator with safety fuse, and the length should be long enough to ensure shotfirer evacuating to a safe place after igniting the fuse and detonating cord.

2. Initiation using a special blasting machine or shock tube starting device

The simplest and safest way of initiating shock tube detonator rounds is using a long shock tube with a length long enough to ensure the shotfirer can evacuate to a safe place to fire the round. The ultrasonic seal at the ends of both starting

Table 4.6 The time range of surface delay connection units provided by some manufacturers

Austin Powder	Shock Star	Delay time (ms)	9	17	25	33	42	67	100	200
		Unit color	Green	Yellow	Red	Orange	White	Light blue	Purple	Black
Dyno Nobel	Nonel Unidet	Delay time (ms)	0	17	25	42	67	109	176	285
		Unit color	Green	Yellow	Red	White	Blue	Black	Orange	Brown
Orica	Excel Connectadet	Delay time (ms)	9	17	25	33	42	65	100	200
		Unit color	Green	Yellow	Orange	Yellow	White	White	Black	Red

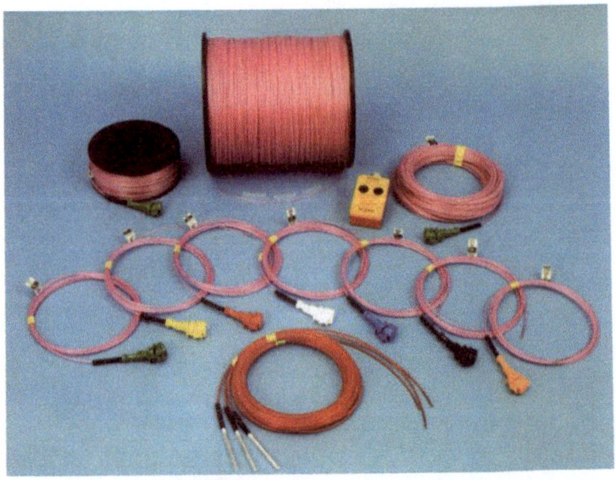

Fig. 4.17 Nonel united system (Dyno Nobel)

Fig. 4.18 Handidet—Orica's non-electric surface and in-hole detonator assembly

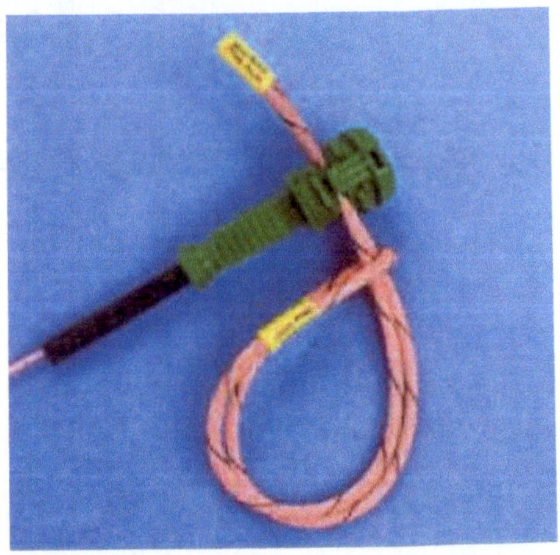

Fig. 4.19 Bunch connecter

connection unit and the long shock tube is cut off and connected together by pushing at least 1 cm into an approx. 4-cm-long outer plastic connecting sleeve, and then the long shock tube is extended to the chosen safe firing point.

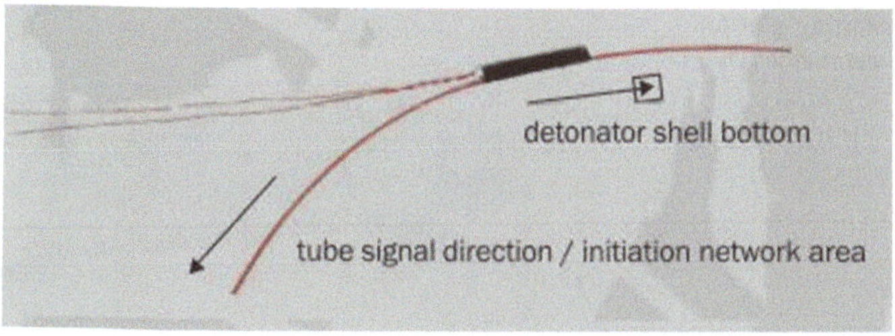

Fig. 4.20 Initiation by means of an electric detonator. Ensure that the detonator bottom is pointed in the direction opposite to the starting shock wave propagation and opposite to the initiation network area

Fig. 4.21 Firing a blasting round using a blasting machine (Dyno Nobel)

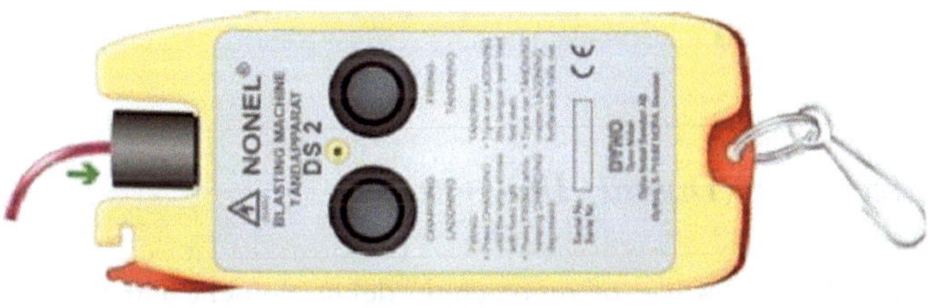

Fig. 4.22 Blasting machine for shock tube detonator (Dyne Nobel)

When the round is ready to be blasted, connect the long shock tube to the blasting machine, by inserting the tube into the chuck as far as possible, then starting the machine, and firing the round (see Fig. 4.21). Figures 4.22, 4.23, and 4.24 show some blasting initiation devices for shock tube detonator system (or by means of remote firing devices).

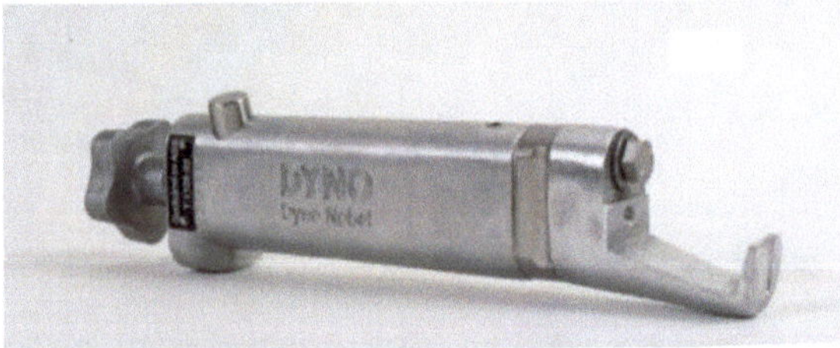

Fig. 4.23 Starting gun for shock tube detonator (Dyno Nobel)

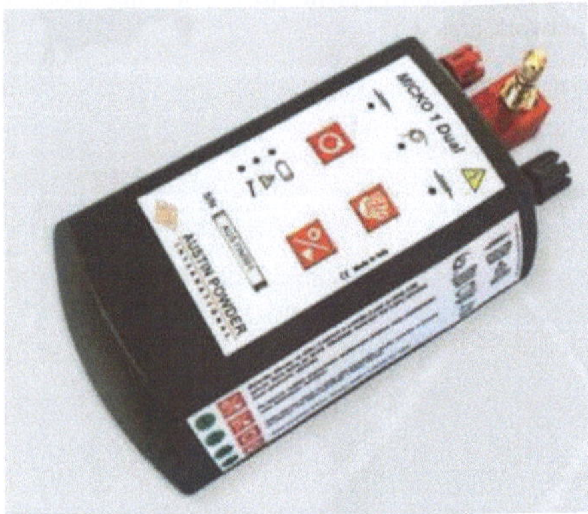

Fig. 4.24 Blasting machine for initiation of both electric and shock tube detonators (Austin Powder)

The advantage of shock tube detonators as a kind of non-electric initiation system versus electric initiation system is perceived to principally be their lack of susceptibility to premature activation by extraneous electrical energy, such as static electricity, stray currents, strong radio or radar signals, or lightning. The principal shortcoming of most non-electrics including shock tube initiation system is the lack of a circuit test capacity.

4.1.6 Electronic (Digital) Detonators

Electric delay detonators use of pyrotechnic delay element and such delays have limitation in terms of accuracy. All the manufacturers indicate variation of about 10 % from the nominal delay period that may lead to possibility of overlap. The electronic detonator is the most recent development in blasting initiation. An integrated circuit chip and a capacitor internal to each detonator control the initiation time. A specially designed blasting machine transmits a selectable signal that is identified by each detonator and determines the detonation timing sequence.

Although the cost of electronic detonator is higher than electric and shock tube detonators, the electronic detonator system has the following advantages:

- Multiple verification of detonators prior to each blast and 100 % verification of reliability of connections in initiation network.
- Delay range of 1 up to 20,000 ms with an increment of 1 ms.
- Precision of 0.01 % of nominal delay time and up to 1000 times more accurate than pyrotechnics.
- Safe and reliable initiation of up to 20,000 units in one blast. It cannot be initiated by foreign energy, i.e., thunder, static electricity, and stray current, only by the specified blasting machine.
- There is a unique ID in each detonator. The ID number is not removable and is readable by the specified logger.

Figure 4.25 shows the structure of the electronic detonator and its integrated circuit board produced by Dyno Nobel and DetNet South Africa (DSA). Figure 4.26 shows the SmartShot detonator and its control and firing devices.

Table 4.7 gives the features of some electronic detonator systems in the international market for reference.

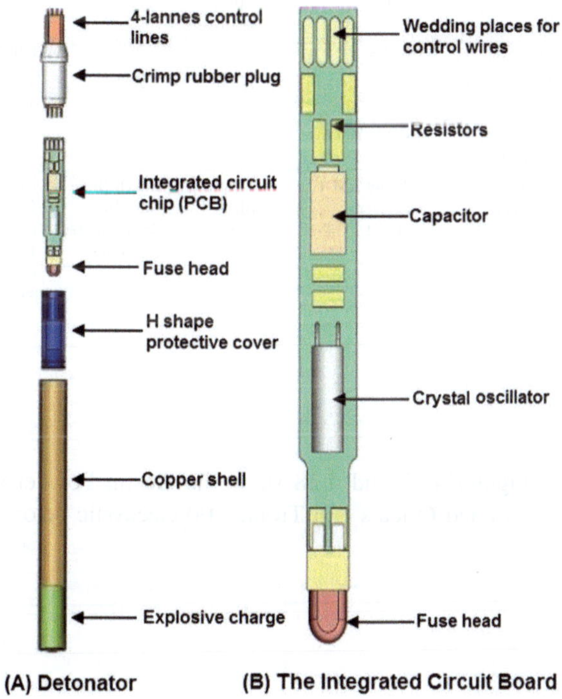

Fig. 4.25 Structure of electronic detonator (SmartShot) (Dyno & DSA). **a** Detonator. **b** The integrated circuit board

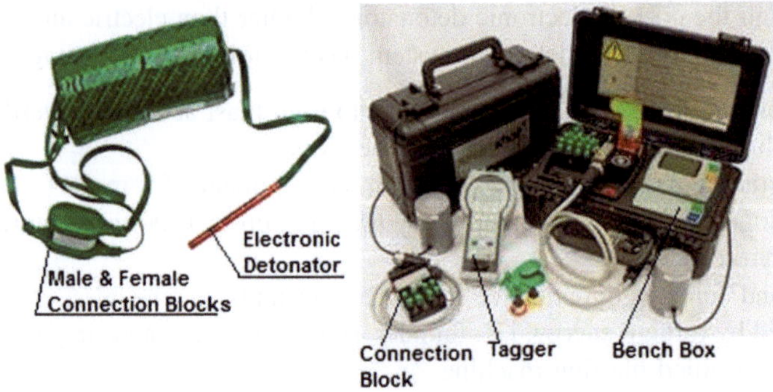

Fig. 4.26 SmartShot electronic detonator and control system (Dyno & DSA)

Table 4.7 Features of some electronic detonator systems

Manufacturer	Austin Powder	Dyno Nobel	Dyno Nobel/DetNet	Orica		
Brand	E*Star	DigiShot Plus	SmartShot	i-Kon II	Uni Tronic 600	eDev II
Max. delay (ms)	10,000	20,000	20,000	30,000	10,000	20,000
Min. delay interval (ms)	1.0	1.0	1.0	1.0	1.0	1.0
Precision (%)	0.005			0.005	0.03	0.01
Max. dets per blast	1,600	7,200	2,400	400–4800	800	800
Detonator strength	Base: 720 mg	#12	#12	Base: 780 mg	Base: 780 mg	Base: 780 mg
Equipment (exclude detonator)	Tester LM-1 Logger DLG1600-2-K Blasting Machine DBM1600-2-k	Tagger Blast Box	Bench Box Base Box String Starter End Plug Tagger	i-Kon Logger i-Kon Blaster CEBS SURBS	Scanner 120/125 Test Box Blast Box 310/310RAU	Blast Box 610 Tester Scanner 125 CEBS Blast Design Software

Figures 4.27 and 4.28 show the Austin Powder's E*Star electronic detonator system and Orica's Uni Tronic 500 electronic detonator system, respectively.

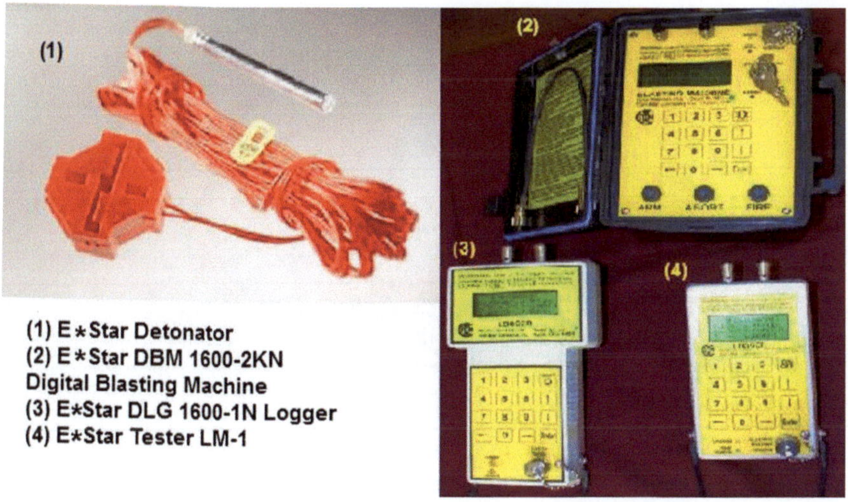

Fig. 4.27 Austin E*Star electronic blasting system. *1* E*Star detonator. *2* E*Star DBM 1600-2KN digital blasting machine. *3* E*Star DLG 1600-1N logger. *4* E*Star tester LM-1

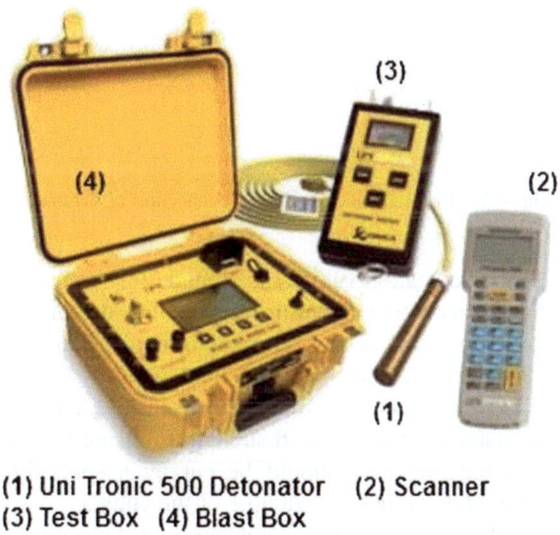

Fig. 4.28 Orica's Uni Tronic electronic detonator system. *1* Uni Tronic 500 detonator. *2* Scanner. *3* Test box. *4* Blast box

4.2 Detonating Cord

Detonating cord (also called detacord, detcord, detonation cord, or detonating fuse) is round, flexible cord containing a core of PETN in varying amounts (1.5, 3, 5, 11, …,40 and 100 g/m) wrapped in a plastic jacket that gives adequate flexibility, waterproofing, tensile strength, and resistance to humidity (see Fig. 4.29).

Fig. 4.29 Detonating cords

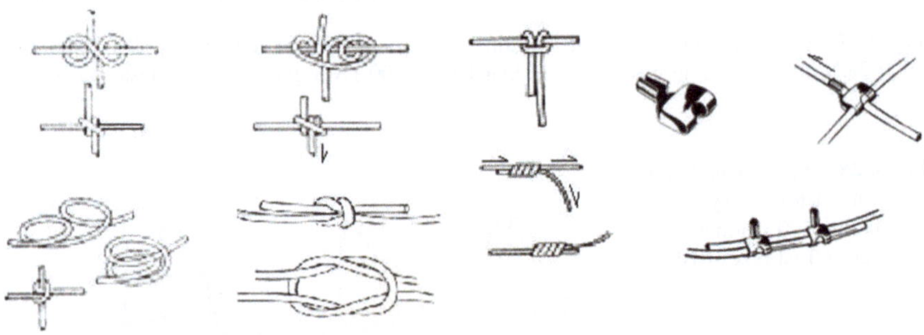

Fig. 4.30 Recommended knots for connection of detonating cords

The detonation velocity is around 7000 m/s. Due to the strong directionality of the detonating cord during propagating detonation, care should be taken when connecting two or more detonating cords. Figure 4.30 gives some recommended connection methods for detonating cords.

Detonating cord is relatively insensitive and requires intimate contact with a detonator of at least No. 6 strength to assure initiation.

Detonating cord should be cut by a specially made cord cutter for safety. Figure 4.31 shows an example of the cutters.

Although the primary application of detonating cords is to propagate the detonation to an explosive charge, they also have other uses. Tables 4.8 and 4.9 give the technical information of detonating cords from some manufacturers and some typical applications.

Fig. 4.31 Cord cutter

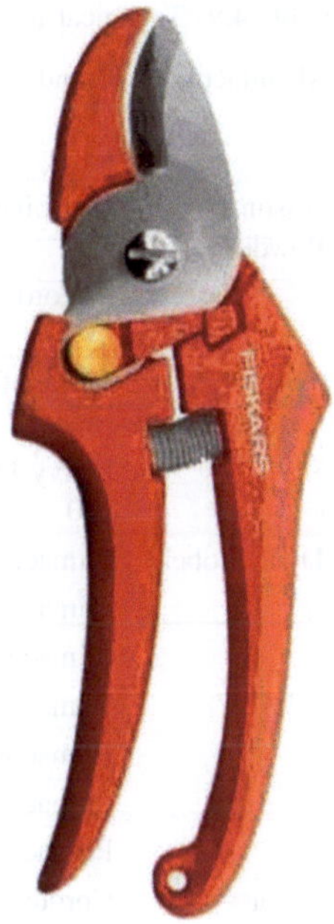

Table 4.8 Application of detonating cord

Core rate (g/m)	Application
1.5–3	Initiation of primers and very sensitive explosives
5, 6	Trunk lines connecting blast (presplitting) holes Used in bunch connecters for initiating shock tube detonators in tunneling
11–20	Initiation of conventional and less sensitive explosives
40	Seismic explosion and also used for lineal charge (double lines) for presplitting
100	Contour blasts and demolition

When detonating cord is used down-the-hole, the ignition train between holes is often detonating cord. In order to provide for delay timing between holes and between rows of holes, the surface detonating cord can be interrupted and a surface delay unit, detonating relay, introduced. A wide range of millisecond delay times are available. Figure 4.32 shows the different types of detonating relays (ms delay).

An in-the-hole delay-initiating device, Slider Primer, is also used in the USA, connecting detonating cord to primers in individual decks of non-cap-sensitive explosives, but it seldom used nowadays.

Table 4.9 Technical information of detonating cords

Manufacturer	Brand	Core load (g/m)	Outside dia. (mm)	Tensile strength (kgf)	VOD (m/s)	Core explosive
Austin Powder	Lite Line	3.2	3.94	104	6500–7000	PETN
	A-Cord	5.3	4.19	104	6500–7000	PETN
	50 Reinforced	10.6	5.0	90.7	6500–7000	PETN
	Heavy Duty 200	42.5	8.5	113.4	6500–7000	PETN
Dyno Nobel	Primacord 1	1.5	3.18	68	7000	PETN
	Primacord 2.5	2.4	2.8	27	7000	PETN
	Primacord 3	3.2	3.66	113	7000	PETN
	Primacord 4Y	3.6	3.61	68	7000	PETN
	Primacord 5	5.3	3.99	68	7000	PETN
	Primacord 8	8.5	4.47	90	7000	PETN
	Primacord 10	10.8	4.7	90	7000	PETN
Orica	Cordtex 3.6 W	3.6	4.0	90	6500–7000	PETN
	Cordtex 5P	5.0	3.7	70	6500–7000	PETN
	40 RDX LS[a]	8.5	4.4	.	6700	RDX
	Cordtex 10P	10.0	4.6	70	6500–7000	PETN
	Cordtex 40	40.0	7.6	70	6500–7000	PETN
China [5]	Low Energy	5.0	4.2	>50	>6000	PETN
	General	11.0	6.0	>50	>6500	PETN
	Seismic	40.0	9.5	>50	>6500	PETN

[a]40 RDX LS detonating cord has a heat resistance rating of 100 h at 145 °C, falling to 1 h at 163 °C

Of serious concern when using detonating cord as the down-the-hole line is the desensitization of some explosives, which is described in Sect. 3.7.8 of Chap. 3: "As the detonation velocity of the detonating cords (usually above 6000 m/s) is much higher than the commercial explosive (usually water gel and emulsion: 3500–5000 m/s, ANFO: 2000 m/s), the shock wave produced by the detonation of in-hole detonating cord compresses the explosive cartridge and increases the density of explosive or breaks the structure of the composition of the explosive (e.g. compressing or breaking the bubbles of the hot spots for water gels and chemically sensitized emulsion explosives) and decreases the sensitivity of the explosives."

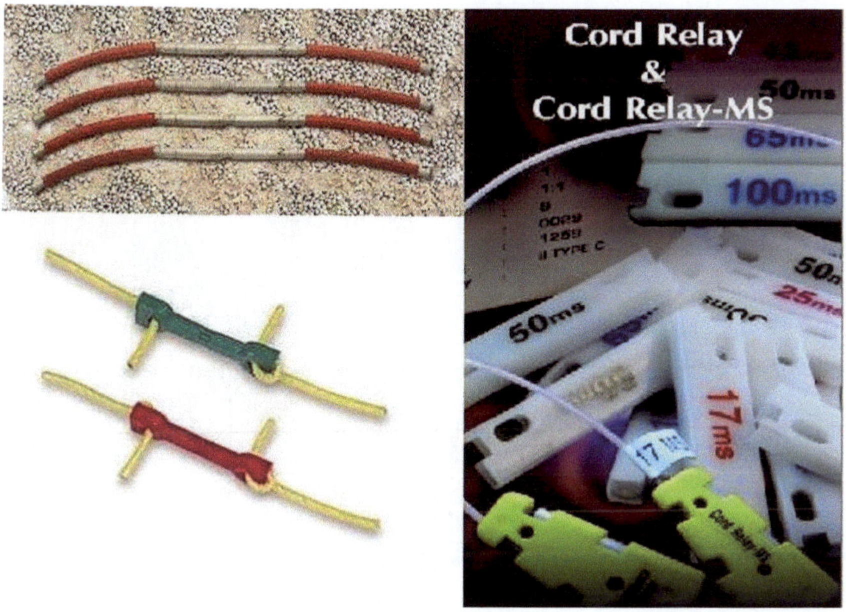

Fig. 4.32 Different types of detonating relays

4.3 Cast Boosters

Most site-mixed and loaded bulk blasting agents, such as ANFO, bulk emulsion, and heavy ANFO, are all non-cap sensitive and need a high strength, high density, and very high velocity of detonation device to initiate them. To meet this demand, compact, high detonation pressure non-nitroglycerin boosters were developed to provide explosives users with complete and convenient non-NG blasting systems.

Cast boosters are explosive units designed to act as primers, comprising a mixture of PETN (or RDX) and TNT, which are inexpensive military high explosives released to the commercial market during the late 1950s through the 1960s and other minor ingredients. Their high energy, high strength, and high VOD (above 7000 m/s) make them suitable for priming the above-said bulk blasting agents. An additional advantage is their lower sensitivity to shock, friction, and impact than NG explosives.

Cast boosters are manufactured in various sizes and weights to meet different requirements and situation of rock blasting. Some small boosters with the size of 15–22 mm in diameter and weight of 8–20 g are specially manufactured for small-diameter drillholes in tunneling (the lower right corner of Fig. 4.33).

There is a kind of slider booster which can be used in explosive columns within one borehole that are separated by decks of inert material in order to enable the explosive columns to detonate within a given time interval, reducing the vibration. The boosters have a plastic tube attached to their side for passing through a light

Fig. 4.33 Cast boosters (Austin)

Fig. 4.34 Slider boosters

loaded detonating cord (2.5–3.8 g/m). A short-lead shock tube detonator with certain delay time is threaded through the tube and inserted into the cap well of the booster (see Fig. 4.34).

References

1. Bhandari S (1997) Engineering rock blasting operations. A.A. Balkema, Rotterdam. ISBN: 90 5410 658 1
2. Persson PA et al (1993) Rock blasting and explosives engineering. CRC Press, Boca Raton. ISBN: 0-8493-8978-X

3. Dick RA et al (1983) Explosives and blasting procedures manual. USBM IC 8925, 1983
4. ISEE (1998) Blasters' handbook, 17th edn. ISBN: 1-892396-00-9
5. Yu Y et al of CSEB (2004) Theory and technology of engineering blasting. Metallurgical Industry Press, Beijing. ISBN: 7-5024-3390-2. 中國工程爆破協會于亞倫主編,工程爆破理論与技朮。冶金工業出版社,2004年, 北京
6. Wang X et al (2010) Handbook of blasting. Metallurgical Industry Press, Beijing. ISBN: 7502415742, 9787502415747. 汪旭光主編,爆破手冊。冶金工業出版社,2010年,北京
7. HKSAR (1997) Chapter 295B, Dangerous Goods (General) Regulation, Government of Hong Kong Special Administrative Region, 1997

Chapter 5
Mechanisms of Rock Breakage by Blasting

5.1 Shock Wave and Stress Wave in Rock Generated by Explosion

5.1.1 Shock Wave and Stress Wave

A change in the local state parameters of a medium (medium density, velocity, state of stresses, etc.) is called perturbations. A propagating perturbation through a medium (solid, liquid, gas, or plasma) (or in some cases in the absence of a material medium, through a field such as an electromagnetic field—that would not be discussed in this book) is called a wave. The interface between the disturbed zone and the non-disturbed zone is called the wavefront. The movement speed of the wavefront is called the velocity of the wave.

A shock wave is a type of propagating perturbations. Like an ordinary wave, it carries energy and can propagate through a medium. Shock waves are characterized by an abrupt, nearly discontinuous change in the characteristics of the medium. Across a shock, there is always an extremely rapid rise in pressure, temperature, and density of the flow. A shock wave travels through most media at very high, supersonic speed.

A propagation of stress (or strain) through a medium is called stress wave or strain wave. The difference from the shock wave is that there usually is not an extremely rapid rise of wave front in stress wave (Fig. 5.1).

5.1.2 Types of Stress Wave

Stress waves in a solid medium are of several types. The stress waves are classified into two groups: body waves and surface waves. The primary wave of body waves is also called the *P*-wave or compression wave or longitudinal wave. It has particle

D. Zou, *Theory and Technology of Rock Excavation for Civil Engineering*,
DOI 10.1007/978-981-10-1989-0_5

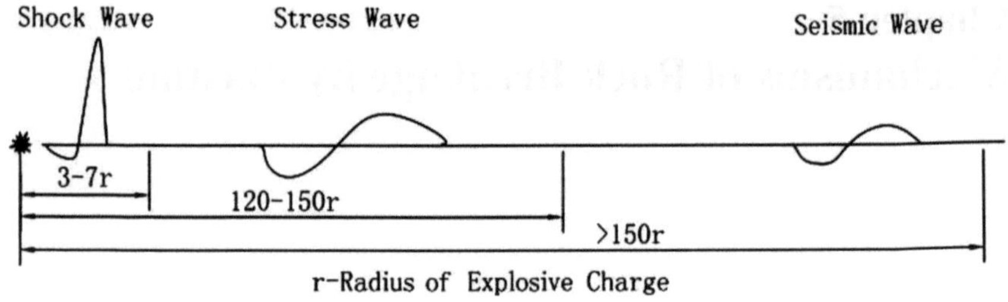

Fig. 5.1 Shock wave, stress wave, and seismic wave generated by explosion in rock mass

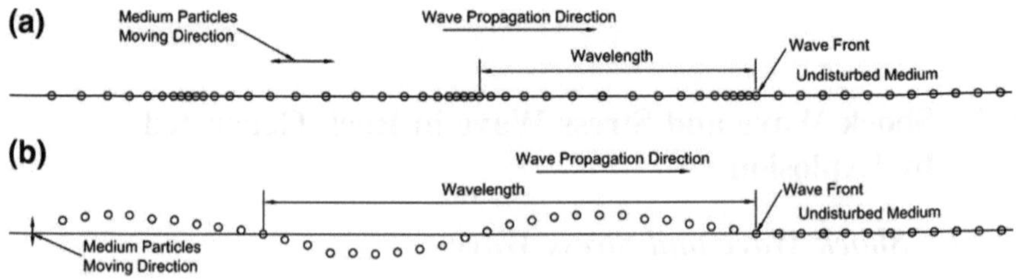

Fig. 5.2 Medium particles moving types of body waves **a** *P*-wave, **b** *S*-wave

motions which are in the radial direction and has the highest propagation velocity. The second type of body wave is called *S*-wave or transverse or shear wave which moves the particles in a direction that is perpendicular to that of the wave propagation (Fig. 5.2). The velocity of the *S*-waves is somewhere between that of the *P*-wave and the surface waves. The surface waves that are usually generated in rock blasting are *R*-waves (Rayleigh wave) and *Q*-waves (Love wave). The *R*-waves are characterized by elliptical particle motions, similar to ocean waves impacting a beach. *Q*-waves are similar as the *S*-waves, the particles move in a direction of perpendicular to that of the wave propagation.

5.1.3 Reflection of Stress Wave from a Free Face

If there is a free face in the medium, the stress waves will be reflected from the free face. A compressive wave will be reflected as a tensile wave (sometimes reflected two waves, a tensile wave and a shear wave depending upon the incident angle of the compressive wave) (see Fig. 5.3).

The characters and effects of the body waves and surface waves generated by the rock blasting will be further discussed in Chap. 11 of this book.

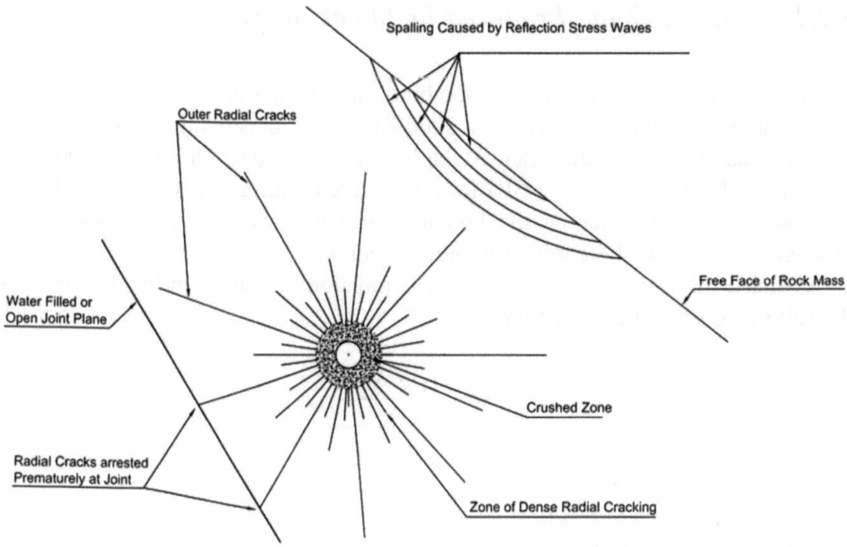

Fig. 5.3 Radial fracturing and spalling caused by reflection of the stress waves

5.2 Mechanism of Rock Breakage by Blasting

During the detonation of an explosive charge inside rock, there are three phases of actions that occur in the rock surrounding the detonated explosive charge:

First phase: A strong impact is produced by the shock wave and crushes the rock closing to the explosive charge.
Second phase: Stress waves damping from shock waves propagate in the rock in a radial direction from the blasthole and break the rock with radial cracks and spalling when the stress waves are reflected from a free face.
Third phase: The gases produced behind the detonation front come into action. The pressure of gases extends fractures and moves the broken rock in a direction towards the free face.

Since the fifties of the last century, many theories have been developed to explain the mechanism of rock breakage by blasting. Most researchers accept the theory: Rock breakage by explosion is the result of both actions of stress waves (including shock wave) and the quasi-static pressure of explosion gases. But even nowadays, it still remains as a debating issue that which action between the stress wave and gas pressure plays the main role during the rock breakage by an explosive charge.

5.2.1 Crushed Zone Produced by Shock Waves

When the charge detonates in the borehole, the detonation wave will propagate along the hole with a velocity of 3000–6000 m/s depending on the type of explosive and the charge diameter. At the front of the detonation wave, the pressure is normally between 5 and 10 GPa for a hole filled with a high explosive. A high pressure will initially act on the borehole wall and working as a shock wave propagates out of the borehole in a radial direction.

As stated in Chap. 3 of this book, the detonation pressure can be expressed by the following simplified equation:

$$\mathrm{PD} = \frac{\rho_e \times \mathrm{VD}^2}{4} \tag{5.1}$$

where:

PD Detonation pressure (kPa);
ρ_e Explosive density (g/cm^3);
VD Detonation velocity of explosive(m/s)

The maximum pressure transmitted to the rock is equivalent of:

$$\mathrm{PT}_m = \frac{2}{1 + n_z}\mathrm{PD} \tag{5.2}$$

where PT_m is also the peak value of the shock wave acting on the rock of borehole wall; n_z is the relationship between the impedance of the explosive and that of the rock:

$$n_z = \frac{\rho_e \times \mathrm{VD}}{\rho_r \times \mathrm{VC}} \tag{5.3}$$

where:

VC Propagation velocity of the wave through rock mass (m/s);
ρ_r Rock density (g/cm^3)

In the first instants of detonation, the pressure in front of the shock wave is much higher than the dynamic compressive strength of the rock, provoking the destruction of its inter-crystalline and inter-granular structure that means the rock is crushed by the shock wave.

The thickness of the so-called crushed zone is about 3–7 times the radius of the explosive charge and increases with detonation pressure of the explosive and with the coupling between the charge and the borehole wall.

According to Hagan [1], almost 30 % of the energy transported by the shock wave is consumed within the crushed zone around the borehole but only contributing a very small volume to the actual rock fragmentation, around 1 % of the total volume corresponding to the normal breakage per blasthole.

5.2.2 Radial Cracking Zone Produced by Stress Waves

After a large amount of energy is consumed in the crushed zone, the shock wave is attenuated to a compressive stress wave. During the propagation of the stress wave, the rock surrounding the blasthole is subjected to an intense radial compression which induces tensile components in the tangential planes of the wave front. When the tangential stresses exceed the dynamic tensile strength of the rock, radial cracks around the crushed zone are formed (Fig. 5.4). The number and length of the radial cracks depend on the intensity of the stress wave out going from the crushed zone, the dynamic tensile strength of the rock, and the attenuation of the stress wave by the rock mass.

Initially, the number of radial cracks is quite large, but only a few of these cracks propagate far because of the stress relaxation spreading from the longest among them. In the absence of a free face, a few cracks become much longer than others. The radial crack propagation velocity is initially of the order of 1000 m/s, gradually decreasing.

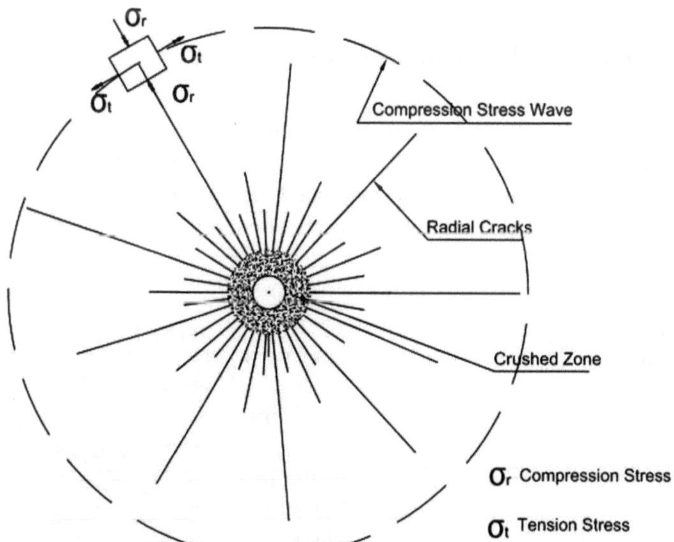

Fig. 5.4 Tensile components in the tangential planes of the compression stress wave front forms radial cracks in the rock

5.2.3 *Reflection of Stress Waves from Free Face*

As the stress wave velocity in hard rock is of the order of 4000–5000 m/s, the radial crack length by the time the stress wave reaches the free face is less than 25 % of the distance to the free face. The compressive stress wave is then reflected back from the free face as two waves, tensile wave and shear wave. Although the relative magnitude of energies associated with the two waves depends upon the incident angle of the compressive stress wave, the fracturing is usually caused by the reflected tensile wave. If the tensile wave is strong enough to exceed the dynamic tensile strength of the rock, which is between 5 and 15 % of the dynamic compressive strength of the rock, spalling will be caused, back towards the interior of the rock. But according to Persson et al. [2], the spalling fracture is not caused in granite at the free face when the explosive powder factor is of the order of 0.5–1.0 kg/m^3 (which is normal in bench blasting).

The experiments showed that some radial cracks grow more rapidly when they interact with the reflected tensile wave even if the energy of the reflected tensile wave is not enough to cause spalling in the free face. It will be most favorable to these cracks if their crack-growing direction is nearly parallel to the reflected tensile wave front, points *A* and *D* in Fig. 5.6, the tensile wave can make the crack get a greater propagation velocity. Other radial cracks which intersect with the wave front of the reflected tensile wave still can get some benefit for increasing their growing velocity from the tensile components of the wave front which is more perpendicular to the crack growing direction, point *B* and *C* in Fig. 5.5.

5.2.4 *Role of the Explosion Gases*

Some researchers, like Langefors and Kihlstrom [3] and Persson [4], have shown that stress wave is not the only mechanism, which is responsible for rock breakage under the action of explosives. They concluded that the energy in the form of stress wave was about 5–15 % of the total theoretical energy of explosive.

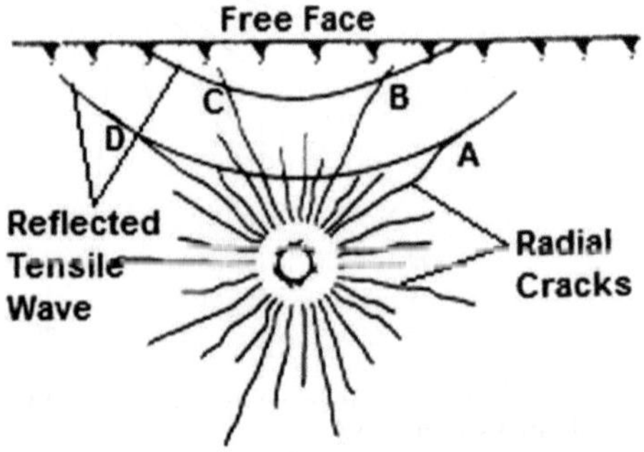

Fig. 5.5 Interaction between reflected tensile wave and growing cracks

Lots of researchers agree that the explosion gases produced by the detonation of explosive in the blasthole play an important role during the breakage of rock mass by blasting. The rolc of the explosion gases expresses three respects:

1. After the stress wave passes, the high pressure of the gases cause a quasi-static stress field around the blasthole.

The peak gas pressure generated within a blasthole is generally taken to be around one-half of the detonation pressure and can be calculated from the equation [5]:

$$P_b = 0.12 f_c^n \rho_{\exp} \mathrm{VOD}^2 \tag{5.4}$$

where:

P_b	Gas pressure (Pa);
VOD	Velocity of detonation of the explosive (m/s);
$\rho_{\exp}$	Density of the explosive;
f_c	Coupling factor, defined as the ratio of the volume of the explosive to the volume of the blasthole (excluding the stemming column); and
n	Coupling factor exponent generally taken to be between 1.2 and 1.3 for dry holes and 0.9 for holes filled with water

For commonly used industrial explosives (ANFO and emulsion explosives), the peak blasthole pressure is between 800 and 4000 MPa for fully coupled explosives. When cartridge explosives are to be used, there is a degree of decoupling between the explosive and the blasthole wall, and peak pressure is reduced relative to those for fully coupled explosives. Assuming normal degree of decoupling, peak blasthole pressure resulting from the use of cartridge explosives is likely to be approximately 1000 MPa. Where blastholes are filled with water, the degree of pressurization increases markedly. Under these conditions, Blasthole pressures with cartridge explosives and normal degree of decoupling can be expected to be around 1500 MPa.

The high pressure of the gases causes a quasi-static stress field around the blasthole. Figure 5.6 is the stress trajectories of the quasi-state stress field plotted by the computer simulation for a burden of 40 mm around a single blasthole.

The quasi-static gas pressure in the hole generates radial compressive stresses. These radial compressive stresses induce tensile components in the tangential planes which are perpendicular to the radial stress trajectories. If the intensity of the tensile components exceeds the tensile strength of the rock, the radial crack would be expected to grow along with the stress trajectories.

The evidence from some experiments also showed that the fracture matrix, either natural or formed by stress waves associated with the detonation of explosive, can extend under the influence of the quasi-static stress field generated by the gas action alone, without penetration of gases into the cracks.

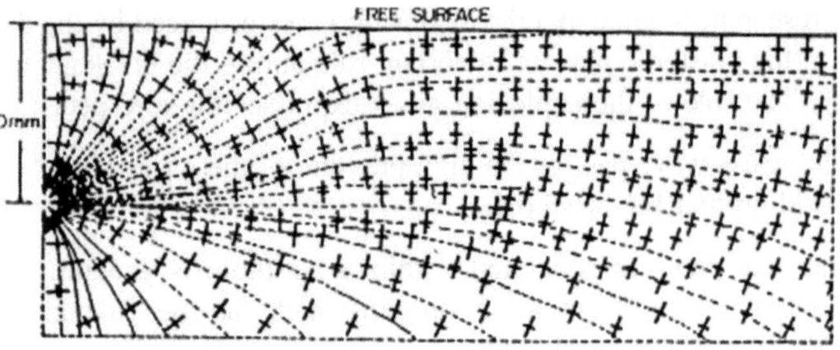

Fig. 5.6 Stress trajectories plotted for a burden of 40 mm with single hole (Reproduced from ref. [7] by permission of Taylor & Francis Group)

2. During or after the formation of radial cracks by the tangential tensile component of the stress wave, the gasses start to expand and penetrate into the fractures, cracks formed by stress wave, and the original fissures in rock mass. The radial cracks are extended under the influence of the stress concentration at their tips.

Some researchers measured gas penetration velocities in the range 5–10 % of the *P*-wave velocity for both sedimentary overburden material and andesites. They suggest, however, that the velocity of penetration will be controlled by rock permeability, rather than by the initial blasthole pressure. Rock mass permeability was considered to be controlled by the in situ joint frequency [5]. At this rate, expected velocity of gas penetration in massive granite could be around 200 m/s.

The extreme gas pressures generated by detonating explosives are dissipated rapidly as the blasthole diameter expands and gas penetrates into the fractures, thereby increasing the volume of the confined gas. One study presented a negative exponential rate of dissipation of the gas pressure [5]. This study also reported peak positive levels of gas pressure up to 300 kPa at distance around 1 m from blastholes, decaying to around 6 kPa at a distance of 8 m. Gas has been observed to flow through and dilate both natural joints, as well as blast-induced fractures, for distances up to 20 m behind the blasthole pattern.

As stated above, the high gas pressure rapidly falls in the blasthole as the gases escape through the stemming, through the radial cracks and with rock displacement, the stored stress energy is rapidly released, generating an initiation of tensile and shear fractures in the rock mass. This fracturing by release-of-load affects a large volume of rock, not only in front of the blastholes but behind the line of the blast cut as well to produce obvious damage to the remaining rock mass.

In bench blasting with long cylindrical charges, during and after the mechanisms of radial fracturing and spalling, the pressure applied by the explosive gases upon the rock mass in front of the explosive column makes the rock act like a beam embedded in the bottom of the blasthole and in the stemming area, producing the deformation and fracturing of the same of the phenomena of flexion (Figs. 5.7, 5.8).

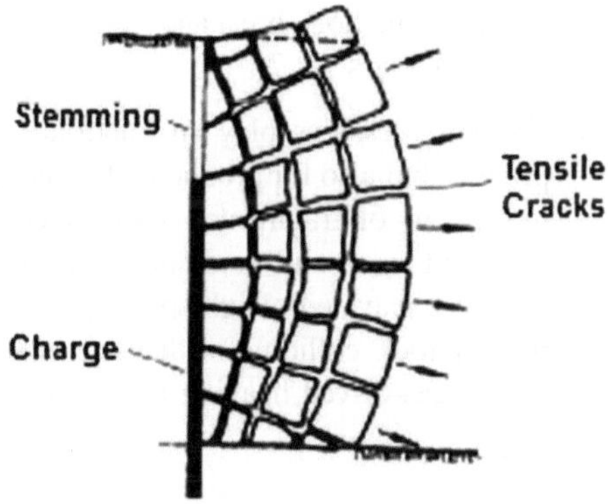

Fig. 5.7 Mechanism of breakage by flexion (Reproduced from ref. [6] with permission from Taylor & Francis Book UK)

Fig. 5.8 Rock breakage by flexion (Reproduced from ref. [6] with permission from Taylor & Francis Book UK)

The degree flexion breakage is controlled by the stiffness property of the burden-rock mass [6, 7].

3. The gas pressures play an important role not only in the rock fragmentation process, but also for producing the movement of the broken rock fragments. The movement of fragmented rock is considered essential in order to introduce sufficient swell and void space that the broken material can be easily and efficiently excavated. High-speed photography of blasting has shown that the broken rock colliding with each other can produce additional fragmentation during gas-driven heaving of rock thrown from the free face.

5.3 Explosive's Energy Distribution During Rock Blasting and Livingston's Blasting Crater Theory

5.3.1 Explosive's Energy Distribution During Rock Blasting

As stated above, during the detonation of an explosive charge inside rock, some of the energy released from the explosive is gainfully utilized and much of it is wasted. The useful part of energy is capable of doing work of fragmentation and displacement of the blasted rock, and the wasteful part of energy can be put to no work, which causes many harmful effects. The total energy released from the explosive charge is characterized by two phases of actions as follows:

First phase: A strong impact is produced by the shock wave and stress wave linked to the ***Strain Energy***, expressed as ***ET***, during a short period of time;

Second phase: The gases produced behind the detonation front come into action, at high temperature and pressure, carrying the ***Thermodynamic*** or ***Bobble Energy***, expressed as ***EB***.

Then, the total energy, expressed as ***ETD***, developed by the explosive can be expressed as the sum of these two components:

$$\mathrm{ETD} = \mathrm{ET} + \mathrm{EB}$$

Summarizing the statement of the section above, the energy distribution developed by the explosive for the rock breakage is described in Fig. 5.9.

5.3.2 Livingston's Crater Blasting Theory

Crater blasting is the basic type of rock blasting. A crate blast is a blast in which a spherical or near-spherical short charge is detonated beneath a rock surface. The detonation of explosive charge breaks the rock and forms a crater-shaped pit which is called a blasting crater, on the rock surface. Crater blasting is a basic form of rock

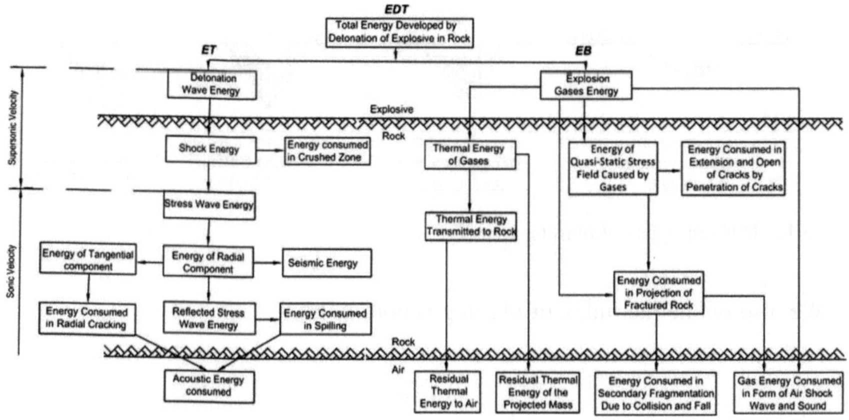

Fig. 5.9 Energy distribution of explosive in rock fragmentation

blasting, and the study of the formation mechanism for revealing the essential nature of rock blasting has great significance. In the study of crater blasting theory, the most practical significance and the most influential is the crater blasting theory developed by C. W. Livingston in 1956. Even today it is still the guidance and basis of theoretical study and practical work of rock blasting in the engineering blasting field.

5.3.2.1 Parameters of Crater Blasting

As shown in Fig. 5.10, the main parameters are as follows:

- *W*—depth of burial (burden), distance from surface to center of charge;
- *r*—radius of crater;
- *R*—active radius of blasting;
- *H*—visible depth of blasting crater; and
- *θ*—opening angle of blasting crater.

Fig. 5.10 Blasting crater formed by spherical explosive charge

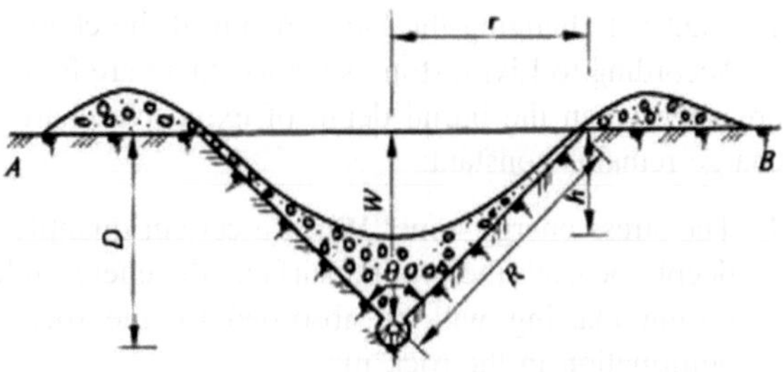

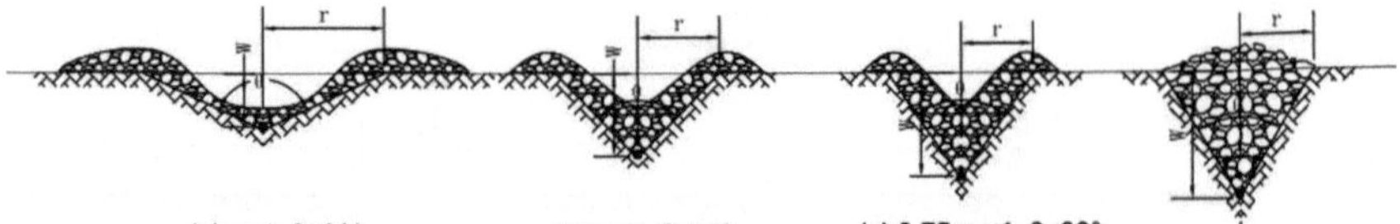

Fig. 5.11 Different types of blasting craters

We also define the index of blasting action, *n,* as:

$$n = \frac{r}{W} \tag{5.5}$$

Basic types of craters blasted are shown in Fig. 5.11.

5.3.2.2 Livingston's Crater Blasting Theory [11, 12]

Under the basis of study and lots of experiments, Livingstone proposed crater blasting theory as the theory of rock fragmentation by blasting in which the criterion of energy balance is adopted. He indicated that the quantity and speed of the energy transferred to the rock by the explosive during the blasting depends on the properties of the rock, performance and quantity of the explosive, the burial depth of the charge, and the initiation method. Under a certain rock condition, the energy released by the explosive charge depends on the quantity of explosive and its detonation velocity.

It is supposed that a certain quantity of explosive charge is buried deeply beneath the ground surface, and most of the energy released by the explosive charge during blasting will be absorbed by the rock mass. When the released energy of explosive charge reaches the saturated state, the remaining energy of explosive charge will cause the breakage, uplift, displacement, and even throwing up of the rock on the surface of the rock mass. But under the saturated state, the rock mass only undergoes elastic deformation and does not break. From the point of view of blasting energy, the blasting result will be same when the burial depth is kept constant and changing the quantity of charge or keeping the explosive charge constant but changing the burial depth of the charge.

According to Livingstone's theory, there are four zones of rock deformation and breakage when the burial depth of explosive charge changes and the quantity of charge remains constant:

1. The stress energy zone. When a certain quantity of explosive charge is buried deeply beneath the ground surface, the energy released by the explosive charge during blasting will be absorbed by the rock mass and only cause elastic deformation in the rock mass.

It has been found that there is a definite relation between the energy of explosive and the volume of the rock that is affected by the blast. This relationship is significantly affected by the placement of the charge.

Livingstone has determined that a strain-energy relation exists, expressed by an empirical equation:

$$W_c = E_s Q^{1/3} \tag{5.6}$$

where:

W_c is the critical distance at which cracks and fracture of the surface above the spherical charge are just noted as the signs of blasting action by the charge;
E_s is the strain-energy factor, which is a characteristic constant in each rock–explosive combination; and
Q is the weight of explosive charge.

2. The shock break zone: When the burial depth of explosive charge is reduced, it means the burial depth $W_b < W_c$, the rock mass will be broken, and uplift forms a blasting crater. The shock break zone and the volume of the produced crater will increase along with the reduction in the burial depth of charge and will reach a maximum value at a certain burial depth.

Equation (5.6) can be written into a different form as

$$W_b = \Delta E_s Q^{1/3} \tag{5.7}$$

where W_b is the distance from the surface to the center of gravity of charge, i.e., the depth of burial; and Δ equals W_b/W_c which is a dimensionless number expressing the ratio of the depth of burial to the critical distance.

When W_b is such that the maximum volume of rock is broken to the fragment size required, this burial depth is called the optimum distance, W_o.

Therefore, $\Delta_o = W_o/W_c$ is called as optimum depth relationship.

3. The fragmentation zone: By continually reducing the burial depth, the volume of crater will decrease. The energy released by the explosive charge is partly used for rock fragmentation and partly used for rock casting and air blast.
4. The air blast zone: When the charge is close to the rock surface, most of the explosive energy is expressed as air blast.

In order to determine the optimum burial depth, a series of tests will be carried out in the same type of rock with the same type and weight of explosive charge. An example curve of the results of crater blasting tests is shown in Fig. 5.12 (in the figure, V is the volume of the produced crater).

Livingston's crater blasting theory is based on a series of tests, closer to practice. In determining rock blastability and their classification, comparison of the performance of explosives and selection of the optimal parameters of rock blasting, and

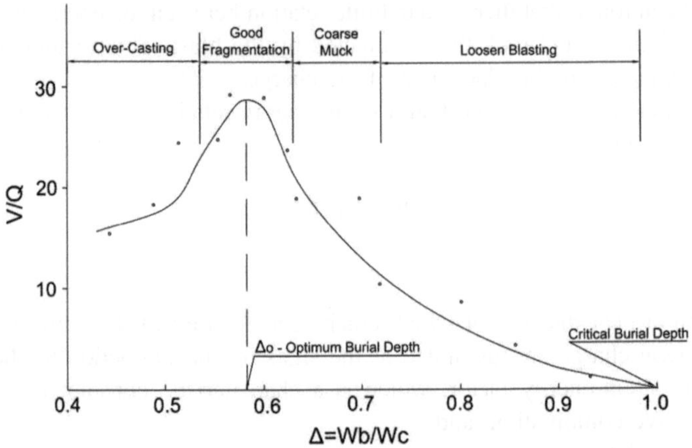

Fig. 5.12 Demonstration curve of the results of crater blasting tests

other aspects of the engineering blasting projects, Livingston's crater blasting theory has been applied successfully. In particular, the use of this theory has achieved good results in the VCR (vertical crater retreat) method in underground mining.

5.4 Rock Classification by Blastability

5.4.1 Rock Blastability and Its Effective Factors

The blastability of rock is the comprehensive reflection of the physical and mechanical properties of rock itself, the performance of the explosive charge, and blasting processes. It is not just a single intrinsic property of the rock, but a combination of a series of inherent properties of the rock, which manifests itself in the blasting process and affects the entire blasting effect.

Of the main factors of rock blastability, one is the internal factor of physical and mechanical properties of the rock itself, on the other hand is the extrinsic factors including performance of explosives and blasting process. The former depends on the geological rock formation conditions, mineral composition, structure, and tectonic effects overtime. It is characterized as physical and mechanical properties of rock, i.e., density or bulk density, porosity, bulking, elasticity, plasticity, brittleness, and rock strength and other properties. The latter depends on the type, form, and weight of the explosive charge, charging structure, initiation methods and initiation intervals, the size of the burden with the free surface, scale of the blast, the direction and the relative position of the free surface, and the charges and so on. Also it includes the effects on the blasting results, i.e., rock fragmentation, rock pile shape, and throw distance. Obviously, the physical and mechanical properties of the rock itself are the most important factors as these are fixed and cannot be changed, whereas the blasting process can be adjusted to suit the rock conditions.

There are two main aspects of the effects of detonation of explosives on rock blasting, one is to overcome the cohesion between rock particles, break the internal structure of the rock, resulting in a fresh fracture surface; the other is to expand and destroy the native or secondary fractures of the rock mass. The former depends on the strength of the rock itself; the latter is controlled by the geological structures of the rock mass. Therefore, the strength of the rock and fissures in the rock are the most fundamental factors influencing rock blastability.

5.4.1.1 The Effect of the Structures (Mineral Composition), Cohesion, and Fissures of Rock on Rock Blastability

Rock consists of solid particles, between which there is a gap, filled with air, water, or other debris. When the rock is acted on by an external load, especially by the explosive shock loading, it will cause changes of state, leading to changes in rock properties.

Minerals are the main components of the rock. When the minerals of the rock are fine-grained, have higher density, and are more angular and interlocked, the rock will be more difficult to be broken by blasting. Mineral density is up to 4 g/cm^3 or more. The bulk density of the rock does not exceed its constituent mineral density. Rock density is generally 1.0–3.5 g/cm^3. With the increase in density, the strength and ability of the rock to resist blasting increases, while the energy consumption for breaking and uplifting rock also increases, which is the reason of more difficult to blast magmatic rocks. For the blastability of sedimentary rocks, in addition to depending on its mineral composition, to a large extent they are influenced by cement composition and particle size. For example, fine-grained sedimentary rocks which have siliceous cement are robust and hard to be blasted; rock with cements containing iron oxides are next; sedimentary rocks containing calcareous and clay-containing cements are less robust and easiest to blasting. The composition and structure of metamorphic rocks are complex, and their blastability is related to the degree of metamorphism. Generally, the rocks which have a high degree of metamorphism and dense texture are relatively strong and harder to be blasted, and vice versa.

Mineral rocks in turn are deposited in different ways and have different chemical compositions and different crystalline lattice structure. Since the mineral composition of the chemical bonds is not the same, the cohesive force of the molecule is different. So, the strength of mineral crystal depends on the internal forces interacted between crystal molecules, crystal structure, and crystal defects. Generally, the cohesive force between the crystals is less than the cohesive force between the molecules within the crystal. Also, the larger the grain size, the smaller the cohesion is; so generally the strength of the fine-grained rock is higher than the strength of coarse-grained rock. And because the cohesion between crystals is smaller than cohesion within the crystal, so damage cracks often appear between the crystal grains. The defects in the form of pores, bubbles, microscopic cracks, and cleavage planes are ubiquitous in the rock; these defects may cause stress concentration. Therefore, microscopic defects will affect the nature of the rock components, and large cracks will affect the overall robustness of the rock, making it easy to blast. Fissures in the rock mass include not only the native fissures and fissures generated by later geological action, but also the secondary fractures generated by production works and previous blasting. They include faults, folds, bedding, cleavage, contact surfaces of different formations, cracks, and other weak faces. These weak faces in the rock mass can have two impacts on to the blastability of rock: On the one hand, a weak surface could allow detonation gas and pressure to leak reducing the effectiveness of explosive energy and affect the blasting result; on the other hand, these weak faces undermine the integrity of the rock and make it easy to break from the weak faces, and they increase the reflective action of the stress wave from the weak face which is favorable to rock fragmentation. However, it must be noted that when the rock mass itself contains many blocks whose dimensions exceed the required size of the excavation works, only a small part of the rock close to the explosive charge will be fully broken, and most of the rock mass which is further from the charge and has been cut by native or secondary fissures will not be fully broken in the blasting process. These blocks are moved and thrown out from the rock mass by detonation gas and blasting vibration to form oversize chunks. These are the two opposing effects of fissured rock during blasting: Some are easy to be broken, and some are prone to produce large fragments. Usually, the rock mass is relatively easy to be fragmented when the blastholes are drilled vertical to the bedding planes or fissures and may be more difficult to be fragmented when the blastholes are drilled parallel or along bedding planes or fissures. Additionally, weathering can break the connection between the various components of the rock, so weathered rock may be fragmented easily by blasting.

5.4.1.2 The Effect of the Density, Porosity, and Broken-Dilatability of Rock on Rock Blastability

Rock density indicates the weight per unit volume of the rock including the internal pores. Rock porosity is the pore volume (including gas or liquid volume) compared with the total volume of the rock. Volume available per unit volume occupied by said rock pores can also be expressed as a percentage. Typically, the rock porosity is of 0.1–50 % (usually igneous rocks of 0.5–2 %, sedimentary rocks of 2.5–15 %). When the rock is pressurized, the porosity is reduced, for example, clay with porosity of 50 % may reduce to 7 % under pressure. As the porosity increases, the shock and stress wave propagation speed in rock mass will be lower.

High-density rock is difficult to be blasted, because a lot of explosive energy is used to overcome gravity in order to fracture the rock, move, and throw it away.

Broken-dilatability is one of the properties of the rock, of which the volume the rock is dilated after broken. The ratio of the volumes of the rock after-breaking and prebreaking is called the coefficient of broken-dilatability. Broken-dilatability is related to the rock structure and the degree of fragmentation. It can be used for estimating the degree of rock fragmentation and calculating the size of the compensation space needed for the rock to move into during the blast.

5.4.1.3 The Influence of Elastic, Plastic, Brittle, and Strength of Rock on Rock Blastability

From the rock mechanics point of view, the properties of rock may appear to be plasticity, elasticity, viscoelasticity, elastic-brittle, and brittle, depending on the different characteristics of the external force and deformation.

The elastoplastic rock will be difficult to blast (such as clay rocks) due to the energy consumption when the applied external load beyond its elastic limit and produces plastic deformation; but the brittle and elastic-brittle rock are easy to blasting (such as brittle coal) as there is almost no residual deformation under the external loads. The plasticity and brittleness of rock are not only related with the nature of the rock itself, but also related with its stress state and loading speed. Rock located deep underground which is equivalent to high pressure often appears plastic, but behaves brittle under impact loading. With the increase in temperature and humidity, the plasticity of rock increases. Generally, in blasting, brittle failure of rock is the main and most useful behavior. But the rock close to the explosive charge suffers plastic failure; although the extent of damage is very small, most of the energy of explosion is consumed in the plastic deformation.

Rock strength is the ability of rock to resist the stresses of pressure, shear, and tension, resulting in rock destruction. Normally, the constant used in material mechanics to indicate the ability of materials to resist these three simple stresses are measured under the uniaxial static loading. However, in blasting, rock suffers the instantaneous impact load, so to deal with rock strength we must consider the new context to emphasize the role of the dynamic triaxial strength index. Only thus the rock blastability can be truly presented.

The dynamic load strength of rock is greater than the static strength. The compressive ultimate strength (σ_p) of rock is the highest, shear (σ_s) strength is the next strongest, and tensile strength (σ_t) is the lowest. Generally, rock has the following relationships: σ_t = (1/10–1/50) σ_p and σ_s = (1/8–1/12) σ_p. Thus, we should as far as possible subject the rock to tensile or shear condition, to facilitate blasting fragmentation and improve the blasting effect.

5.4.1.4 The Influence of Blasting Parameters and Process on Rock Blastability

Blasting parameters and process are other important factors affecting the blasting of rock. According to the theory of crater blasting, the overall effects of surface blasting depend on the burden status of each blasthole and explosive power. Therefore, burden and spacing (or afforded area) of blastholes, stemming quality, and the initiation interval between holes have a direct effect on the rock fragmentation, i.e., the blastability.

The direction of blastholes' burden towards free face is the dominant direction of rock breakage and movement. The size of burden and spacing of blastholes control the rock amount to be blasted and degree of fragmentation of the rock as well. If the burden and spacing is too large, it may produce large fragments or leave a rock foot or rock wall, indicating the deterioration of the quality of the rock blasting.

The quality of hole stemming affects directly the utilization of explosive energy. Good stemming can prevent premature escape of explosive gases, extending the acting duration of explosive energy and is favorable for rock blasting fragmentation.

Blasthole initiation sequence and time intervals have an important effect on the blastability of rock and the blasting result. Under the conditions of reasonable time intervals and firing sequences of blastholes, the first firing blastholes break their burden rock and move it a distance, creating a new free face for the follow-up firing holes, greatly improving the blastability of rock and blasting results.

The commonly used explosive charging structures in engineering blasting include coupled charging, decoupled charging, continuous charging, and decking charging. They can be selected to use according to the engineering requirements and geological characters of rock mass structure. When coupled charging is used, the blasthole bears high explosion pressure and violent shock action, and the rock is damaged significantly. When decoupled charging is used, the peak value of explosion pressure acting on the blasthole wall is reduced, then the damage to the rock surrounding the hole is decreased as well, that is favorable for protecting the remained rock body behind the blastholes. Decking charging can improve the fragmentation of the upper part of the bench rock due to the explosive charging height being increased in the blastholes.

5.4.1.5 The Influence of Explosive Performance on Rock Blastability

When different explosives are used in the same type of rock mass, the indexes of rock blastability may vary greatly, which is caused by the effective factors, such as explosive density, detonation velocity, and pressure of explosion gasses, etc.

The high-temperature and high-pressure gases generated by explosive detonation act on the blasthole wall, sparking an intense shock and stress waves in the rock, which cause rock deformation and failure. Explosion gas pressure is a very important factor of the performance of explosive on rock fragmentation. The explosion pressure depends on the characteristics of explosives and is proportional to the product of the density of explosive and the square of its detonation velocity. Explosive detonation velocity is higher; the higher the explosion gas pressure, the more violent the blasting action, and the more effective the rock fragmentation. So that, for hard and dense rock, the explosives with high density and high detonation velocity should be used due to their high explosion pressure. Conversely, for weak and easily blasted rock, the explosives with low density and low detonation velocity should be selected to use for its low explosion pressure and long acting duration. With a longer duration of action of explosion pressure, stress waves can be more effective in influencing the development and expansion of the initial fracture, therefore, improving the blasting results.

5.4.1.6 The Influence of the Matching of Wave Impedance of Both Explosive and Rock mass on Rock Blastability

The wave impedance of rock is the product of the density of rock, ρ_r, and propagation speed of *P*-wave in rock, C_p. It reflects the resistance in the rock to the stress wave-driving movement of rock particles. Wave impedance of rock has a direct effect on the efficiency of propagation of the stress wave in the rock. The wave impedance of the explosive is the product of the density of explosive, ρ_e, and detonation velocity of explosive, D_e. When the wave impedance ratio of explosive and rock $R = \frac{\rho_e D_e}{\rho_r C_p} = 1$, there is no wave reflection between explosive and rock, that is called "full matching." In this condition, the transmitted explosive energy to the rock is the maximum, and the blasting gets a better result. Therefore, wave impedance ratio R is an important parameter for the selection of explosives. However, the wave impedance of general industrial explosives are quite different from the wave impedance of rocks. Getting an explosive which is completely matched with the rock is very difficult or uneconomic, and in fact not all the rocks need a strong stress wave. In general, for a dense and hard rock with high elastic modulus and small Poisson's ratio, an explosive with high detonation velocity and density should be selected in order to ensure a strong stress wave passing into the rock, resulting in the initial fissures. For moderate hard rocks, the explosive with a moderate detonation velocity and density should be chosen. For a rock with well-developed joints and fissures, or a soft and plastic rock, the explosive with low detonation velocity and density can be chosen as the stress wave is attenuated fast and acts to a lesser degree, so that stress wave plays a secondary role in rock fragmentation in these kinds of rocks.

5.4.2 *Criterion of Rock Classification by Blastability*

So far, many dozens of classification methods on rock blastability have been published by different countries, within which some typical points and methods will be discussed in the next section. Throughout all published methods of rock classification of blastability, the criterion for classification can be divided into six categories.

5.4.2.1 Criterion of Mechanical Strength of Rock

Ultimately, for the rock material, its damage occurring under blasting load can be attributed to the problem of strength. Therefore, using rock strength parameters as the classification criterion of rock blastability has some merits. The problem is that the stress state of rock under explosive load is extremely complex, and rock strength characteristics change greatly. The strength index of rock samples measured in the laboratory, even if the index is measured in triaxial dynamic tests, cannot reflect the real properties of the complex rock mass under the explosion stress state.

5.4.2.2 Criterion of Powder Factor of Explosives

In the classification system of rock blastability, it is defined that rocks are divided into different classes according to the powder factor of explosives to be used under the standard conditions. This criterion is simple, intuitive, and thus became the first applied classical criteria. The basic principles were recognized in Chinese and overseas blasting circle. It is also easy to be applied in mines and working sites. Although the powder factor (explosive unit consumption) is a relatively common indicator, it is still not ideal to use it as the unique criterion of classification of rock blastability. Because rock blastability is a complex function related with many parameters, especially with the parameters and process conditions during blasting.

5.4.2.3 Criterion of Parameters of Engineering Geology

Engineering geological parameters are the important internal factors to affect rock blastability. Sometimes the effects of joints and fissures of rock mass to the blasting results are much greater than that of other rock mechanical parameters. Therefore, many researchers proposed their rock blasting classification systems based on the geological classification of rock mass, including the geological features of joints and fissures, while ignoring the effects of the inherent physical and mechanical properties of the rock itself or the blasting parameters and process conditions on the rocks blastability.

5.4.2.4 Criterion of Speed of Elastic Wave in Rock mass

The propagation velocity of elastic sound wave in the rock is the function of elastic properties of rocks and rock density, and closely linked with the strength characteristics of the rock, developed degree of joints and fissures, condition of contained water, and other aspects of rock mass. And the quantitative data of propagation velocity of elastic sound wave in the rock can be measured directly with instruments. Thus, it is more scientific and practical than the aforementioned criteria engineering geology.

5.4.2.5 Criterion of Energy

Energy criterion is the most common criteria. The energy released by explosives in the instant of explosion passes to the rock, causing the rock deformation and destruction. The energy consumed for the damage of different rocks is different; thus, it can be used to determine the degree of difficulty of rock blasting. Therefore, the criterion of energy should be used as a criterion in rock classification by blastability.

5.4.2.6 Criterion of Critical Fracture Velocity of Rock

Rock failure is the result of the particle movement, separation, and even movement. The critical velocity at the time of breakage is a function of a series of parameters of the rock: rock density, longitudinal wave velocity, Poisson's ratio, elastic modulus, compressive and tensile strength of rock mass, mean dimension of natural structure unit, and a coefficient representing the properties of filling of fracture and their degree of opening. This criterion encompasses both considerations: the physical and mechanical properties of rocks and the effects of the geological structure of the rock mass. It seems an ideal criterion, but to measure the critical velocity of rock cracking is not easy, so it is difficult for practical application.

5.4.3 Rock Classification by Blastability

Blastability can be defined as the blasting characteristics of the rock mass subjected to a specified blast design, explosive characteristics, and specified legislative constrains depending on the site specifics. In other words, blastability indicates how easy it is to blast a rock mass under a specific condition. To determine the blastability of rock mass and the classification, different approaches have been made and various criteria have been taken by different researchers during the past decades and a general view of them are aimed in this book.

5.4.3.1 Rock Blastability Classification by Protodyakonov (М.М. Протодьяконов) (1926) and Sukhanov (А.Ф. Суханов) (1940)

In the early 1920s, Russian Professor A. A. Protodyakonov raised the concept of "Rock Sturdiness" firstly in the world and carried out a systematical study on it. He adapted a coefficient to describe sturdiness of the rock. The coefficient is called

Protodyakonov's Coefficient, "f" of a rock, commonly used in Russia and China. The definition of "*f*," its determination method, and the classification of rock according to the Protodyakonov's coefficient "*f*" have been discussed in Chap. 1 of this book.

Protodyakonov addressed that "Rock sturdiness in all aspects of performance is insistent." In practice, rock's drillability, blastability, and stability are not entirely consistent, some easy drilling but difficult blasting, some difficult drilling but easy blasting, and the strength of a small rock sample (e.g., 7 × 7 × 7 cm) under static uniaxial compressive load does not characterize the blastability under the impact load of explosion. Furthermore, as the measured data have a large variability, typically 15–40 %, some up to 80 %, its classification method using a simple index cannot represent the actual rock blastability required by rock blasting engineering as the above-mentioned shortcomings despite the classification method have obtained wide applications in mining industry.

In light of Protodyakonov's classification system, another Russia's researcher, Prof. A. F. Sukhanov (А. Ф. Суханов), raised his new classification system in 1930s. He addressed that there is no practical significance to classify rock ruggedness using an abstract coefficient. He suggested using the specific mining methods to determine the rock ruggedness during drilling and blasting. He adapted the explosives quantity (kg/m^3) consumed for blasting 1 m^3 rock (powder factor) or drillhole length needed to blast 1 m^3 rock (unit borehole length) (m/m^3) to characterize the rock blastability and provided a series of standard test conditions. He divided rock into 16 classes according to the powder factor and unit borehole length. If the powder factor and unit borehole length are higher, the rock is more difficult to be blasted, otherwise, it is easy to be blasted. It must be noted that the powder factor is not a constant and is a variable as well affected by many factors. Therefore, Sukhanov proposed a series of correction factors for non-standard conditions, which is very cumbersome, which also affected the authenticity of rock blasting. Furthermore, powder factor does not reflect the blasting fragmentation which is an important blasting result. Therefore, Sukhanov's classification method does not accurately characterize rock blastability Table 5.1 from [14].

5.4.3.2 Rock Blastability Classification by Hanukaev (1969)

In the 1950s, former Soviet Union's researcher, Prof. A.N. Hanukaev (А.Н. Ханукаев), pointed out that fissures in the rock mass have the most important significance on rock blastability. He stated the wave impedance of rock, ρv_r, the product of rock density ρ, and the velocity of elastic longitudinal wave in rock v_r, reflects the integrity of the rock mass. In 1969, Hanukaev proposed his classification system of rock blastability using this index, ρv_r, together with the characterisation of rock fissures, which is shown in Table 5.2 (from [11, 13]).

Table 5.1 Comparison between Protodyakonov's classification and Sukhanov's classification

Protodyakonov's classification			Sukhanov's classification			Representative rocks
Sturdiness coefficient (*f*)	Class	Sturdiness level	Blastability	Class	Powder factor with #2 Rock AN (kg/m^3)	
20	I	Most sturdy	Most difficult for blasting	1 2 3	8.3 6.7 5.3	Dense microcrystalline quartzite No amide quartzite ultimate secret The most dense quartzite and basalt
18 15 12	II	Very sturdy	Very difficult for blasting	4 5 6	4.2 3.8 3.0	Extreme close andesite and diabase Quartz porphyry Extreme close siliceous sandstone
10 8	III IIIa	Sturdy	Difficult for Blasting	7 8	2.4 2.0	Dense granite, sturdy iron ore Tight sandstone and limestone
6 5	IV IVa	Fair sturdy	Fair difficult for blasting	9 10	1.5 1.25	Sandstone Sandy shale
4 3	V Va	Moderate	Moderate	11 12	1.0 0.8	Not rugged sandstone and limestone Shale, dense shale rock
2 1.5	VI Via	Fair weak	Fair easy for blasting	13 14	0.6 0.5	Soft shale Anthracite
1.0 0.8	VII VIIa	Weak	Easy for blasting	15 16	0.4 0.3	Dense clay, Soft coal rock Pumice, Tuff
0.6 0.5 0.3	VIII IX X	Soil Loose Quicksand	No need blasting			

5.4.3.3 Rock Blastability with Livingstone's Strain-Energy Factor (1956)

As we stated in Sect. 5.3.2.2, US researcher C.W. Livingstone created the crater blasting theory in 1956 and raised the conception of strain-energy factor, E_s:

Table 5.2 Harukaev's rock classification of blastability

Grade of fissures	Average spacing of fissures (m)	Grade of sturdiness of Rock	Natural fissure's area in 1 m^3 rock	*P*'s Sturdiness coefficient (*f*)	Density g/cm^3	Wave impedance ratio ρv_r * 10^5 g/cm^3*cm/s	Content of natural blocks (%)			Powder factor kg/m^3	Class of rock blastability
							>300 mm	>700 mm	>1000 mm		
Extreme Fissured	<0.1	Weak	33	<8	<2.5	<5	<10	Close to 0	0	<0.35	Easy
Strongly fissured	0.1–0.5	Moderate sturdy	33–9	8–12	2.5–2.6	5–8	<70	<30	<5	0.35–0.45	Moderate
Moderate fissured	0.5–1.0	Sturdy	9–6	12–16	2.6–2.7	8–12	<90	<70	<40	0.45–0.65	Difficult
Slightly fissured	1.0–1.5	Very sturdy	6–2	16–18	2.7–3.0	12–15	100	<90	<70	0.65–0.9	Very difficult
Very slightly fissured	>1.5	Sturdiest	2	= or >18	>3.0	>15	–	–	100	= or > 0.9	Extreme difficult

Table 5.3 Strain-energy factor E_s of rock in a Canada mine

Rock	E_s	Explosive	Rock	E_s	Explosive
Magnetite (1)	4.6	60 % granular dynamite	Ferric ore	3.4	ANFO
Magnetite (2)	4.4	60 % granular dynamite	Quartzite (2)	3.4	ANFO
Magnetite (3)	4.3	Slurry	Rock salt	3.2	ANFO
Granite	4.2	60 % granular dynamite	Frozen topsoil	2	ANFO
Quartzite (1)	3.7	Slurry	Frozen rock	1.8	Slurry

$$E_s = \frac{W_C}{Q^{1/3}} \tag{5.8}$$

where:

W_c is the critical distance;
E_s is the strain-energy factor; and
Q is the weight of explosive charge.

When Q remains constant, the greater the value of W_c, the greater the E_s, and the more difficult the rock to blast. So, the strain-energy factor, E_s, can be used as a index of blastability of rock. Table 5.3 shows the test values of E_s in a mine of Canada. The values of E_s were used to determine the rock blastability and the optimal powder factor of explosives in blasting.

5.4.3.4 Rock Blastability Index Developed by Lilly (1986) and Ghose (1988) [8]

Australia researcher, P. Lilly, developed a Blastability Index (BI) based on five rock mass parameters as follows: rock mass description (RMD), joint density (JPS), orientation (JPO), specific gravity (SGI), and hardness (H). The five parameters and their ratings for the BI are shown in the table below: Table 5.4

The value of the Blastability Index is given as:

$$\mathrm{BI} = 0.5(\mathrm{RMD} + \mathrm{JPS} + \mathrm{JPO} + \mathrm{SGI} + \mathrm{H}) \tag{5.9}$$

This index can be closely related to powder factor. To use Lilly's blastability index, it is required to establish a site-specific relationship between this Blastability Index and the powder factor. This can be established either with the help of historical blast records or from trial blast results.

A.K. Ghose in 1988 also proposed a rock blastability system for in use coal mining and correlated the powder factor with Blastability Index, BI. However, the blastability is limited for surface blasting only and is given by Tables 5.5 and 5.6

Table 5.4 Rock mass parameters and their ratings from [8]

Rock mass parameters	Rating
RMD—rock mass description	
Powdery or friable	10
Blocky	20
Totally massive	50
JPS—joint plane spacing	
Close (< 0.1 m)	10
Intermediate (0.1–1 m)	20
Wide (> 1 m)	50
JPO—joint plane orientation	
Horizontal	10
Dip out of face	20
Strike normal to face	30
Dip into face	40
SGI—specific gravity influence	
SGI = 25SG-50, where SG is in tonnes/m^3	
H—Hardness	
H—hardness in Moh's Scale (1–10)	

Table 5.5 Assigned ratio for the parameters of Blastability Index (from [8])

Parameters		Ranges				
Density (t/m^3)	Data	<1.6	1.6–2.0	2.0–2.3	2.3–2.5	>2.5
	Rating	20	15	12	6	4 < 0.2
Discontinuity spacing (m)	Data	<0.2	0.2–0.4	0.4–0.6	0.6–2.0	>2.0
	Rating	35	25	20	12	8
Point load strength index (MPa)	Data	<1	1–2	2–4	4–6	>6
	Rating	25	20	15	8	5
Joint plane orientation	Data	DIF	SAF	SNF	DOF	HOR
	Rating	20	15	12	10	6
Adjustment factor 1		Highly confined Reasonably free				−5 0
Adjustment factor 2		Hole depth/Burden > 2 Hole depth/Burden = 1.5–2 Hole depth/Burden < 1.5				0 −2 −5

DIF dip into face; *SNF* strike normal to face; *HOR* horizontal; *DOF* dip out of face; and *SAF* strike at an angle acute to face

$$BI = (DR + DSR + PLR + JPO + AF1 + AF2) \tag{5.10}$$

where:

DR Density ratio;
DSR Discontinuity spacing ratio;

Table 5.6 Relationship between Blastability Index and powder factor (from [8])

Blastability Index, BI	30–40	40–50	50–60	60–70	70–85
Powder factor (kg/m^3)	0.7–0.8	0.6–0.7	0.5–0.6	0.3–0.5	0.2–0.3

PLR Point load strength index ratio;
JPO Joint plane orientation ratio;
AF1 Adjustment factor 1;
AF2 Adjustment factor 2

5.4.3.5 Rock Blastability Index Developed by Rakishev [9]

In 1982, a scholar from former Soviet Union, Prof. B.R. Rakishev (Б.Р. Ракишев), raised a concept of "critical velocity" of rock fracture. He stated that rock failure is the result of the particle movement, separation, and even throwing away, and the "critical velocity V''_{cr} at the time of rock fracture is a function of a series parameters of the physical and mechanical properties, wave impedance, and the weak faces of the rock mass. He developed a formula to express this relationship:

$$V_{cr} = k\sqrt{g \times d_n} + \frac{0.1\sigma_c + \sigma_t}{\rho_o \times c} \tag{5.11}$$

where:

V_{cr} Critical velocity at the time of rock fracture, m/s;
k A coefficient representing the properties of filling of the fracture and their degree of opening;
g Gravitation acceleration, m/s^2;
d_n Mean dimension of a natural structure unit, m;
σ_c Compressive strength of rock, MPa;
σ_t Tensile strength of rock, MPa;
ρ_o Rock density, kg/m^3;
c longitudinal wave velocity,m/s.

He divided rocks into five classes of their blastability, as shown in Table 5.7, according to the measured data of the critical velocity of rock fracture.

5.4.3.6 Rock Classification by Blastability Developed by Northeastern University, China (1985) [10]

In 1985, scholars of Northeastern University developed their rock classification system by blastability. They stated that energy criterion is the most fundamental and

Table 5.7 Correlation of rock blastability with critical fracture velocity, from [9]

Critical velocity of rock fracture V_{cr} (m/s)	Class of blastability
$V_{cr} < 3.6$	EB (Easily blasted)
$3.6 \leq V_{cr} < 4.5$	MB (Moderately easily blasted)
$4.5 \leq V_{cr} < 5.4$	DB (Difficult to blast)
$5.4 \leq V_{cr} < 6.3$	VDB (Very difficult to blast)
$V_{cr} \geq 6.3$	EDB (Exceptionally difficult to blast)

universal norm of rock failure and also reflects the basic principles of rock blasting in nature. Explosive release of energy is absorbed by rock and consumed by destroying rock that is in line with the energy conservation guidelines. Rock wave impedance ratio characterizes the physical and mechanical properties of rock and also reflects the significant effects of rock joints, crack, and other structurally weak interfaces to rock fragmentation by blasting. Combination of both respects for judging blasting results must undoubtedly be scientific, easy to test, and practical.

Based on the data collected from site tests and laboratories' examination for 63 different rocks in 13 mines, they defined the "Index of Rock Blastability, *N*." *N* is expressed by the equation of:

$$N = 67.22 - 38.44 \ln V + K + 2.03 \ln(\rho C_p) \tag{5.12}$$

where:

V Volume of rock in blasting crater test, m^3;
ρ Rock density, g/cm^3;
C_p Elastic longitudinal wave in rock mass, m/s;
K Index of rock fragmentation and expressed as:

$$K = \ln \frac{K_1^{7.42}}{K_2^{4.75} \times K_3^{1.89}} \tag{5.13}$$

where

K_1 Proportion of big rock fragments (>300 mm), %;
K_2 Proportion of small rock fragments (<50 mm), %;
K_3 Average pass rate, %

Rock is classified into five classes according to its value of "Index of Rock Blastability, N," which is given in Table 5.8.

Table 5.8 Rock classification by blastability using index of rock blastability, *N* from [10]

Class		*N*	Blastability	Representative rock
I	I_1	<29	Very easy	Phyllite, crushed sandstone, argillite crushing of dolomite
	I_2	29.001–38		
II	II_1	38.001–46	Easy	Breccia, green mudstone, beige dolomite
	II_2	46.001–53		
III	III_1	53.001–60	Moderate	Actinolite quartzite, lamprophyre, marble, gray dolomite
	III_2	60.001–68		
IV	IV_1	68.001–74	Difficult	Magnetite quartzite, amphibolite long, gneiss
	IV_2	74.001–81		
V	V_1	81.001–86	Very difficulty	Skarn, granite, light-colored sandstone
	V_2	>86		

References

1. Hagan TN (1977) Rock breakage by explosives. 6th symposium on gas dynamics of explosives and reactive system. Stockholm, pp 329–340. ISBN: 0 08 025442 X
2. Persson PA et al (1994) Rock blasting and explosives engineering. CRC Press, Inc., Boca Raton. ISBN: 0-8493-8978-X
3. Longefors U, Kihlstrom B (1973) The modern technique of rock blasting. Wiley, New York. ISBN: 0-470-99282-4
4. Persson PA et al (1977) The basic mechanism of rock blasting. In: Proceedings of 2nd congress international society rock mechanics. Belgrade, vol III, pp 19–33
5. Blastronics Pty Ltd. (2000) Method of assessment and monitoring of the effects of gas pressures on stability of rock cuts due to blasting in the near-field. GEO Report No. 100. Geotechnical Engineering Office, Civil Engineering Department, The Government of the Hong Kong Special Administrative Region
6. Jimeno CL, et al (1995) Drilling and blasting of rock. A. A. Balkema, Rotterdam. ISBN: 90 5410 199 7. Copyright ©1995 A. A. Balkema
7. Bhandari S (1977) Engineering rock blasting operations. A.A. Balkema, Rotterdam. ISBN: 90 5410 658 1
8. Dey K, Sen P (2003) Concept of blastability-an update. Indian Mining Eng J 42(8 and 9):24–31
9. Rakishev BR (1982) A new characteristics of the blastability of rock in quarries. Soviet Mining Sci 17:248–251
10. 鈕強 (1984) 我國礦山岩石爆破性分級的研究. <爆破 > 1984(1). Niu Q (1984) The study of rock blastability classification for mines in China (in Chinese). Blasting 1
11. 中國工程爆破協會于亞倫主編 (2004) 工程爆破理論与技术。冶金工業出版社,2004年, 北京。Yalun Y et al of CSEB (2004) Theory and technology of engineering blasting. Metallurgical Industry Press, Beijing. ISBN: 7-5024-3390-2
12. 汪旭光主編 (2010) 爆破手冊。冶金工業出版社 2010年, 北京。Wong XG et al (2010) Handbook of blasting. Metallurgical Industry Press, Beijing. ISBN: 978-7-5024-5345-9
13. 哈努卡耶夫 AH (1974) 礦岩爆破物理過程 (Ханукаев АН (1974) *Физические процессы при отбойке горных пород взрывом*) 。劉殿中譯,李國喬校。冶金工業出版社,1980年, 北京。
14. 林韻梅等. 岩石分級的理論與實踐. 冶金工業出版社, 1996, 北京。 Lin Y et al (1996) Theory and practice of rock classification (in Chinese). Metallurgical Industry Press, Beijing

Chapter 6
Blasting Assessment Report

Rock excavation by blasting can adversely affect the stability and integrity of slopes, retaining walls, structures, buildings, services, and utilities through ground vibration and other effects. The transport, storage, and use of explosives for blasting also pose a safety hazard to the public. A blasting assessment should be undertaken in the design stage to assess such adverse effects and hazards and to propose measures to demonstrate the practicality of safely carrying out of blasting works. In this chapter, we will discuss the contents of the Blasting Assessment Report (BAR) and how to prepare it according to the practice in Hong Kong. As an Annex, the "Contents of a Blasting Assessment" given in the GEO Circular No. 27, Geotechnical Control of Blasting, issued by GEO, CEDD of HKSAR is attached at this Chapter for reference.

6.1 Desk Study

The desk study is a very important stage of preparing the Blasting Assessment Report (BAR). The desk study involves the collection and review of information required for the planning of the proposed blasting works and the assessment of any impact and risk to the surrounding area.

6.1.1 Gathering Information from Relevant Authorities

The Following information should be collected from local relevant authorities:

- Topographic maps at scales of 1:20,000 down to 1:500;
- Geological maps at a scale of 1:20,000 with memoirs (1:5000 maps are available for some areas);

D. Zou, *Theory and Technology of Rock Excavation for Civil Engineering*,
DOI 10.1007/978-981-10-1989-0_6

- Aerial photographs at various scales depending on the scale of the features being studied;
- Records of past works including ground investigation data and site history; and
- Laws and regulations relevant to the proposed project and blasting works.

6.1.2 Identification of the Area Which Will Be Affected by the Blasting Works

To carry out a blasting assessment, various sensitive receivers shall be identified surrounding the blast area. These sensitive receivers include slopes, retaining walls, natural terrain, boulders on natural terrain, structures, buildings, services, utilities, railways, roads, and any other installations and properties in the vicinity that are likely to be affected by the proposed blasting works through flyrock, ground vibration, air overpressure, and other effects generated by the blasts.

The commonly used guideline to identify the influence area by blasting is ground vibration.

For identifying the area of influence by blasting works, the maximum charge weight per delay which may be used for the project and the allowable PPV for the critical sensitive receiver should be known, then the distance between the blasting area to the boundary of influence area shall be assessed.

Usually, 100–300 m distance for the zone of influence by blasting is adopted for surface blasting projects in (or near) the urban area according to the scale of blasting works and the sensitivity of the vicinity area.

According to GEOGUIDE 4 issued by the Hong Kong Government [1], as a general guide, blast vibration from subsurface works are normally not potentially damaging at distance of more than 50 m and only exceptionally at distances of more than 100 m.

6.1.3 Contact and Collect Necessary Information from All Units Within the Affected Area by the Excavation Work

All necessary information should be collected from the following related parties within the zone of influence by blasting:

- Allowable PPV (and PD, if any), layout plan, service facilities, and special requirements from all public utilities, such as Water Supplies Department, Drainage Services Department, Electricity supply companies, Gas supply company, Telecommunication companies, Internet service companies, and other public and private utility units;

- Allowable PPV, based on school and kindergarten schedule of activities including entering and leaving time of students, exam date, and time;
- Allowable PPV and any special requirements from nearby hospitals, sanatorium, and nursing homes;
- Allowable PPV and special requirement, if any, from nearby buildings, residential estates, private houses, public or private structures, and facilities;
- Allowable PPV and special requirements from the relative companies or authorities of railways, subways (MTR), and roads;
- Allowable PPV and any special date that needs to be avoided from blast or any special requirements from the management of temples, religious architectures, and historical buildings or structures;
- Allowable PPV and special requirements from the related units or authorities of the dangerous goods store, such as explosive magazine, gas tanks, and oil stations;

6.1.4 Collect Necessary Information About Existing Geotechnical Features, Meteorological, and Hydrogeological Data

The information of existing geotechnical features, including slopes and retaining walls, and hydrogeological data can be collected from the relative department of the government, such as the Geotechnical Engineering Office in Hong Kong.

The meteorological information including the rainfall of the site area may be collected from the local meteorological department.

6.2 Field Investigation and Condition Survey

6.2.1 Investigation, Including Photography and Surveying, of the Current Condition of All Facilities Within the Zone of Influence by Blasting

All buildings, houses, structures, public utilities, railways, subways, and roads within the zone of influence by blasting should be investigated in situ, including photography and surveying, and record all information of current conditions of them.

6.2.2 Field Investigation of All Geotechnical and Geological Features Within the Zone of Influence by Blasting

All existing slopes, including man-made slopes and natural slopes, and other geotechnical features should be investigated in situ, including field mapping, surveying, photography, and sampling if necessary (Fig. 6.1).

All geological features, natural terrain, and boulders on natural terrain should be investigated in situ, including photography and surveying.

All retaining walls should be investigated in situ, including photography and surveying.

6.2.3 Records of Existing Defects

During the field investigation, attention must be taken on the existing defects on sensitive receivers, such as cracks on the wall of the buildings, houses, retailing walls, structures and facilities, potential damage or failures or instability of the slopes, any damage or defects or potential defects on any utilities and facilities. All the existing problems should be recorded in detail including photography and

Fig. 6.1 Boulders above natural slope

surveying. If it is necessary, a notarization should be carried out for some important features.

6.3 Analysis of the Potential Influence of Blasting Works to the Environment

6.3.1 Environment Impact Assessment (EIA) for the Transport and Storage of Explosives

The Report of Environment Impact Assessment (EIA) for the transport and storage of explosives is an independent report from the Blasting Assessment Report (BAR). Application together with the EIA report must be submitted to the environment protection authority for approval and issuance of the permit. According to the Chapter 499 of the Environment Impact Assessment Ordinance of Hong Kong Government, the assessment shall include the following.

The applicant shall investigate alternative construction method to avoid the use of explosives. If explosives to be used for the construction activities and the explosive storage or blasting location is in close proximity to populated areas and/or potentially hazardous installation sites, the applicant shall carry out hazard assessment as follows:

(1) Identify hazardous scenarios associated with the transport, storage, and use of explosives and then determine a set of relevant scenarios to be included in a quantitative risk assessment (QRA);
(2) Execute a QRA of the set of hazardous scenarios determined in (1), expressing population risks in both individual and societal terms;
(3) Compare individual and societal risks with the criteria for evaluating hazard to life stipulated in Annex 4 of the Technical Memorandum; and
(4) Identify and assess practically effective and cost-effective risk mitigation measures.

The methodology of hazard assessment shall be agreed and approved by the authority.

6.3.2 Potential Risk of Blasting Flyrock

Flyrock is the uncontrolled propelling of rock fragments produced in blasting and constitutes one of the main sources of material damage and harm to people.

The details of flyrock formation and control will be discussed in Chap. 11, Part II of this book.

If the surface blasting works is close to a public or residential area, a safety clearance distance for flyrock should be considered in the blasting assessment.

6.3.3 *Influence of Blasting Vibration*

6.3.3.1 Influence of Blasting Vibration and Its Prediction

Blasting vibration acceptance criteria depend on the type of structure, technical installation, and occupancy as well as the dominant frequency of the vibration. Among the concerning factors, peak particle velocity (PPV), due to its closer relation to damage potential (for residences) than acceleration (PPA) or displacement (PPD) for most cases, is generally accepted as the best component to assess the behavior of blast induced ground vibration on various installation.

Upper limits on blast vibration can be set using standard values which have been shown through experience to be acceptable. The safe vibration limits used in various countries vary but all follow the same pattern. A peak particle velocity of 50 or 70 mm/s is often the basic values. In Hong Kong, a value for peak velocity of 25 mm/s has been used for many years. The limits are maximum values normally acceptable for a building.

On the other hand, a peak particle vibration of 5 mm/s has usually been adopted as the lower bound limit which is used as the safe limit of the historical or religious constructions in Hong Kong.

The vibration that results from a blast may be calculated using a formula as follows:

$$\mathrm{PPV} = K(R/Q^{d})^{-b} \tag{6.1}$$

where

PPV Predicted peak particle velocity in mm/s;
K a constant which determined by the characters of ground that vibration passes through and affected by the explosive's power;
Q Maximum charge weight per delay interval in kg;
R Distance in m between the blast and the measuring point;
d Charge exponent, usually $d = 1/2$ for a cylindrical charge or $d = 1/3$ for a charge of symmetrical sphere;
b Attenuation exponent.

In the formula, K and b are the site-specific parameters related to local geological conditions and explosive strength and estimated by means of regressive analysis based on the collected data from trial blasts of the determined site. Table 6.1 gives some ranges of d, K, and b from different sources for reference.

Table 6.1 Ranges of constant *d*, *K* and *b* from different sources

Constants	USA	China	Mines division, Hong Kong
d	1/2	1/3	1/2
K	1140 for average value, 1725 for upper bound, and 4316 for heavy confined	50–150 hard rock, 150–250 moderate rock, 250–350 soft rock	644
b	1.6	1.3–1.5 hard rock, 1.5–1.8 moderate rock, 1.8–2.0 soft rock	1.22
Remark	From [7]	From [6]	From [14]

6.3.3.2 Analysis of Stability for Slopes and Other Geotechnical Works and Geological Features

A number of approaches may be used to assess the stability of rock and soil slopes subject to blasting vibration. These include the pseudo-static approach, dynamic analysis, reliability approach, the empirical approach, and the energy approach. There are also lots of computer codes to help the assessment work.

In Hong Kong, the Energy Approach for rock slopes and Pseudo-static Approach for soil slopes have been used as a guideline for many years and the method of assessment is published in the GEO Report No. 15, Assessment of Stability of Slopes Subjected to Blasting Vibration, by the Geotechnical Engineering Office, Civil Engineering Department in 1992 (reproduced from ref. [2]) by permission of GEO, CEDD, HKSAR Gov.

The Energy Approach tackles the problem by a consideration of the blasting vibration energy transmitted to the potential failure wedge (modeled as a rock block, see Fig. 6.2) resting on a rock slope, as well as the energy dissipation at the rock joint. The stability and the downslope displacement of rock block are assessed using equations based on the principle of conservation of energy. Different rock joint models may be incorporated in the analysis to calculate the energy dissipated at the joint. Analysis using the empirical rock joint formulae developed by Barton has been carried out for a range of situations (Fig. 6.3).

The peak particle velocity of the rock block is a key parameter in the Energy Approach. It is a good measure of the vibration energy induced by blasting. Therefore, application of PPV value in the analysis is likely to give more realistic predictions than the use of PPA's.

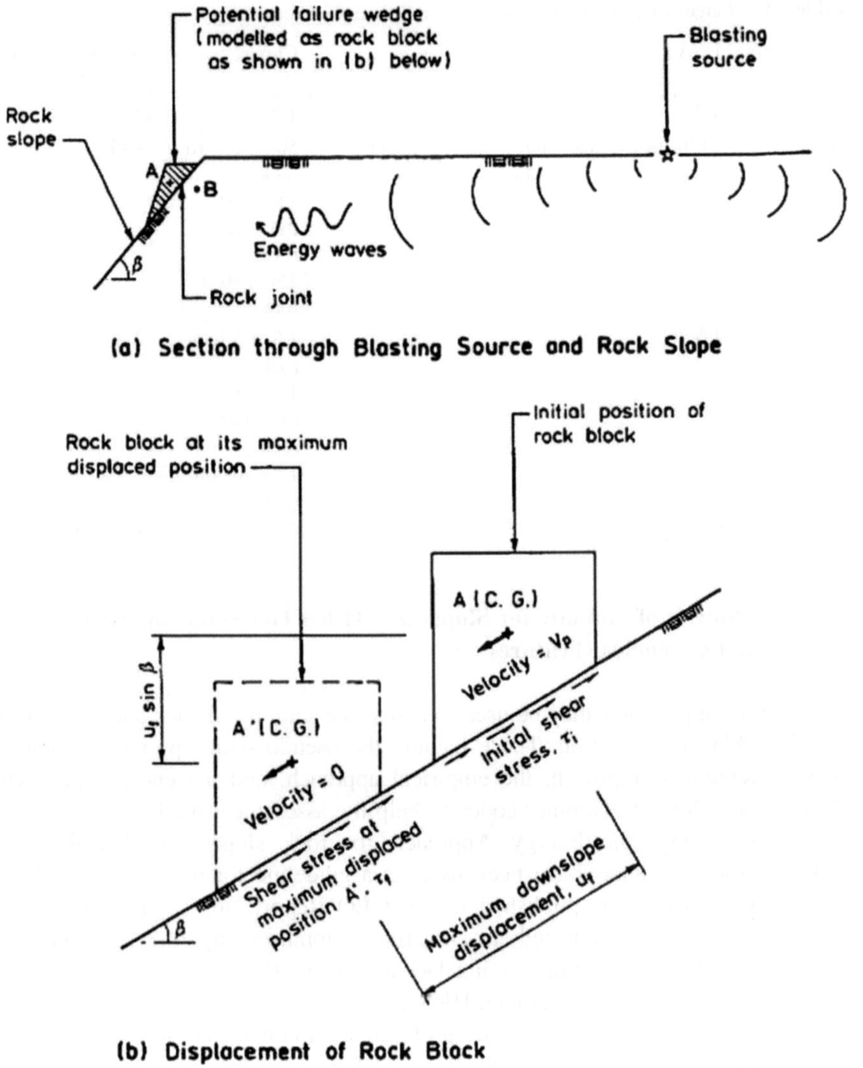

Fig. 6.2 Rock block system subject to blasting vibration (Reproduced from Ref. [2]) by permission of GEO, CEDD, HKSAR Gov.

When these approaches are applied for assessment, detailed field investigation of ground conditions and structural form is required, including the groundwater condition. Typical frequency of vibration adopted for assessment is 30 Hz.

For stability assessment of retaining wall subjected by blasting, the blast induced vibration is treated as a seismic load and the Mononobe–Okabe (M–O) Method and EUROCODE 8, Part 5 are often used.

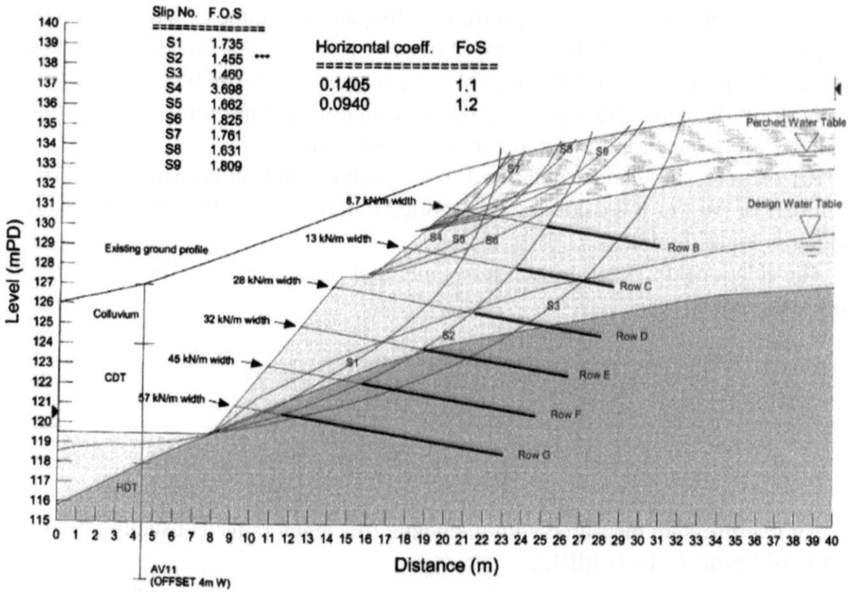

Fig. 6.3 An example of analysis graph for assessment of slope subjected to blasting vibration

Since the practices in Hong Kong shows that the assessment results for soil slopes using the Pseudo-static Approach are usually conservative, GEO issued a Technical Guidance Note No.28 (TGN 28), New Control Framework for Soil Slopes Subjected to Blasting vibration, in 2010 [3]. TGN 28 recommends that as an alternative to using GEO Report No.15, the PPV limit of 25 mm/s for the slopes that pose negligible risk to life under the condition of some risk control measures shall be implemented by the geotechnical site supervision staff (Reproduce from ref. [2]) by permission of GEO, CEDD, HKSAR Gov.

6.3.4 Influence of the Air Overpressure Produced by Blasting

Blast induced air overpressure (defined as pressure above normal atmospheric pressure) is the air pressure wave generated by an explosion. More commonly, this is termed as air blast. Air overpressure is considered to be one of the most detrimental side effects due to generation of noise and even causing some potential damages to residences.

Air blast or air overpressure (AOP) is an energy transmission in the form of pressure waves from the blast site within the atmosphere. The pressure waves consist of energy over a wide spectrum of frequencies, some of which are audible and hence may be sensed in the form of noise, but many are at inaudible frequencies of <20 Hz and beyond 20,000 Hz. These frequency ranges can be sensed by people in the form of a pressure impact known as concussion. The combination of effects of noise and concussion is known as air overpressure.

Air overpressure can be measured in two ways: either in pressure units (kPa or millibars or psi), or more commonly reported in terms of decibels (dB), the scale of which is logarithmic.

The relationship between pressure units and decibels can be presented by the following equations:

$$\mathrm{dB} = 20 \times \mathrm{Log}_{10}\left(\frac{\mathrm{kPa}}{6.9}\right) + 170.75 \tag{6.2}$$

$$\mathrm{psi} = 10^{\left(\frac{\mathrm{dB}-170.75}{20}\right)} \tag{6.3}$$

Air overpressure measured with the whole frequency range is usually reported in terms of linear decibels (dBL).

Noise measurements are usually made with standard sound meters that have built-in filters called "weighting scale". In the meters, only the air waves which fall within the frequency range of 20–20,000 Hz are measured. An "A-weighting dB" is a measure of noise pressure level, dBA, designed to reflect the acuity of the human ear. The A-weighted sound level is also called the noise level. The sound pressure level (SPL) can be expressed with the air overpressure as follow:

$$\mathrm{SPL(dBA)} = 20 \times \mathrm{Log}_{10}\left(\frac{P}{P_0}\right) \tag{6.4}$$

where

P is the measured overpressure in kPa or psi;

P_0 is a reference pressure level—the "threshold of hearing," representing the lowest sound audible, to the ear. $P_0 = 2 \times 10^{-8}$ kPa or 2.9×10^{-9} psi and represents a level of zero dB

Air blast characteristics are not easy to predict. Factors such as climate and topography intervene which, along with the actual blast design, can give different results in each case. Some researchers have found the air overpressure induced by blasting is proportional to $(D/W^{1/3})^{-1.45}$ [4], where D is the distance between the blasting charge and the point of observation and W is the maximum instantaneous charge weight per delay (MIC).

6.3.5 Arrangement of the Monitoring Points for All Sensitive Receivers

Within the area which will be influenced by the blasting works all sensitive receivers shall be identified. The monitoring points for blasting vibration and air overpressure shall be arranged for the sensitive receivers.

6.4 Proposals for Elimination or Controlling the Adverse Influences

6.4.1 Protective Measures for Prevent Blasting Flyrock

According to different circumstances one or two or three of the following protective measures should be applied for prevention of the flyrock in blasting:

(a) Ground covers for the horizontal rock surface to be blasted with wire mesh or mats;
(b) All the horizontal rock surface to be blasted is fully covered with blasting cages
(c) Vertical blasting screens (movable/fixed) are erected at the sides of the blasting area which are facing to the public or properties.

In some very special situation, a roof-over structure may be required to be built to prevent any flyrock from harming the public.

The details of the flyrock control and protective measures will be further discussed in Chap. 11, Part II of this book.

6.4.2 Limitation of Blasting Vibration for All Protective Objects and Its Controls

For protection of structures, buildings, and utilities from any damage by ground vibration generated in blasting, some restrictions of ground vibration by blasting are specified in some regulations or specifications. The following tables give some figures which were adapted in different countries for reference.

6.4.2.1 Limits of Blasting Vibration Usually Used in Hong Kong (Reproduced From ref. [5]) by Permission of GEO, CEDD, HKSAR Gov.)

(Table 6.2)

Table 6.2 Vibration limits commonly adopted in Hong Kong (OAP 2014) from [5]

Current Hong Kong vibration limits	PPV limits (mm/s)	PD limit (mm)
MTRC		
Railway structures/ permanent way	25	0.2
Q relays	40	0.2
Insulators and overhead lines	50	0.2
Overhead line mast	10	ns
WSD		
Water retaining structures/water tunnels	13	0.1
Water mains/pipes	25	0.2
Towngas		
Towngas transmission/distribution network		
All installations/pipes	25	0.2
Gas governors	13	0.1
Gas Tunnels	13	0.1
Off-take stations/above ground gas pipes	13	0.1
Towngas production facilities		
Gas holders	5	0.1
Naphtha tanks	5	0.1
Towngas production facilities (continued)		
Switch rooms/control room containing electrical	5	0.1
Switchgear (with protection relays)		
Gas production plant & corresponding facilities	5	0.1
Gas production plant columns and stacks	5	0.1
Water tanks & vessels etc.	5	0.1
Major gas and naphtha pipelines	15	0.2
DG Stores	15	0.2
Buildings	15	0.2
Boundary wall	15	0.2
CLP		
Power stations	6.28–11	0.02–0.1
Major substations	13	0.1
Minor substations	25	0.2
Underground cable joints	13	0.1
Underground cable and pylon foundations	25	0.2
Cable tunnel	13	0.1
HKE		
Transmission/distribution facilities (electrical protection equipment, transformer & HV/LV switchgear)	0.2 g	0.02
Transmission (275/132 kV) cable and joint	12 mm/s/0.223 g	0.1
Building structures of primary substation	0.07 g	ns

(continued)

Table 6.2 (continued)

Current Hong Kong vibration limits	PPV limits (mm/s)	PD limit (mm)
Submarine cable landing point and associated structures	0.07 g	ns
Pylon foundations	0.07 g	ns
Cable tunnel/ portal	0.07 g	ns
Power station building & the associated facilities	0.07 g	ns
DSD		
All structures	25	ns
Highways department		
All structures and road drains	25	0.2
Buildings		
Schools/residential buildings/private property	13–25	0.2
Historical buildings and monuments	5–10	0.2
Green concrete (GEO Report No. 102—time based) [15]		
≤4 h age	10	ns
6–8 h age	20	ns
10–12 h age	30	ns
18 h age	40	ns
24 h age	70	ns
3 days age	100	ns
7 days age	125	ns
>28 days age	150	ns
Green concrete (GEO Report No. 102—compressive strength based)		
3.0 MPa	77	ns
5.0 MPa	88	ns
10.0 MPa	106	ns
20.0 MPa	126	ns
30.0 MPa	140	ns
40.0 MPa	151	ns
50.0 MPa	160	ns
Hong Kong tramways		
All facilities	25	0.2
Shell		
LPG utilities/ equipment/ structures	5	–

6.4.2.2 Safety Limitation of Blasting Vibration in P. R. China [6]

(Table 6.3)

Table 6.3 Safety limitation of blasting vibration in China (GB6722-2014), (ref. [6])

No	Protective objects	Allowable PPV (cm/s)		
		$f \le 10$ Hz	10 Hz $< f \le 50$ Hz	$f > 50$ Hz
1	Loess cave dwelling, adobe house, and rubble house	0.15–0.45	0.45–0.9	0.9–1.5
2	General civil construction	1.5–2.0	2.0–2.5	2.5–3.0
3	Industrial and commercial building	2.5–3.5	3.5–4.5	4.2–5.0
4	General ancient buildings and monuments	0.1–0.2	0.2–0.3	0.3–0.5
5	Running hydropower stations and power plants and the equipment in the central control room	0.5–0.6	0.6–0.7	0.7–0.9
6	Water tunnel	7–8	8–10	10–15
7	Traffic tunnel	10–12	12–15	15–20
8	Mine's drift	15–18	18–25	20–30
9	Permanent high rock slopes	5–9	8–12	10–15
10	Green mass concrete (C20): age from batching			
	0–3 days	1.5– 2.0	2.0–2.5	2.5–3.0
	3–7 days	3.0–4.0	4.0–5.0	5.0–7.0
	7–28 days	7.0–8.0	8.0–10.0	10.0–12

Note 1. In the table, particle velocity is the maximum among the three components of the vibration; the vibration frequency is the frequency of predominant pulse

2. The frequency range is determined according to the site measured waveform or selected by the following data: chamber blasting $f < 20$ Hz; open deep hole blasting $f = 10$–60 Hz; open shallow hole blasting $f = 40$–100 Hz; underground deep hole blasting $f = 30$–100 Hz; and underground shallow hole blasting $f = 60$–300 Hz

3. Blasting vibration should be monitored with three orthogonal components of the particle motion

6.4.2.3 Blasting Vibration Regulation in USA

The only existing federal blast vibration Regulations are those of Office of Surface Mining (OSM) in the USA, which were published in 1983. The regulation allowed the choice of one method among three options. The methods 1 and 2 are shown in Table 6.4.

Table 6.4 US Code of Federal Regulations (CFR) 30 §715.19 and §816.67

Distance to blast area		OSM Method 1: Max. particle velocity		OSM Method 2: recommended scale distance (SD) when instrumentation is not available	
ft	m	in/s	mm/s	ft/lbs^2	m/kg^2
0–300	0–90	1.25	32	50	22.30
301–5000	90–1500	1.00	25	55	24.50
>5000	>1500	0.75	19	65	29.00

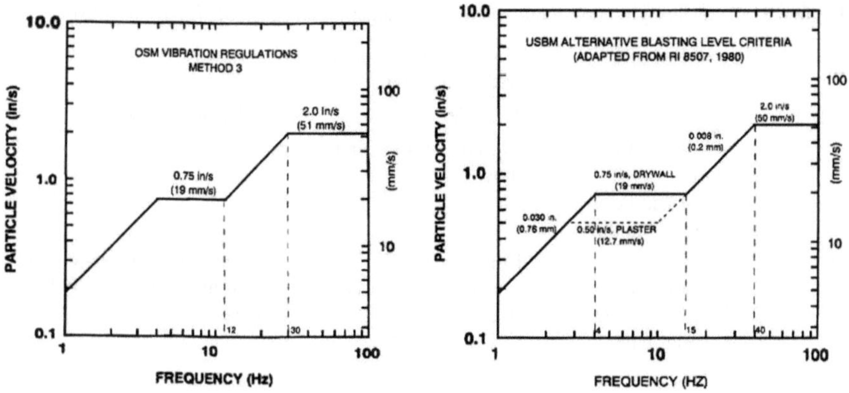

Fig. 6.4 OSM regulation and USBM recommended guideline (ref. [7])

In OSM method 3, OSM has prepared an alternate limit in graphical form, which allows the particle velocity to increase with an increase in frequency. This guide may be selected as an alternate to Method 1 or Method 2 but will require monitoring of particle velocity and frequency. See Fig. 6.4.

A similar graphical guideline was also published in RI 8507 of US Bureau of Mines in 1980 by Siskind, D.E. et al. as a recommendation. Although USBM did not have regulatory authority, however, some portion or all of the guideline has been incorporated into a number of state codes, regulations, ordinances, and project specifications. Basically, it consists of a recommended limit of 2.0 ips (50.8 mm/s) for frequencies above 40 Hz, 0.75, and 0.5 ips (19.1 and 12.7 mm/s) for low frequency vibrations, and transition lines to connect the two sets of data. The transition lines are equivalent to constant displacements. [7, 8].

6.4.2.4 Transient Vibration Guide Values for Cosmetic Damage in UK

In the UK, British Standard 7385: Part 2-1993 [9] is a definitive standard in terms of relevant blast vibration damage criteria for the prevention of structural damage, against which the likelihood of building damage from ground vibration can be assessed.

The guide values from this standard for transient vibration, judged to result in a minimal risk of cosmetic damage from residential buildings and industrial buildings, are presented in Table 6.5.

Table 6.5 Transient vibration guide values for cosmetic damage—BS 7385.2 (ref. [9])

Line	Type of building	Peak component particle velocity in frequency range of predominant pulse	
		4–15 Hz	15 Hz and above
1	Reinforced or framed structures Industrial and heavy commercial buildings	50 mm/s at 4 Hz and above	
2	Unreinforced or light framed structures Residential or light commercial type buildings	15 mm/s at 4 Hz increasing to 20 mm/s at 15 Hz	20 mm/s at 15 Hz increasing to 50 mm/s at 40 Hz and above

Note 1 Values referred to are at the base of the building
Note 2 For line 2, at frequencies below 4 Hz, a maximum displacement of 0.6 mm (zero to peak) should not be exceeded

Table 6.6 Recommended peak particle velocity—AS 2187.2 (ref. [10])

Type of building or structure	Peak particle velocity (mm/s)
Houses and low-rise residential buildings; commercial buildings not included below	10
Commercial and industrial buildings or structures of reinforced concrete or steel construction	25

6.4.2.5 Blasting Vibration Criteria Recommended in Australia

The vibration velocity "damage" criteria recommended in the Australian Standards Explosives Code, AS 2178.2-1993 (refer to [10]), vary according to the type of building and are defined in terms of peak particle velocity (PPV), as shown in Table 6.6.

The standard goes on to say that the likelihood of damage in residential areas starts to increase at ground vibration levels above 10 mm/s (peak particle velocity). Structures which may be particularly susceptible to ground vibration should be examined on an individual basis. Peak particle velocity may not be the appropriate criterion for determination of damage. In the absence of a particular site-specific study which may determine the appropriate damage criterion, then peak particle velocity is suggested as the damage criterion and a maximum level of 5 mm/s is recommended for blast design purposes, as experience has shown that damage is unlikely to occur at a ground vibration level below this level.

6.4.2.6 Blasting Vibration Criteria Recommended in Scandinavian Countries (Reproduced from ref. [11] by Permission of SANDVIK)

The blasting vibration criteria in all countries, Finland, Sweden, and Norway, are similar. The Finnish limit values for peak particle velocity can be calculated accordingly:

$$\mathrm{PPV} = F_k \times v_1 \tag{6.5}$$

where

F_k Structural coefficient (Table 6.7);
v_1 Peak particle velocity as a function of the distance (R) for structures and buildings that have been founded on different materials/vertical component

Table 6.7 shows a categorization of seven classes for several structures and buildings. The structural coefficients F_k of curing concrete are also given in the table and are based on past experience as well as a series of tests. The F_k values given for electrical cables and pipelines, as well as for rock masses, are quite certain based on experience. Table 6.8 shows the values for peak particle velocity (v_1) as a function of the distance (R) for structures and buildings that have been founded on different materials, based on the fact that the dominating frequency is lower when the distance is greater. The recommended peak particle velocities reduce with increasing distance, as the low frequencies are more dangerous for structures than high frequencies.

When there is a great distance between the blasting location and the structure (approx. over 50–70 m), the limit v_1 values are conservative. In certain cases, more

Table 6.7 Structural coefficients F_k ($V = F_k \times V_1$, where $V_1 = Fd(v_0)$ (ref. [11])

Structural categories (structures in good condition)	Structural coefficient F_k
1. Heavy structures such as bridge and piers.	2.00[a]
2. Concrete and steel building, rock caverns with shotcrete reinforcement	1.50[a]
3. Office and commercial buildings of brick and concrete. Wood-frame houses on concrete or stone foundation	1.20[a]
4. Brick and concrete residential buildings with no light concrete, limestone-sand brick etc. Rock caverns with no shotcrete reinforcement Curing concrete >7 days old[a], Electrical cables etc.	1.00
5. Building with light concrete structures. Curing concrete 3–7 days old[a]	0.75
6. Very vibration sensitive buildings, such as museum, churches, and other buildings with high vaults and great spans. Buildings of limestone-sand bricks. Curing concrete <3 days old[a]	0.65
7. Old historical buildings at the point of collapse such as ruins.	0.50

[a] Values over one (1) are permitted only when a blasting or vibration specialist is present

Table 6.8 Permitted peak particle velocity (vertical component) V_1 ($V_1 = Fd(v_0)$ as a function of distance (R) for structures and buildings founded on different materials (the structural coefficient is $F_k = 1$) (ref. [11])

Distance R (m)	Soft moraine, sand, shingle, and clay	Moraine, slate, soft limestone, and soft sandstone	Granite, gneiss, quartzite, hard sandstone, hard limestone, and diabase
	Wave velocities c		
	1000–1500 m/s	2000–3000 m/s	4500–6000 m/s
	Permitted peak particle velocity v_1 (mm/s)		
1	18	35	140
5	18	35	85
10	18	35	70
20	15	28	55
30	14	25	45
50	12	21	38
100	10	17	28
200	9	14	22
500	7	11	15
1000	6	9	12
2000	5	7	9

economical blasting can be achieved by measuring all the three components plus the time history of the vibration to carry out the frequency analysis. This allows for use of higher limit values.

If the structure is not a residential building with a structural coefficient of 1, the permissible limit value for peak particle velocity should be calculated with the equation:

$$\mathrm{PPV} = F_k \times v_2 \tag{6.6}$$

where

F_k Structural coefficient;
v_2 particle velocity versus frequency, $F_d(v_0)$, Fig. 6.5

6.4.2.7 Blasting Vibration Criteria Recommended in Germany (Reproduce from Ref. [11] by Permission of SANDVIK)

According to German Standard, the criteria is listed for three classes of structures and triaxial measurements for frequencies under 10 Hz, from 10 to 50 Hz, 50 to 100 Hz, and over 100 Hz (Table 6.9; Fig. 6.6). Measurements should be taken in the foundation and on the roof of the building.

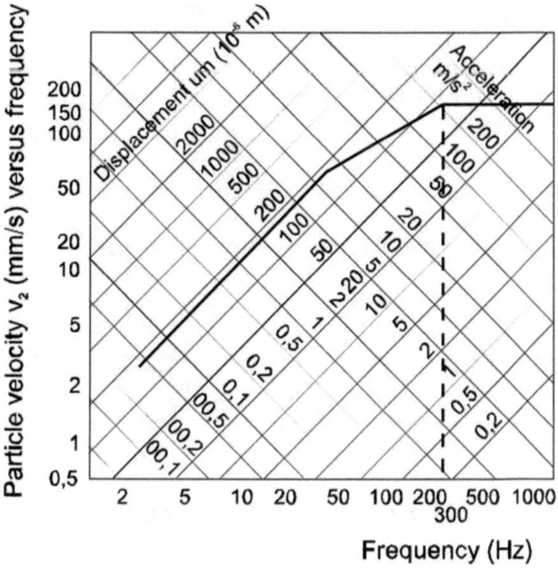

Fig. 6.5 particle velocity v_2 (mm/s) versus f (Hz) (Finnish proposal) (Reproduced from Ref. [11] by permission of Sandvik)

Table 6.9 Criteria for particle velocity v based on effects of short term vibration (ref. [11])

Line	Type of structure	Criteria for particle velocity v in mm/s			
		Base			Top of the highest floor
		Frequencies			
		<10 Hz	10–50 Hz	50–100 Hz[a]	All frequencies
1	Commercial and Industrial structures	20	20–40	40–50	40
2	Residences (also under construction)	5	5–15	15–20	15
3	Structures which are especially vibration sensitive (for example historical buildings)	3	3–8	8–10	8

[a] For frequencies over 100 Hz should at least criteria for 100 Hz be used

The methods for reducing the effects of ground vibration induced by surface and underground blasts will be discussed in Chap. 11 in Part II and Chap. 22 in Part III of this book, respectively.

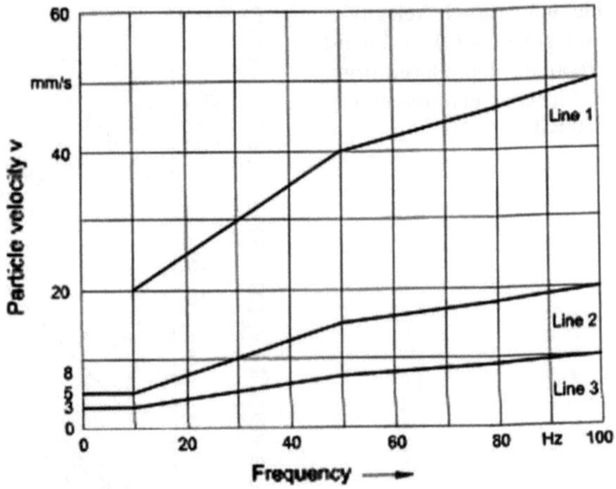

Fig. 6.6 Vibration in Structures, DIN 4150, part 3/11/1986, Germany (Reproduce from Ref. [11] by permission of Sandvik)

6.4.3 *Limitation of Air Overpressure*

Air overpressure (or airblast) can cause discomfort to persons and, in some cases, damage to structures. There are two kinds of limitation or guideline in various countries for air overpressure induced from blasting. One kind of limitation or guideline is used for controlling structural damage and another for controlling the noise induced from blasting. The current legislation, standards, guidelines, and criteria of some countries are introduced below.

6.4.3.1 Limits and Guidelines Specified in USA [7]

For Construction and quarry blasting specifications, there has been a long history of using 140 dB as an overpressure limitation to avoid structural damage. In recent years, it has become more common to make use of more restrictive limitations that were developed for surface mining operation and to apply them to all forms of blasting. A common overpressure limitation is the 134 dB recommended in Bureau of Mines RI 8485 in 1980. The overpressure of 134 dB is one-half of the overpressure of 140 dB that has served previously as a long-term common standard. In Fact, neither of these has been shown to cause windows breakage or structural damage. Table 6.10 shows the overpressure limits recommended by USBM RI 8485.

Table 6.10 Overpressure limits recommended by Bureau of mines for surface mining (USBM RI 8485)

134 dB—0.1 Hz High pass measuring system
133 dB—2.0 Hz High pass measuring system
129 dB—6.0 Hz High pass measuring system
105 dB—C-weighting scale on a sound level meter[a] (events less than or equal to 2-s duration)

[a] C-weighted scale is the least sensitive at low frequency

6.4.3.2 Limits and Guidelines Specified in Safety Regulations for Blasting of P.R. China, GB 6722-2014 (Refer to [6])

In Sect. 13.3 of the regulations, the limits and guidelines for the safety of human and structure for blasting induced air overpressure are given below.

- When the explosives charge of one blast do not exceed 25 kg in surface blasting, the safe distance between the charge and the shelter which is a safe protection from air overpressure for operating workers can be calculated using the following formula:

$$R_k = 25\sqrt[3]{Q} \tag{6.7}$$

where:

R_k Distance between the charge and the shelter, m;
Q Explosive charge quantity equivalent to TNT in one blast, taking the maximum quantity per delay when second delay is used or the total charge quantity of this blast when ms delay is used, kg;

- The allowable limit of blasting overpressure is 0.02×10^5 Pa for non-operating workers and 0.01×10^5 Pa for the operating workers in the shelter.
- The relationship between the extents of damage of the structure due to air overpressure is list in Table 6.11.

In the Sect. 13.4 of the regulations, the control standards of the blasting noise are given in Table 6.12 below.

6.4.3.3 Limits and Guidelines Specified in Australia and New Zealand

Item 10.2 together with the item J3.3 of Appendix J, Ground Vibration and Airblast, of Australia Standard, AS 2187.2-2006 specified the guidelines of airblast:

> Airblast can cause discomfort to person and in some cases damage to structures. Appropriate limit levels for airblast for local conditions may be required by the relevant authority. A limit of 120 dB for human discomfort is commonly used and 133 dB to avoid structural damage is generally appropriate.

The criteria normally recommended for blasting in Australia and New Zealand, based on human comfort, are contained in the Australian and New Zealand Environment Council (ANZEC) guidelines. The ANZEC for the control of blasting impact at residences are [12]:

- The recommended maximum level for airblast is 115 dB Linear.
- The level of 115 dB Linear may be exceeded on up to 5 % of the total number of blasts over a period of 12 months. However, the level should not exceed 120 dB Linear at any time.

Table 6.11 The relationship between the degree of damage of the structure due to air overpressure [6]

Class of damage	1	2	3	4	5	6	7
Damage degree	Basically no damage	Very light damage	Light damage	Moderate damage	Serious damage	Very serious damage	Complete damage
Overpressure $\times 10^5$Pa	<0.02	0.02–0.09	0.09–0.25	0.25–0.40	0.40–0.55	0.55–0.76	>0.76
The extent of damage to buildings							
Glass	Occasional damage	A small part was broken chunks, most were small	Most were broken into small pieces to smash	Smash	–	–	–
Wooden doors and windows	No damage	Small amount of casement damage	Lot of casement damage, doors, and window frames destruction	Casement falling and inverted, window frames, doors extensive damage	Doors, casements destroyed, casement falling	–	–
Brick out wall	No damage	No damage	Small cracks appear, the width is less than 5 mm, slightly inclined	Larger cracks appear, width 5–50 mm, obviously tilt, small cracks appear in brick stack	Large cracks appear, larger than 50 mm, the heavily skewed, larger cracks appear in brick stack	Partial collapse	Most or all of collapse

(continued)

Table 6.11 (continued)

Class of damage	1	2	3	4	5	6	7
Wooden roof	No damage	No damage	Wooden roof deformation, fracture occasionally	Wooden roof and purlins fractured, wooden frame support loosen	Wooden purlins broken, timber roof bars occasionally broken, bearing dislocation	Partial collapse	Fully collapsed
Tile roof	No damage	Small amount of movement	Large amount of Movement	Most movement, even tilt	–	–	–
Reinforced concrete roof	No damage	No damage	No damage	Small cracks occur less than 1 mm	1–2 mm wide cracks appear, can continue to use after repair	Cracks greater than 2 mm appear	Load-bearing brick walls collapsed, reinforced concrete load-bearing columns severely damaged
Ceiling	No damage	Small amount of plaster fall	Large amount of plaster fall	Wooden keel partially destroyed, sagging seam appear	Slump	–	–
Interior wall	No damage	Small amount of plaster from lath wall falling	Lots of plaster from lath wall falling	Small cracks appear on the brick interior wall	Large cracks appear on the brick interior wall	Serious cracks on the brick interior walls appear and even partially collapse	Most of brick interior walls collapse
Reinforced concrete column	No damage	No damage	No damage	No damage	No damage	Tilt	Have a greater tilt

Table 6.12 Blast noise control standards (unit: dB(A)) [6]

Sound environment category	Corresponding region	Different time control standards	
		Day time	Night
0	Recuperation area, health care area with serious patients or living area. Farmed animals zone (hibernation)	65	55
1	Residential, general health, culture, education, scientific research and design, and the administrative offices which are needed to keep quiet	90	70
2	Commercial finance, fair trade as the main function, or residential, commercial, industrial mix, the need to maintain a quiet residential area. Noise-sensitive animals concentrated breeding areas, such as the chicken farm	100	80
3	The area of industrial production, warehousing and logistics as the main function, and need to prevent serious industrial noise impact on the surrounding environment	110	85
4	Border guard officers, non-noise-sensitive animals concentrated breeding areas, such as pig farms	120	90
Working Area	Mining, water conservancy, transportation, railway, infrastructure engineering and construction field, and explosion processing area	125	110

6.4.3.4 Limits and Guidelines Specified in Canada

The overpressure limits recommended BY USBM RI 8485 (Table 6.7 of this Chapter) are widely used in Canada, but in some provinces, the more strict limits are used in recent years.

The Ontario Ministry of the Environment uses a standard limit of 128 dB, reduced to 120 dB if routine monitoring is not undertaken. [13]

6.4.3.5 Limits and Guidelines Specified in Hong Kong

Hong Kong Government also applies the limit of 120 dBL as a target to restrict the blasting works in the urban area [4].

6.4.4 Equipment for Monitoring

The instrument for monitoring ground vibration and air overpressure should be listed in the BAR and conform with the specifications which are given in the

Table 6.13 Minimum general specification requirements, based on ISEE (2000) [5]

Frequency range	2–250 Hz, within zero to −3 dB of an ideal flat response
Accuracy	± 5 % or ± 0.5 mm/s, whichever is larger, between 4 and 125 Hz
Transducer density	<2.4 g/cc
Digital sampling	1000 samples/s or greater, per channel
Operating temperature	−12 to 49 °C

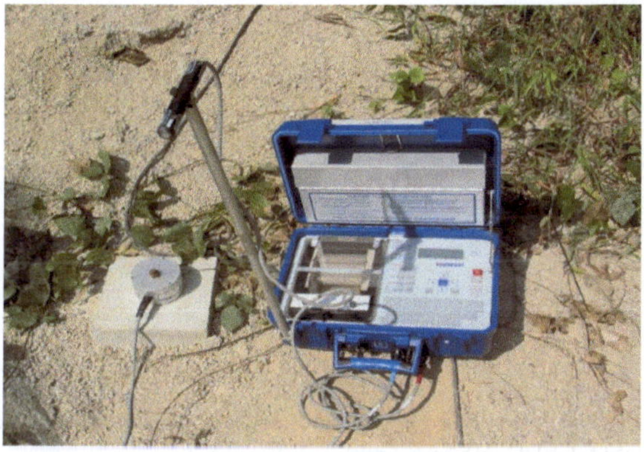

Fig. 6.7 Vibration and air overpressure monitoring

Appendix D of ISEE Blaster's Handbook (18th Edition), Blasting Seismograph Standard. The Mines Division of Hong Kong Government also published a Guidance Note on Vibration Monitoring which is quoted in Table 6.13 below. The Monitoring station and mounting details for horizontal geophones is also quoted in Figs. 6.7 and 6.8 (reproduced from ref. [5]) by permission of GEO, CEDD, HKSAR Gov.

6.4.5 Levels of Alert, Action, and Alarm of Blasting Vibration and AOP and Corresponding Requirements and Measurement

A list of the alert, action, and alarm including limits of ceasing works shall be specified in the BAR, including blasting vibration and air overpressure limits. The following tables of the AAA levels, control procedure, and requirements which were used successfully in Hong Kong are for reference only (Tables 6.14 and 6.15).

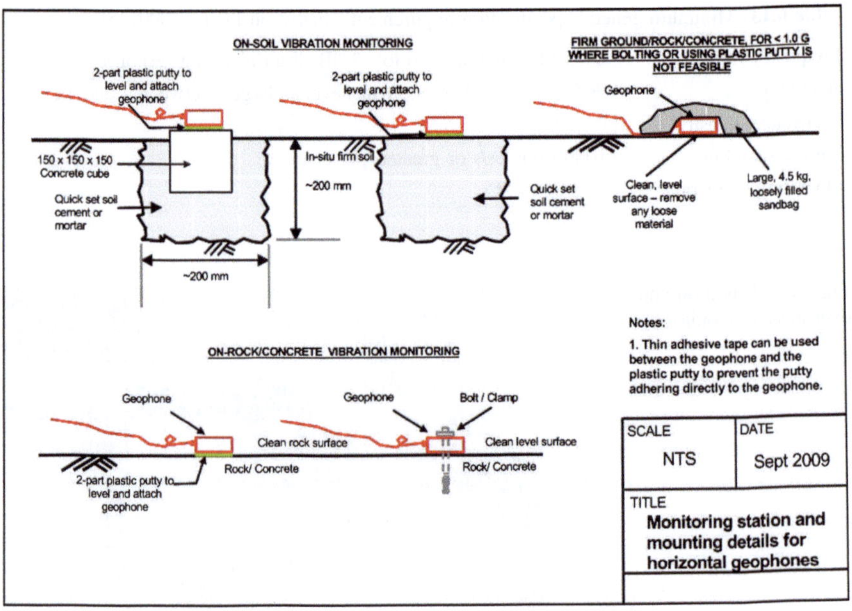

Fig. 6.8 Monitoring station and mounting details (Reproduced from Ref. [5]) by permission of GEO, CEDD, HKSAR Gov)

Table 6.14 Blast vibration control procedure

Control level	Sensitive receiver PPV limit	Control procedure/requirements
ALERT	PPVc90 %	• Inform the RE, Blasting Engineer, and SDO of CEDD, GEO(MD) of occurrence
		• Review monitoring data and recording equipment for accuracy
		• Check secure placement of transducers
		• Review Blast pattern and parameters
		• Inspect sensitive receivers affected
		• Prepare action plan
		• Continue Blasting
ACTION	PPVc95 %	• Inform the RE, Blasting Engineer, and SDO of CEDD, GEO(MD) immediately
		• Review/modify blast pattern and blast parameters
		• Review and implement action plan if necessary
		• Detailed inspection of all affected sensitive receivers
		• Plan and prepare remedial works if necessary
		• Resume blasting

(continued)

Table 6.14 (continued)

Control level	Sensitive receiver PPV limit	Control procedure/requirements
ALARM	PPVc100 %	• Report to the RE, Blasting Engineer, SDO of CEDD, GEO (MD) immediately
		• Assess the possible blast related reasons for the vibration exceeding/non-compliance
		• Immediate inspection of affected sensitive receivers
		• Implement action plan
		• Stop blasting if signs of instability/damage to sensitive receivers is observed
		• Undertake remedial work as necessary
		• Blasting Specialist to review and modify the blast design with the contractor to comply with the limits
		• Modify the blast vibration assessment parameters to reduce the PPV and estimated charge weight per delay
		• The engineering geologist shall review the blast area and the vibration monitoring area that exceeded the allowable limit and identify whether any geological feature was contributing the excessive vibration level
		• Seek approval to resume a period of trial blasting after the various investigation have been carried out

Table 6.15 Air overpressure control procedure

Control level	Sensitive receiver PPV limit (dBL)	Control procedure/requirements
ALERT	115	• Review monitoring data and recording equipment
		• Check placement of microphone
		• Check microphone is wind shielded
		• Measure background wind dBL level
		• Continue blasting
ACTION	118	• Carry out all Alert action
		• Review/modify blast pattern and blast direction
		• Prepare action plan
		• Resume blasting
ALARM	120	• Report occurrence to the RE, CEDD, GEO immediately
		• Carry out all Alert actions

(continued)

Table 6.15 (continued)

Control level	Sensitive receiver PPV limit (dBL)	Control procedure/requirements
		• Implement action plan if required
		• Stop blasting if public complaints are received
		• Review/modify blast parameters and blast direction
		• Seek approval to resume a period of trial blasting

6.5 Methodology of the Excavation by Blasting

6.5.1 Arrangement of Excavation Process

A working plan to show the excavation procedure and sequence, especially the blasting works for the project should be induced in the BAR. The time schedule of the blasting works including the period over which blasting will be carried out, number of blasts per day and per week, and the blasting time daily shall be included as well.

6.5.2 Working Scale, Excavation Method, and Main Equipment

The exact locations and areas that will be excavated by both methods of non-blast and blasting shall be shown in both plan and section drawings. The amount of rock to be blasted and the amount of explosives to be used per blast and per day shall be included in the BAR. Main equipment of drilling, charging explosives, moving, and installing protective measures (cages or screens, if any) for blasting works shall also be listed.

6.5.3 Explosives to Be Used in the Project and the Outline of Blasting Design

All explosive products to be used for the project shall be listed in the BAR. A proposal of the arrangement for delivery of explosives to the site and the safe location for parking and unloading explosives from the delivery vehicles shall be specified.

An outline of the blasting design shall be presented, which should include the range of blasting parameters (hole diameters, hole depth, powder factor, stemming, burden, and spacing), basic design of hole patterns, initiation methods, and arrangement of the protective measures. The BAR shall show that the proposed blasting design could be safely carried out and the proposed limits and any other constraints could be satisfied.

6.5.4 Security Measures, Scope of Safe Evacuation, and Evacuation Program

Proposals of protective and precautionary measures, i.e. flyrock prevention structures (blasting mats, cages, and blasting screens, if necessary for surface blasting and blasting door for underground blasting) shall be included in the BAR.

The scope of safe evacuation for public and site staff shall be outlined. The roads and other facilities which may need to be closed during blasting shall be considered. A detail program of evacuation, road closing proposal, and warning system shall be proposed by the contractor in the method statement of blasting. If roads are required to be closed, agreement for this must be obtained from relevant parties, such as the Transport Department or Police.

6.5.5 Assessment of the Feasibility of an On-Site Explosive Magazine, and Its Arrangement and Security Measures

If it is necessary to set up an on-site explosive magazine, an assessment report of its feasibility shall be proposed. An Environment Impact Assessment (EIA) Report may need to be submitted to the Environment Protection Authority for approval in advance of the BAR.

The drawings to indicate the location and arrangement of the magazine shall be included. The security measures and safe operation regulation and procedure shall be proposed as well.

Annex: Contents of a Blasting Assessment

(Quoted from "Geotechnical Engineering Office, GEO Circular No. 27, Geotechnical Control of Blasting", 13.7.2006. Reproduced by permission of Geo, CEDD, HK SAR Gov.)

(a) Site plans clearly indicating the proposed areas of blasting and locations of all sensitive receivers including streets, structures, foundations, railways, public utilities, water mains, drains, sewers, gas mains, and other services, geotechnical features such as slopes, retaining walls, boulders, tunnels, and caverns that may be damaged or destabilised by the proposed blasting works.
(b) A report containing the results of a study, including the site topography, geology, ground, groundwater and surface water conditions, and the physical site constraints, sensitive receivers, and site history.
(c) A report containing examination of the conditions of the sensitive receivers on and adjacent to the site.
(d) A report containing an assessment of the effects of blasting works to demonstrate that the proposed blasting would not cause any injury to persons or damage to property and sensitive receivers.
(e) Proposals of preventive measures to be carried out for sensitive receivers, if considered necessary.
(f) A list of the alert and cease works limits to be specified for the implementation of blasting works, including blasting vibration limits and air overpressure limits, to ensure that the blasting works to be carried out would not cause any injury to persons, damage to sensitive receivers, significant disruption to traffic, or undue nuisance to the public. The limits proposed shall take into account the existing conditions of all sensitive receivers. The source of the limits and documentary evidence of consultation and agreement, where appropriate, with the key stakeholders (e.g. owners or maintenance agents) of the sensitive receivers shall be provided.
(g) An outline of the blast design to demonstrate that the blasting works could be safely carried out and the proposed limits and any other constraints could be satisfied.
(h) A document setting out methods to be employed, working procedures and sequences for all blasting works, and the safety management system.
(i) Particulars of the site inspections, surveys, and monitoring to be carried out to check and measure the effects of blasting, including plans showing the locations of the monitoring stations, the performance criteria, and the alert and cease works limits.
(j) Proposals of protective and precautionary measures to be taken, including any evacuation and closure of public areas (such as roads and other facilities) and warnings needed to protect the sensitive receivers and the safety of the public and workers.
(k) Proposals of the arrangement for delivery of explosives to the site to demonstrate the practicability of completing the blasting works and the rock excavation needed within the construction period.
(l) If an on-site explosive store is considered necessary, a report containing an assessment of its feasibility and proposed arrangement.

References

1. Stomme B (1992) Guide to cavern engineering, GEOGIDE 4, Geotechnical Engineering Office, Civil Engineering Department, Hong Kong
2. Wong HN, Pang PL R (1992) Assessment of stability of slopes subjected to blasting vibration. GEO Report No.15. Geotechnical Engineering Office, Civil Engineering Department, Hong Kong
3. GEO Technical Guideline Note No. 28 (TGN 28) (2010) New control framework for soil slopes subjected to blasting vibration. Geotechnical Engineering Office, Civil Engineering and Development Department, HKSAR
4. Richards AB (2008) Prediction and control of air overpressure from blasting in Hong Kong. GEO Report No. 232. Geotechnical Engineering Office, Civil Engineering and Development Department, HKSAR
5. Guidance Note on Vibration Monitoring, Mines Division, Geotechnical Engineering Office, Civil Engineering and Development Department, HKSAR, January 2015
6. 爆破安全規程, GB6722-2014, 中華人民共和國國家質量監督檢驗檢疫总局发布, 2014-12-05 发布,2015-7-1实施. Safety Regulation for Blasting, GB6722-2011. Issued by The State Administration of Quality Supervision, Inspection and Quarantine, Issued on 5 December, 2014, Enforced from 1 July 2015
7. ISEE (1998) Blaster's handbook. 17th edn. ISBN: 1-892396-00-9
8. Siskind DE (2000) Vibrations from blasting, ISEE 2000, ISBN: 1-892396-11-4
9. BRITISH STANDARD, BS 7385-2 (1993) Evaluation and measurement for vibration in buildings—Part 2: Guide to damage levels from groundborne vibration, ISBN 0 580 22188 1
10. Australian Standard, AS 2187.2 (1993) Explosives—storage, transport and use, Part 2 Use of explosives, Standards Australia, ISBN: 0 7262 8229 4
11. SANDVIK & TAMROCK (1999) Rock excavation handbook for civil engineering. Editor in Chief: Matti Heiniö
12. Environment Australia (1998) Noise vibration and airblast control. Department of the Environment of Australia. ISBN 0642 545103
13. McCabe NB, Bonsma I (2014) Noise and vibration impact assessment. Wabush 3 Pit, West Labrador
14. Li UK, Ng SY (1992) Prediction of blast vibration and current practice of measurement in Hong Kong. In: Proceeding of the conference "Asia Pacific—quarying the Rim". Hong Kong
15. Kwan AKH, Lee PKK (2000) A study of effects of blasting vibration on green concrete. GEO Report No. 102. Geotechnical Engineering Office, Civil Engineering and Development Department, HKSAR

Part II
Surface Excavation

Chapter 7
Non-blasting Excavation

Non-blasting methods for rock excavation may be required in some situations:

- The excavation site is very close to the public area; any flyrock which may be caused by blasting is a major concern even if protective measures are adopted;
- There are some very sensitive receivers to ground vibration induced by blasting, like some important buildings or facilities, some monuments or some precise instrument, in the adjacent area of the excavation site;
- Nearby slopes or natural terrain where blasting may affect the stability or cause rockfalls which may threaten the safety of people or structures;
- Small quantity of rock excavation, so no need to apply for a blasting permit from the local authority, which may consume time and effort.

7.1 Manual Splitting and Hydraulic Rock Breaker

7.1.1 Manual Splitting

Manual splitting is usually used for small-scale rock excavation. Holes are first drilled along a line on the rock at a certain spacing. The distance depends upon the drillhole diameter and rock characters. Small holes (40–50 mm) with small spacing are used for small splitting tools and large holes (76–89 mm) with larger spacing are used for larger splitting tools. The tools which are used to split the rock consist of one to three pieces, and one or two of them have a wedge shape (see Fig. 7.1). The wedge splitter is pushed into the drillhole by hand hammer (8 or 12 lbs) for small hole splitting, or hydraulic hammer for large holes.

D. Zou, *Theory and Technology of Rock Excavation for Civil Engineering*,
DOI 10.1007/978-981-10-1989-0_7

Fig. 7.1 Spliters for manual splitting

7.1.2 *Hydraulic Hammer Breaker*

Hydraulic hammer breaker is a common method used for non-blasting excavation of rock. The main working components of a hydraulic hammer breaker include cylinder, piston, control valve, and chisel (breaking tool) (see Fig. 7.2). The impact energy has a wide range, from <500 J to more than 21,500 J, and can be selected according to the work purpose. The breaker body is usually installed on an excavator which works as a mobile carrier (see Fig. 7.3). Tool selection is an important productivity factor in rock breaking. A wedge-shaped chisel is used as the breaking device, so the sharpness of the wedge is essential for achieving a good breaking productivity. Sharpening the chisel is recommended in abrasive conditions.

The usual way to use a hydraulic hammer breaker is to have the excavator located under the bench. Bench height should be as high as can be excavated in one phase (3–5 m). Extremely high benches may collapse. Low benches (<1 m) are difficult to work, and the root section of a low bench is often the most difficult area to excavate (Fig. 7.3; [1]). Working on the top of high bench is dangerous, with the severe risk of the bench collapsing [1].

Sometimes a rock drill is used to help the excavation by hydraulic hammer beaker for creation of free face in the rock or to guide the direction of the development of cracks produced by the breaker.

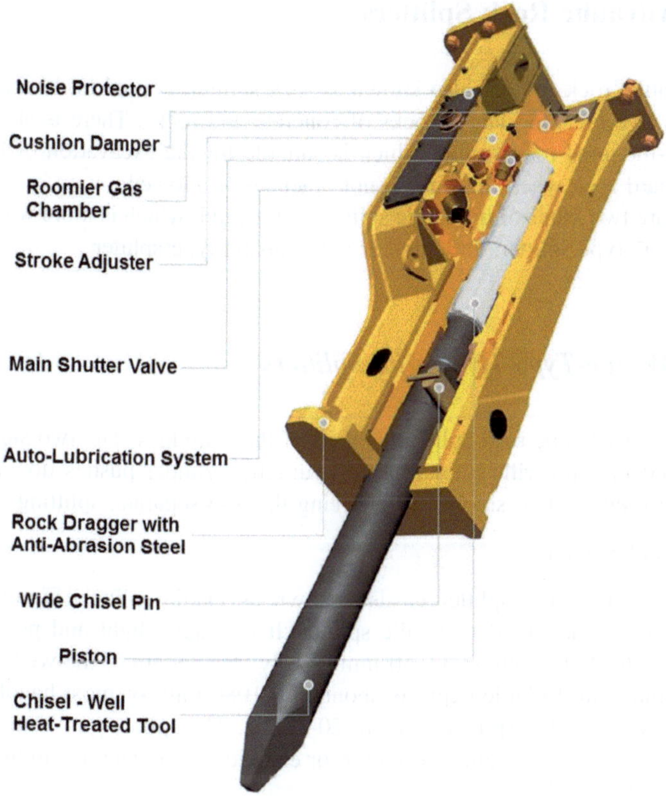

Fig. 7.2 Main components of a hydraulic rock breaker (reproduced with permission from Balavto Ltd. Ajdovščina)

Fig. 7.3 Root of the bench (*Arrow*) is the most difficult point of breaking with a hydraulic breaker. (reproduced from [1] with permission from SANDVIK)

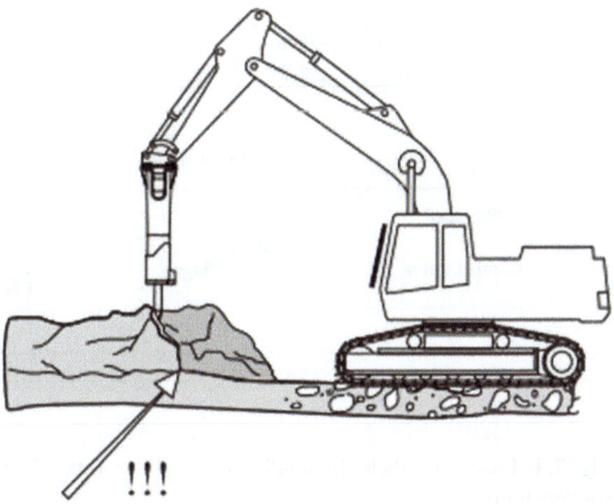

7.2 Hydraulic Rock Splitters

The hydraulic rock splitter, also known as rock splitter, is used in demolition jobs which involve breaking large blocks of concrete and rocks. There is also a larger excavator-mounted rock splitter which is suitable for the excavation of large volumes of hard rock where blasting is not practical or allowed.

There are two kinds of hydraulic splitters. One kind, which is more widely used, is the wedge-type splitter, and another is the piston-type splitter.

7.2.1 Wedge-Type Hydraulic Splitters

Wedge-type hydraulic rock splitters consist of three wedges. The two side-wedges are inserted in a predrilled hole, and a hydraulic cylinder pushes down a center wedge between the two side-wedges forcing them to separate, splitting the rock.

- Handheld Splitters

Handheld hydraulic splitter is also known as Darda splitter. Figure 7.4 is a photograph of a handheld hydraulic splitter. It is usually light and portable. The diameter of drillhole is about 32–50 mm usually. The length of the wedge is about 150–500 mm and drillhole depth is about 270–1080 mm for most handheld splitters. The weight of the splitter is about 20–31 kg. The splitter can be powered by gasoline engine, diesel engine, air motor, or electric motor. The maximum splitting force is about 220–500 tons.

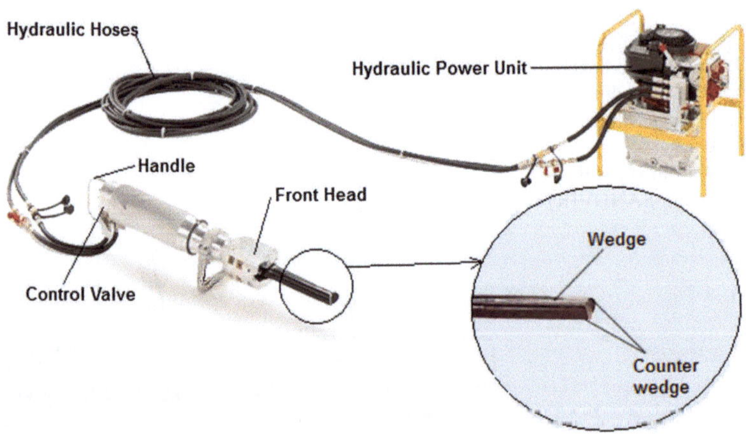

Fig. 7.4 Darda handheld hydraulic rock splitter (https://www.crowdersupply.com/darda-rock-splitters.htm)

Fig. 7.5 Multi-gun hydraulic rock splitter (courtesy of Wuxi Joyray International Corp.)

Fig. 7.6 Multi-gun hydraulic rock splitter is working (Courtesy of China Flying Technology Ltd.)

- Multi-guns Hydraulic Rock Splitters

For higher productivity of rock excavation, multi-guns of hydraulic splitter unit is usually used. Figure 7.5 shows an unit of the multi-guns splitters, and Fig. 7.6 shows how the splitters work.

- Excavator-Mounted Hydraulic Rock Splitter

A heavy hydraulic splitter can be mounted in an excavator. Figure 7.7 shows an excavator-mounted splitter breaking a large rock block. The diameter of the drill-hole for the splitter is 75–90 mm, and recommended hole depth is about 1300 mm. The theoretical maximum splitting force is about 24,000 KN.

Fig. 7.7 Excavator-mounted hydraulic splitter

7.2.2 *Piston-Type Hydraulic Rock Splitter*

Piston-type hydraulic splitter consists of a power unit and multiple piston-type splitting rods. In the power unit, the hydraulic pump is driven by a gasoline engine/diesel engine/air motor/electric motor. The size (diameter) of the piston-type splitting rod is 80–95 mm. The length of the splitting rods is variable according to the requirement of the splitting work but 700–800 mm of length is usually used. The diameter of drill hole is around 90–105 mm and depth is 1.0–2.5 m. The diameter and depth of drillholes depend on the size of the splitting rods used. Hole spacing in a row is about 400–600 mm depending on the strength of the rock. Figure 7.8 shows one of the products in the market. The rods are inserted into the drillholes and hydraulic pressure pushes all the pistons out from the rods, which fractures the rock with splitting force of 1000 tons up to 2500 tons within few minutes.

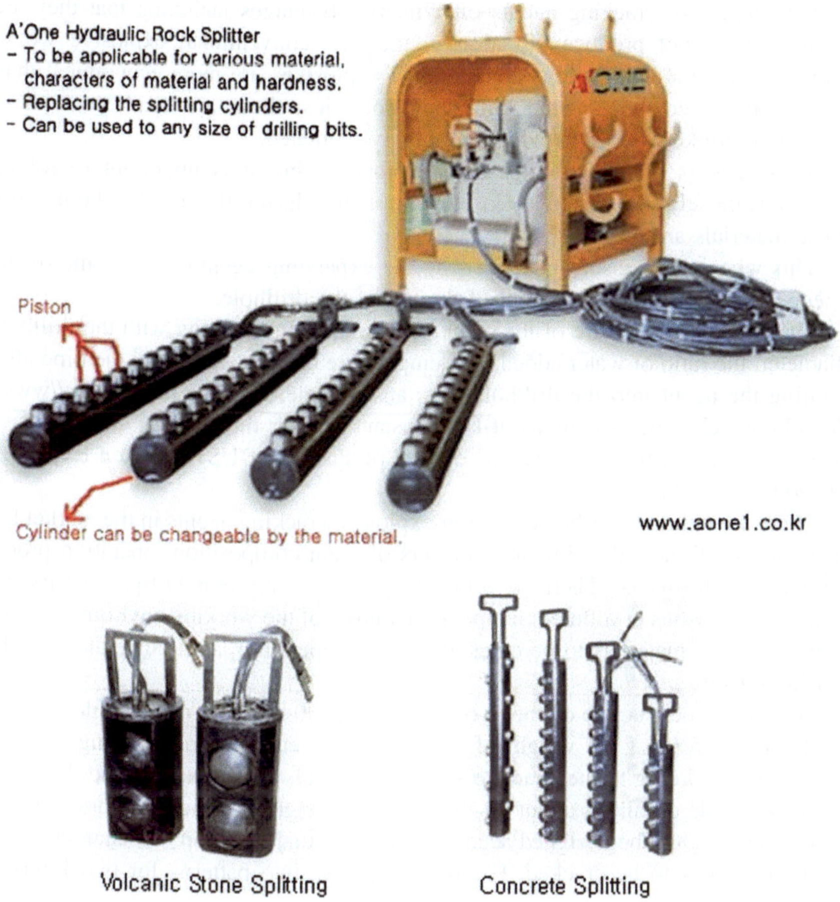

Fig. 7.8 Piston-type hydraulic splitter manufactured by A'One Machinery Co. Ltd. Korea

7.3 Non-explosive Cracking Agent

Non-explosive cracking agents are also called as chemical (demolition) cracking agent or chemical expanding agent or soundless (silent) cracking agent or expansive mortar. Non-explosive cracking agent is a powder product consisting mainly of calcium oxide and small quantities of additives like Al_2O_3, Fe_2O_3, SiO_2, and MgO. Non-explosive cracking agents are an alternative to explosives and gas pressure blasting products for demolition and rock excavation. To use non-explosive cracking agents, holes are drilled in the rock like they would be drilled for use with conventional explosives. A slurry mixture of the non-explosive cracking agent and water is poured into the drill holes. Over some hours, the slurry expands, cracking the rock in a pattern somewhat like the cracking that would occur from conventional explosives.

Non-explosive cracking agents offer many advantages including that they are silent and do not produce vibration in the way conventional explosive would although in some applications conventional explosives are more economical than non-explosive cracking agents. In many countries, these are available without restriction unlike explosives which are highly regulated.

These agents are much safer than explosives, but it is important to follow directions closely in order to avoid blown-out shot during the first few hours after these materials are placed.

This when the pressure generated by the expanding agent causes some of the agents to be violently ejected out of the top of the drillhole.

The expansive pressure of the cracking mortar changes along with the drillhole diameter, the ratio of water added, the temperature of rock/water, and the time after pouring the agent into the drillholes. As an example, Fig. 7.9 (from http://www.demolitiontechnologies.com/about-DTI-Bustar) shows the change in expansive pressure along with these factors of the product of BUSTAR Non-Explosive Demolition Agent.

There are hundreds of brands of non-explosive cracking agents in the market but they almost all have the similar characters of main composition, operation procedure, and performance. There are different types for every brand of product for use in different seasons or different temperature ranges of the working environment, i.e., temperature of materials to be cracked, water temperature, and temperature of the cracking agent.

For most products, the diameter of drillholes is 30–50 mm, ratio of added water is about 28–35 % of the weight of the cracking agent, drillhole spacing is about 30–70 cm according to the characters of the materials to be cracked, and the depth of the drill hole usually is about 80–105 % of the height of bench (or boulder). The hole pattern should be designed according to the actual situation and characteristics of the materials to be cracked. Figure 7.10 shows two patterns for two kinds of working situations.

When a non-explosive cracking agent is properly used within the parameters as noted in the Supplier's instruction for use, no ejection due to heat generation (blown-out shot) occurs, because of the mortar's strong adhesion and resistance to the upper surface of the hole. But all necessary protective measures like protective glasses, rubber gloves, and dust-proof mask should be equipped for all operation workers and all safe working procedures must be carried out strictly (Fig. 7.11).

7.4 Other Non-explosive Excavation Methods

To overcome the negative effects of blasting using high explosives, flyrock, ground vibration, and air overpressure, many attempts have been done in recent decades to develop non-explosive rock excavation techniques. These methods include

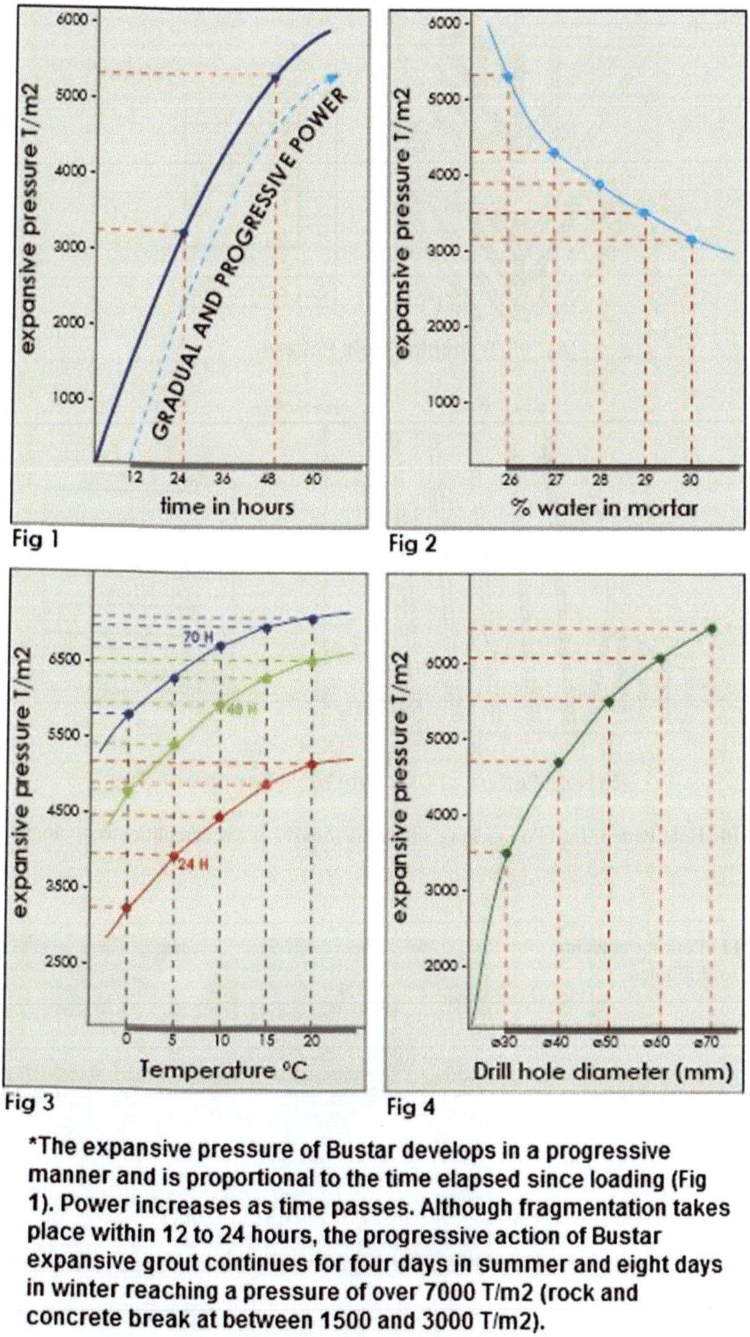

Fig. 7.9 Relationship between expansive pressure and the working parameters (*Source* www.demolitiontechnologies.com)

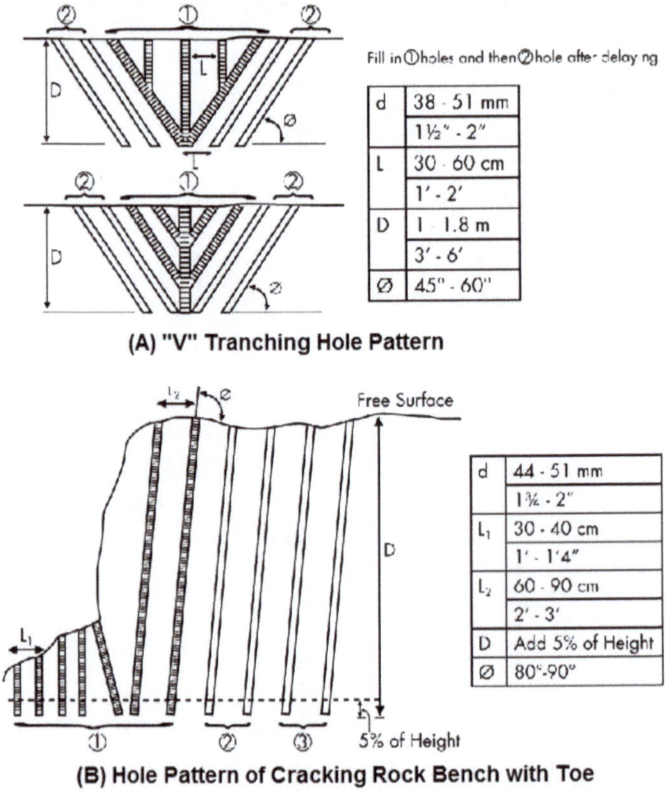

Fig. 7.10 Hole Pattern for two cracking situations (*Source* www.demolitiontechnologies.com)

Fig. 7.11 Pouring cracking mortar to drillholes

mechanical, hydraulic, gas pressure, and pulse plasma. But due to some disadvantages, such as high costs, complex equipment, low productivity, or still containing some low explosives, these techniques have not been widely used in rock excavation. Some techniques are briefly introduced below for information.

7.4.1 Controlled Foam Injection [2]

In the controlled foam injection (CFI) method, a viscous foam is injected into the bottom of a predrilled hole in the rock to be broken by means of a barrel incorporating an inexpensive and highly effective hole-bottom sealing method (Fig. 7.12).

The high-pressure foam penetrates into the fractures of the rock and breaks the rock. Pressure needed to break rock with CFI method are significantly less than are required in methods based upon the use of small explosive or propellant charges. The ability to tailor the viscosity and the stored gas energy of the foam to specific rock breakage characteristics results in highly controlled and predictable breakage. The general features of a device to use controlled foam injection (CFI) are illustrated in Fig. 7.13 (Refer to: [2]).

Fig. 7.12 Rock broken by the expansion of cracking agent

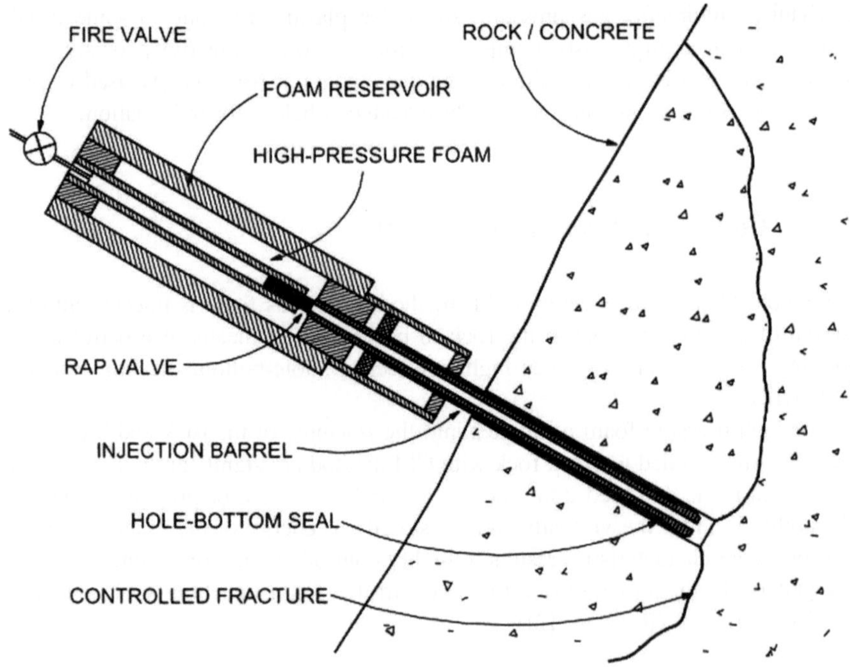

Fig. 7.13 Basic hardware and geometry for controlled form injection (CFI) fracture of rock (Ref. [2])

7.4.2 *Gas Pressure Rock-Breaking Products*

7.4.2.1 Cardox: CO_2 Non-detonating Expansion System [3]

Cardox has been used since the 1960s. Cardox CO_2 systems are comprised of reusable steel tubes filled with liquid carbon dioxide, which when energized by a small electrical charge from the firing head, instantly converts into gas. The CO_2 volume builds up until the pressure inside the tube causes a rupture disk at the end to burst and release the gas at a controlled pressure of up to 275 MPa (40,000 psi). These pressures build within milliseconds in the hole which is drilled in the rock and generates forces capable of breakage of the surrounding material along natural and/or induced fractures while heaving it away from the mass of rock. The structure of the cardox tube is shown in Fig. 7.14.

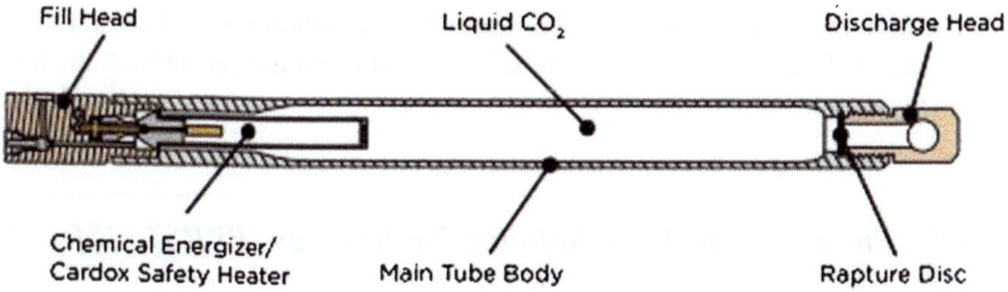

Fig. 7.14 Cut section of a cardox tube (*Source* www.cardox.com)

7.4.2.2 GasBlaster™: A Gas Pressure Blasting Product [4]

Gas pressure rock-breaking products using pyrotechnic charges have been around for many years.

The process replaces high-explosive methods with a slow-burn "deflagrating" low-explosive one that "heaves" the material apart in a much safer and gentler manner. This more benign process reduces environmental and safety concerns by significantly reducing fly rock, noise levels, shock wave, vibration, dust levels, nitrate gasses, and damage to the surrounding environment. Exclusion zones are also reduced.

GasBlaster™, which is patented in UK in 2012, is a sealed plastic tube with a solid high-tech plastic core centrally connecting the fixed (blue) end stops (Fig. 7.15). When the pyrotechnic charge (filled around the core) is ignited (via standard detonators), the initial deflagration pressure forces the two (red) high-tech plastic collars up the face of the (blue) conical retaining end stops. This action locks (self stems) the charge in the drill hole (now drillable at any angle and cartridge length only if required). Very quickly thereafter, the charge releases its full energy potential, breaking the core and bursting the sleeve to create the desired heaving effect to break rock.

GasBlaster™ can be used for rock excavation and demolition works. In particular beneficial for marine applications, the technology also reduces shockwaves, which helps safeguard marine life.

Fig. 7.15 Tube of GasBlaster™

The UN hazardous material Classification of GasBlaster™ is 1.4S, and it is necessary to obtain a transport, storage and application license/permit from the local authority in most countries.

7.4.3 Pulse Plasma Rock Splitting Technology (PPRST) [5]

Pulse plasma rock splitting technology (PPRST) is the technology application of plasma and particle accelerator science for rock-breaking operations. It is titled "Electro Power Impactor (EPI) Method." The core equipment used in this technology is an electro power impactor (EPI) unit that generates a high pulse electric energy. This energy is supplied to the electrolyte cells filled with powders of aluminum and copper oxide. In a millisecond, the electrolyte attains plasma state and generates high heat and impact wave (pulse). This heat wave gets transmitted to the surrounding rock bed and, in turn, breaks the rock with weak noise and vibration. The cracked rock, thus generated, can then be easily excavated using conventional methods Fig. 7.16.

The reaction of the powder in the cell is only a thermal process and no gas is generated:

$$2Al + 3CuO \rightarrow Al_2O_3 + 3Cu + 1.197\ kJ$$

Fig. 7.16 Overall of pulse plasma rock fragmentation technology (*Source* Kapra & Associates, Korea)

Kinds of the Cell

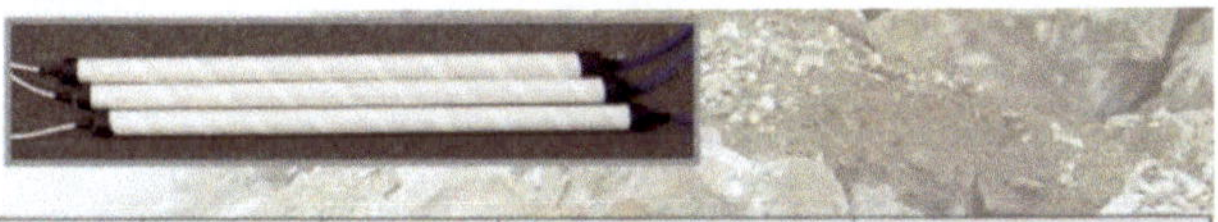

Diameter(mm)	Length	Weight	Energy	Hole (Light Rock)	Remarks
Cell : 34 (Hole : 51)	600mm	925g	3,784kJ	1.0m x 1.0m x 2.4m	Near-safety zone
	800mm	1,234g	5,048kJ	1.1m x 1.1m x 2.7m	General zone (Std)
	1000mm	1,542g	6,308kJ	1.3m x 1.3m x 3.0m	Far-safety zone

Fig. 7.17 Cells and their specification (*Source* Kapra & Associates, Korea)

The operation can be divided into three parts, including a generator for power supply, the EPI for storage and discharge, and a set of cells for plasma reaction. The generator can supply 20 kW of electricity to the EPI. The maximum storage and discharge amount of energy in the EPI is 268.92 kJ and 134,460 kW, respectively. After setting up the circuit, the operator uses a remote control for rock-breaking operation. Figure 7.16 is an overall of the technology, and Fig. 7.17 is the cell and their specification.

References

1. Matti H (1999) Rock excavation handbook for civil engineering. Sandvik & Tamrock
2. Young C (1999) Controlled-foam injection for hard rock excavation. In: Proceedings of the 37th US rock mechanics symposium. Vail
3. Cardox International Limited (2016) Brochures: civils. Cardox: UK Main page—for blockage clearance, build up clearance and quarrying. http://www.cardox.co.uk/uk_main.htm. Accessed 15 April 2016
4. Controlled Blasting Solutions Ltd (2016) The GasBlast™ technology. http://www.controlledblastingsolutions.com/product/. Accessed 15 April 2016
5. KAPRA & Associates (2016) Pulse plasma rock fragmentation technology. http://www.kapra.org/catalog.pdf. Accessed 15 April 2016

Chapter 8
Bench Blasting

Bench blasting is the method with which the rock excavation is carried out by drilling blastholes downward into rock from an upper working level, with a vertical or inclined rock face along one or more sides of the area to be blasted. It is the most common and basic method of surface rock excavation.

8.1 Geometrical Parameters of Bench Blasting

8.1.1 Blasthole Inclination, β

Both vertical blastholes and inclined blastholes can be used in bench blasting (see Fig. 8.1).

As the inclined holes are drilled in parallel with the front face of the bench, the holes of the first row have a uniform burden along the length of the blastholes. So blasting with inclined holes has the advantages of better fragmentation, displacement and swelling of the muck pile, and less backbreak to the newly formed slope behind the blast area. But they also have disadvantages such as increasing the deviation of drilling, lower drilling efficiency, increased drilling length, more wear on the drilling tools, and increasing the difficulty of explosives charging especially in blastholes with water. Inclined blastholes also increased the potential risk of flyrock especially in the case of rugged bench face as there is an upward component of rock movement.

Horizontal blastholes are very rarely used due to difficulty with drilling and charging as well as the greater potential risk of flyrock in surface rock blasting.

D. Zou, *Theory and Technology of Rock Excavation for Civil Engineering*,
DOI 10.1007/978-981-10-1989-0_8

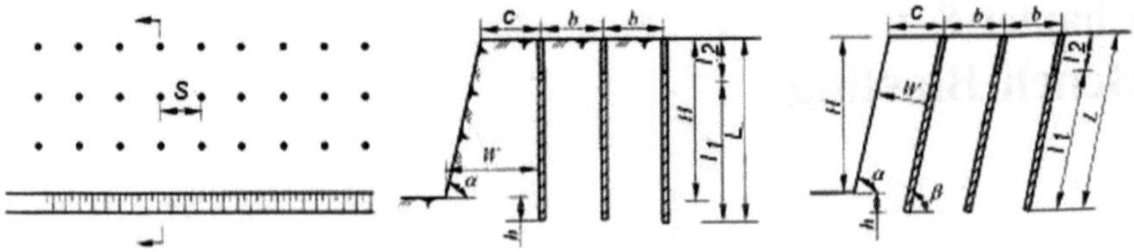

H - *Bench Height* *W* - *Burden of the Holes of First Row* *L* - *Depth of the Blastholes*
l_1 - *Charge Length* l_2 - *Stemming Length* *h* - *Subdrilling* α - *Angle of Bench Face*
β - *Clined Angle of Blastholes* *S* - *Spacing between Blastholes in a Row*
b - *Distance Between Rows* *c* - *Safe Distance between the bench edge and first row of Blastholes*

Fig. 8.1 Geometrical parameters of bench blasting

8.1.2 Blasthole Diameter, D

The diameters which are used for rock blasting in construction excavation projects usually fall within the range of 38–150 mm.

The diameter of blasthole is selected for a given excavation project depending upon the following factors:

- The blasting scale and productivity required of the excavation project.
  For small and shallow hole blasting, 38–50 mm hole diameters are used and 76–125 mm for moderate to large and deep hole blasting.
- Degree of fragmentation required to suit the loading and transportation equipment. The blasting fragmentation with small diameter blastholes is better as there is more optimum distribution of explosives in the rock mass being blasted than with large diameter holes.
- Height of bench and configuration of charges. Small diameter blastholes are usually used in lower bench height.

8.1.3 Height of Bench, H

The height of the bench usually depends on the scale of the excavation project and the equipment used for drilling and loading. Sometimes, it is also limited by the design requirements of the project. Usually, lower height of bench (equal or less than 5 m) is suitable for small blastholes and 10–15 m high bench is used for moderate to large-scale rock blasting.

In bench blasting, the ratio of the bench height to the burden, H/W in Fig. 8.1, is an important concept—the stiffness of the bench, and it has great influence on the results of blasting. When the stiffness of bench, H/W, is large, it is easy to displace and deform rock in front of the blastholes. If $H/W < 1$, the fragmentation will be bad and the rock body in front of the blastholes, especially in the lower part of the bench, is difficult to be pushed forward and may cause vertical flyrock. If $H/W = 1$, the fragmentation will be large, with overbreak and toe problems. When $H/W = 2$, these problems are attenuated, and are completely eliminated when $H/W =$ or >3.

8.1.4 Burden, W

Burden is one of the most critical parameters in bench blasting. The burden, W, is the minimum distance from the axis of a blasthole to the free face of the bench. But when vertical blastholes was used, the burden is measured usually from the toe of the bench to the axis of blasthole, see Fig. 8.1. The value of the burden is controlled by the rock blastability, the diameter of the blasthole, and the performance of explosive charged. There are numerous formulas that have been suggested to calculate the burden. Most formulas utilize charge volume, charge weight, or hole diameter as the basic parameter to calculate the burden. Some common formulas are introduced herewith for reference:

1. Langefors' (Sweden) Formula [1]

$$W_{\max} = \frac{d}{33}\sqrt{\frac{\rho \mathrm{PRP}}{\bar{c}f\left(\frac{S}{W}\right)}} \tag{8.1}$$

where

$W_{\max}$ Maximum burden (m);
d Diameter in the bottom of blasthole (mm);
ρ Loading density of explosive, (kg/dm^3);
PRP Weight strength of the explosive, for dynamite: PRP = 1.0, ANFO: PRP = 0.84, and TNT: PRP = 0.97;
$\bar{c}$ $\bar{c} = c + 0.75$ kg/m^3 for $W_{\max} = 1.4$–15.0 m and c is the rock constant, normally in surface blasts and with hard rock $c = 0.40$ kg/m^3;
f Degree of fixation, 1.0 for vertical holes and 0.95 for holes with inclination 3:1;
S/W Ratio of spacing and burden.

Langefors' formula can be simplified to [2]:

$$W_{\max} = 1.47\sqrt{l_b} \quad \text{for Dynamite;} \tag{8.2}$$

$$W_{\max} = 1.45\sqrt{l_b} \quad \text{for Emulsion cartridge;} \tag{8.3}$$

$$W_{\max} = 1.36\sqrt{l_b} \quad \text{for ANFO.} \tag{8.4}$$

where l_b is the required charge concentration (kg/m) of the selected explosive in the bottom part of the blasthole. Hole inclination is assumed to be 3:1, and the rock constant c is 0.4. Bench height $H \geq 2W_{\max}$. For other values of hole inclination and rock constant, $W_{\max}$ is corrected by factor R_1 and R_2 which are listed below:

Inclination	Vertical	10 : 1	5 : 1	3 : 1	2 : 1	1 : 1
R_1	0.95	0.96	0.98	1.00	1.03	1.10

Rock Constant c	0.3	0.4	0.5
R_2	1.15	1.00	0.90

In practice, actual operating burdens are frequently around 10 % less than the maximum values calculated using the above formula (8.1)–(8.4).

Considering the deviation in drilling, the practical burden can be calculated using the following formula:

$$W = W_{\max} - e' - d_b H \text{ (m)} \tag{8.5}$$

where

H bench height (m),
e' Collaring error (m/m), and
d_b Blasthole deviation (m).

2. Baron's Formula (Барон. Л. И.) (Former Soviet Union) [3]

For the vertical blastholes, Baron's formula of burden is:

$$W = d\sqrt{\frac{7.85\rho L\tau}{qmH}} \tag{8.6}$$

where

W Burden, m;
d Diameter of blastholes, dm;
ρ Loading density of explosive, (kg/dm^3);
L Depth of blastholes, m;
τ Charging percentage (height of charge/depth of blasthole) %;
q Designed powder factor, kg/m^3;
m Ratio of spacing and burden; and
H Bench height.

3. Tamrock's Method (Finand, [4])

Tamrock in its handbook of "Surface Drilling and Blasting," edited by Jukka Naapuri in 1987–1988, offered a method to estimate the value of burden. The author supposed that for given drillhole diameter, rock type, and blastability, there is an optimum burden (suitable fragmentation and toe conditions). The optimum burden is normally found to lie in the following range and depends in particular on the properties of the rock (Fig. 8.2).

W = 25–40d, where W in m and d in mm, or
W = 2.5–3.5d, where W in ft and d in inch.

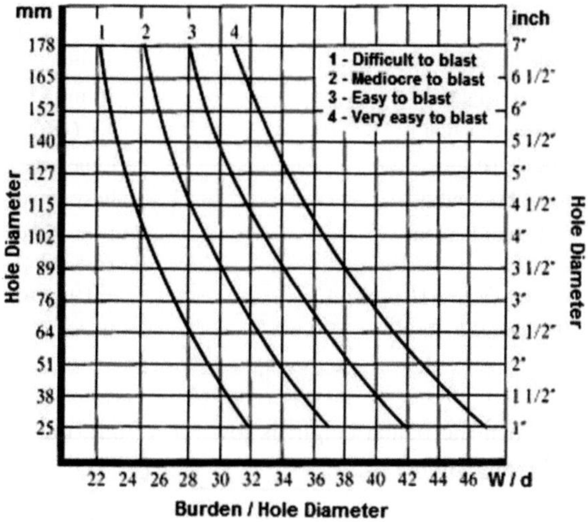

Fig. 8.2 Maximum burden/hole diameter ratio (*W*/*d*) for different hole diameters as a function of rock blastability. Hole inclination is 15:20° from vertical; for vertical holes decrease burden by 5–10 %. (Reproduce from Ref. [4] by permission of Sandvik)

The selected value of burden from Fig. 8.2 should be corrected with the factors given in Table 8.1.

Table 8.1 Correction factors for burden values (*Source* [4])

Bench height m	Hole diameter mm								
	45	51	64	76	89	102	115	127	152
4	0.92	0.85	0.73	0.66	–	–	–	–	–
5	1.00	0.97	0.86	0.77	–	–	–	–	–
6	0.97	1.0	0.98	0.89	0.91	0.89	–	–	–
7	0.95	0.97	1.00	0.98	0.94	0.93	–	–	–
8	0.92	0.95	0.98	1.00	0.97	0.96	0.93	0.93	0.93
9	–	0.93	0.96	0.98	0.98	0.99	0.96	0.96	0.96
10	–	0.90	0.94	0.96	1.00	0.99	0.98	0.98	0.98
11	–	0.88	0.92	0.93	0.98	1.00	0.99	0.99	0.99
12	–	0.85	0.90	0.91	0.97	1.00	1.00	1.00	1.00
13	–	–	0.88	0.89	0.95	0.99	0.99	1.00	1.00
14	–	–	0.84	0.87	0.94	0.97	0.98	1.00	1.00
15	–	–	0.82	0.86	0.92	0.97	0.97	0.99	0.99
16	–	–	–	0.84	0.91	0.96	0.97	0.98	0.98
17	–	–	–	0.82	0.89	0.94	0.96	0.97	0.98
18	–	–	–	–	0.87	0.94	0.95	0.96	0.97
19	–	–	–	–	0.86	0.93	0.94	0.95	0.96
20	–	–	–	–	0.84	0.93	0.94	0.94	0.95
21	–	–	–	–	–	0.92	0.93	0.93	0.94
22	–	–	–	–	–	0.90	0.91	0.93	0.93
23	–	–	–	–	–	–	0.89	0.91	0.92
24	–	–	–	–	–	–	0.88	0.90	0.91

Example	
Hole diameter	89 mm
Bench height	16 m
Hole inclination	3:1 (18°)
Rock type	Limestone
Rock blastability	Easy to very easy to blast
Optimum burden	$W = 36d$ (from Fig. 8.2) $W = 3.2$ m
Correct factor ($H = 16$ m)	=0.91 (from Table 8.1)
Burden ($H = 16$ m)	$W = 0.91 \times 3.2$ m = 2.9 m

When the bench height is low ($H < 70d$), the burden can be calculated by the following equation:

$$W = W_{\max} - 0.1 - 0.03H \tag{8.7}$$

where

$W_{\max}$ Maximum burden (m);
0.1 Collaring error (m); and
$0.03H$ Alignment error (m) and H—bench height, (m).

4. Bhandari's (India) Experimental Method for Optimum Burdens [5]

For a given charge, rock type, and spacing, there is an optimum burden (for which the volume of suitably fragmented and loosened rock is maximum and toe conditions are acceptable). Sushil Bhandari, an Indian professor, in his book "Engineering Rock Blasting Operations," divided it into two concepts: optimum breakage burden and optimum fragmentation burden.

At the optimum breakage burden, the volume or mass of rock broken is maximized, but the fragmentation obtained is not essentially acceptable as it has greater fines and some large boulders. This was shown by laboratory scale blasting where it was found that optimum fragmentation burden was 30–40 % less than the optimum breakage burden. At the optimum fragmentation burden, quantity of rock broken is less, but the fragmentation obtained is uniform and large boulders are absent. He also suggested that larger spacing at optimum fragmentation burden can be utilized as compared to that for optimum breakage burden; hence, there is no loss of total quantity of rock breakage.

Based on the Livingstone crater blasting theory (see Sect. 5.3.2 of Chap. 5), Bhandari suggested to use a modified crater blasting method called bench fragmentation method to get both optimum breakage burden and optimum fragmentation burden. In this method, separate single holes at different burdens with constant explosive charge (mass or volume) are blasted to a bench geometry. Preferably, the same explosive and hole diameter as in production blasting are used with same method of priming and blasting. After each blast, the results in terms of volume of breakage and fragmentation are recorded. From these results, the optimum breakage burden or the optimum fragmentation burden can be obtained.

5. Some Empirical Methods to Estimate the Burden

Tables 8.2 and 8.3 are some empirical methods to estimate the value of burden in bench blasting for reference.

8.1.5 Spacing, S

The distance between adjacent blastholes, measured perpendicular to the burden, is defined as the spacing. Usually, spacing of blastholes is calculated as a function of burden, but it is also affected by the initiation timing between blastholes and the initiation sequence.

When a small spacing is adopted, a crack may form between two adjacent holes and allow premature emission of explosive gases, producing large blocks in front of the blastholes and toe problems. But excessive spacing between blastholes may cause inadequate fracturing between holes, along with toe problems and an irregular face with overhang in the new bench face.

The spacing is selected according to widely held concept that since the break angle made by the charge to the bench face is near 90° (Fig. 8.3) thence spacing larger than two times the burden are not possible. So for decades, the spacing to burden ratio has been between 1 and 2.

Table 8.2 Design parameters in bench blasting with hole diameters *d* of 65–150 mm (Reproduce from Ref. [7] by Permission of Taylor & Francis Book UK)

Design parameter	Uniaxial compressive strength of rock (MPa)			
	Low <70	Medium 70–120	High 120–180	Very high >180
Burden—*W*	39*d*	37*d*	35*d*	33*d*
Spacing—*S*	51*d*	47*d*	43*d*	38*d*
Stemming—*l*2	35*d*	34*d*	32*d*	30*d*
Subdrilling—*h*	10*d*	11*d*	12*d*	12*d*

Table 8.3 Recommended burden/hole diameter values for bench blasting by USBM (Ref. [3])

Explosive	Rock	Density (g/cm[3])	*W*/*d*
ANFO 0.85 g/cm^3	Light	2.20	28
	Medium	2.70	25
	Heavy	3.20	23
Slurry/dynamite 1.20 g/cm^3	Light	2.00	33
	Medium	2.70	30
	Heavy	3.20	27

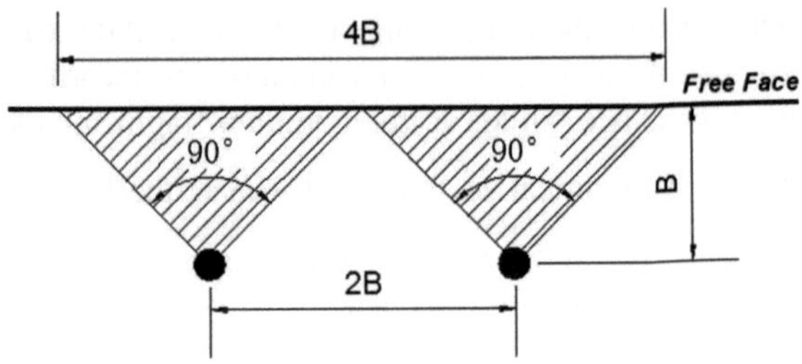

Fig. 8.3 Widely supposed break angle near 90° from blasthole to free face

In 1963, Langefors et al. demonstrated from laboratory model scale tests that ratio exceeding three for simultaneously fired charges with reduced burden in a single row gave better fragmentation. For the same model tests with multiple rows of charges fired together, but rows of holes delayed relatively resulted in good fragmentation up to spacing to burden ratio of eight. The method suggested has been approved by the practice of production blasting in surface mines. Hence, this technique became popular in early 1970s and is known as "Swedish Wide Spacing Technique."

Zou explored the mechanism of improving rock fragmentation using wide spacing blasting technique in his paper [6]. His study using the BMMC computer simulation model, which will be discussed in Sect. 12.2.4 of Chap. 12 of this book, indicated that:

In the wide spacing technique, along with the decrease of burden, the action of reflected strain waves is strengthened, hence changing the distribution of the energy of strain waves in the rock being blasted. It not only increases the energy density of strain waves in the range near the free face, but also makes the distribution of energy more uniform (see Fig. 8.4); therefore, rock fragmentation is effectively improved.

In practice of rock excavation by blasting for construction, as there are some differences from the mining and quarry blasting, the selection of blasthole spacing should be according to the purpose of blasting and requirement to fragmentation. Wide spacing with small burden can result in better fragmentation and looser muck pile, but also may be accompanied with toe problems in the middle between two adjacent holes and an irregular face in the new bench face and even increasing the potential risk of flyrock; hence, $S/W = 1.5$ to 2 and less than 3 is considered adequate if good fragmentation is required. Small spacing can improve the roughness of the newly formed bench surface, especially if the presplitting technique is used, but more large blocks may be produced. For normal rock blasting, $S/W = 1.25$ to 1.5, or selected referring to Table 8.2, is usually adopted. When blasting is adjacent to the final slope without presplitting, $S/W = 1$ to 1.2 should be considered to keep a relative smooth rock face of final slope. If large blocks of rock are required, S/W can be further reduced to 0.8

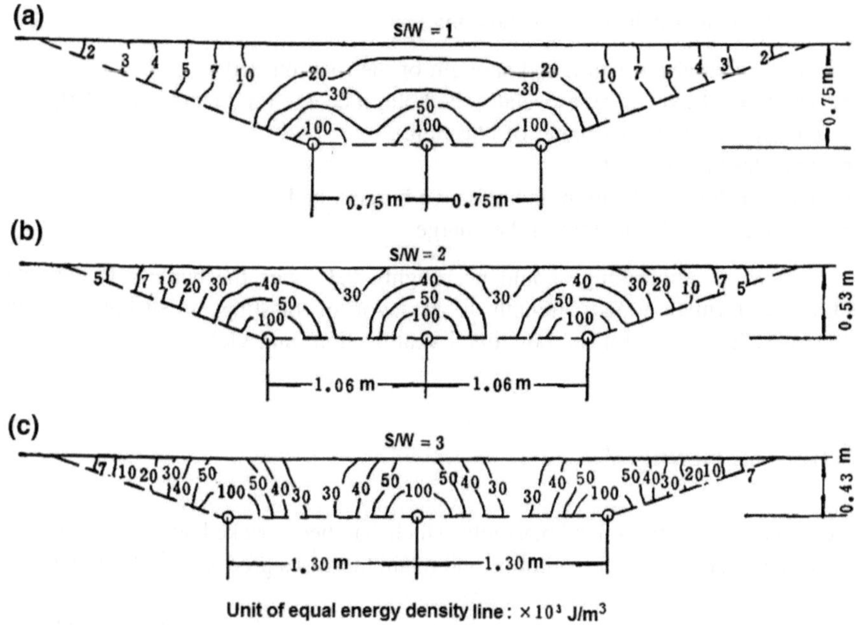

Fig. 8.4 Distribution of average energy density of strain energy on the middle plane section of the burden when S/W = 1, 2, 3. (*Source* [6])

8.1.6 Subdrilling, H

To avoid leaving a hump or toe in bench floor after blasting, the blastholes are drilled below the floor level. This is termed as subgrade drilling or subdrilling (lettered as h in Fig. 8.1).

If the subdrilling is small, the rock will not be completely sheared off at floor level, which will result in a hump or toe appearance and will increase the difficulty in excavating the muck pile. However, if the subdrilling is excessive, the following problems will occur:

- An increase in drilling and blasting costs;
- An increase in ground vibration level;
- Excessive breakage in the top part of the underlying bench, causing drilling problems, and even cause upward flyrock when blasting the underlying bench if the broken fragments on the bench top are not cleared; and
- Worse fragmentation in the upper part of bench as the charge is emphasized in the lower part of the blasthole if the height of charge column or powder factor is not increased.

An optimum subdrilling depends on:

- The structural formation and strength of the rock blasted;
- The type of explosive, especially the bottom charge (and more particularly, the energy generated per meter of blasthole),
- The blasthole diameter,
- The blasthole inclination, the effective burden, and
- The location of initiators in the charge.

In practice, subdrilling should be roughly 0.3 times the burden. If the rock structural formation and strength is in favor of blasting, subdrilling can be decreased (even to 0 for the horizontal stratification) and vice versa.

8.1.7 Stemming, L_2

Stemming is the portion of blasthole which has been packed with inert material above the charge in order to confine and retain the gases produced by the explosion, thus improving the fragmentation process.

The type and length of stemming have no significant effect on the characteristics of the explosion generated strain wave. But, a stemming column of suitable length and material can reduce premature venting of high pressure explosion gases to the atmosphere; hence, it always enhances fracture and displacement of bench rock by gas energy.

If insufficient stemming length or unsuitable stemming materials are used, it will cause noise, airblast, and danger of flyrock as the explosion gases prematurely vent to the atmosphere and also affect the rock fragmentation and displacement. On the other hand, if the stemming is excessive, there will be a large quantity of boulders produced from the top part of the bench, poor swelling of the muck pile, and increasing ground vibration.

An optimum stemming depends on both suitable length of stemming and suitable stemming materials.

In practice, the optimum stemming length varies between $25d$ to $40d$ (or $0.75W$ to $1.2W$) but should consider the following factors:

- Rock characteristics, the stemming length should increase as the quality and competence of the rock decrease;
- In multiple row blasts, special care should be taken when stemming the front row blastholes, especially when bench face irregularities are present or vertical holes are used, as they cause great differences in burden dimensions from top to toe of the bench. If the mean direction of rock movement tends more and more toward the vertical with successive rows, a longer stemming column is often used in the last row to reduce overbreak (Fig. 8.5); and

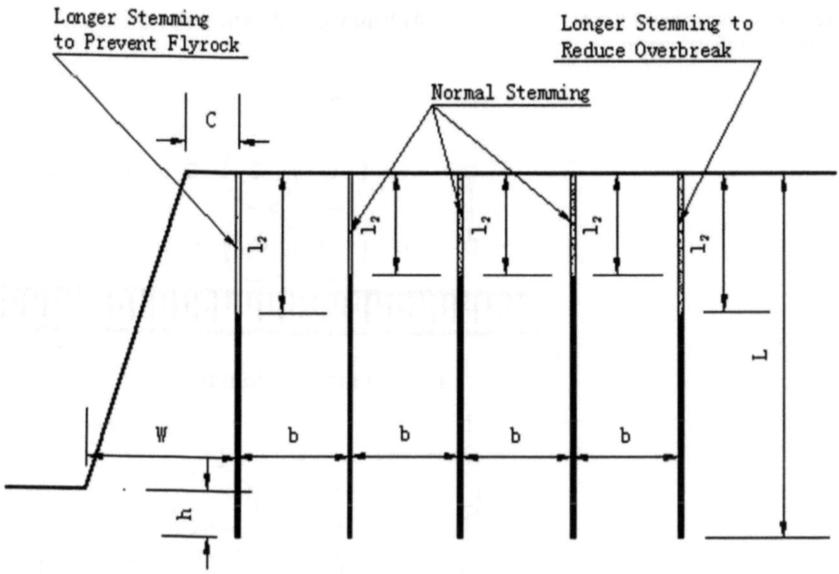

Fig. 8.5 Longer stemming in front row and last row of blastholes

- Special care must be taken when the blasting location is close to a public or residential area, and more stemming length (>0.85W to 1.0W) should be adopted to reduce noise, airblast and prevent flyrock.

In practice, it has been established that effective stemming is achieved with granular particle materials with the size range between 1/17d and 1/25d as they exhibit high frictional effects and free flowing when stemming. Drill cuttings, which are usually located near the hole collar and convenient to use, and soil are not suitable for stemming due to their low friction and may cause heavy dust during blasting.

8.1.8 Blasthole Pattern

In bench blasting, there are two types of blasthole patterns, quadrangular pattern (including square and rectangular) and staggered pattern (Fig. 8.6).

Owing to the ease of marking the collaring point and suitable for covering with some protective measures (like blasting mats or blasting cages), the quadrangular pattern is often used in bench blasting. However, the most effective is staggered pattern, especially those drilled on an equilateral triangular grid, as they give

Fig. 8.6 Comparison of three types of blasthole pattern

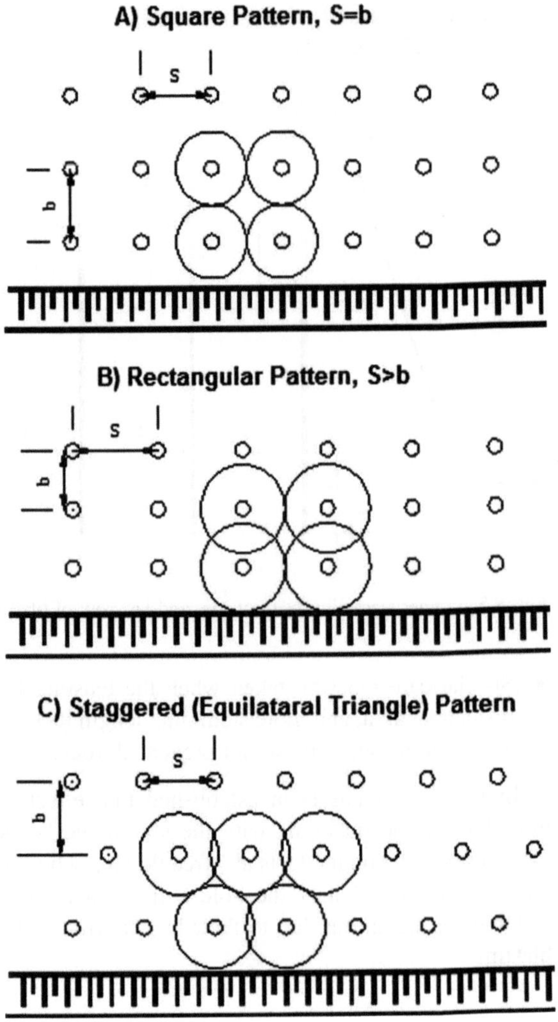

optimum distribution of explosive energy in the bench rock, hence result a better fragmentation, especially in hard and integral rock mass.

For improving bench top fragmentation, when blasthole pattern is larger than 2.5 × 2.5 m in a hard rock, some additional small charges can be used together with the main blasting charges. Figure 8.7 shows two proposals.

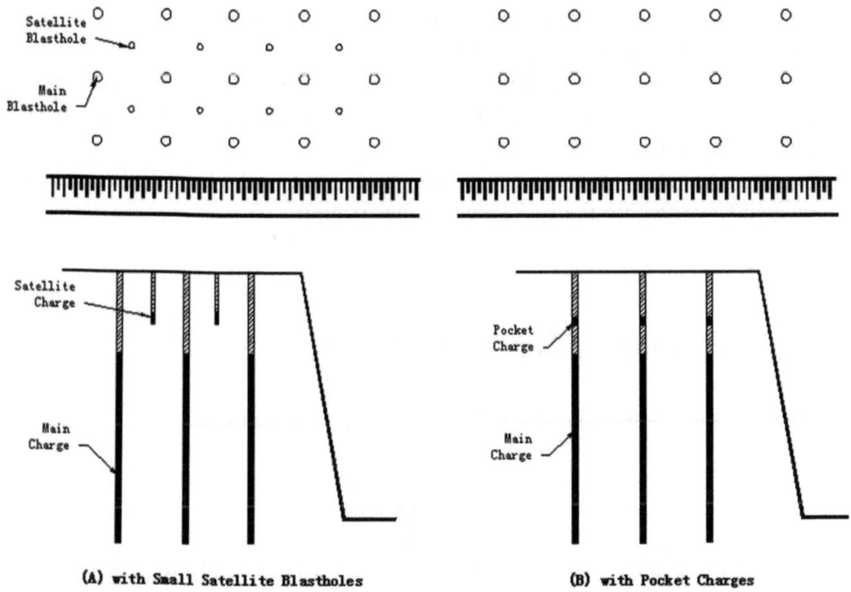

Fig. 8.7 Improving bench top fragmentation with additional small charge

8.2 Specific Charge (Powder Factor) and Explosive Charging Calculation

8.2.1 Specific Charge (Powder Factor)

Two terms are often used to relate explosive mass and rock mass to be blasted: specific charge and powder factor, q. Specific charge or powder factor is defined as the quantity of explosive (in kg or lb) necessary to fragment 1 m^3, ft^3, or 1 tonne of rock, and expressed as kg/m^3, lb/ft^3, or kg/t. In this book, we use the unit of kg/m^3.

Specific charge is one of the important parameters of bench blasting. The specific charge or powder factor varies between 0.1 and 0.7 kg/m^3 for normal bench blasting. The value is affected by the following factors:

- The blastability of rock blasted. Tables 5.1, 5.2, and 5.6 of Chap. 5 have shown the relationship between the powder factor and rock blastability;
- The fragmentation to be required to the blasted rock. Higher specific charge will get better fragmentation of blasted rock;
- The swelling and throwing (cast) distance required. As the specific charge of a blast is increased, the displacement and swelling of blasted rock also increase;
- The performance of explosive to be used. If the rock characteristics are similar for getting similar blasting results (fragmentation, swelling, and throwing distance), the specific charge required of a explosive with high energy and high

density (e.g., emulsion explosives) is lower than that of a explosive with low energy and low density (e.g., ANFO);

- The distribution of explosive in the blastholes and rock mass. The specific charge of a blast is increased by a poor charge distribution in the rock mass or blastholes; and
- Degree of fixation of the rock mass to be blasted. A higher specific charge is needed to overcome the resistance of the fixation.

8.2.2 *Calculation of Explosive Charging*

According to the definition of specific charge or powder factor:

$$q = \frac{Q}{V}\ \text{kg/m}^3 \tag{8.8}$$

where Q is the explosive weight and V is the solid volume of the rock blasted.

8.2.2.1 Charge Weight for a Blasthole

For the blast of single row of blastholes or the first row of blastholes of a multiple row blast, the explosive charge of one hole, Q_1, is calculated according to the following formula:

$$Q_1 = q \times S \times W \times H \tag{8.9}$$

where

q the specific charge, in kg/m^3;
S the spacing between two adjacent holes, in m;
W the burden at the toe of the bench face, in m; and
H the height of the bench, in m.

For multiple row blast, the explosive charge weight per hole of all blastholes from the second row is calculated as:

$$Q_2 = k \times q \times S \times b \times H \tag{8.10}$$

where

k A coefficient for considering the resistance of blasted rock of front rows, $k = 1.1\text{–}1.2$;
b Distance between rows, in m.

8.2.2.2 Linear Charge Concentration

Linear charge concentration L_e is often used in blasting terminology especially when dealing with any type of perimeter blasting. The term *loading density* (used in the USA) is equivalent to linear charge concentration.

For a fully coupled charged blasthole, the expression of linear charge concentration L_e simply gives the amount of explosive used per 1 m loaded blasthole:

$$L_e = \frac{\pi d^2 \rho_e}{4} \text{ kg/m} \tag{8.11}$$

where

d the blasthole diameter, in m;
ρ_e loading density (or called degree of packing) of explosive, in kg/m^3.

For a decoupling charged blasthole, the expression of linear charge concentration L_e should include a coefficient of decoupling:

$$L_e = k_d \frac{\pi d^2 \rho_e}{4} \text{ kg/m} \tag{8.12}$$

where

k_d coefficient of decoupling,
$k_d = \frac{d_e^2}{d^2}$, d_e is the diameter of the explosive column (or cartridge).

8.3 Charge Configuration

8.3.1 Continuous Charge

As shown in Fig. 8.8a, the continuous charge of blastholes is usually used for most bench blasting as the charging operation is most simple and convenient, especially for short blastholes.

8.3.2 Continuous Decking Charges (Fig. 8.8b)

Due to the higher resistance of rock breakage in the lower part of the bench to overcome the shear strength which is much greater than the tensile strength, more explosive energy is needed in the bottom part than the column part of the blastholes, especially when vertical blastholes are used in bench blasting. This means that explosives of high density and strength should be used in the bottom charge, such

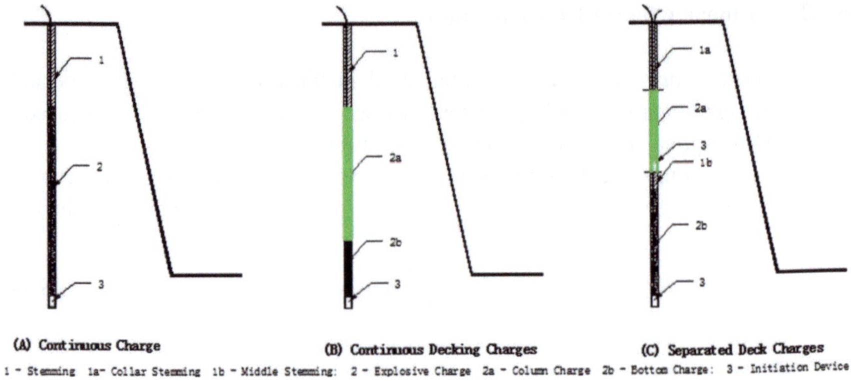

Fig. 8.8 Charge configuration

as high-density emulsion or heavy ANFO, and medium strength, low density explosives in the column charges, such as ANFO or low-density emulsion. The bottom charge should have a length of at least $0.6W$ (burden), so that its center of gravity is above or at the same level as the bench floor. According to Langefors, extension of the bottom charge to a length more than equal to the burden value does not contribute appreciably to breakage at the bench toe level, therefore suggesting that the lower charge should be between 0.6 and $1.3W$ [7].

The continuous deck charges with selected explosives have the following advantages [7]:

- Increased drilling productivity as a consequence of a larger pattern and a smaller length of subdrilling.
- Better bottom breakage which eliminates toe problem and favors the loading operation.
- Lower drilling and blasting costs, especially in hard rock.
- Lower powder factor due to improving the efficiency of use of explosive energy.

8.3.3 Isolated Deck Charges (Fig. 8.8c)

In the deep hole bench blasting, the charge configuration of two or three decks which are isolated by a middle section of stemming and have different firing times is often used. The length of middle stemming should be not shorter than $10d$ (diameter of blasthole) to prevent the sympathetic detonation or desensitization of the later initiated explosive charge by the first initiated explosive charge. The delay duration between decks usually adopts 25 ms when shock tube (Nonel) detonator initiation system is used (e.g., 450, 475, and 500 ms delay for DYNO's products).

The isolated deck charges have the following advantages:

- Reduce explosive charge weight per delay. It is an effective measure to reduce the ground vibration.
- Enhance the explosive column height and improve the fragmentation of the top part of bench rock.
- Lower drilling and blasting costs, especially in hard rock.

8.4 Firing Method and Firing Sequence

8.4.1 Electric Detonator Initiation System

8.4.1.1 Formulae for Electric Firing Circuits

The formulae for simple series and single parallel circuits are given below.

- Series Circuit [8]:

With reference to Fig. 8.9, the blasting circuit resistance R_o can be expressed by

$$R_o = \sum_{i}^{n} R_i + R_f + R_c \tag{8.13}$$

where

R_i the resistance of detonator i in ohms;
R_c the resistance of the connection wire in ohms; and
R_f the resistance of the firing line, in ohms.

Ohm's law gives the voltage required from the power supply to provide the necessary current I if the blasting circuit resistance is known:

$$U_o = IR_0 \tag{8.14}$$

where

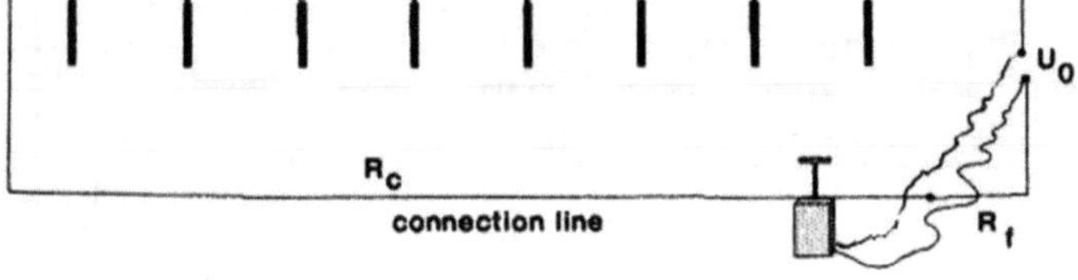

Fig. 8.9 Series circuit

U_o the voltage of the power supply in volts and
I the current in amperes.

The power supply must have sufficient capacity to feed the circuit with the required current and voltage. The needed power P, in watts, of the series circuit can be calculated:

$$P = U_o I \tag{8.15}$$

or

$$P = \frac{U_o^2}{R_o} \tag{8.16}$$

or

$$P = I^2 R_o \tag{8.17}$$

- Parallel Circuit

For the parallel circuit (Fig. 8.10), the blasting circuit resistance is

$$R_o = R_{\mathrm{f}} + \frac{1}{\sum_i^n \frac{1}{R_{\mathrm{i}}}} \tag{8.18}$$

and the total current flowing is

$$I_o = \sum_i^n I_i \tag{8.19}$$

- Parallel–Series Circuit

Parallel–series circuit is the common type of circuit used in the blasting (Fig. 8.11). The main advantage of the parallel–series circuit is the large number of detonators which can be fired from a blasting machine without a large input voltage requirement. In a balanced parallel–series circuit, the blasting circuit resistance is:

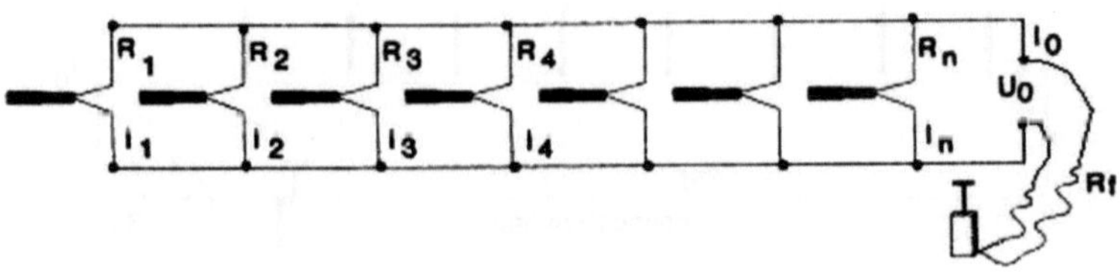

Fig. 8.10 Parallel circuit

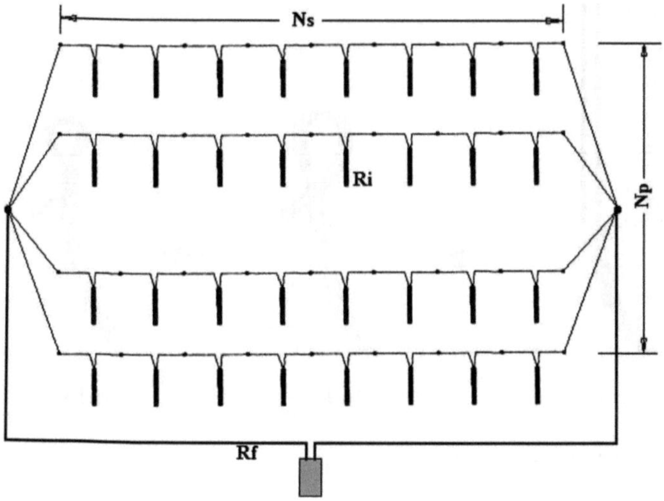

Fig. 8.11 Parallel–series circuit

$$R_o = R_f + \frac{R_i \times N_s}{N_p} \tag{8.20}$$

where

N_s number of detonators in a series;
N_p Number of parallel series.

8.4.1.2 Recommended Wire Splices

The connection of the detonator wires between themselves or to the buswire should be done in accordance with the recommended patten in Fig. 8.12.

8.4.1.3 Firing Sequence Design

In surface bench blasting, millisecond delay detonators are usually used with multi-row blastholes as it results a better fragmentation, smaller ground vibration, and high productivity. The firing sequence varies according to the site conditions.

- Blasts with one free face

In Fig. 8.13, the different initiation sequences available for multiple row blastholes with square or staggered patterns are given.

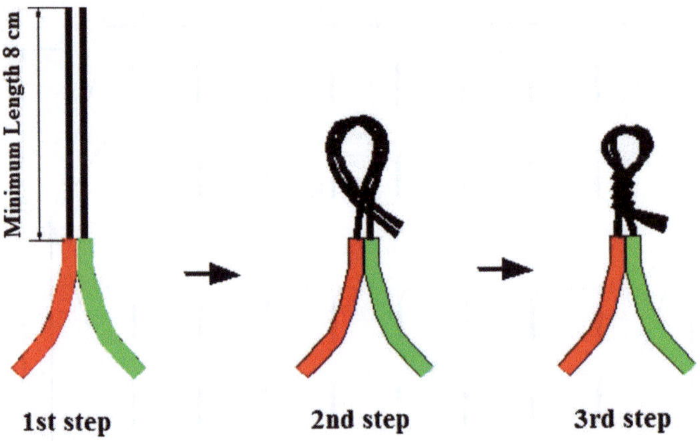

Fig. 8.12 Recommended wire splices

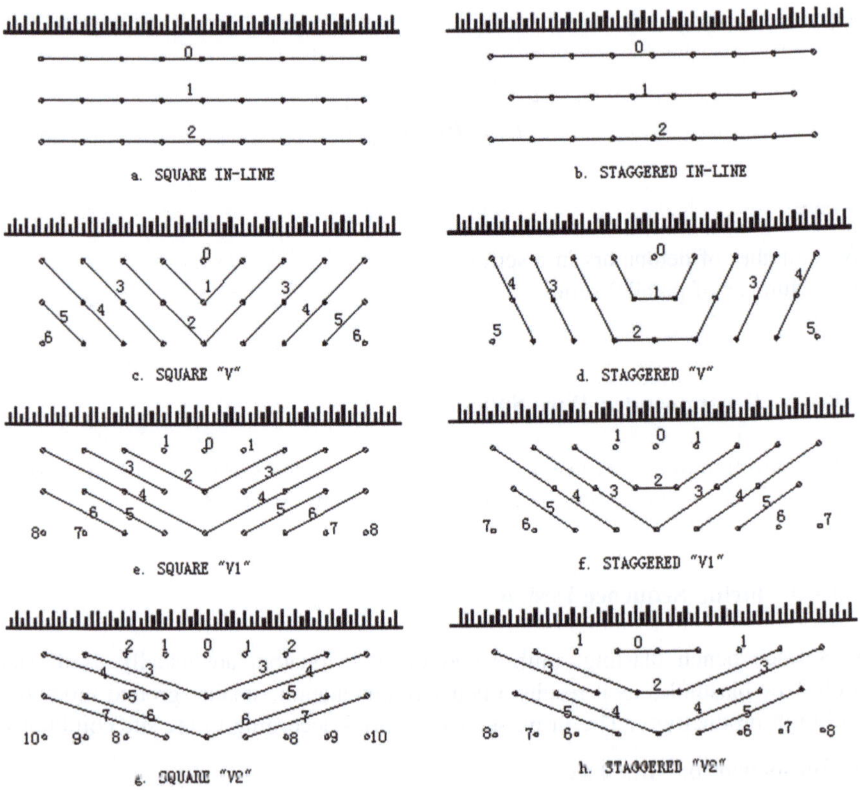

Fig. 8.13 Firing sequences for square and staggered pattern with electric MS delay detonators (Reproduced from Ref. [9] by permission of ASCE)

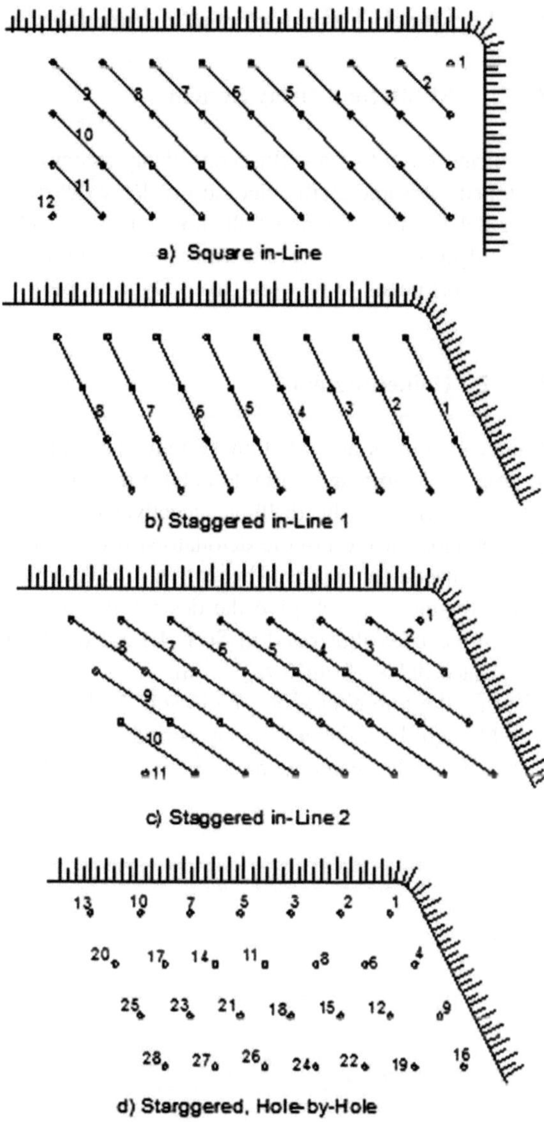

Fig. 8.14 Firing sequences for square and stagged pattern with two free faces

- Blasts with two free face

Blasting with two free faces, Fig. 8.14, is the most frequent geometric configuration in bench blasting.

8.4.2 *Shock Tube Detonator Initiation System*

8.4.2.1 Millisecond Delay System

The firing circuit of a millisecond delay system of shock tube detonators can be designed the same as the electric ms delay detonator which is described in above Sect. 8.4.1.3, but to connect all shock tubes of each blasthole the connector blocks (0 ms delay) (Fig. 8.15.) or a trunk line of low load detonating cord (3.6–5.0 g/m) (Fig. 8.16.) needs to be used (from Dyno Nobel).

8.4.2.2 Unidet System

Unidet system is an initiation system that employs a uniform delay time in the in-hole detonators and variable delay times in the connector units on the surface. The delay time in the drillhole usually has a longer delay time which normally enables most of the in-hole detonators to be initiated from the surface before any rock displacement begins. This is then supplemented by delay times in the surface connector units, which give the desired initiation sequence. The principle of this system has been illustrated in Sect. 4.1.5.2 of Chap. 4.

Surface delays from 0 to 200 ms are available, which give great flexibility in adapting the initiation sequence to suit the burden and rock characteristics.

Although the number of delay periods is less than the ms delay system which is of benefit to the manufacturers, the shotfirers can use them to design various and

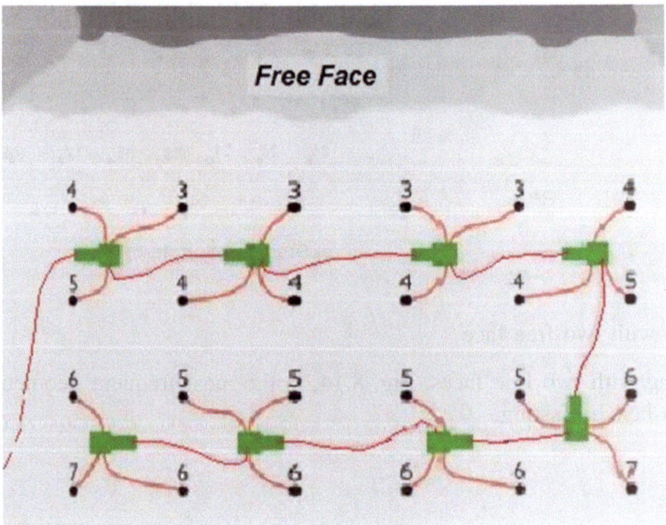

Fig. 8.15 Connection using connector (0 ms delay) blocks

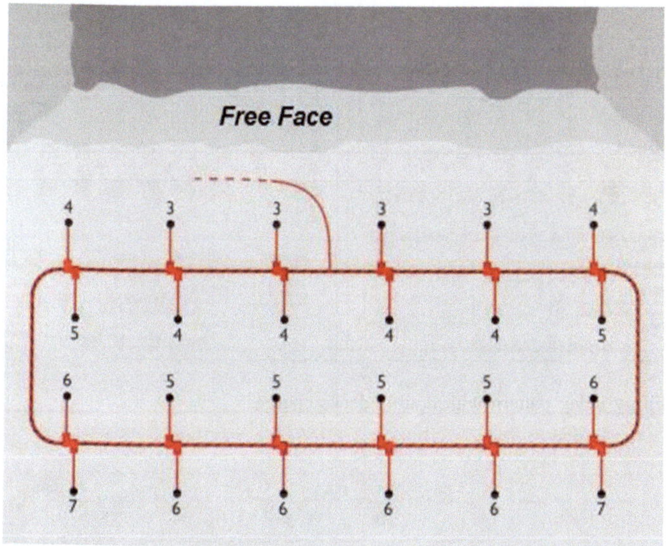

Fig. 8.16 Connection using detonating cord

flexible firing patterns to suit the practical situations. Figure 8.17 shows two firing patterns which are commonly used in bench blasting. But attention must be paid as there are some restrictions (see the lower left corner of each drawing of Fig. 8.17) to the number of rows or holes in a row (in length) in different pattern models; otherwise, there may be some holes firing at the same time (overlap).

Figure 8.18 shows two firing patterns in which double deck charges are used and each deck charge has its own delay time so that the charge weight per delay is greatly reduced to suit some sites where there are restrictions on ground vibration. Figure 8.19 (from Dyno Nobel) is a firing pattern with 3 deck charges in each hole. When this kind of deck charge is used, more attention must be paid to the number of rows or holes in a row to prevent any overlapping.

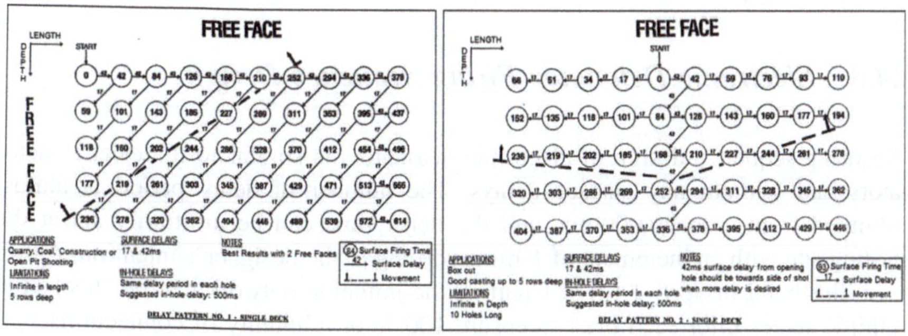

Fig. 8.17 Firing delay pattern with single deck charge

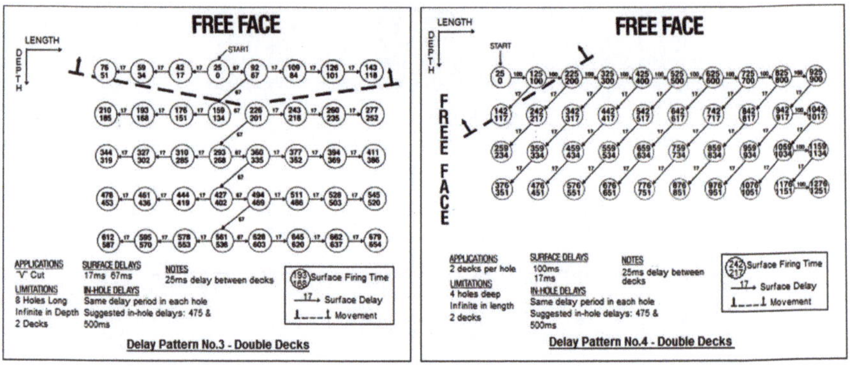

Fig. 8.18 Firing delay pattern with double decks charge

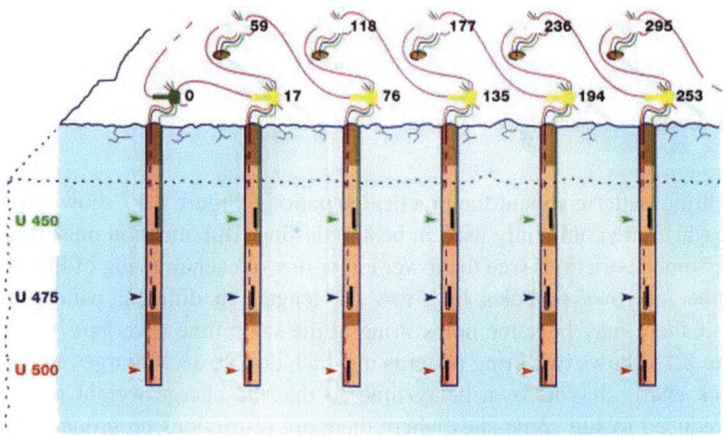

Three-deck charge in each hole for sites where there are restrictions on vibration. Each sub-charge has its own delay time. There is a 101 ms delay between rows and 59 ms between the holes in the rows (with the exception of holes 1 and 2 in the first row).

Fig. 8.19 Firing pattern—three decks charge per hole

8.4.3 Electronic Detonator Firing Sequence Design

The firing sequence design of electronic detonators is similar to the electric detonators and without any surface delays. The number of delay period is almost unlimited from 1 to 20,000 ms, and the delay time can be arbitrarily set at the blasting site with an increment of 1 ms according to the designer's intention. As we have illustrated in Sect. 4.1.6 of Chap. 4, the initiation network can be checked at multiple stages prior each blast to ensure 100 % of reliability of connections.

Figure 8.20 is an example which is the firing sequence design of a blast in KWP quarry in Hong Kong using SmartShot electronic detonators manufactured by Dyno-DSA (see Photo 4.15 of Chap. 4). The shotfirer loaded the primer with the electronic detonator into each blasthole first, then connected all the leg wires of the detonators from No. 1 to No. 46 according to the design (Fig. 8.20), and connected the end leg wires of No. 46 detonator to an active end plug (Fig. 8.21 is the blast bench surface after completion of network connection). After checking all connections are ready, the shotfirer connected the leg wires of No. 1 detonator to the Tagger and programmed the delay time of each detonator one by one from No. 1 to No. 46 according to the design, Fig. 8.20. After verifying all timing setting and connections, explosive charging and stemming follow. The shot was fired after finally verifying the network with the bench box.

8.5 Delay Timing

The optimum blasting results should meet the following objectives:

- Adequate rock fragmentation, swelling, and displacement.
- Control over the flyrock and overbreak.
- Minimum level of ground vibration and airblast.

The delay timings between holes and rows play a very important role in fulfillment of these objectives.

An adequate delay timing for certain rock characteristics should allow the strain waves to have enough time to break the rock and the explosion gas to push the broken rock to accelerate up to a velocity that assures an adequate horizontal displacement.

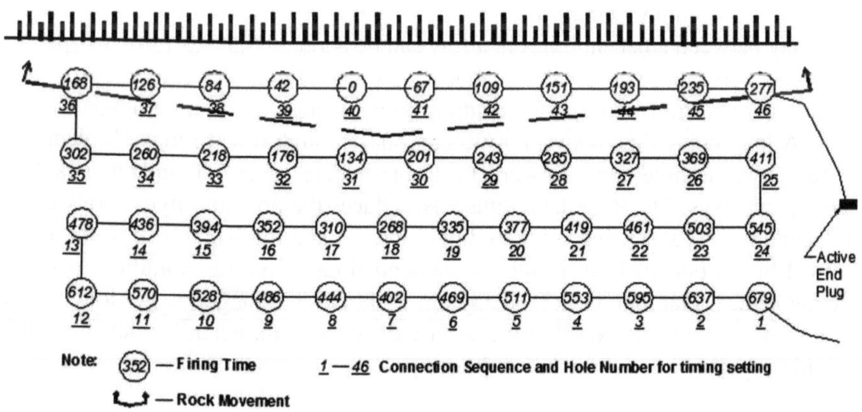

Fig. 8.20 Firing sequence design of electronic detonators (SmartShot of Dyno-DSA) in KWP Quarry

Fig. 8.21 Bench surface after completion of firing connection with SmartShot electronic detonators in KWP Quarry, 2013

If the delay between rows is too short, the movement of burden is restricted which causes excessive burden in the second and subsequent rows and the rock tends to move vertically because of insufficient relief (Fig. 8.22). This causes poor fragmentation, tight muck piles, and high ground vibrations and flyrock, and backbreak along the new face may persist, jeopardizing the stability of the slope or opening. If the delay time between rows is too great, it may cause airblast, cutoffs of surface delay network, and even flyrock if the burden is small in the first rows.

There are lots of literatures to discuss the adequate delay timing. Table 8.4 (from [5]) gives some recommendations by different researchers.

According to my experience of construction blasting practice for decades, the adequate delay timing should be:

Delay between adjacent holes in a row can be adopted 5.5–8.3 ms/m of spacing, and delay between rows should be about 2–3 times of that between holes in a row, about 14–28 ms/m of burden for medium to hard rocks and a little longer for soft rock. A long delay time between rows can offer enough time to move broken rock of the previous fired rows forward to accommodate space for broken rock from subsequent rows. The long delay time also reduces the potential flyrock risk due to the tendency of later firing rows to move upward and the backbreak to the newly formed bench (slope) face. In the recommended delay timing condition, it is also recommended that the in-hole delay time should be longer, e.g., longer than 450 ms, to prevent any cutoff of the surface delay network when Unidet shock tube initiation system is used for bench blasting.

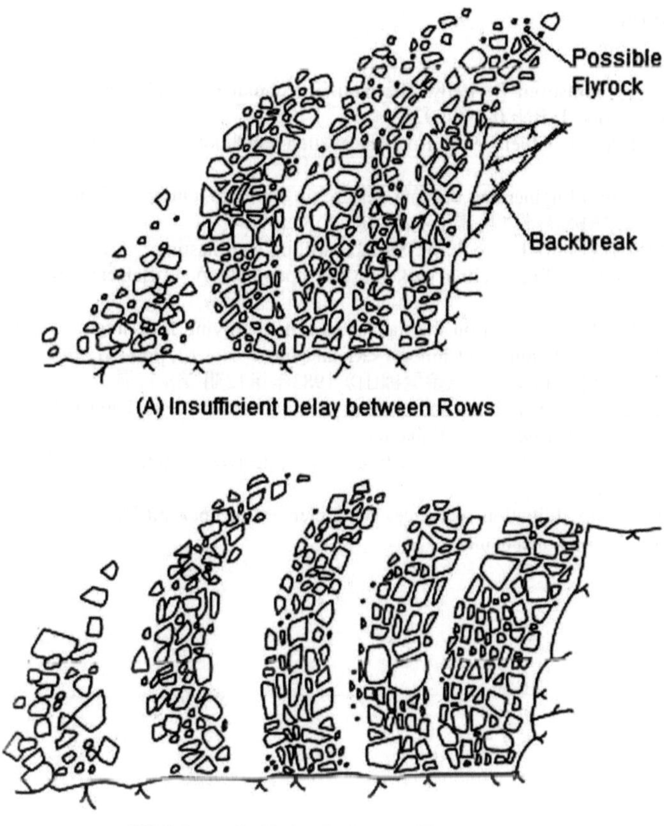

Fig. 8.22 Effect of insufficient delay time (**a**) and effect of adequate delay (**b**) between rows (Reproduced from Ref. [2] by permission of Sandvik)

Table 8.4 Delay intervals recommended by different researchers [5]

1. Langefors and Kihlström (1973)	2–5 ms/m of burden (from experience in production blasting)
2. Bergmann et al. (1974)	1–2 ms/ft (3.3–6.6 ms/m) of burden (from model blasting experiment)
3. Hagan (1977)	8 ms/m of burden for long collars, low powder factor (kg/m^3), soft, densely fissured, low density rock, and 4 ms/m of burden for short collars, high powder factor (kg/m^3), dense, tough, massive rock
4. Lang and Favreau (1972)	1.5–2.5 ms/ft (5–8.3 ms/m) of burden (from high speed photography in iron ore mines)
5. Andrews (1981)	1–5 ms/ft (3.3–17 ms/m) of spacing between adjacent holes in a row, and 6.6–50 ms/m between rows
6. Winzer (1978)	3.4 ms/ft (11 ms/m) of relief between holes and about 8.6 ms/ft (28.7 ms/m) along the echelon
7. Anderson et al. (1981)	8.4 ms/ft of B_e recommendation for an optimum breakage and forward movement

References

1. Langefors U, Kihlstrom B (1978) The modern technique of rock blasting, 3rd edn. Halsted Press, New York. ISBN: 0-470-99282-4
2. Heiniö M, Editor in Chief (1999) Rock excavation handbook for civil engineering. Sandvik & Tamrock
3. Huang SJ (1996) Engineering blasting design. (in Chinese), Ordnance Industry Press, Beijing. 黄紹鈞: 工程爆破設計。兵器工业出版社,北京,1996
4. Naapuri J (ed) (1988) Tamrock. In: Surface drilling and blasting
5. Bhandari S (1977) Engineering rock blasting operation. A.A. Balkema, Rotterdam. ISBN: 905410 658 1
6. Zou D (1983) Investigation on the mechanism of improving rock fragmentation using wide spacing blasting technique (in Chinese) <Metal Mines> No. 12, pp 8–11. 邹定祥:大孔距爆破改善破岩效果机理的探討。《金属礦山》1983年第12期,第8-11頁。
7. Jimeno CL et al (1995) Drilling and blasting of rock. A.A. Balkema Publishers. ISBN: 90 5410 199 7. Copyright ©1995 A. A. Balkema
8. Persson PA et al (1994) Rock blasting and explosives engineering. CRC Press. ISBN: 0-8493-8978-X
9. Hagan TN (1975) Initiation sequence—vital element of open pit blast design. In: 16th U.S. symposium on rock mechanics

Chapter 9
Trench Blasting

Trench blasting is also called ditch blasting. It is a part of the construction works for the excavation of various types of pipelines, such as water mains, sewerage and drainage system, gas mains, oil pipes, electric cable, optical fiber and normal communication and Internet cable. Trench excavation with explosives presents a series of particular characteristics which are different from bench blasting. As most trench blasting work is within or near urban areas, particular concern on the blasting safety is an important issue.

9.1 Blasthole Pattern and Firing Sequence

9.1.1 Blasthole Diameter

It is important to choose correct blasthole diameters for trench blasting, since these affect the drilling and blast cost and also the total cost of making the trench. Large diameter reduces the drilling cost but increases the cost of excavation and back-filling due to the increased overbreak. Small-diameter blastholes (32–45 mm when trench depth is <1.5 m) are drilled using hand-held drills in small urban operations. For larger trench excavations (50–76 mm when trench depth is >1.5 m), drilling rigs may be used.

9.1.2 Drilling Pattern

The drilling patterns of trench blasting depend upon the size of the trench. The holes are usually placed in rows with no slope at the sides. When blasting smaller trenches, the hole should have an inclination of 2:1–3:1 (26.5°–18.5° from the

D. Zou, *Theory and Technology of Rock Excavation for Civil Engineering*,
DOI 10.1007/978-981-10-1989-0_9

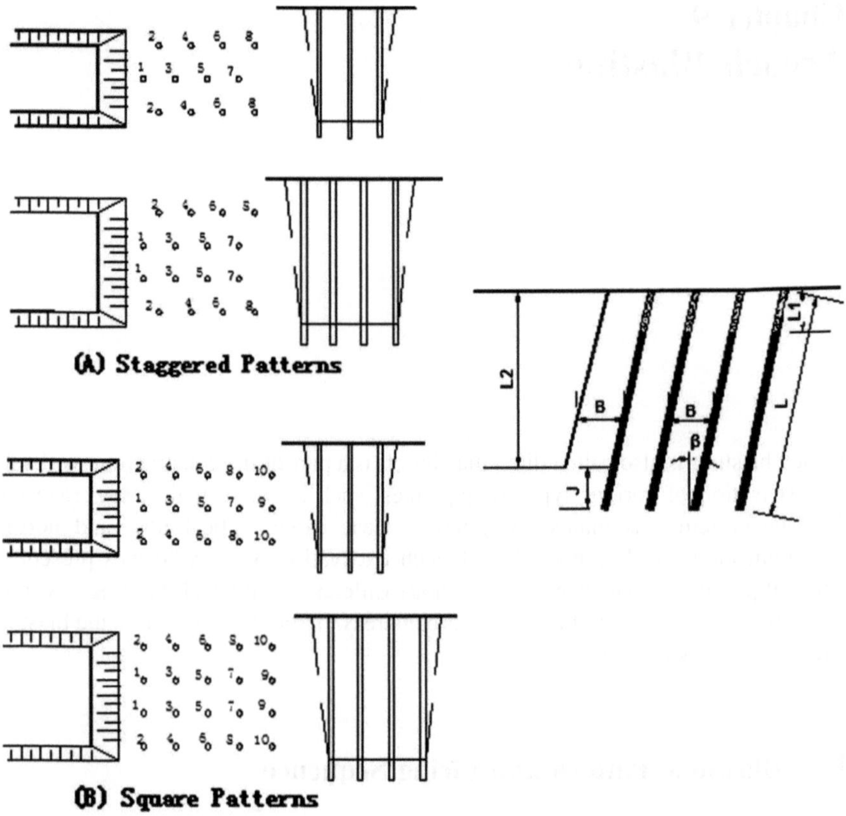

Fig. 9.1 Samples of blasthole patterns and firing sequence of trench blasting. **a** Staggered patterns. **b** Square patterns

vertical, as shown in Fig. 9.1). For the larger and longer trench, it is often more feasible to use vertical holes.

The major parameters of trench blasthole patterns are given below for general rock conditions:

D Diameter of blastholes, D = 32–45 mm for trench depth $L_2 \leq 1.5$ m; D = 50–76 mm for trench depth $L_2 > 1.5$ m;

B Burden, $B = 26D$ for $D < 50$ mm; $B = 24D$ for $D > 50$ mm;

S Spacing, $S = W$ when (trench width) $W < 0.75$ m; $S = W/2$ when W = 0.75–1.5 m; $S = W/(2–3)$ when $W > 1.5$ m;

J Subdrilling, $J = 0.5B$ with minimum value of 0.2 m;

L_1 Stemming, $L_1 \geq B$;

L Length of blastholes, $L = L_2/\cos\beta + J$, where **β**—angle of blasthole inclination.

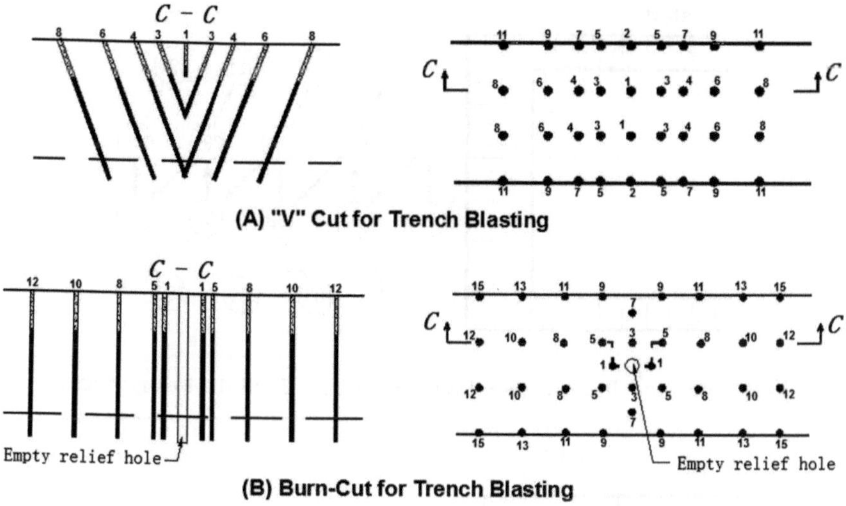

Fig. 9.2 Cut-hole types for trench blasting. **a** "V" cut for trench blasting. **b** Burn-cut for trench blasting

Square or rectangular and staggered patterns are used in trench blasting (see Fig. 9.1).

If there is no free face for starting trench blasting, some cut holes are needed to form a free face for trench excavation. Figure 9.2 shows two types of cut-hole patterns. When burn-cut pattern is used, it is better to adapt one or more larger uncharged holes for relief.

9.1.3 Firing Sequence

The firing sequence in trench blasting should be so designed that it can produce a good fragmentation and rock breakage while, at the same time, maintaining overbreak at a minimum. Usually, the millisecond delay electric or shock tube initiation system is used in trench blasting and the central holes are always fired earlier than the both sides' holes. Figures 9.1, 9.2, and 9.3 show some examples of the firing sequences.

9.2 Explosive Charging

There are two types of trench blasting techniques to be used—the conventional and the controlled (or called smooth wall) techniques. The conventional one applies explosive charging, almost the same for all blastholes (Fig. 9.4a). The controlled

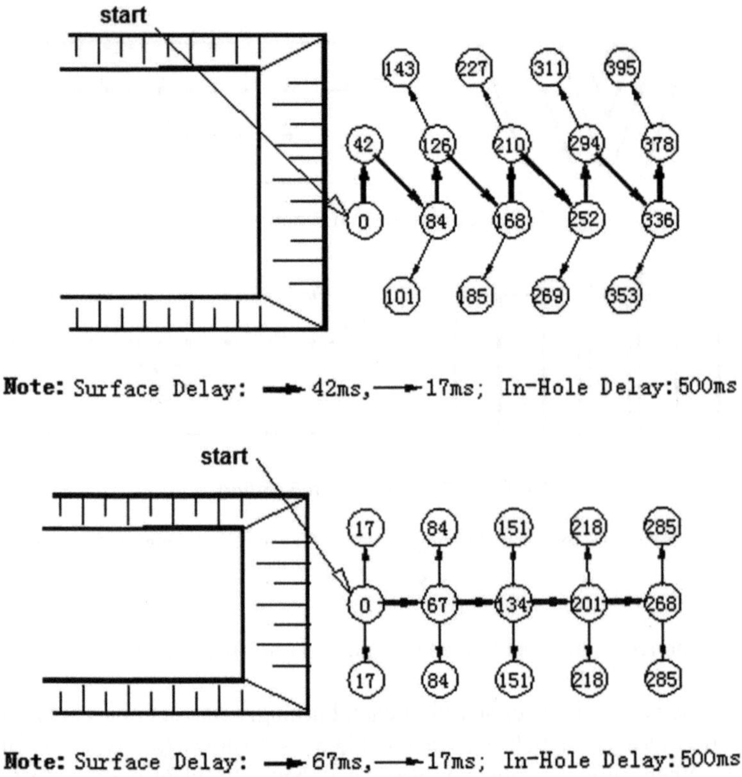

Fig. 9.3 Firing sequence of trench blasting with shock tube detonator united system

technique (Fig. 9.4b) has holes placed in rectangular pattern with the center holes charged more heavily and the side holes charged only lightly. In trench blasting, it is important to minimize overbreak, as this results in more materials to be removed and costly refilling work. Controlled trench blasting technique makes it easy to minimize overbreak and cut the rock along the designed lines of excavation.

As the rock is more confined in the lower part of the trench, for both methods, strong explosive (higher power and higher density) is usually charged in the bottom of blastholes as the bottom charge and weaker explosives are used as the column charge. The concentration of explosive in the column charge is considerably less than that in the bottom charge, about 25–35 %. For the length of the bottom charge, one can refer to Table 9.1 below, where L_2 is the depth of the trench to be excavated, in meters.

Compared to bench blasting, charging in the column parts of the blastholes is lighter in order to reduce overbreak beyond the designed contour of the trench although the overbreak is difficult to avoid completely, even in solid rock formations. ANFO is a popular column charge explosive for trench blasting dry blastholes as it has low density and less strength. Where water is encountered,

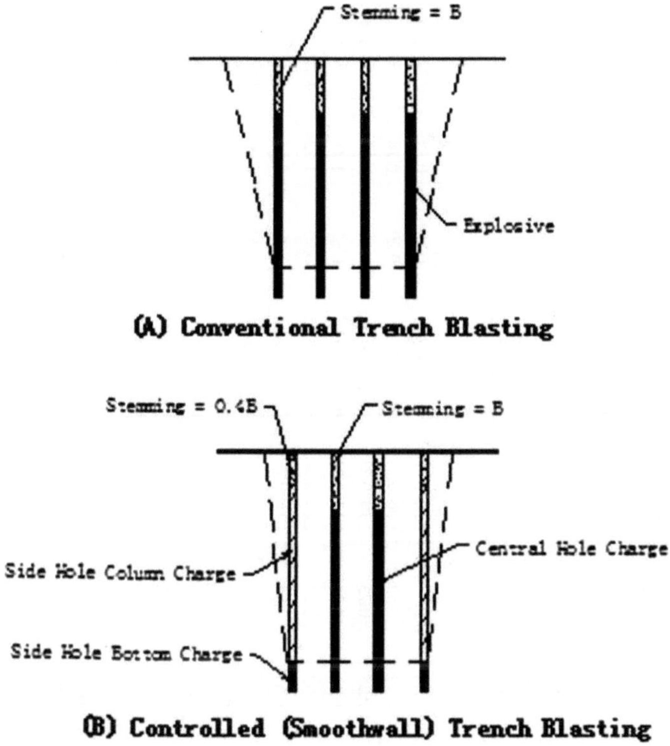

Fig. 9.4 Explosive charging method for trench blasting. **a** Conventional trench blasting. **b** Controlled (Smoothwall) trench blasting

Table 9.1 The length of bottom charge for trench blasting

Conventional blasting	Smooth wall blasting
All the blastholes $0.4 + (L_2 - 1)/5$	Central blastholes: $1.3[0.4 + (L_2 - 1)/5]$ Contour blastholes: $0.7[0.4 + (L_2 - 1)/5]$

Reproduce from Ref. [1] with the permission from Taylor & Francis Book UK

water-resistant cartridge explosives with a smaller diameter than the bottom charge should be used. The bulk emulsion explosive with less energy can be also used as the column charge when the diameter of the blastholes is not less than 50 mm.

As the rock is more confined in trench blasting, higher powder factors are used than with conventional bench blasting, and drilling patterns are closer. The following tables give the blasting parameters for reference. Tables 9.2, 9.3, and 9.4 are compiled from Table 6 to 9b of [2] for hole diameters of 32 and 38 mm. If the surrounding environment is not very restricted and allows adequate throw, backbreak and ground vibration, some larger diameter blastholes, such as 50 or 76 mm, may be adopted to increase the productivity and reduce the cost. The parameters in Table 9.5 can be referred.

Table 9.2 Blasting parameters for trench excavation with conventional blasting method

Trench design		Geometric parameters				32 mm hole diameter					38 mm hole diameter				
Bottom width (m)	Depth (m)	Hole number per row	Subdrilling (m)	Hole length (m)	Stemming (m)	Burden (m)	Bottom charge (kg)	Column charge (kg)	Charge per hole (kg)	Powder factor (kg/m^3)	Burden (m)	Bottom charge (kg)	Column charge (kg)	Charge per hole (kg)	Powder factor (kg/m^3)
0.7	0.5	3	0.25	0.75	0.35	0.50	0.24	0.03	0.27	2.31					
0.7	1.0	3	0.25	1.30	0.60	0.80	0.32	0.08	0.4	1.71					
0.7	1.5	3	0.30	1.90	0.90	0.80	0.40	0.13	0.53	1.51					
0.7	2.0	3	0.30	2.40	0.80	0.80	0.48	0.25	0.73	1.56					
0.7	2.5	3	0.35	3.00	0.80	0.80	0.56	0.38	0.94	1.61					
0.7	3.0	3	0.35	3.50	0.70	0.80	0.65	0.50	1.15	1.64					
1.0–1.5	1.0	3	0.55	1.60	0.90	0.80	0.32	0.08	0.4	1.20–0.80	0.90	0.46	0.12	0.58	1.74–1.16
1.0–1.5	1.5	3	0.55	2.15	1.15	0.80	0.40	0.13	0.53	1.06–0.71	1.10	0.58	0.20	0.78	1.56–1.04
1.0–1.5	2.0	3	0.55	2.65	1.05	0.80	0.48	0.25	0.73	1.10–0.73	1.10	0.69	0.40	1.09	1.64–1.09
1.0–1.5	2.5	3	0.55	3.20	1.00	0.80	0.56	0.38	0.94	1.13–0.75	1.10	0.81	0.60	1.41	1.69–1.13
1.0–1.5	3.0	3	0.55	3.70	0.90	0.75	0.64	0.50	1.14	1.14–0.76	1.10	0.92	0.80	1.72	1.72–1.15
1.0–1.5	3.5	3	0.55	4.25	0.85	0.70	0.72	0.63	1.35	1.16–0.77	1.10	1.04	1.00	2.04	1.75–1.17
1.0–1.5	4.0	3	0.55	4.75	0.85	0.60	0.72	0.75	1.47	1.10–0.74	1.10	1.04	1.20	2.24	1.68–1.12
2.0	1.0	4	0.55	1.60	0.90	0.90	0.32	0.08	0.4	0.80	0.90	0.46	0.12	0.58	1.16
2.0	1.5	4	0.55	2.15	1.15	1.00	0.40	0.13	0.53	0.71	1.10	0.58	0.20	0.78	1.04
2.0	2.0	4	0.55	2.65	1.05	1.00	0.48	0.25	0.73	0.73	1.10	0.69	0.40	1.09	1.09
2.0	2.5	4	0.55	3.20	1.00	0.95	0.56	0.38	0.94	0.75	1.10	0.81	0.60	1.41	1.13
2.0	3.0	4	0.55	3.70	0.90	0.85	0.64	0.50	1.14	0.76	1.10	0.92	0.80	1.72	1.15
2.0	3.5	4	0.55	4.25	0.85	0.85	0.72	0.63	1.35	0.77	1.10	1.04	1.00	2.04	1.17
2.0	4.0	4	0.55	4.75	0.85	0.75	0.72	0.75	1.47	0.74	1.10	1.04	1.20	2.24	1.12
2.5–3.0	2.0	4	0.55	2.65	1.05						1.10	0.69	0.40	1.09	0.87–0.73
2.5–3.0	2.5	4	0.55	3.20	1.00						1.10	0.81	0.60	1.41	0.90–0.75
2.5–3.0	3.0	4	0.55	3.70	0.90						1.10	0.92	0.80	1.72	0.92–0.76

(continued)

Table 9.2 (continued)

Trench design		Geometric parameters				32 mm hole diameter					38 mm hole diameter				
Bottom width (m)	Depth (m)	Hole number per row	Subdrilling (m)	Hole length (m)	Stemming (m)	Burden (m)	Bottom charge (kg)	Column charge (kg)	Charge per hole (kg)	Powder factor (kg/m^3)	Burden (m)	Bottom charge (kg)	Column charge (kg)	Charge per hole (kg)	Powder factor (kg/m^3)
2.5–3.0	3.5	4	0.55	4.25	0.85						1.10	1.04	1.00	2.04	0.93–0.78
2.5–3.0	4.0	4	0.55	4.75	0.85						1.10	1.04	1.20	2.24	0.90–0.75
2.5–3.0	4.5	4	0.55	5.30	0.80						1.00	1.05	1.40	2.45	0.87–0.73
2.5–3.0	5.0	4	0.55	6.00	0.80						1.00	1.38	1.60	2.98	0.95–0.79

Reproduce from Ref. [2] with the permission from Sandvik

Table 9.3 Blasting parameters for trench excavation with controlled blasting method using Φ32 mm holes, hole inclination 3:1, bottom charge 0.6 kg/m, column charge 0.16 kg/m for side holes, and 0.5 kg/m for center holes

Trench design		Geometric parameters						Explosive charges						
Bottom width (m)	Depth (m)	Hole number per row	Subdrilling (m)	Hole length (m)	Burden (m)	Stemming (m)		Side holes			Center holes			Powder factor (kg/m³)
						Side holes	Center holes	Bottom charge (kg)	Column charge (kg)	Charge per hole (kg)	Bottom charge (kg)	Column charge (kg)	Charge per hole (kg)	
0.7	0.5	3	0.25	0.75	0.45	0.10	0.25	0.20	0.05	0.25	0.30		0.30	5.08
0.7	1.0	3	0.25	1.30	0.70	0.30	0.60	0.25	0.10	0.35	0.35	0.05	0.40	2.24
0.7	1.5	3	0.30	1.90	0.70	0.30	0.60	0.30	0.20	0.50	0.40	0.35	0.75	2.38
0.7	2.0	3	0.30	2.40	0.70	0.30	0.60	0.35	0.25	0.60	0.45	0.50	0.95	2.19
0.7	2.5	3	0.35	3.00	0.70	0.30	0.60	0.40	0.35	0.75	0.50	0.75	1.25	2.24
0.7	3.0	3	0.35	3.50	0.70	0.30	0.60	0.45	0.40	0.85	0.60	0.95	1.55	2.21
1.0–1.5	1.0	3	0.55	1.60	0.70	0.30	0.70	0.25	0.15	0.40	0.35	0.15	0.50	1.86–1.24
1.0–1.5	1.5	3	0.55	2.15	0.70	0.30	0.70	0.30	0.20	0.50	0.40	0.40	0.80	1.71–1.14
1.0–1.5	2.0	3	0.55	2.65	0.70	0.30	0.70	0.35	0.30	0.65	0.45	0.60	1.05	1.68–1.12
1.0–1.5	2.5	3	0.55	3.20	0.70	0.30	0.70	0.40	0.35	0.75	0.50	0.80	1.30	1.60–1.07
1.0–1.5	3.0	3	0.55	3.70	0.65	0.30	0.70	0.50	0.40	0.90	0.60	1.00	1.60	1.74–1.16
1.0–1.5	3.5	3	0.55	4.25	0.65	0.30	0.70	0.60	0.45	1.05	0.70	1.20	1.90	1.76–1.17
1.0–1.5	4.0	3	0.55	4.75	0.55	0.30	0.60	0.70	0.55	1.25	0.80	1.40	2.20	2.14–1.42
2.0	1.0	4	0.55	1.60	0.80	0.30	0.80	0.30	0.15	0.45	0.40	0.10	0.50	1.19
2.0	1.5	4	0.55	2.15	0.80	0.30	0.80	0.35	0.20	0.55	0.45	0.30	0.75	1.08
2.0	2.0	4	0.55	2.65	0.80	0.30	0.80	0.50	0.25	0.75	0.60	0.45	1.05	1.13
2.0	2.5	4	0.55	3.20	0.75	0.30	0.80	0.55	0.30	0.85	0.65	0.65	1.30	1.15
2.0	3.0	4	0.55	3.70	0.75	0.30	0.80	0.60	0.40	1.00	0.80	0.80	1.60	1.16
2.0	3.5	4	0.55	4.25	0.70	0.30	0.80	0.70	0.45	1.15	0.90	1.00	1.90	1.24
2.0	4.0	4	0.55	4.75	0.70	0.30	0.70	0.80	0.50	1.30	1.00	1.20	2.20	1.25

Reproduce from Ref. [2] with the permission from Sandvik

Table 9.4 Blasting parameters for trench excavation with controlled blasting method using Φ38 mm holes, hole inclination 3:1, bottom charge 0.9 kg/m, column charge 0.25 kg/m for side holes, and 0.7 kg/m for center holes

Trench design		Geometric parameters						Explosive charges						
						Stemming (m)		Side holes			Center holes			
Bottom width (m)	Depth (m)	Hole number per row	Subdrilling (m)	Hole length (m)	Burden (m)	Side holes	Center holes	Bottom charge (kg)	Column charge (kg)	Charge per hole (kg)	Bottom charge (kg)	Column charge (kg)	Charge per hole (kg)	Powder factor (kg/m³)
1.0–1.5	1.0	3	0.55	1.60	0.80	0.35	0.85	0.35	0.20	0.55	0.50	0.15	0.65	2.19–1.46
1.0–1.5	1.5	3	0.55	2.15	1.00	0.35	0.85	0.45	0.30	0.75	0.55	0.50	1.05	1.70–1.13
1.0–1.5	2.0	3	0.55	2.65	1.00	0.35	0.85	0.50	0.45	0.95	0.65	0.75	1.40	1.65–1.10
1.0–1.5	2.5	3	0.55	3.20	1.00	0.35	0.85	0.55	0.55	1.10	0.70	1.10	1.80	1.60–1.07
1.0–1.5	3.0	3	0.55	3.70	1.00	0.35	0.85	0.70	0.65	1.35	0.85	1.35	2.20	1.63–1.09
1.0–1.5	3.5	3	0.55	4.25	1.00	0.35	0.85	0.85	0.75	1.60	1.00	1.60	2.60	1.66–1.10
1.0–1.5	4.0	3	0.55	4.75	1.00	0.35	0.75	1.00	0.85	1.85	1.15	1.95	3.10	1.70–1.13
2.0	1.0	4	0.55	1.60	0.80	0.35	0.95	0.45	0.15	0.60	0.55		0.55	1.44
2.0	1.5	4	0.55	2.15	1.00	0.35	0.95	0.55	0.30	0.85	0.65	0.30	0.95	1.20
2.0	2.0	4	0.55	2.65	1.00	0.35	0.95	0.60	0.40	1.00	0.75	0.60	1.35	1.18
2.0	2.5	4	0.55	3.20	1.00	0.35	0.95	0.70	0.50	1.20	0.90	0.90	1.80	1.20
2.0	3.0	4	0.55	3.70	1.00	0.35	0.95	0.80	0.60	1.40	1.10	1.10	2.20	1.20
2.0	3.5	4	0.55	4.25	1.00	0.35	0.95	1.00	0.70	1.70	1.25	1.35	2.60	1.23
2.0	4.0	4	0.55	4.75	1.00	0.35	0.85	1.20	0.80	2.00	1.45	1.60	3.05	1.26
2.5–3.0	2.0	4	0.55	2.65	1.00	0.35	0.95	0.65	0.40	1.05	0.85	0.50	1.35	0.96–0.80
2.5–3.0	2.5	4	0.55	3.20	1.00	0.35	0.95	0.80	0.50	1.30	1.05	0.75	1.80	0.99–0.83
2.5–3.0	3.0	4	0.55	3.70	1.00	0.35	0.95	1.05	0.55	1.60	1.35	0.90	2.25	1.03–0.86
2.5–3.0	3.5	4	0.55	4.25	1.00	0.35	0.95	1.15	0.65	1.80	1.50	1.15	2.65	1.02–0.85
2.5–3.0	4.0	4	0.55	4.75	1.00	0.35	0.95	1.20	0.75	1.95	1.60	1.45	3.05	1.00–0.83
2.5–3.0	4.5	4	0.55	5.30	0.90	0.35	0.95	1.20	0.90	2.10	1.60	1.80	3.40	1.09–0.91
2.5–3.0	5.0	4	0.55	6.00	0.90	0.35	0.95	1.20	1.05	2.25	1.60	2.30	3.90	1.09–0.91

Reproduce from Ref. [2] with the permission from Sandvik

Table 9.5 Blasting parameters for trench excavation with conventional blasting method using Φ50 mm holes, hole inclination 3:1, and 3 holes per row

No.	Trench depth (m)	Hole length (m)	Burden (m)		Bottom charge (kg/hole)		Column charge (kg/hole)
			Maximum	General	Width of trench bottom (m)		(charge concentration: about 0.4 kg/m)
					1.0	1.5–2.0	
1	0.6	0.9	0.6	0.6	0.15	0.20	
2	1.0	1.4	0.8	0.8	0.20	0.25	0.20
3	1.5	2.0	1.4	1.1	0.30	0.40	0.35
4	2.0	2.5	1.4	1.1	0.40	0.55	0.50
5	2.5	3.1	1.4	1.1	0.50	0.65	0.75
6	3.0	3.6	1.4	1.1	0.60	0.75	0.90
7	3.5	4.1	1.4	1.1	0.75	0.95	1.10
8	4.0	4.6	1.4	1.1	0.9	1.15	1.30

Reproduce from Ref. [3] with the permission from Metallurgy Industry Press

9.3 Blasting Safety in Trench Excavation

As stated in the beginning of this chapter, due to the strong confinement of the rock in trench blasting, higher powder factor, closer drilling patterns, and inclined blastholes are used in most situations, and then higher ground vibration and greater rock throw may be caused, especially when cut-hole blasting is used when there is no free face (Fig. 9.5). Some preventive and protective measures must be carried out during trench blasting, especially within or near the urban area:

Fig. 9.5 Trench blasting with ground cover of rubber tyre mats

- Use smaller blastholes, reduce the charge weight per delay, and properly increase the delay time between rows to reduce the ground vibration.
- Heavy ground cover to prevent flyrock, such as heavy blasting mats, blasting cages, or cover with thick soil or sand after completion of charging and connecting the initiation network.

References

1. Jimeno CL et al (1995) Drilling and blasting of rock. A.A. Balkema Publishers. ISBN 90 5410 199 7. Copyright ©1995 A. A. Balkema
2. Tamrock (1988) Surface drilling and blasting. In: Naapuri J (ed)
3. 汪旭光主編 (2010) 爆破手冊。冶金工業出版社,2010年, 北京. Guang WX et al (2010) Handbook of blasting. Metallurgical Industry Press, Beijing. ISBN: 978-7-5024-5345–9

Chapter 10
Contour Blasting Tcchnique for Surface Excavation

Contour blasting is also called controlled blasting as its main purpose is to minimize the damage to the rock (overbreak) beyond the boundary of the designed contour of the excavation (Fig. 10.1). These techniques are used in both surface and underground excavations by blasting. But it has to be emphasized that these techniques are not a cure-all, it depends primarily on the geology of the rock formation being blasted.

10.1 Types of Contour Blasting

There are several different techniques for contour blasting that have been developed since the 1950s:

- Presplitting;
- Smooth blasting;
- Cushion blasting, and
- Line drilling.

For construction excavation, presplitting and smooth blasting are the most common techniques to form the permanent slopes of the excavated area, and these will be discussed in detail in the following section. Line drilling is used sometimes when the rock formation is suitable, e.g., the homogeneous rock, or where the weakness planes in the rock mass are almost parallel to the designed slope face. Cushion blasting is seldom used in construction excavation and mostly used in large open pits and quarries.

D. Zou, *Theory and Technology of Rock Excavation for Civil Engineering*,
DOI 10.1007/978-981-10-1989-0_10

Fig. 10.1 Comparison between the results achieved using control blasting (*on the left*) and normal bulk blasting for a surface excavation in gneiss

10.2 Presplitting and Smooth Blasting

10.2.1 The Characters of Presplitting and Smooth Blasting Holes

Presplitting and smooth blasting involve a single row of closely spaced holes drilled along the designed contour line. In construction projects, the holes are usually the same diameter as the bulk blasting holes (50–101 mm). All the presplitting or smooth holes are lightly loaded and decoupled with well-distributed explosives.

The main difference between presplitting and smooth blasting is the firing sequence of the contour holes relative to the bulk blastholes in front of them. The charges in presplitting holes are detonated simultaneously, prior to the bulk charges in the main area to be excavated. On the contrary, all charges of the contour holes in smooth blasting are fired either together with the bulk holes or after them.

10.2.2 The Mechanism of Presplitting and Smooth Blasting

The main function of the presplitting and smooth blasting holes is to create a fracture plane along the designed contour of the excavation area when the holes are detonated almost simultaneously. The mechanism of the formation of the fracture plane can be explained as the following three aspects:

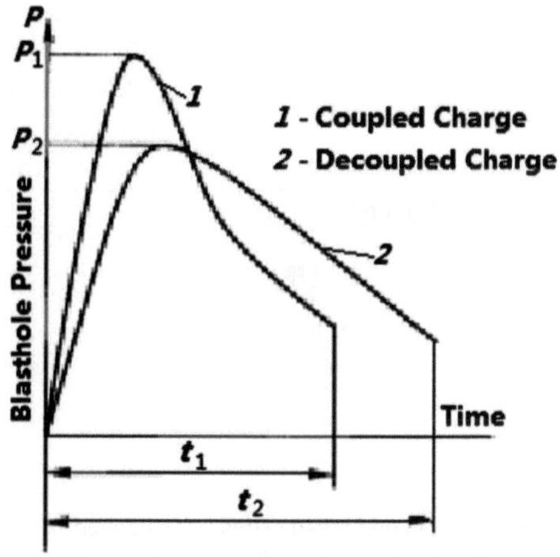

Fig. 10.2 Blasthole pressure–time curve

- Function of decoupling

As the presplitting or smooth blasting holes are loaded with decoupled charges, the effective pressure of the shock waves and gas on the borehole wall is cushioned (Fig. 10.2) due to the buffer function of the annulus air or other inert material, and it can expressed as:

$$P_b = P_e \times \left(\frac{V_e}{V_b}\right)^{1.2} \tag{10.1}$$

where

P_b Effective pressure on the borehole wall;
P_e Detonation pressure of the explosive;
V_e Volume of the explosive; and
V_b Volume of the borehole.

This effect reduces the crushing and radial cracking of the rock around the boreholes.

- Effect of stress superposition/guiding (or dominating) effect causing crack by close empty hole.

When the charges of two adjacent holes are fired simultaneously, the strain waves generated by the two charges collide in the central point between the two holes creating the effect of superposition (Fig. 10.3). The complementary tangential tensile components of the strain waves, if they exceed the dynamic tensile breaking strength of the rock, produce a new crack and favor the propagation of the radial crack in the designed direction of the cut face.

This theoretical explanation of the mechanism of splitting is based on the assumption that the adjacent two charges are detonated simultaneously or close to simultaneously, but lots of practical evidence shows that a good crack along the

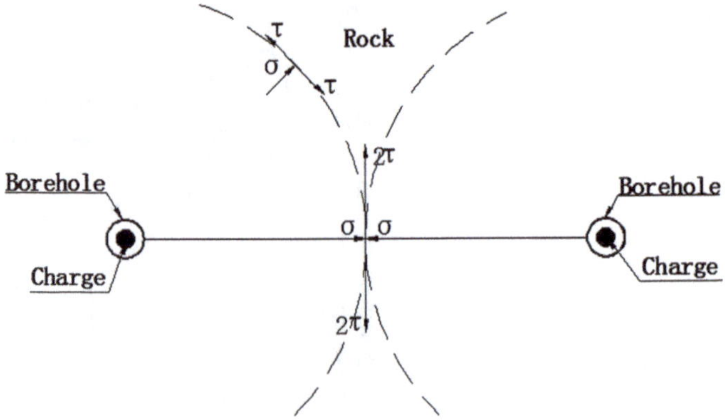

Fig. 10.3 Double tensile stresses generated by the collision of the strain waves produced by the simultaneous detonation of two adjacent charges

Fig. 10.4 Slope face formed by presplitting

axis of presplitting holes can still be formed even if the presplitting holes are detonated with 25 even 50 ms delay between holes. Langefors and Kihlström gave another explanation of the mechanism of formation of cracks in contour blasting in their book according to their experiments [1]: "When an elastic material with an empty circular hole is in a state of tensional stress, for instance from a detonating charge in an adjacent hole, it can be shown by calculations that there is a three-fold increase of the stress at the two points of the empty hole that are closest and farthest from the loaded one. This gives an effect causing cracks at this empty hole if it is close enough to the charged one. These cracks tend to connect the holes."

- Gas penetration and extension crack

The cracking between boreholes produced by the strain waves is subsequently extended and widened by the expanding gases to form a fracture plane in accordance with the designed contour. This split or crack in the rock forms a discontinuous zone which minimizes or eliminates overbreak from the subsequent primary blast and produces a smooth finished rock wall (Fig. 10.4).

10.3 Parameters of Presplitting and Smooth Blasting

10.3.1 Theoretical Approaches

- Paine et al. in the 1960s raised a theoretical approach to calculate the parameters of presplitting and smooth blasting based on the two essential prerequisites: no crushing on the borehole wall and adjacent two holes firing simultaneously to form a crack through the two holes [2].

According to the first prerequisite, the pressure acting on the wall of the borehole should meet the condition: $P_b < \sigma_c$. According to formula (10.1):

$$P_b = P_e \times \left(\frac{V_e}{V_b}\right)^{1.2} \quad (10.1)$$

Referring to Fig. 10.3, the compressive stress, P_M ($P_m = \sigma$ in Fig. 10.3), at the middle point where the strain waves from two adjacent holes collide can be expressed as:

$$P_M = P_b\sqrt{\frac{2r}{S}} \times e^{-\frac{KS}{2C}} \quad (10.2)$$

where

r The radius of the borehole;
S Spacing between two holes;
K Time coefficient;
C Velocity of the strain wave; and
E The base of the natural logarithm.

The tangential tensile component of the strain waves τ is:

$$\tau = E\varepsilon_0 = -\mu\sigma = -\mu P_M \tag{10.3}$$

where

E Young's modulus;
ε_0 The strain in the tangential direction of the strain wave;
μ Poisson's ratio.

As it is difficult to get the time coefficient K, Duvall suggested to replace the last item of the Formula (10.2) $e^{-\frac{KS}{2C}}$ with $e^{-\frac{\alpha S}{2r_e}}$, thus r_e

$$\tau = -\mu P_b \sqrt{\frac{2r}{S}} \times e^{-\frac{\alpha S}{2r}} = -\mu P_b \sqrt{\frac{2r}{S}} \times e^{-\frac{\alpha S}{d}} \tag{10.4}$$

where

α The absorption constant of rock;
r_e Radius of explosive charge, and
D Diameter of explosive charge.

When $\tau > \sigma_T$, where σ_T is the tensile strength of the rock, a crack through the two holes will be formed.

- Feshenko and Eristov (Фещенко, А. А. and Эристов, В. С., scholars of the former Soviet Union) (refer to [2]) raised the formulas of the linear charge concentration and hole spacing for presplitting:

$$\Delta_e = 0.1\pi r^2 \frac{\sigma_c \Delta\left(2.5 + \sqrt{6.25 + \frac{1400}{\sigma_c}}\right)}{100 Q_u} \tag{10.5}$$

$$S = 1.6 \left[\frac{\left(\frac{\sigma_c}{\sigma_t}\right)\mu}{1-\mu}\right]^{2/3} d_h \tag{10.6}$$

where

Δ_e Linear charge concentration;
R The radius of the borehole;
Q_u Detonation heat of explosive;
μ Poisson's ratio; and
S Spacing between two holes;
σ_c Compressive strength of rock;
σ_t Tensile strength of rock; and
d_h Diameter of borehole.

10.3.2 Empirical Approaches

- Hole diameter and decoupling

In surface excavation, the most frequent diameters used for contour blasting are in the 35–76 mm range and drilled with the same drill rigs used for the bulk blast holes.

As discussed above, the decoupled charges of presplitting or smooth blasting holes can reduce the crushing and radial cracking of the rock around the boreholes. The adequate coefficient of decoupling, *d/D* where *d* is the diameter of charge and *D* is the diameter of blasthole, is 2–5 [2].

- Hole spacing and burden [3]

In smooth blasting, the spacing between the boreholes is usually 15–16 times the borehole diameter and the burden (the distance between the boreholes and the free face created by the previously detonated blastholes) is 1.25 times the spacing.

For presplitting, the spacing is normally 8–12 times the borehole diameter and the burden can be considered infinite.

- Linear charge concentration

– Persson et al. empirical formula [4] gives the minimum required linear charge concentration for smooth blasting and presplitting as a function of the hole diameter:

$$\Delta_e = 90d_h^2 \tag{10.7}$$

where

Δ_e Linear charge concentration of ANFO-equivalent explosives in kg/m;
d_h Diameter of borehole, in m

– Empirical formulas of Gezhouba` Dam Construction Bureau, China [2]:

$$\Delta_e = 9.318\sigma_c^{0.53}r^{0.38}\ \text{kg/m} \tag{10.8}$$

Under the rock condition of $\sigma_c = 10 - 150\ \text{MP}_a$ and $r = 23 - 85$ mm.

$$\Delta_e = 0.595\sigma_c^{0.5}\ S\,\text{kg/m} \tag{10.9}$$

Under the rock condition of $\sigma_c = 20 - 150\ \text{MP}_a$ and $S = 45 - 120$ cm.

where

σ_c Compressive strength of rock, in MPa;
R The radius of the borehole, in mm; and
S Spacing between two holes in cm.

Table 10.1 Presplitting parameters (Gustafsson 1981) (reproduced from Ref. [7] with permission from Taylor & Francis Group)

Hole diameter (mm)	Charge concentration (kg/m)	Charge type	Spacing (m)
25–32	80 g	Detonating cord	0.30–0.60
25–32	0.30	Pipe charge 17 mm	0.35–0.60
40	0.30	Pipe charge 17 mm	0.35–0.50
51	0.60	Pipe charge 17 mm	0.40–0.50
64	0.46	Charge 25 mm	0.60–0.80

Table 10.2 Parameters of presplitting and smooth blasting offered by Sandvik and Tamrock, (reproduced from Ref. [9] withe permission of Sandvik)

Hole dia.		Presplitting			Smooth blasting		
mm	In	Charge concentration (kg/m)	Hole spacing (m)	Specific drilling (m/m^2)	Charge concentration (kg/m)	Hole spacing (m)	Burden (m)
32	$1\frac{1}{4}$	0.13–0.21	0.45–0.7	2.22–1.43	0.21	0.6–0.8	0.9–1.1
38	$1\frac{1}{2}$	0.21	0.45–0.7	2.22–1.43	0.21	0.6–0.8	0.9–1.1
51	2	0.38–0.47	0.5–0.8	2.00–1.25	0.38–0.47	0.7–1.0	0.9–1.4
64	$2\frac{1}{2}$	0.38–0.55	0.5–0.9	2.00–1.43	0.38–0.55	0.7–1.3	0.9–1.6
76	3	0.55–0.71	0.7–0.9	1.67–1.11	0.55–0.71	1.0–1.3	1.2–1.7
89	$3\frac{1}{2}$	0.90–1.32	0.7–1.1	1.43–0.91	0.90–1.32	1.2–1.5	1.7–2.0
102	4	0.90–1.32	0.7–1.1	1.43–0.91	0.90–1.32	1.2–1.5	1.7–2.0

Table 10.3 Presplitting parameters offered by MIMR, China, Ref. [8]

Hole diameter (mm)	Spacing (m)	Charge concentration (kg/m)	Charge type
32	0.3–0.5	0.15–0.25	#2 Rock AN
42	0.4–0.6	0.15–0.30	
50	0.5–0.8	0.20–0.35	
80	0.6–1.0	0.25–0.50	
100	0.7–1.2	0.30–0.70	

- Some empirical data from different approaches for reference

1. R. Gustafsson's data for presplitting, 1981 (Table 10.1).
2. Data offered by Sandvik and Tamrock (Table 10.2).
3. Data offered by Maanshan Institute of Mining Research, China (Table 10.3).

10.4 Blast Design, Charging, and Initiation for Presplitting and Smooth Blasting

10.4.1 Blast Design of Presplitting and Smooth Blasting

Figure 10.5 is a typical blast design of presplitting with 76-mm diameter blastholes. Usually, there is a row of buffer holes in the front of the presplitting holes with 1/3 to 1/2 charge of the production holes. In hard rock, a row of additional shallow holes is needed to help break the top rock between the presplitting holes and buffer holes.

The firing of presplitting holes can be done together with the production holes, but the detonation should be from 90 to 120 ms (at least 50 ms) in advance of the production holes [5, 6].

The blasthole pattern of smooth blasting is similar to the presplitting, but the distance (burden) between the contour holes and the holes of buffer row is larger than presplitting and the spacing between smooth holes are also a little bit larger (refer to tables above). The main difference between presplitting and smooth blasting is that with smooth blasting all holes are detonated after all production blastholes in front of them, with a delay about 75–150 ms.

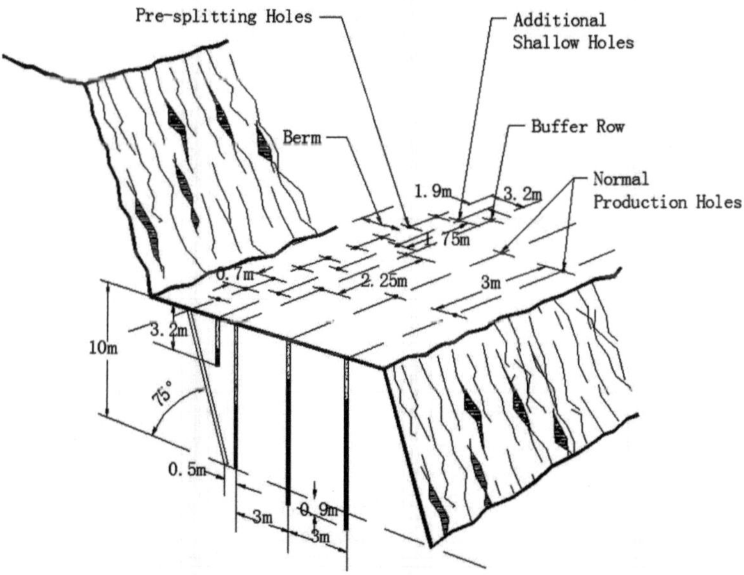

Fig. 10.5 Typical design of presplitting, hole diameter: 76 mm

10.4.2 Explosive Charging for Presplitting and Smooth Blasting Holes

- Conventional Charges

Small diameter cartridges (22–32 mm) are tied and placed with certain separation (according to the designed linear concentration) on a detonation cord (5 or 11 g/m) then fixed on a long bamboo stick down-loaded in the blastholes (Fig. 10.6a) for the blasthole diameter of 50–76 mm.

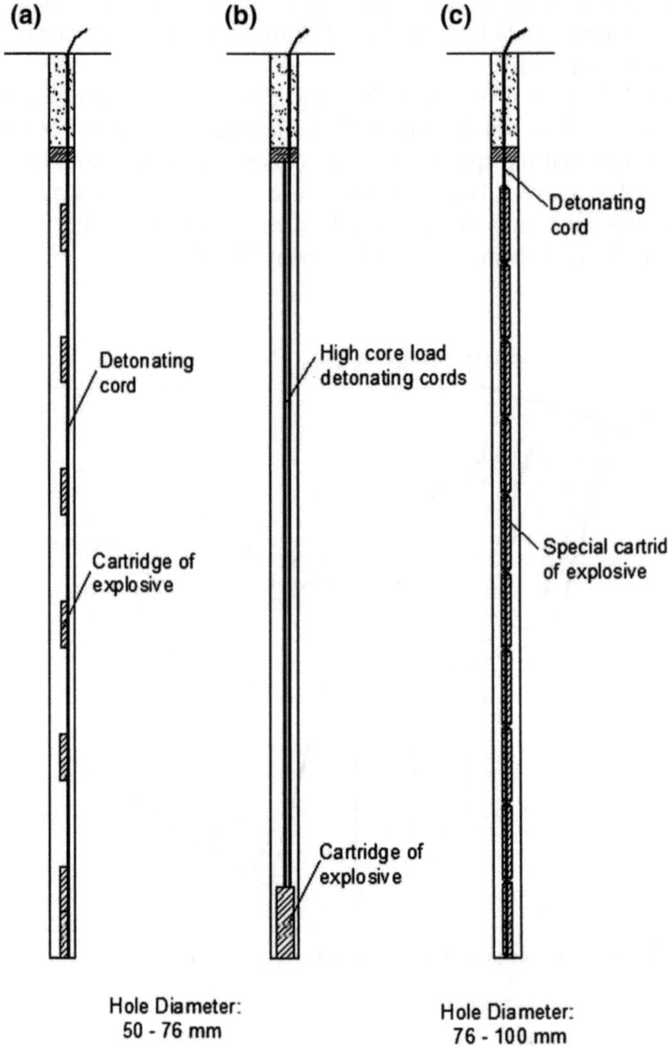

Fig. 10.6 Typical charge loads for contour blasting with different hole diameters

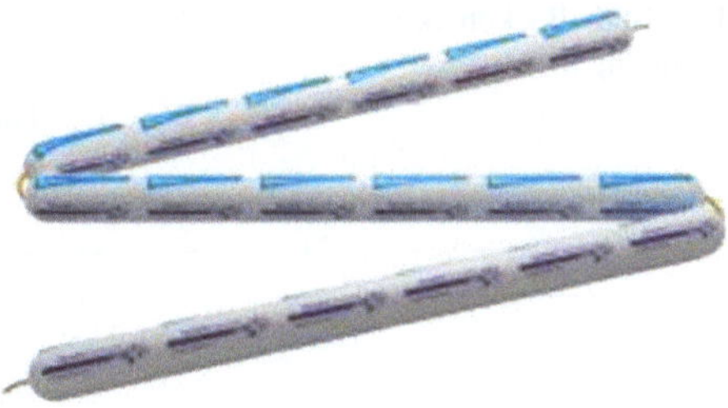

Fig. 10.7 Orica's product of SenatelTM PowersplitTM for presplitting/smooth blasting. Diameter: 26/32 mm, length per clip: 400 mm

- Detonating Cord

Usually two lines of high core load detonating cord (40, 60, or 100 g/m) with a cartridge of emulsion explosive (32–60 mm diameter, according to the blasthole diameter and length) in the bottom are loaded in the blastholes, see Fig. 10.6b.

- Special Cartridges

The manufacturers of explosives supply some specially designed cartridges for contour blasting to facilitate and speedup the charging of blastholes (see Fig. 10.6c). Figure 10.7 shows one of these kinds of products. The emulsion cartridges are internally traced with 10 g/m detonating cord that ensures fast and complete detonation.

10.4.3 Stemming and Initiation of Contour Blasting Holes

10.4.3.1 Stemming

Some researchers believe that no stemming is needed for presplitting or smooth blasting as the stemming may cause the damage to the designed berm of the slope especially where the detonating cord passes through the stemming (see Fig. 10.8) and the crack plane cannot be formed between adjacent two holes in the stemming area, resulting in "nose of ox" that can be found in tunnel blasting with smooth blasting.

According to our experience of presplitting for surface excavation, it is believed that:

- The damage of the berm usually is caused by the unfavorable discontinuities in the rock mass to be blasted, even when no stemming is adopted.
- If there is no stemming, the explosion gas will escape to the atmosphere prematurely and affect the formation of the crack plane between holes.
- No stemming or insufficient stemming will produce airblast and flyrock; this is never permitted when the blasting site is within or near the residential area.

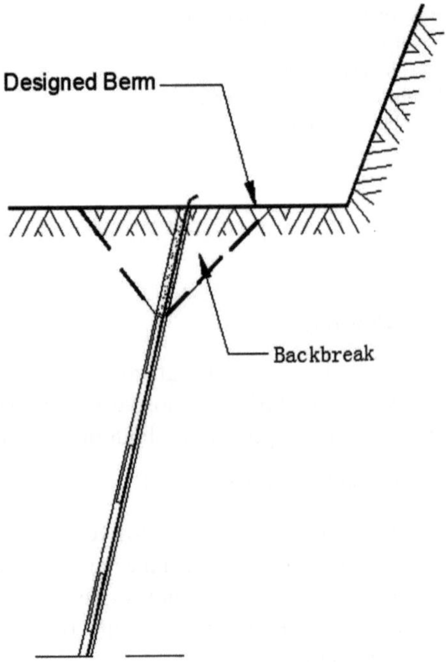

Fig. 10.8 Backbreak as stemming

- But excessive stemming is also not recommended as it will seriously affect the quality of the formed slope face and seriously damage the designed berm of the slope.

  So it is recommended that:

- Stemming column should be 12 times the hole diameter but not more than 1.2 m (4 ft).
- Some plastic plugs, like VARI-STEM plugs (Fig. 10.9), are recommended to be used for contour blasting with stemming. The use of the plug together with granular particle materials can effectively improve the stemming effect, reduce the length of stemming to 10 times the hole diameter, and greatly reduce the airblast produced by the contour blasting.

10.4.3.2 Initiation of the Contour Blast

- Initiation with a trunk line of detonating cord

The common method of initiation of the contour blastholes is to use a trunk line of detonating cord (see Fig. 10.10). For reducing the airblast and noise produced by the detonating cord, a light load core of 0.5 g/m of detonating cord is usually used as the trunk line. Due to the very high detonating velocity of detonating cord (about

Fig. 10.9 VARI-STEM plug inserted into blasthole

6700 m/s), all the blastholes connected to the trunk line are considered as initiated simultaneously. To reduce the ground vibration, the blastholes can be divided in groups, and shock tube delay detonators or detonating relays are used to connect each group one by one.

- Initiation with In-Hole Detonators

When a trunk line of detonating cord is used to initiate contour blastholes, airblast and loud noise are often produced by the detonating cord, even a light loaded cord (5 g/m) and when ground cover is used. To decrease the adverse effect, an initiation system with in-hole detonators is used for initiation of contour blasting. Although electric delay detonators without surface connectors can be used, shock tube detonators are most commonly used, such as the Unidet system (Fig. 10.11).

When instantaneous detonators (o ms delay) are used for all the in-hole detonators, the blastholes in the same group (up to 5 or 6 holes as limited by the number of shock tubes connected by the surface connector) are initiated simultaneously. The delay time between groups is controlled by the surface delay connectors.

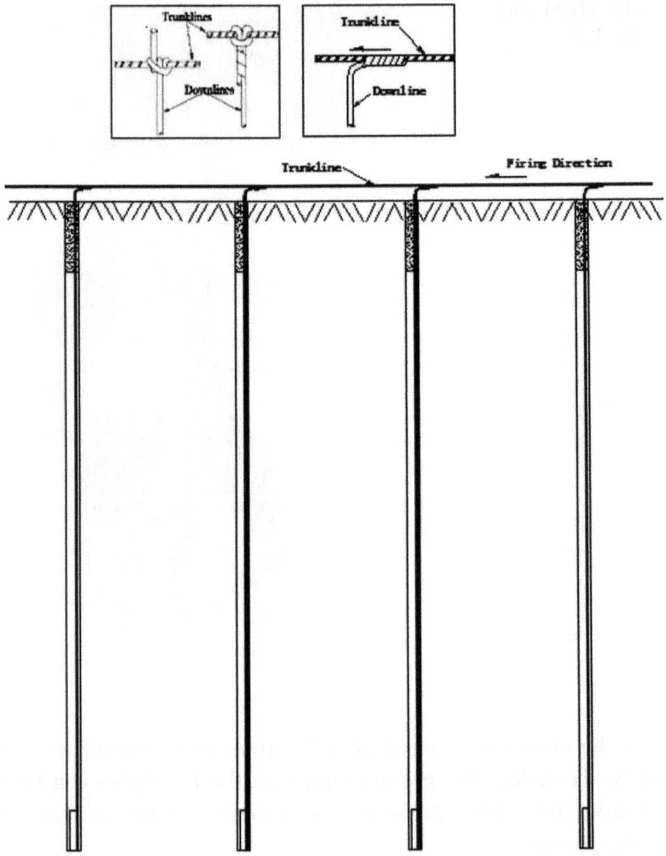

Fig. 10.10 Initiate contour holes simultaneously with a trunk line of 5 g/m detonating cord

10.5 Cushion Blasting and Line Drilling

10.5.1 Cushion Blasting

In cushion blasting, a single row is drilled along the perimeter of the excavation. Drillhole size is the same as the production holes and varies between 50 and 100 mm for construction excavation.

Similar to smooth blasting, the spacing between holes is normally larger than that used in presplitting. The burden is designed so that it is greater than the spacing, $B = 1.2\ 1.3S$. In general, subdrilling is not necessary. Table 10.4 gives the approximate values of parameter for cushion blasting for reference.

Decoupling is used when charging. The explosive are well distributed with certain separation along a detonating cord (usually 11 g/m) (Fig. 10.12a) or loaded in the bottom of the hole (Fig. 10.12b). In the early years, the inert materials, like

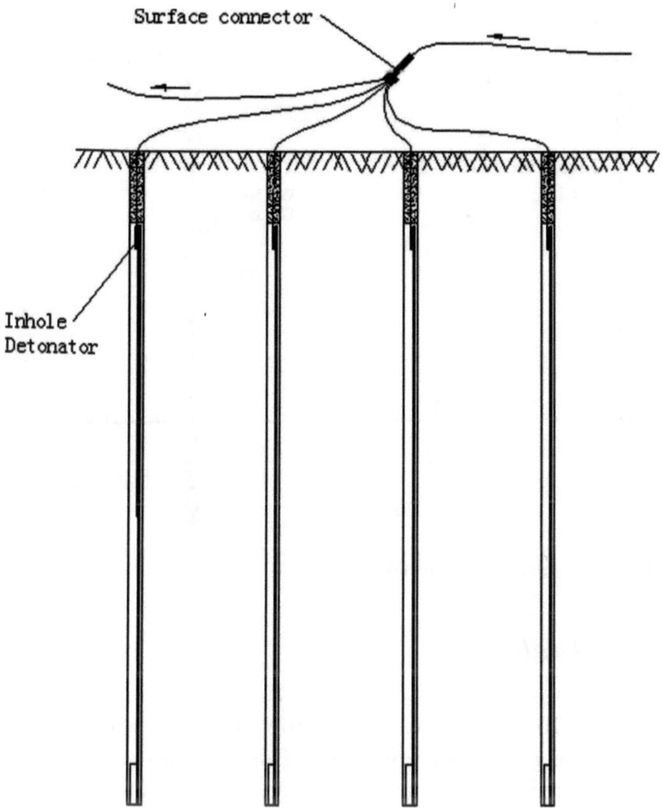

Fig. 10.11 Initiate contour holes using both in-hole and surface shock tube detonators

Table 10.4 The approximate values of parameters for cushion blasting (reproduced from Ref. [9] with the permission of Sandvik)

Drillhole diameter (mm)	Charge concentration emulsion or Dynamex (kg/m)	Burden (m)	Hole spacing (m)	Stemming (m)
50–64	0.12–0.35	1.20	0.90	1.20
75–89	0.20–0.70	1.50	1.20	1.50
102–114	0.35–1.10	1.80	1.50	1.80
127–140	1.10–1.50	2.10	1.80	2.10
152–165	1.50–2.20	2.70	2.10	2.70

drill cuttings, sands, or soil, are placed in the void space around the charges (Fig. 10.12a), but in recent years they are replaced by air except in the collar part of the blastholes. The holes are fired after the main excavation is removed. A minimum delay between the holes is desirable.

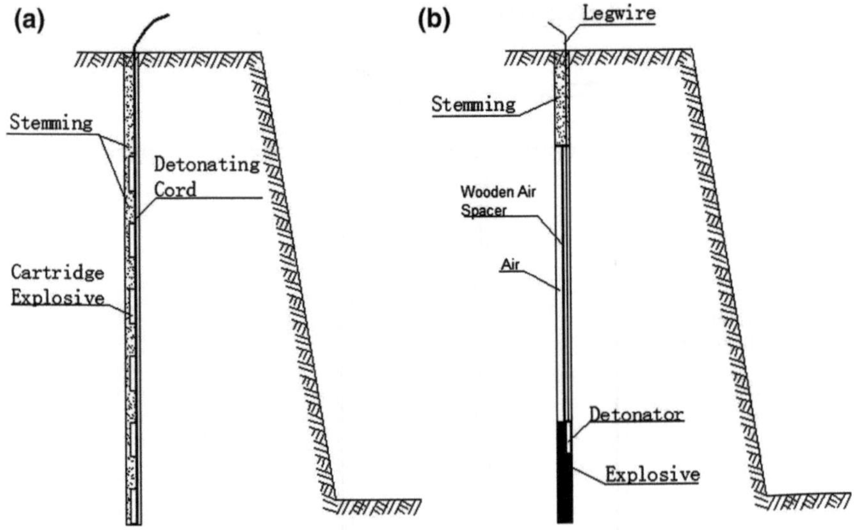

Fig. 10.12 Charging structures of cushion blasting

10.5.2 Line Drilling

Line drilling (Fig. 10.13) involves a single row of closely spaced, unloaded, small diameter holes along the excavation line. This provides a weak plane that the primary blast can break. The diameter of line drilling holes are generally 51–76 mm (2″–3″) and are spaced at 2–4 times the hole diameter along the excavation line. Holes larger than 76 mm are seldom used in line drilling since higher drilling costs cannot be offset by increased spacing.

Drilling precision is very important to obtain good results, as well as the homogeneity of the rock because, if not, the natural fissures of the rock tend to create planes of weakness more easily than that caused by the drilled holes.

Blastholes directly adjacent to the line drill holes, which act as the buffer holes, are generally loaded lighter and are more closely spaced than the primary blastholes. The distance between the line holes and the directly adjacent blastholes is usually 50–75 % of the normal burden.

Line drilling is applicable in areas where even light explosive loads associated with other controlled blasting techniques may cause damage beyond the excavation limit. The explosion gases produced by the blastholes can escape to the atmosphere through the unloaded line holes, and the strain waves produced by the blast are also reflected from the line holes, which reduces shattering and stressing in the finished wall. The ground vibration induced by blasting beyond the excavation limit can be obviously reduced.

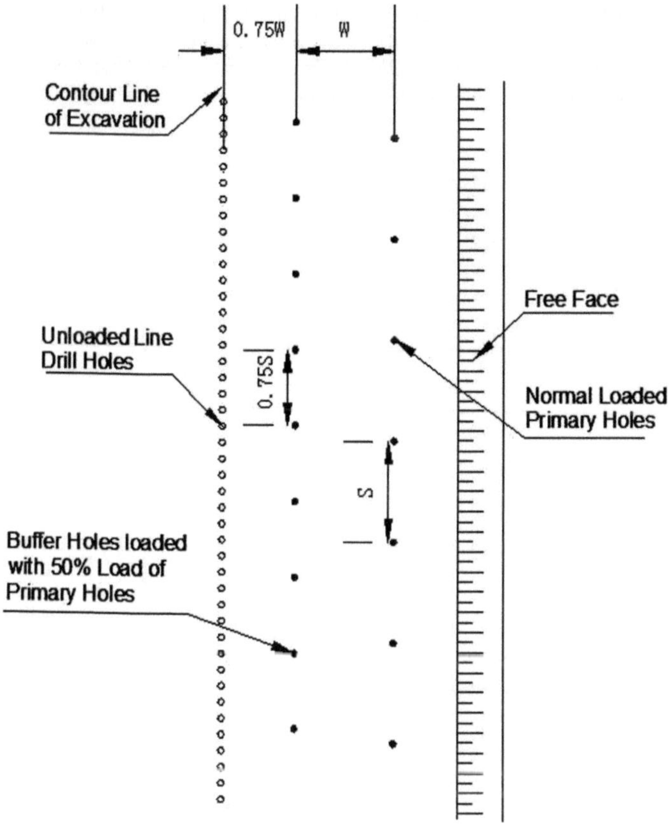

Fig. 10.13 Typical pattern of line drilling

References

1. Langefors U, Kihlstrom B (1978) The modern technique of rock blasting, 3rd edn. Halsted Press, New York. ISBN 0-470-99282-4
2. Huang SJ (1996) Engineering blasting design. Ordnance Industry Press, Beijing. ISBN: 7-80132-004-2. (in Chinese) 黄紹鈞 (1996): 工程爆破設計。兵器工业出版社,北京,1996
3. Hook KE, Brown ET (1980) Underground excavation in rock. Institution of Mining and Metallurgy, London. ISBN 9780415461658
4. Persson PA et al (1994) Rock blasting and explosives engineering. CRC Press, Boca Raton. ISBN 0-8493-8978-X
5. Jimeno CL et al (1995) Drilling and blasting of rock. A.A. Balkema Publishers, Amsterdam. ISBN 90 5410 199 7
6. Wong XG et al (2010) Handbook of blasting. Metallurgical Industry Press, Beijing. ISBN: 978-7-5024-5345-9. 汪旭光主編(2010), 爆破手冊。冶金工業出版社,2010年, 北京。
7. Bhandari S (1977) Engineering rock blasting operation. A.A. Balkema, Rotterdam. ISBN 905410 658 1

8. Gao S et al (1985) Study on the cracking mechanism of pre-splitting. In: Proceedings of soil and rock blasting, 2nd series. Metallurgical Industry Press, Beijing (in Chinese). 高士才等(1985), 預裂爆破成縫機理的研究, "土岩爆破文集" 第二輯, 冶金工業出版社,1985, 北京。
9. Matti H, Editor in Chief (1999) Rock excavation handbook for civil engineering. Sandvik & Tamrock

Chapter 11
Blasting Safety for Surface Blasting

One of the largest single problems that are faced by a blasting contractor is blasting safety. The principal disturbances created by blasting are flyrock, ground vibration, and airblast. All of them can, under some circumstances, cause damage to structure nearby and, apart from this, be the source of serious conflict with the inhabitants who live close to the operation.

In order to solve these problems, it is necessary to have more highly qualified blasting engineer and blasting supervisors so that they can reduce the level of disturbances at a reasonable cost. Another issue to take into account is the job of information and public relations, which is becoming more necessary, undertaken by the directors of the project. Moreover, as well as careful blast design, careful monitoring of blast effects, and meeting with all those potential affected inhabitants to explain the care used to protect their property and safety is necessary. In situations where complaints persist, continued attention to blast design, effective monitoring, and good record keeping may take on even greater importance. If the complaint is taken to the authorities, records detailing blast design and monitoring records can be very important when discussing the problem.

11.1 General Roles of Blasting Operation

(Reproduced form Ref. [1] with the permission from Sing Tao Publishing Group Ltd.)

- No person except registered shotfirers can handle and prepare the explosive products.
- No person shall smoke at or near any blasting site while explosives are being removed from a store thereat or while charges are being prepared for blasting or are being laid.
- No person except the registered shotfirer in charge can set off the shot.

D. Zou, *Theory and Technology of Rock Excavation for Civil Engineering*,
DOI 10.1007/978-981-10-1989-0_11

- Any blasting operations, including storage, transportation, loading, firing, and destroying of any explosive products should be in accordance with the regulation, permit conditions and restrictions imposed by the government.
- No person except the registered shotfirers can handle any misfire. When a misfire is being handled by the shotfirer, the evacuation procedure should be remained effective.
- No blast shall be fired unless effective and adequate precautions are taken to prevent any fragments being projected in a dangerous manner, in particular going beyond the site boundary.
- An approved evacuation procedure should be carried out strictly. For a period lasting from 5 min prior to blasting until all charges have been fired, warning gongs shall be beaten continuously so as to be audible at a distance of 150 m from the blast and red flags shall be displayed continuously at all points of access to the place of blasting and at a distance of 150 m from the blast.
- Any person who, after the commencement of the warning signals referred to in above paragraph, enters or, upon request being made to him by any public servant or any person engaged in the blasting, refuses to leave the blasting area shall be guilty of an offense.
- Any person who finds any remaining explosive products in the site should report to the registered shotfirer immediately, and no person except the registered shotfirers can handle and destroy the remaining explosives.
- A strict and effective managing system is one of the most important guarantees to the safety and project progress. The principle of an effective management system should be that "strictly carry out every step of the procedure of blasting works and clarify the duty and responsibility of each step and every person in the working procedure." An example of the flowchart of the procedure of blasting works for a major site formation project is shown in Fig. 11.1 (Refer to [1]).

11.2 Flyrock and Its Control

11.2.1 Flyrock and Its Cause of Formation

Flyrock, also called rock throw, is the uncontrolled propelling of rock fragments produced in blasting and constitutes one of the main sources of material damage and harm to people.

Flyrock usually appears on the front free face or top free face of the blasted bench. Figure 11.2 shows the schematic crater effects that cause flyrock, and Fig. 1.40 in Chap. 1 shows a flyrock accident occurring on a bench top.

Some of the common causes of flyrock are as follows:

- Adverse geological conditions of rock mass.

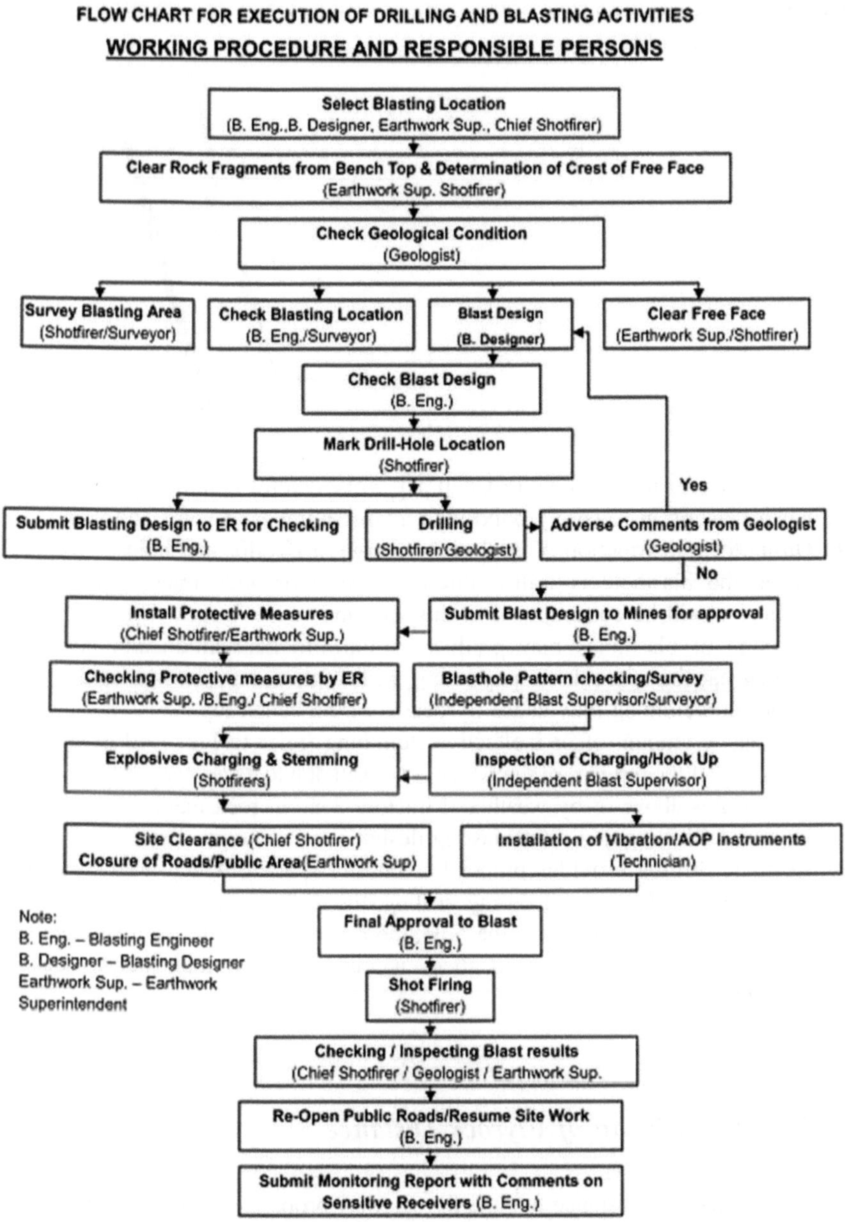

Fig. 11.1 Blasting management procedure

Intensely fissured and jointed rocks facilitate the appearance of flyrocks more than massive and homogeneous rocks, especially when the face is very irregular due to the unfavorable joints, causing reduced burden in front of explosive

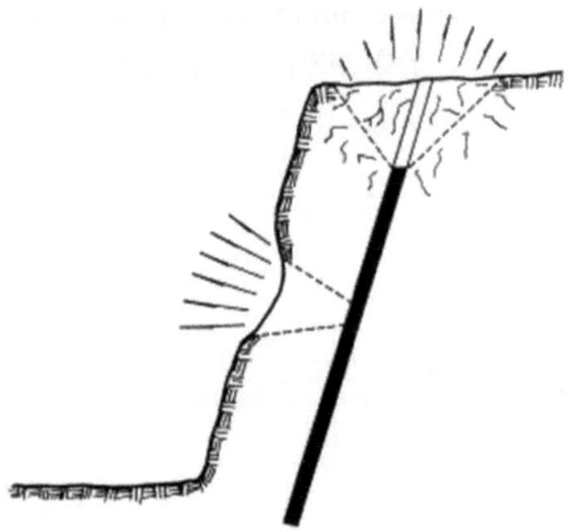

Fig. 11.2 Hackneyed crater effects to cause flyrock in bench blasting

column (refer to Fig. 1.38 in Chap. 1). Very careful control should be taken when blasting in Karstic ground with a large number of voids and vugs.

- Un-uniform distribution of explosives charge or overloading holes.
  When the blasthole is drilled through a void or wide fracture, the loaded explosive will be concentrated in this part and cause flyrock. Un-uniform distribution of blastholes may produce un-uniform distribution and inadequate burden causing flyrock. Top priming with high energy booster and without sufficient stemming may cause flyrock from the bench top.
- Inadequate amount of or ineffective stemming (refer to Fig. 1.40 in Chap. 1), especially if the top part of the bench is broken due to excessive subdrilling from the benches above or by weathered fracture zone in rock mass.
- Inaccurate drilling and poor blasting design.
  Accurate drilling provides proper burden and spacing which are essential for better results. However, if during drilling, the driller is not guided properly for the position and direction of the hole, the drill may deviate from its calculated position and inclination, and then, it may cause flyrock during blasting (Fig. 11.3).

11.2.2 Calculation of Flyrock Distance

How to estimate the distance that the flyrock may propel is an important issue, as it is the base for determining the safe distance of blasting within which all people should be evacuated.

Nils Lundborg et al of the Swedish Detonic Research Foundation did effective and efficient research works through the model and field experiments and theoretical analysis in the 1970s [2].

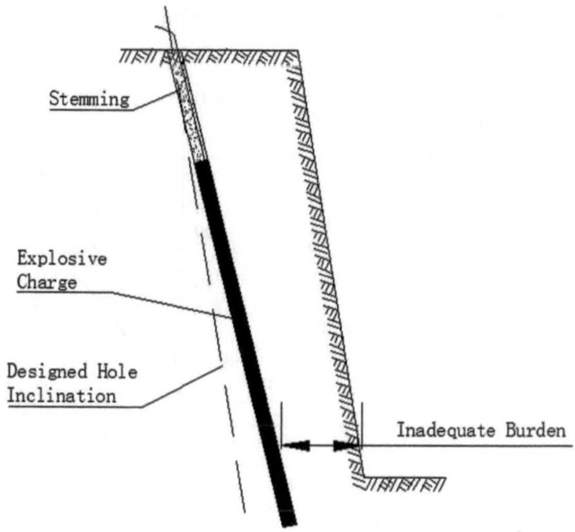

Fig. 11.3 Insufficient burden caused due to deviation of drillholes results in flyrock

For bench blasting, Lundborg et al. [2] had, in his different papers using impulse laws and air resistance, calculated the relative throw distance, L, at blasting with different hole diameters, d:

$$L = 260d^{2/3} \tag{11.1}$$

He also obtained the simple relation as follows:

$$d = k\frac{\rho}{2600}\Phi v \tag{11.2}$$

where

- d diameter of the blasthole in inch;
- L maximum flyrock distance in meter;
- ρ density of the stone (flyrock) in kg/m^3;
- Φ diameter of the stone in meter;
- v velocity of the stone in m/s; and
- k a constant, which can be determined experimentally. It was taken $k = 0.1$ in Lundborg's experiments.

Figure 11.4 shows the calculated maximum throw distance as a function of the diameter of the stones with the hole diameter as a parameter. Figure 11.5 shows the throw distance as the function of the specific charge (powder factor) at the hole diameters of 1 and 4 inches.

American researcher, Roth [3], also attempted to obtain the throw range of flyrock. In his approach, the critical variable in all flyrock throw range calculation was the estimation of the initial flyrock velocity, V. But on the comparison between

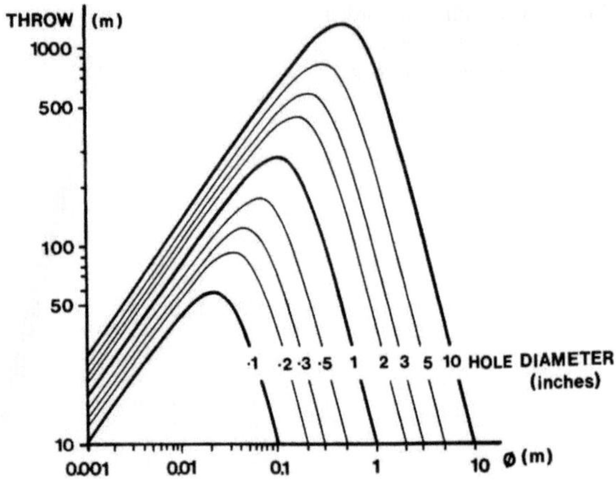

Fig. 11.4 Calculated maximum throw distance as a function of the diameter of the stones with the hole diameter as a parameter (courtesy of SveDeFo, Ref. [2])

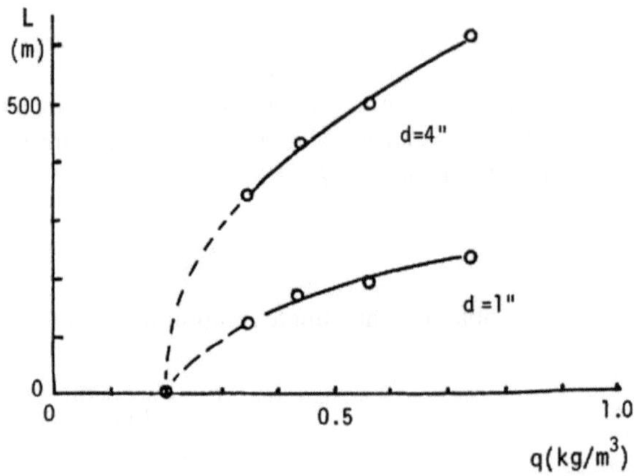

Fig. 11.5 Throw distance as function of the specific charge at 1′ and 4′ hole diameters (courtesy of SveDeFo, Ref. [2])

calculated results and field-measured velocity data, the former values were found to be 1.6 times greater than the observed velocity.

Australia's A. B. Richards and A. J. Moore developed a methodology for quantification of flyrock distances relative to the explosive confinement conditions based on their recent flyrock investigations combined with the research works carried out by Lundborg (1981), Workman et al. (1994), and St George et al. (2001) [16].

In their paper [16], three key mechanisms of flyrock due to the lack of confinement of energy in the explosive column were identified. An illustration of each mechanism is shown in Fig. 11.6.

Based on Workman et al's formula, Richard et al. established the following formulas to calculate the maximum flyrock throw distances from a blast:

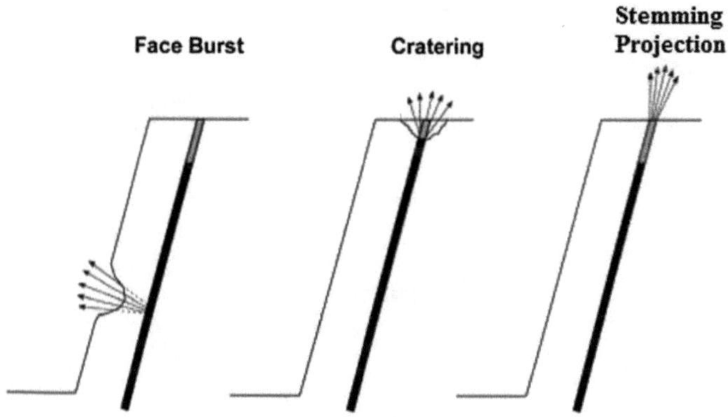

Fig. 11.6 Three key mechanisms of flyrock

Face burst is as follows:

$$L_{\max} = \frac{k^2}{g}\left(\frac{\sqrt{m}}{B}\right)^{2.6} \tag{11.3}$$

$$\text{Cratering:} \quad L_{\max} = \frac{k^2}{g}\left(\frac{\sqrt{m}}{SH}\right)^{2.6} \tag{11.4}$$

$$\text{Stemming Ejection:} \quad L_{\max} = \frac{k^2}{g}\left(\frac{\sqrt{m}}{B}\right)^{2.6} \sin 2\theta \tag{11.5}$$

where

θ	drillhole angle;
$L_{\max}$	maximum throw (m);
m	charge mass/m (kg/m);
B	burden (m);
SH	stemming height (m);
g	gravitational constant; and
k	constant relative to rock characters.

The empirically determined constant k range is 13.5 for soft competent rock, such as coal overburden, and 27 for hard competent rock, such as basalt or granite.

Because the throw distance increases greatly with small changes to burden or stemming height, it is considered reasonable that the maximum throw distance be doubled to determine the minimum clearance distances to plant and equipment and this be doubled again to determine the minimum clearance distance for personnel.

The clearance distance to the sides of a blast face should also be considered. The shape of the recommended clearance zone is shown in Fig. 11.7.

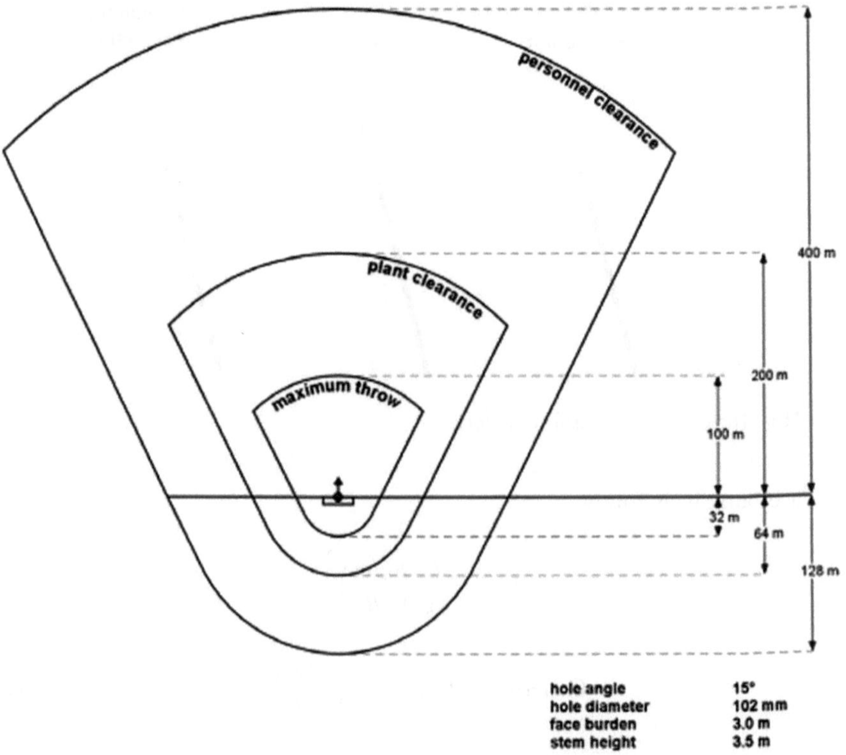

Fig. 11.7 Recommended minimum clearance zone—102-mm (4′) blastholes (courtesy of ISEE, Ref. [16])

11.2.3 Precaution and Protective Measured for Flyrock

11.2.3.1 Measures for Reducing the Risk of Flyrock

For precaution against flyrock in blasting the following should be kept in mind:

- Carefully inspect the geological conditions of the rock mass to be blasted and make a proper blasting design to suit the conditions.
- Keep a proper burden of blast holes specially the holes close to the free face.
- Have good stemming using free-flowing materials.
- Take adequate protective measures to protect nearby properties and public area.

11.2.3.2 Evacuation Distance and Evacuation Procedure

To guarantee the safety of personnel and equipment, an effective evacuation procedure must be carried out during blasting.

The range of evacuation is determined according to the safe distance which may be specified in the blasting assessment and method statement which is proposed by the blasting engineer according to the regulations or specifications and site condition.

The safe distance considered adequate for blasting in hilly terrain in Hong Kong takes into account of the relative differences of blast elevation and usually downslope sensitive receivers or area of concern. A formula is used in Hong Kong for calculating the safe distance from flyrock:

$$D_s = 150 \times \frac{d}{75} + (H_1 - H_2) \tag{11.6}$$

where

D_s safe distance for flyrock, m;
d diameter of blastholes, mm;
H_1 blast location elevation, mPD;
H_2 flyrock sensitive receiver elevation, mPD.

For example, if $d = 75\,\text{mm}$, $H_1 = 170\,\text{mPD}$, and $H_2 = 122\,\text{mPD}$, then the safe distance from flyrock is $D_s = 198\,\text{m}$, say 200 m. All people within the area which has a distance less than 200 m from the blasting area should be evacuated, and beyond this the normal activities can be carried out during the blast.

People must be prevented from entering into the evacuation area during and immediately after the shot. It requires the proper placement of guards and barricades on all accesses to the blast site. It also requires constant and effective communication between the blasters and the persons responsible for the site control.

All equipment must be moved to a safe distance from the shot and keep there until after the shot is fired.

If the blasting site is close to public roads, approval for road closure should be obtained from the local authorities and carried out during blasting with the help of the police.

11.2.3.3 Protective Measures for the Prevention of Flyrock

According to the different circumstances, one or more of the following protective measures should be applied for the prevention of flyrock in blasting:

Fig. 11.8 Ground cover with wire mesh mats

(A) The ground cover

(a) General ground cover
The horizontal rock face to be blasted is fully covered with gunny sack (or canvas sheets) and mats of wire mesh then weighed down with sufficient sandbags (Fig. 11.8).

(b) Heavy covers

(1) Rubber tire blasting mats (Fig. 11.9);

Fig. 11.9 Rubber tire blasting mats covering

Fig. 11.10 Blasting cages with counterweight

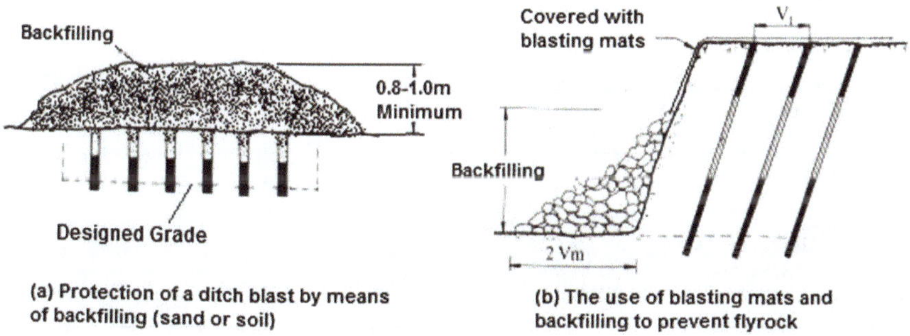

Fig. 11.11 Use of backfilling to prevent flyrock in trench blasting

(2) Blasting cages (Fig. 11.10); and
(3) Backfill (Fig. 11.11).

(C) Vertical blasting screens (movable/fixed) shall be erected at the sides of the blasting area which are facing to the public or properties (Fig. 11.12).

(D) Roofover

In some special situation, a roofover structure may be required to built-up to prevent flyrock from harming public (Fig. 11.13).

Fig. 11.12 Vertical screens and cages to prevent flyrock

Fig. 11.13 Roofover protective structure for MTRC Yau Tong Station of Hong Kong. *Left* external view; *Right* inner view

11.3 Ground Vibration and Its Control

11.3.1 Ground Vibration Produced by Rock Blasting

The most common type of blasting damage is caused by ground vibrations. When an explosive detonates in a blasthole, it generates intense stress waves moving in the rock. Ground vibrations from blasting are acoustic (as opposed to electromagnetic) waves that propagate through the earth. They are also termed "seismic" waves because their propagation characteristics are similar to the ground motions produced by earthquakes. Ground vibration from blasting have much lower peak amplitudes and higher dominant frequencies than earthquake vibration, primarily because of their lower energies and smaller propagation distance.

11.3.1.1 Propagation Velocity and Particle Velocity

As mentioned in Chap. 5, a propagating perturbation through a medium (solid, liquid, gas, or plasma) is called wave. The interface between disturbed zone and the non-disturbed zone is called the wavefront. The movement speed of the wavefront is called the velocity of the wave. A seismic wave such as a P-wave will travel at different velocities through different materials, and the different waveforms travel at different velocities through a single material. These are called wave propagation velocities, wave transmission velocities, or simply seismic wave velocities.

We are interested also in another kind of velocity, that is, the velocity at which the ground surface oscillates as a wave passes under it. That velocity is called a particle velocity, and its maximum value is called peak particle velocity—PPV.

As seismic waves travel through an area, they generate particle motions that we call vibration.

11.3.1.2 Type of Seismic Waves

Seismic waves in geologic media are of several types. These waves have different particle motions and travel at different wave velocities through the material. The waves have different characteristics that affect the way structures and people respond to them. The waves in one group are called body waves and consist of primary waves and shear waves which travel through the "body" of the material, as well as including the surface. The waves in another group are called surface waves which can also take different forms but travel only along the surface (see Fig. 11.15).

The primary wave is also called the P-wave or compressional wave. It has particle motions which are in the radial direction and has the highest propagation velocity (and therefore arrive first).

Next to arrive are the "secondary" waves (shear or S-waves, or transverse waves) with particle motions perpendicular to the radial line. These are of two types, horizontally and vertically oriented particle motions. P- and S-waves are relatively high frequency and are collectively called "body waves."

Slowest and last to arrive are the low-frequency surface waves. The surface waves that are usually generated in rock blasting are as follows: Rayleigh waves (R-waves) and Love waves (Q-waves). Other types of surface waves are the channel waves and the Stoneley waves which are not important as they supply very little information. The most significant is typically the Rayleigh wave with

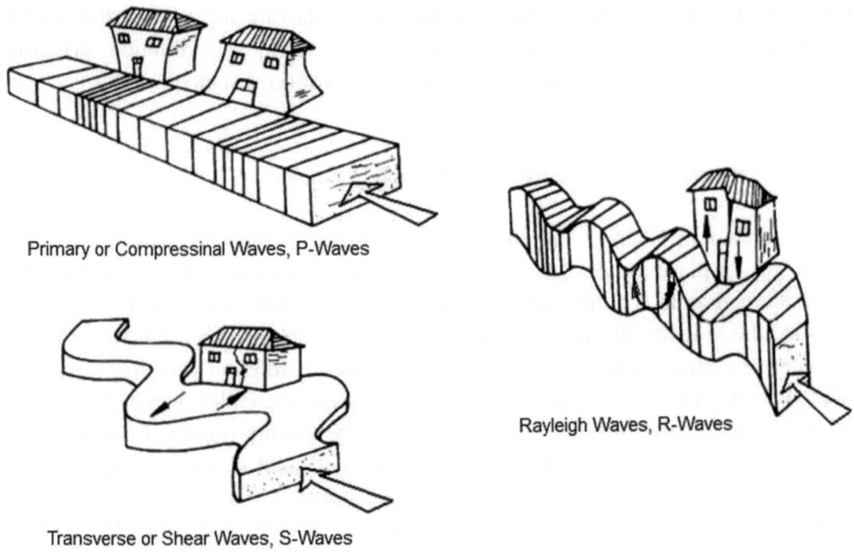

Fig. 11.14 Three types of seismic waves (reproduced from Ref. [6] with the permission from Sandvik)

retrograde elliptical particle motions, similar to ocean waves impacting a beach (Fig. 11.14).

In homogeneous rock, seismic wave velocities are expressed with following equations [5]:

$$\text{P-waves}: \quad c_p = \left[\frac{E(1-\nu)}{\rho(1-2\nu)(1+\nu)}\right]^{1/2} \text{(m/s)} \tag{11.7}$$

$$\text{S-waves}: \quad c_s = \left(\frac{G}{\rho}\right)^{1/2} = \left[\frac{E}{\rho^2(1+\nu)}\right]^{1/2} \text{(m/s)} \tag{11.8}$$

$$\text{R-waves}: \quad c_R \approx c_S \frac{0.86+1.14\nu}{1+\nu} \quad \text{(m/s)} \tag{11.9}$$

where E is the modulus of elasticity (Pa), G is the shear modulus of the material (Pa), ρ is the density of the material (kg/m^3), and ν is Poisson's ratio.

The distribution of the energy transported by the different types of waves has been studied by several researchers such as Miller and Purey (1955), and Vorob'ev (1973), who have come to the conclusion that the Rayleigh waves carry between 70 and 80 % of the total energy [4].

11.3.1.3 Parameters of Seismic Waves

For convenience of study, the seismic waves generated by blasting can be simplified as harmonic motion type waves, as shown in Fig. 11.15.

The basic parameters for analysis are as follows:

- amplitude (A), maximum displacement of a particle from its rest position;
- velocity (v), velocity at which a particle moves;
- acceleration (a), velocity per unit time, i.e., $a = v/t$; and
- frequency (f), complete number of oscillations or cycles per second. The frequency is the inverse of the period T, $f = 1/T$.

At any time t, the displacement of the particle is $u = a\sin(\omega t)$, where $\omega = 2\pi f$.

The relationships between displacement, velocity, and acceleration of the particle are as follows:

$$u = A\sin(\omega t) \tag{11.10}$$

$$v = \frac{\mathrm{d}u}{\mathrm{d}t} = A \times \omega \times \cos(\omega t) \tag{11.11}$$

$$a = \frac{\mathrm{d}v}{\mathrm{d}t} = -A \times \omega^2 \times \sin(\omega t) \tag{11.12}$$

When only the maximum absolute values of these parameters are taken into account, the above relationships are converted into:

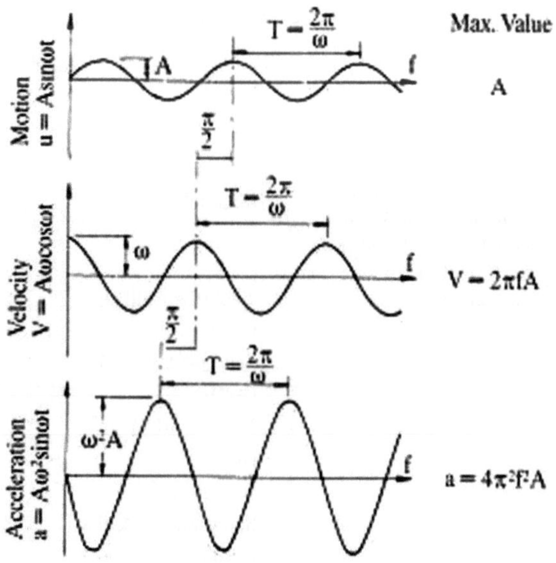

Fig. 11.15 Basic terms for determining wave motion (reproduced from Ref. [6] with the permission from Sandvik)

$$v = A \times \omega = 2\pi f A \tag{11.13}$$

$$a = A \times \omega^2 = 4\pi^2 f^2 A \tag{11.14}$$

In most situations, the maximum acceleration is expressed in g's, where g is the acceleration of gravity, $g = 9.8\ \text{m/s}^2$, usually simplified as $g = 10\ \text{m/s}^2$. For example, $a = 0.25$, $g = 2.5\ \text{m/s}^2$.

11.3.2 *Factors which Affect the Magnitude of Ground Vibration*

The parameters which affect the characteristics of vibration can be classified in two groups: controllable and uncontrollable. Controllable factors include blast geometry, type and amount of explosive used, priming, and initiation. Uncontrollable include distance, geological condition, and initiation timing errors.

11.3.2.1 Site Geology

The site geology and rock geomechanics have great influence on vibration generated by blasting. The propagation of the ground vibration is strongly influenced by the lithology of the rock mass. The strength, density, and porosity of rock affect the propagation velocity of the waves significantly as shown in Table 11.1.

Harder rock will have higher wave velocity and frequency. When the rocky substratum is covered by soil overburden, this usually affects the intensity and frequency of vibration. Soil usually has less elasticity modulus than the rock and, for this reason, the wave propagation velocity diminishes in this type of material.

Table 11.1 Velocity of seismic waves in different materials (reproduced from Ref. [6] with the permission from Sandvik)

Material	Velocity of seismic waves (m/s)	
	S-waves	P-waves
Clay or silt (dry)	−200	400–600
(wet)	−200	1300–1600
Sand or gravel (dry)	200–400	400–700
(wet)	200–400	1400–1700
Moraine (dry)	200–700	700–1500
(wet)	200–700	1400–2000
Broken rock	800–1200	1900–2500
Sandstone or schist	1200–1600	2500–3400
Granite or gneiss	2000–2500	4000–4800

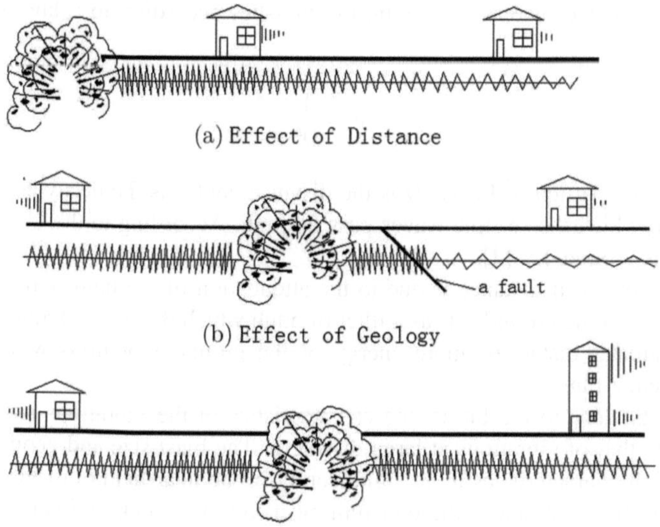

Fig. 11.16 Effects of site conditions on blast vibration

The vibration frequency also diminishes, but the displacement increases significantly.

In the area around a blast, the influence of rock characteristics on the rate of ground motion reduction with distance varies. The magnitude of the vibrations decreases rapidly with distance if soil overburden is present because a large amount of the energy is used up in overcoming friction between particles and in displacing them (Fig. 11.16).

The presence of discontinuities such as joints, fractures, faults, and weak zones in the path of ground vibration waves also acts to scatter the peak vibrations. Some components of the ground motion, such as the high-frequency wave components, are lost as the wave crosses a discontinuity and the propagation direction of the vibration wave may also be influenced by the complex geological structures, giving different attenuation indexes or laws of propagation.

Whenever the burden rock consists of highly fractured or jointed strata, there exists a possibility of escape of gases through them, and this may cause high air overpressure but also decreases the ground vibration as part of the energy has been lost with the escaped gases.

11.3.2.2 Distance from the Blasting

As the distance from the blasting increases, the particle velocity and frequency of ground vibrations decrease due to the absorption, dispersion, and dissipation of the elastic wave.

As the distance increases, vibration diminishes according to a law of the following type:

$$v = \frac{1}{D^b} \tag{11.15}$$

where v is the particle velocity, D is the distance, and b is the decay coefficient of the ground which the seismic waves pass through. According to the US Bureau of Mines, b is around 1.6 [4].

Another effect of distance is due to the attenuation of the high-frequency wave components, as the ground acts as a filter through which the lower frequencies pass. Thus, at greater distances, more energy of the ground vibrations will be in the low-frequency range.

At points close to the blasts, the characteristics of the vibrations are quite different from that of vibrations at large distances. The blast size and geometry affect the vibration characteristics more significantly in the near-field blast vibration, but for far-field blast vibration, the transmitting medium of rock and soil overburden dominates the wave characteristics (Fig. 11.17 from [7]).

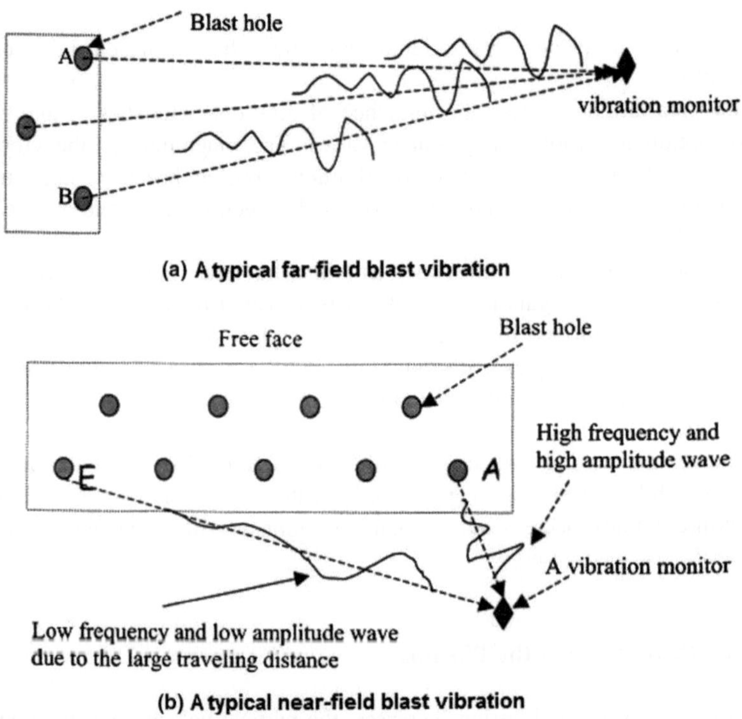

Fig. 11.17 Near-field and far-field blasting vibrations (reproduced from Ref. [7] with the permission from the author, Mr. Ruilin Yang, Orica)

11.3.2.3 Type, Amount, and Coupling of Explosives

The amplitude of the pressure pulse is directly proportional to the strength of explosive and amount of explosive used.

Under the same site condition, the ground vibration produced by blasting using high-strength explosives, e.g., dynamite or emulsion explosives, is higher than using low strength explosives (low density and detonation velocity), e.g., ANFO blasting agent.

The more energy the explosive releases, then the higher the ground vibration will be. Charge weight per delay is the most important factor which controls the intensity of ground vibration. The larger the quantity of charge detonated per delay, the higher the vibration. The relationship that exists between vibration intensity and the charge weight is expressed as follows:

$$V \propto Q^a \tag{11.16}$$

where V is the particle velocity and Q is the charge weight per delay. The investigations carried out by US Bureau of Mines show that the value of a is around 0.8.

The coupling of the explosives charge to the rock affects how much of the energy is transferred to the rock, hence the intensity of the vibration, as well the effectiveness of blasting in fracturing the rock. As it is stated in Chap. 10, when the explosive is loaded in blastholes with decoupling technique, the effective pressure of the shock waves and gas on the borehole wall is cushioned (Fig. 10.1) due to the buffer function of the annulus air or other inert materials. This effect reduces the crushing and radial cracking of the rock around the boreholes and the ground vibration as well.

An interesting and sometimes confusing issue is the powder factor. Some engineers propose to reduce the powder factor of the blast and reduce the amount of explosive, but the result usually is not what they expected. In fact, if the powder factor is lower than the optimum level, the vibration levels measured will be higher as a consequence of the confinement and poor spatial distribution of explosive, causing lack of displacement and swelling (see Fig. 11.18).

11.3.2.4 Geometric Parameters of Blasting Design

- Burden, bench height, and the stiffness of bench

If the burden is excessive, the resistance to rock movement and swelling increases, and more explosive energy will transfer to ground vibration. As it is discussed in Chap. 8, it was addressed that the stiffness of bench $H/B > 2$ should be maintained. If H/B = or <1, the rock body in front of the blastholes, especially in the lower part of the bench, will be difficult to be pulled forward, and as a consequence, the problems of bad fragmentation, toe problem, and backbreak will appear and high ground vibration will be caused as well.

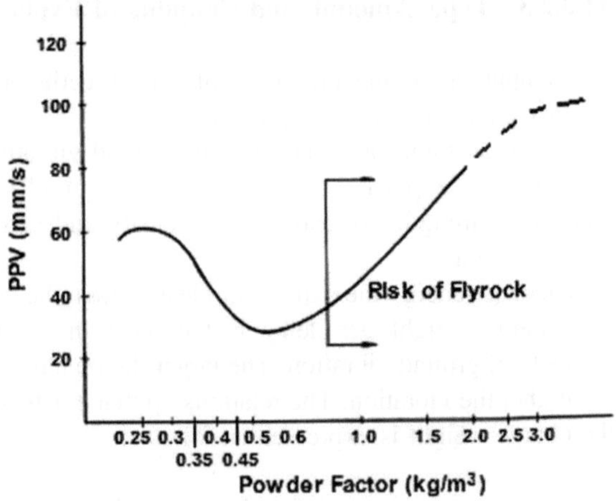

Fig. 11.18 Powder factor influence on vibration intensity

- Spacing

As discussed in Chap. 8, wide spacing and small burden can improve the distribution of energy of strain waves in the bench rock, resulting in improved fragmentation, and also reducing the ground vibration due to the reduction in the confinement of the burden rock.

- Subdrilling

Excessive subdrilling will clearly increase the ground vibration due to the higher resistance for shearing the toe rock of the bench.

- Shape of blasting area and firing direction

The vibration caused by a blast with less rows and rectangular shape with long free face is obviously less than the vibration caused by a blast with more rows and short free face. Ground vibration caused in the trench blasting is much higher than that in the normal bench blasting due to the strong confinement (Fig. 11.19).

The firing direction also affects the ground vibration intensity along different directions of propagation paths. In Fig. 11.15, the vibration intensities recorded at the three monitoring points show $A < B < C$ due to the recoil effect of rock movement.

11.3.2.5 Detonator Timing Scatter

As discussed in Chap. 4, for the pyrotechnical delay detonators, there is always an unavoidable scatter in the firing times of different detonators with the same nominal firing time. This can be due to small variations in the length, the packing density, and the composition of the pyrotechnic delay charges, or to change in the burning rate occurring from aging during storage. The burning rate may change due to slow

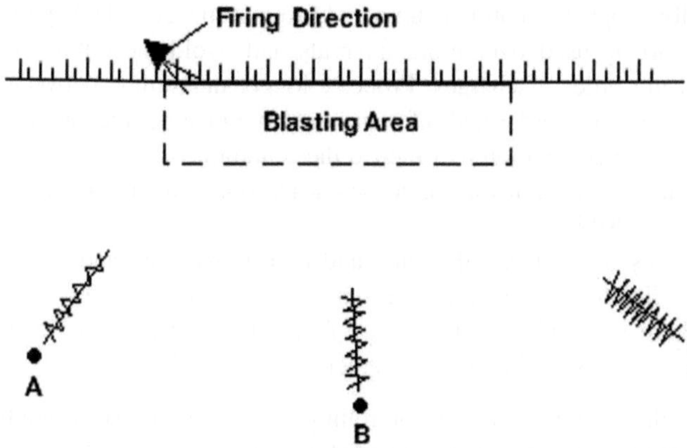

Fig. 11.19 Effect of firing direction on the vibration intensity along different propagation paths

oxidation reactions within the pyrotechnic charge itself, often accelerated by moisture. The timing scatter increases with the total time it takes to detonate the cap, that is, its delay number or nominal firing time. The timing scatter of the detonators in a blast causes timing overlay of some blastholes which should be detonated in different firing time according to the designed firing sequence, and there is a great chance that two or more charges will detonated simultaneously, or even in reverse firing order. That means the charge weight per delay is greatly increased; as a result, the ground vibration is also increased.

It has become a common practice in various regulations, criteria, and project specifications to consider the maximum charge weight per delay that explosive quantity which detonates within any given 8-ms time interval. In other words, the delay interval should be equal to or greater than 8 ms in order that the two charges can be considered separate charges. This conclusion was made by US Bureau of Mines (USBM RI 6151, Duvall et al. 1963). Although in the later time, the same group of Bureau of Mines' researchers found 5-ms delay also to be effective, the 8 ms delay intervals as a criterion is still accepted and applied by most countries in the world.

11.3.3 Controlling Ground Vibration During Blasting

The first important thing for controlling ground vibration is an optimal blasting design:

- As stated in the above section, an optimal value of burden is very important as it can get the best fragmentation, proper swelling, and the lowest ground vibration. Too large burden and choke blasting must be avoided strictly. The optimal burden can be determined through trial blasts.

- A carefully adopted initiation pattern and firing sequence is the key to spread the explosive to be used over more intervals and avoid the timing overlap and reverse firing order of charges. Properly longer delay time between rows can avoid reverse firing order and offer an inner free face between adjacent rows to improve the fragmentation and reduce the vibration.
- Accurate drilling of holes needs to be carried out to avoid uneven burden and excessive subdrilling.
- Carefully design the firing direction and rock movement, and try to avoid the backward directing from facing the sensitive vibration receivers.
- Ensure there is a sufficient free face of the blasted rock mass in front of the direction of rock swelling and movement.

The most effective measure for controlling the vibration level is controlling the explosive charge weight per delay. The following methods can be used to reduce the charge weight per delay:

- Reduce the hole number and diameter;
- Reduce the bench height or divide a high bench into more benches;
- Use decked charge by dividing the necessary drillhole charge level into more ignition intervals through inter-stemming; and
- Use decoupled charge with the cartridges of smaller diameter than the hole diameter.

11.3.4 Prediction and Restriction Criteria of Ground Vibration by Rock Blasting

11.3.4.1 Prediction of Ground Vibration Induced by Blasting

Like all force field, seismic waves die out or decay with distance. The decay is often called attention. The waves die out in a fairly regular manner, which makes them predictable with acceptable accuracy and allows restrictions on blasting vibrations to be regulated by means of either mathematical expressions or actual measurement of the vibrations with portable seismographs.

General formula for prediction of ground vibration in far-field locations

As discussed in Sect. 6.3.3 of Chap. 6, the vibration that results from a blast can be calculated using a formula of form:

$$\text{PPV} = K\left(R/Q^{d}\right)^{b} \qquad (11.17)$$

where

PPV Predicted peak particle velocity in mm/s;

K a constant which is determined by the characteristics of the ground that the vibration passes through and affected by the explosive's power;
Q Maximum charge weight per delay interval in kg;
R Distance in m between the blast and the measuring point;
d Charge exponent, usually $d = 1/2$ for a cylindrical charge or $d = 1/3$ for a charge of symmetrical sphere;
b Attenuation exponent

In the formula, *K* and *b* are the site-specific parameters related to local geological conditions and explosive strength and estimated by means of regression analysis based on the collected data from trial blasts of the determined site. Table 6.1 gives some ranges of *d, K,* and *b* from different sources for reference.

Regression analysis is usually used to validate the experimental formula from the collected data of the blasts in the site. Figure 11.20 is an example of the graph of regression analysis of a surface construction blasting site.

In this graph, it is shown that the vibration recorded in the underground tunnel is obviously lower than that recorded in the ground surface for the same blasts at the same scaled distances. The same result was also concluded by other researchers, e.g., Figure 11.21, from "Surface blasting above the old Jenny Mine, Kentucky"

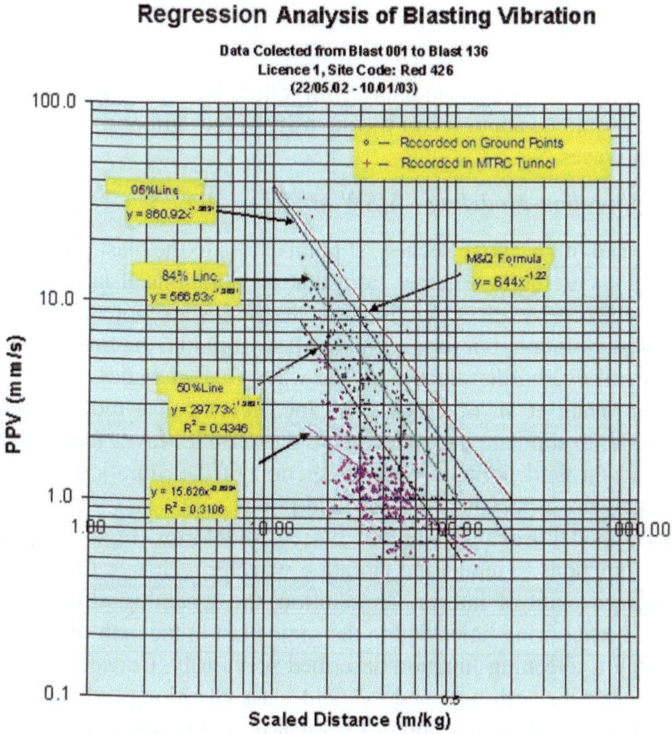

Fig. 11.20 Regression analysis chart of a blast site in Hong Kong

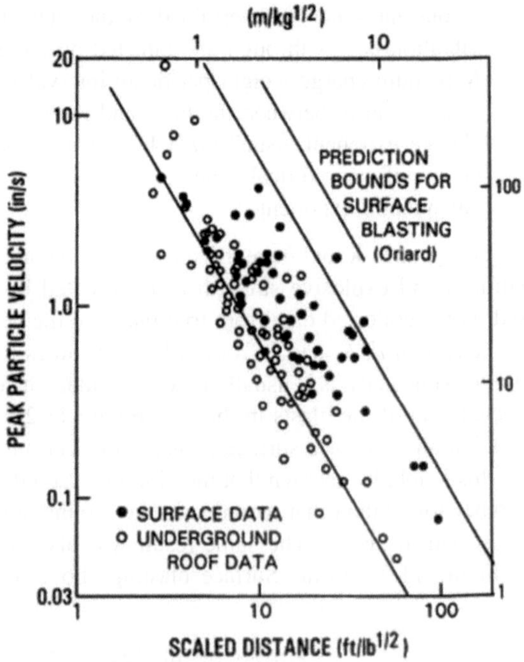

Fig. 11.21 Data plot shows the fact that the underground roof vibration showed a reduction to about 40 % of the recorded on the surface (courtesy of ISEE, Ref. [8])

page 495 of [8]. This is consistent with other experiences that body waves underground would be about 1/2 of the intensity of the surface motion generated by the Rayleigh waves.

Near-Field Vibration Modeling—MSW Model

As it is indicated in the above section, at points close to the blast, the characteristics of the vibrations are quite different from that of vibrations at large distances. The blast size and geometry affect the vibration characteristics more significantly in the near-field blast vibration, but for a far-field blast vibration, the transmitting medium of rock and soil overburden dominates the wave characteristics.

Dr. Ruilin Yang et al. of Orica USA Inc. developed a model, named MSW model, in 2007, to simulate the processes of the seismic waves from a blast to both near-field and far-field points for the prediction of the vibration intensity at any vibration receivers, especially the near-field points.

The MSW model uses multiple sets of seed waveforms (Figs. 11.22 and 11.23) and transfer functions to model the vibration waveform change from different blast holes to a given point of interest. In addition, the screening effect of the broken ground from earlier firing holes within the same blast in the path of vibration is also modeled using a screening function described previously. Consequently, the MSW model is suitable for both near and far-field blast vibration predictions.

To model the vibration at a point of interest from a production blast, a signature wave (three triaxial components) is selected for a blasthole according to the nearest

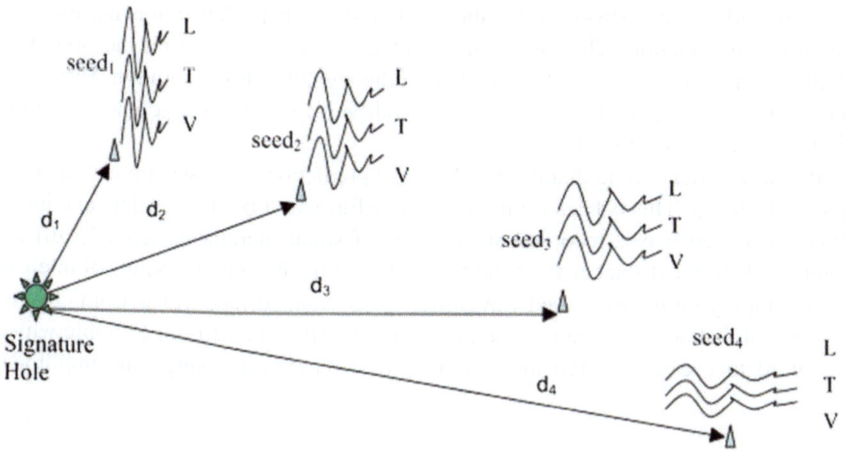

Fig. 11.22 Triaxial seed waveforms (L. V. T) measured at different distances from a signature hole at the site (reproduced from Refs. [10, 11] with the permission from the author, Mr. Ruilin Yang, Orica)

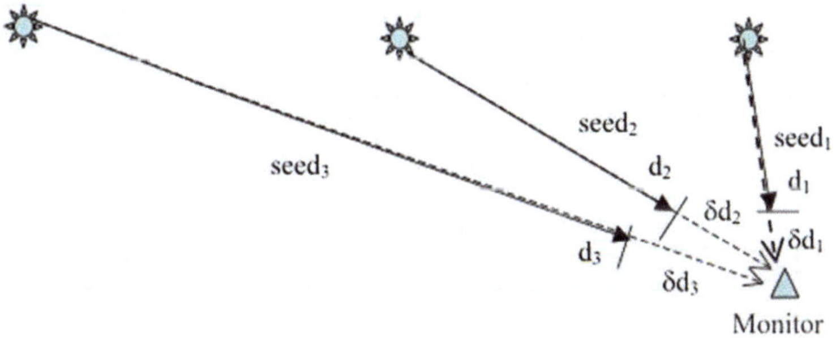

Fig. 11.23 A set of signature waveforms is selected for each charge according to the distance match (reproduced from Ref. [11] with the permission from the author, Mr. Ruilin Yang, Orica)

smaller distance of the signature wave (d_{sd}) compared to the distance (d_{hl}) from the blasthole to the monitor in the production blast (as shown in Figs. 11.22, 11.23). The change in waveform over the distance difference $\delta_d = d_{hl} - d_{sd}$ is modeled with the Kjartansson transfer function—a constant Q model [9]. The Kjartansson transfer function was successfully used for monitoring both amplitude and frequency attenuations of seismic waves at lower strains typical to the far-field regime. It assumes that the rock exhibits linear viscoelastic behavior. However, in the near-field, particularly in highly nonlinear soft ground, the Kjartansson transfer function is not suitable for modeling amplitude attenuation. On the other hand, it may still be a useful choice for modeling frequency attenuation for near-field blast vibration since it is one of the simplest models for the frequency attenuation and the latter is the major phenomenon for the waveform change over a small distance (δ_d).

Consequently, in the MSW model, the Kjartansson transfer function is not used for amplitude attenuation. The vibration amplitude is determined from the nonlinear charge weight scaling law established from the signature hole vibration. However, the frequency attenuation that produces a modified wave shape is adopted from the Kjartansson transfer function.

It is worth mentioning that the MSW model prediction is not sensitive to the rock quality factor Q. This is because the wave transformation by the transfer function is minimal if seed waves recorded at distances of small increments (e.g., 15 m) are comparable to the distances from the blast holes of the blast to the points of interest.

By employing multiple seed waveforms, *p*-, *s*-, and surface waves from charges at different distances can be included in the model (Fig. 11.24 is an example which was used in a case research in a quarry [10]). Waveform changes in amplitude,

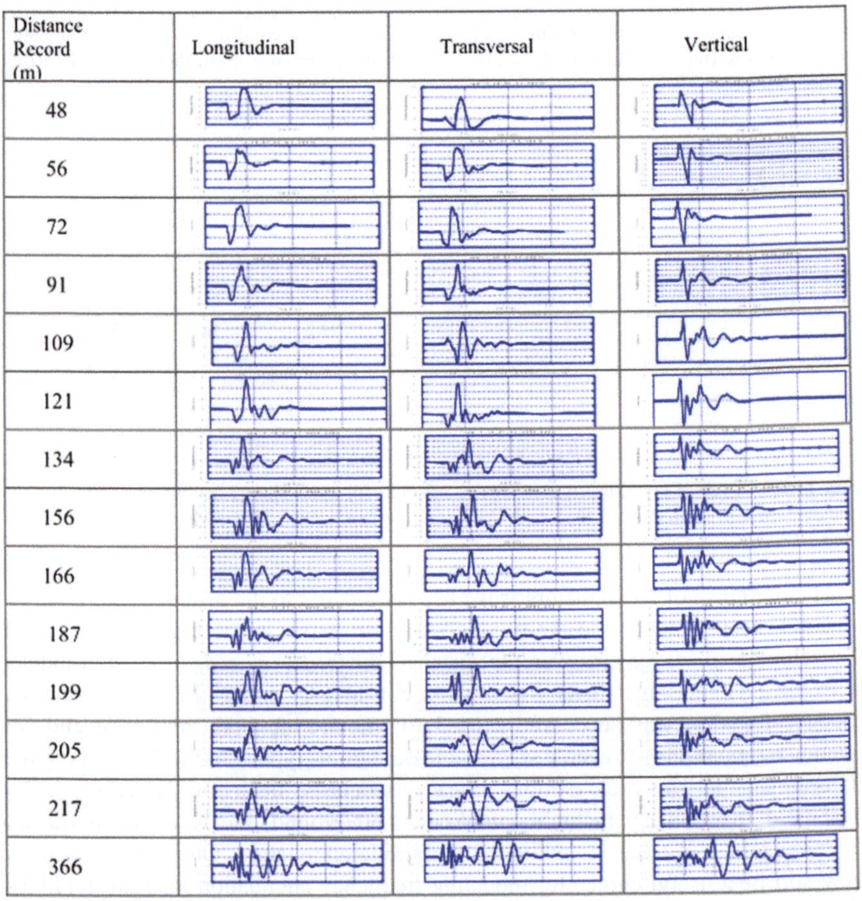

Fig. 11.24 The multiple sets of seed waveforms used as input to the model for a surface blasting site (reproduced from Ref. [11] with the permission from the author, Mr. Ruilin Yang, Orica)

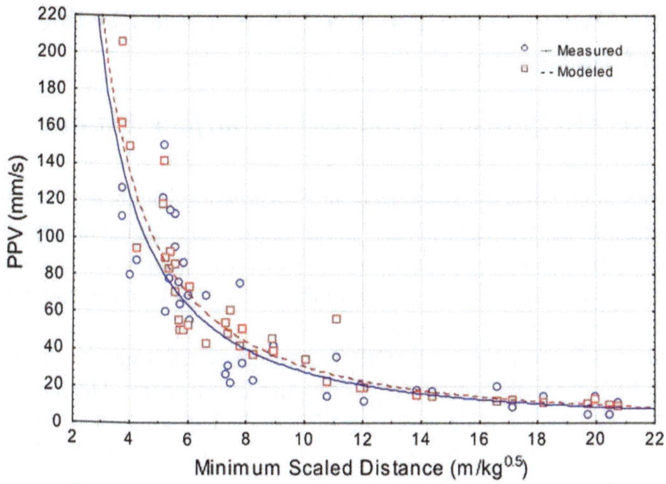

Fig. 11.25 Comparison of the model predictions against the field measurements (reproduced from Ref. [11] by permission from the author, Mr. Ruilin Yang, Orica)

frequency, and duration due to the mixture of wave types and frequency attenuation with distance are automatically taken into account by the multiple seed waveforms. In addition, some geological effects on different seed waveforms can also be input to the model.

MSW model has been successfully used in both surface and underground blasting for the prediction of ground vibration in both near-field and far-field situations. Figure 11.25 is the comparison between the modeling results and the measured records as a case study in an open pit (from [11]).

11.3.4.2 Restriction Criteria of Ground Vibration by Rock Blasting

For the protection of structures, buildings, and utilities from any damage by ground vibration generated in blasting, some restrictions of ground vibration from blasting are specified in some regulations or specifications. In Sect. 6.4.2 of Chap. 6, the restriction criteria proposed in different countries are given for reference.

11.3.5 Instrumentation for Monitoring Blasting Vibration

The International Society of Explosives Engineers (ISEE) published "Blasting Seismograph Standard—Performance Specifications for Blasting Seismographs" in the "ISEE Blasters' Handbook^{TM}" in 2000 [12]. Based on it, Mines Division of the Hong Kong Government also published the "Guidance Note on Vibration Monitoring (Jan. 2015)" [13].

11.4 Air Overpressure Produced by Surface Blasting and Its Control

11.4.1 Air Overpressure Produced by Surface Blasting

As discussed in Chap. 6, blast-induced air overpressure (defined as pressure above normal atmospheric pressure) is the air pressure wave generated by explosions. More commonly, this is termed as airblast. Air overpressure is considered to be one of the most detrimental side effects due to the generation of noise and even causing some potential damage to residences.

Airblast or air overpressure (AOP) is an energy transmission in the form of pressure waves from the blast site within the atmosphere. The pressure waves consist of energy over a wide spectrum of frequencies, some of which are audible and hence may be sensed in the form of noise, but most at inaudible frequencies at less than 20 Hz and beyond 20,000 Hz. These frequency ranges can be sensed by people in the form of a pressure impact known as concussion. The combination of effects of noise and concussion is known as air overpressure.

The maximum excess pressure above normal atmospheric pressure of this wave is known as the peak air overpressure, generally measured in decibels linear (dBL).

An "A-weighting dB" is a measure of noise pressure level designed to reflect the acuity of the human ear, which does not respond equally to all frequencies. The ear is less efficient at low and high frequencies than at the middle speech-range frequencies. Noise measurement is usually made with standard level meters that have built-in filters called "weighting scales." These filters distort the true pressure readings and reduce the effects of the low and high frequencies with respect to the medium frequencies. The A-weighted sound level is also called the noise level.

As mentioned above, air overpressure contains a considerable amount of low-frequency energy which can eventually produce direct damage on structures; however, high-frequency air waves are more common and are observed due to the movement of windows, dishes, doors, etc. Figures 11.26 and 11.27 show data collected by the author in a construction blast site [14].

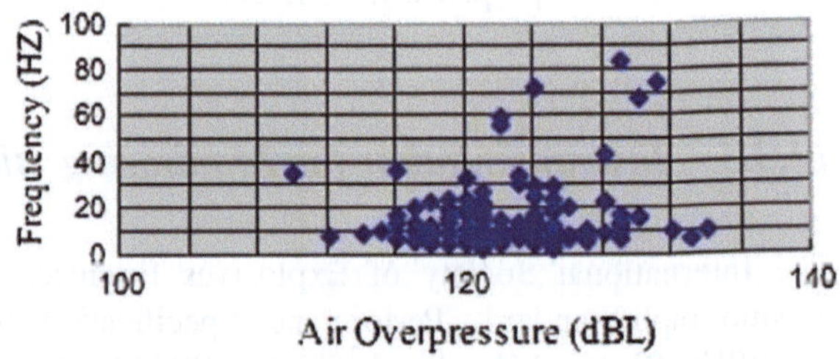

Fig. 11.26 Frequency distribution of AOP

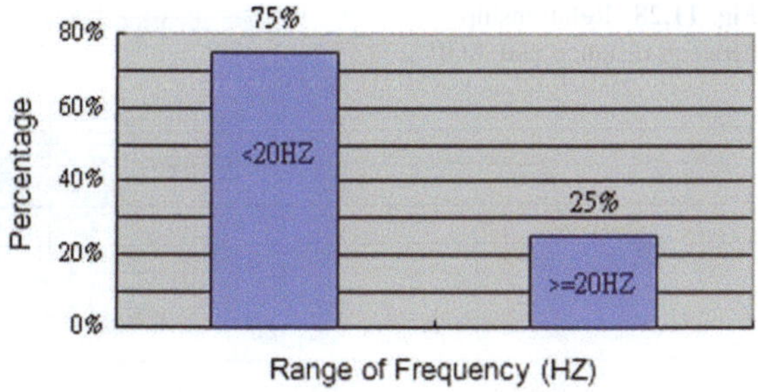

Fig. 11.27 Frequency distribution of AOP

11.4.2 Factors which Affect the Air Overpressure [14]

There are many factors which affect the air overpressure. They can be classified in two groups: uncontrollable and controllable. These influencing factors are listed below:

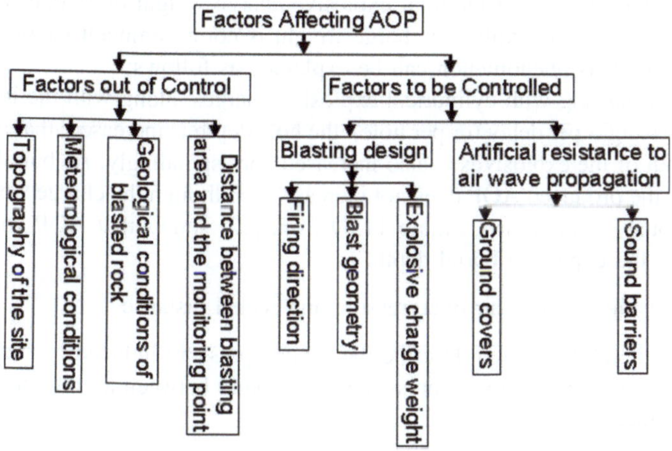

Zou, D. (the author of this book) carried out some research works in a large site formation project in the urban area of Hong Kong and summarized some experiences of the construction blasts, refer to [14]:

- Distance in between blasting area and monitoring points

Both inside and outside of the residential buildings' four monitoring points were set up and two of them were chosen for each blast. The monitoring points inside the building were placed close to the window at almost the same elevation to the blasting level. The relationship between the distance and the recorded air overpressure is shown in Fig. 11.28.

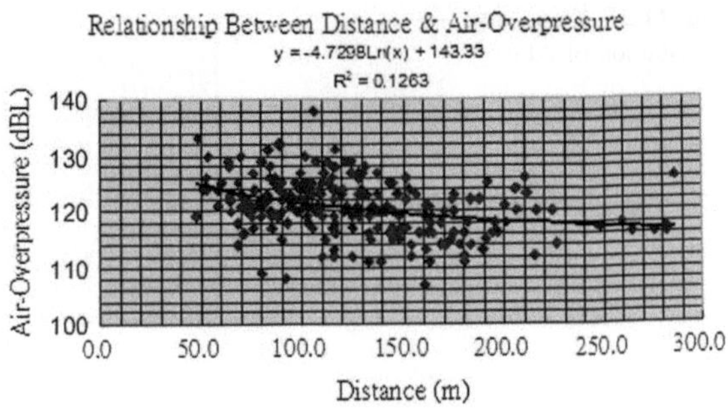

Fig. 11.28 Relationship between distance and AOP

Although the data appear scattered due to the effects of other influencing factors, the obvious decreasing trend of the overpressure along with the increasing distance can be seen.

- Effect of the explosive charge weight, hole depth, and hole number

The statistical data show that air overpressure does not necessarily increase along with the increase in the total explosive charge weight or charge weight per delay. That means the explosive charge weight is not a dominant factor of the air overpressure. This phenomenon can be explained as follows:

In bench blasting with cylindrical explosive charges, along with the increase in the charge weight per delay (or per hole), the hole depth is increased; that means the burial depth of the explosives is also increased correspondingly. As burial depth is increased, the produced AOP does not increase even though the charge weight fired simultaneously is increased. Figure 11.29 from [12] shows that AOP will greatly affected by the explosive burial depth.

- Relationship between air overpressure and scaled distance

In most literatures which describe the air overpressure induced by blasting, the following formula is used to express the relationship between air overpressure and scaled distance:

$$\mathrm{AOP} = a\left(\frac{D}{Q^{1/3}}\right)^{b} \tag{11.18}$$

where,

AOP	Air overpressure;
D	Distance between blasting area and monitoring point;
Q	Maximum charge weight per delay;
a and b	Constants related with the site condition and environment factors.

According to this formula, the regression analysis is carried out and shown in Fig. 11.30.

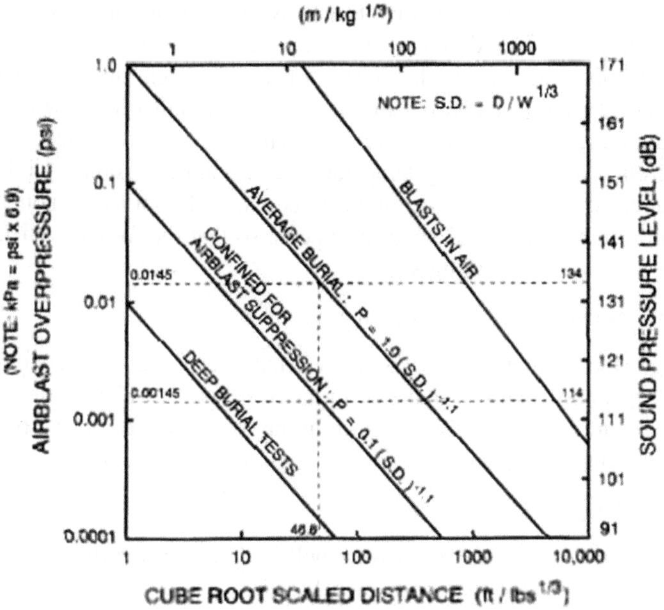

Fig. 11.29 AOP versus charge burial (courtesy of ISEE, Ref. [12])

Fig. 11.30 Air overpressure versus scaled distance

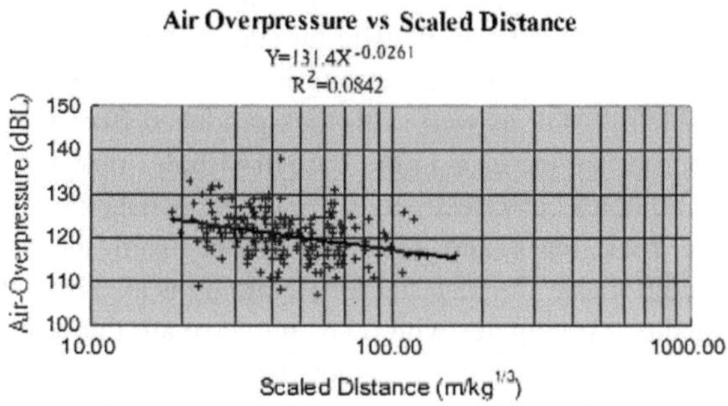

From the graph, we note that the coefficient of determination is very small, $R^2 = 0.0842$, due to the large scatter of the data. It means that it is impossible to predict the overpressure with any degree of accuracy using the formula similar to predicting ground vibration as there are too many unpredictable and uncontrollable effects of atmospheric conditions and the location conditions (refer to [4, 15]).

- Meteorological conditions

Because air overpressure is transmitted through the atmosphere, meteorological conditions such as wind speed and direction, temperature, humidity, and atmospheric pressure will all affect the intensity of the air overpressure.

Wind changes the angles of wave fronts. Wind gradients are highly directional. The wave intensity and duration are found to be enhanced in the downwind direction. Temperature affects the density of the air. High humidity increases the absorption of the air. Usually, the air overpressure is higher in the winter than that in the summer. In some literatures, a phenomenon usually mentioned, that is called "focusing effects," is caused by temperature and wind inversions. As the blasting site and affected area of construction projects is usually not large enough, it is not easy to form temperature and wind inversions, so that the focusing effects seldom appear in construction working sites.

- Geological condition of the rock mass

Geological condition of the blasted rock mass, especially the condition of the joints and fractures, plays a vital role in air overpressure. If there are some open joints or fractures reaching the blast face in the rock mass, explosive pressure will vent out into the atmosphere and produce high air overpressure.

- Blasting design

Blasting direction—Air overpressure will be larger in the direction of the free face and rock movement than any other direction.

Blasting geometry—The main geometrical parameters of blasting are burden, spacing, hole length, subdrilling, and stemming length. But in choosing these parameters, all aspects such as rock fragmentation, requirement of site formation, reducing flyrock, AOP, and ground vibration should be considered properly not only for reducing AOP. The burden of the first row holes and the stemming length are the most important among these parameters to the blasting-induced AOP. For reducing AOP as well as flyrock produced from the front free face and hole collar, appropriate increase in the burden of holes near the free face and stemming length of all holes can effectively increase the resistance of explosive gas venting out from the rock mass and rock movement from the front face and benchtop face. Additionally, the material for stemming is also important for ensuring the stemming quality. Aggregates with 3–10 mm size are the best stemming materials. Excessive levels of AOP are often associated with stemming ejection which commonly occurs when drill cuttings are substituted for stemming aggregates.

11.4.3 Estimation of Air Overpressure

In some literatures which describe the air overpressure induced by blasting, the formula of (11.18) was recommended to be used for the estimation of the intensity of air overpressure. But as it was stated above, air overpressure characteristics are not easy to predict. Factors such as climate and topography intervene which, along with the actual blast design, can give different results in each case.

11.4.4 Criteria for Limiting Air Overpressure in Surface Blasting

As it was discussed above, air overpressure (or airblast) can cause discomfort to persons and, in some case, damage to structures. There are two kinds of limitation or guidelines in varies countries for air overpressure induced from blasting. One kind of limitation or guideline is used for controlling structure damage and another for controlling the noise induced from blasting. The current legislation, standards, guidelines, and criteria of some countries have been introduced in Sect. 6.4.3 of Chap. 6.

11.4.5 Measures for Reducing Air Overpressure Produced by Surface Blasting

Air overpressure should be controlled with regard to not only potential damage, but also adverse human response. The following measures can be adopted at the blasting site:

1. All holes should be adequately stemmed. A minimum stemming height no less than thirty borehole diameters is recommended to prevent "blowouts" and resultant airblast.
2. Assure proper front row burden (recommend no less than 28 hole diameter).
3. Assure proper timing sequence.
4. Do not allow any exposed explosives including detonating cord or surface connectors. If they cannot be avoided, at least 30 cm thickness of soil or sand should be placed on them. As stated in Sect. 10.4.3.1 of Chap. 10, when presplitting is carried out, for the elimination of surface detonating cords, an instantaneous or 17-ms delay non-electric detonator is connected to down-hole detonating cords to replace the surface trunk line detonating cord, together with the plastic plugs that approved to be effective to reduce the air overpressure produced by presplitting (see Fig. 11.26 and Photo 11.6) (Figs. 11.31 and 11.32).
5. Limit the charge weight of explosives per delay.

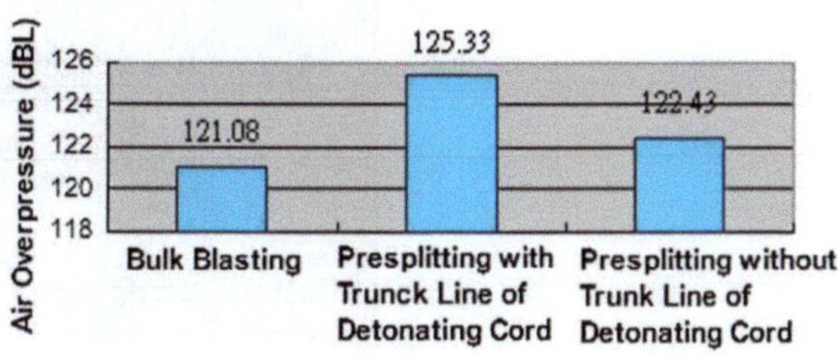

Fig. 11.31 Average AOP in different blasting methods

Fig. 11.32 Covering on the presplitting holes

6. When possible, delay the blast to direct rock movement (direction of main free face of a blast) away from critical areas, especially the residential area.
7. Avoid adverse environmental conditions such as blasting when wind is blowing toward residential areas, or under an atmospheric temperature inversion.
8. Use artificial resistance to the air waves propagation.
   When air waves propagate, some artificial resistance may reduce their intensity. The following measures had been taken in blasts of this area. Figure 11.33 shows the blast which is very close to residential buildings. For reducing the AOP, on the top and the side facing to residential buildings, blasting cages which cover the top blasting area are all covered with thick rubber tire mats. Figure 11.34 shows the vertical screens which are surrounding the blasting area to prevent any flyrock are also covered with canvas as sound barriers.
9. Maintain good public relations by notifying neighbors before blasting operations start and before each blast.

Fig. 11.33 Thick rubber tire mats covering on the top and side of the cages which is facing to the residential buildings

Fig. 11.34 Vertical screens covered with canvas as sound barriers are surround the blasting area

References

1. Zou D (2007) Engineering blasting in Hong Kong, Sing Tao Press, 2007, Hong Kong. ISBN: 978-962-672-682-2
2. Lundborg N (1985) The probability of flyrock, Report DS 1985:5, SveDeFo, 1985
3. Roth J (1979) A model of the determination of flyrock rane as a function of shot conditions, US Bureau of Mines, Report NTIS, PB81-222358, 1979
4. Jimeno CL et al (1995) Drilling and blasting of rock. A.A. Balkema Publishers. 1995. ISBN 90 5410 199 7
5. Persson PA et al (1994) Rock blasting and explosives engineering. CRC Press. 1994. ISBN: 0-8493-8978-X
6. Tamrock (1988) Surface drilling and blasting. Edited by Jukka Naapuri, 1988
7. Yang R (2007) Near-field blast vibration monitoring, analysis and modeling. In: Proceeding of the 33rd annual conference ISEE, Nashville, Tennessee, USA
8. Oriard LL (2002) Explosives engineering, construction vibrations and geotechnology, ISEE. ISBN: 1-892396-13-0
9. Kjartansson E (1979) Constant Q wave propagation and attenuation. J Geophys Res 84:4737–4748
10. Yang R et al (2010) An integrated approach of signature hole vibration monitoring and modeling for quarry vibration control. In: Blasting-Sanchidrian (ed) Rock Fragmentation. Taylor & Francis Group, London, ISBN 978-0-415-48296-7
11. Yang R, Scovian DS (2010) A model for near and far field blast vibration based on multiple seed waveforms and transfer functions for waveform change and amplitude attenuation. Blast Fragm J 4(2):91–116 (ISEE)
12. ISEE (2010) Blaster's handbook, 18th edn. Cleveland, Ohio, USA, ISBN: 9781892396198
13. Guidance Note on Vibration Monitoring, Mines Division, GEO, CEDD, HKSAR, Jan 2015. http://www.cedd.gov.hk/eng/services/mines_quarries/mgd_mdgn.html
14. Zou D (2007) Study on air overpressure produced by blasting near high rise residential buildings. In: Proceeding of the 33rd annual conference ISEE, Nashville, Tennessee, USA
15. Scottish Executive (2000) PAN 50 Annex D: controlling the environment effects of surface mineral working. ISBN: 0748092307
16. Richards AB, Moore AJ (2002) Flyrock control—by chance or design. In: Proceedings of the 28th annual conference on explosives and blasting technique, ISEE, New Orleans, USA

Chapter 12
Blasting Models and Computer-Aided Design for Bench Blasting

12.1 Introduction

12.1.1 Blasting Models

Rock blasting model includes two kinds of models: empirical models and theoretical models. The former is built on the basis of empirical formula or statistical data for dealing with specific engineering design and parameter optimization problems within a certain range; the latter is based on the mechanism of rock blasting generally applicable to a variety of blasting calculation and theoretical analysis.

Due to the rapid development of computer technology in recent decades, rock blasting theoretical models take the computer technology and variety of software as a platform, and use the numerical methods to calculate and simulate the physical processes of rock blasting based on a variety of known rock blasting fragmentation theories. Therefore, the computer simulation of the rock blasting process has become another method of study in addition to laboratory model tests and field trials blasting, and it has undergone rapid development and wide application in recent decades.

Summarizing the computer simulation models of rock blasting, they include the following areas:

- The models for studying the developing progress of crack, fracture, and damage in rock caused by blasting refer to [1, 2];
- The models for predicting the blasting results, including fragmentation, rock movement, and the shape of muck pile after blasting for optimize blasting parameters refer to [3–6];
- The models for predicting the ground vibration, backbreak, and potential flyrock for reducing the adverse effect of rock blasting refer to [7, 8].

D. Zou, *Theory and Technology of Rock Excavation for Civil Engineering*,
DOI 10.1007/978-981-10-1989-0_12

With the deepening of the understanding of the mechanism of rock blasting, the simple traditional analytical methods on many issues with the actual situation have large discrepancies. A greater number of new analysis methods such as fractal geometry, fuzzy mathematics, mutation theory, and other applications bring the theory of rock blasting into a new situation.

However, these methods are still in the development stage, they may also be coupled and mutual penetration each other in the future.

In this chapter, some models for bench blasting, including empirical models and theoretical models, will be introduced briefly.

12.1.2 *Computer-Aided Blasting Design*

Application Research of Computers in Engineering Blasting began in the early 1980s. In the early stage, it was mainly for surface bench blasting. Along with the development of expert systems and CAD technology, it has developed as system software combined with blasting design, parameter optimization, blasting simulation results analysis, cost analysis, and data management.

A complete system design package consists of several modules: the raw data input module (including site topography, site geology, drills, and explosives), blasting design optimization module (including blasting parameters calculation and optimization, firing pattern, and initiation sequence), blasting effects analysis module (including rock fragmentation distribution, rock pile shape, throwing range, blasting vibration, and air blast prediction), and blasting cost analysis and data management module.

In this chapter, some computer-aided design system software for surface bench blasting works (open pit mine) will be briefly introduced.

12.1.3 *Size Distribution of Rock Fragments in Blasting*

As some blasting models and CAD for blasting often use indexes or distribution functions to describe the degree of rock fragmentation in blasting, we need to have some knowledge about them before we introduce the blasting models and CAD design systems in this chapter.

12.1.3.1 Description of the Degree of Rock Fragmentation in Blasting

Engineering blasting, especially in mining production blasting, the index which directly reflects the good and bad effects primarily is the fragmentation after blasting. Because the degree of the rock fragmentation directly affects the follow-up operation processes—the productivity of loading, transport, crushing and others, production costs and total production costs, and also determines the use rates of stone blasted in construction of dams, embankments, and road engineering and sales prices. Therefore, there is an important significance to study rock blasting fragmentation and its rules, not only for the study of the mechanism of rock blasting but also for the realization of "blasting optimization."

There are usually two different methods to describe quantitatively the rock fragmentation after blasting. One description method is of using a single index. The most commonly used are "the rate of unqualified blocks" (or rate of oversize fragments) indicator and the "mean size of fragments" indicators. The former reflects the percentage of large blocks which have a major impact on subsequent operations (loading, transport, and crushing), and typically must be further broken, and the latter reflects the average degree of rock fragmentation after blasting. Some researchers found that there is a nearly linear relationship between the "rate of unqualified block" and "mean size of fragments" by statistics. In theoretical research, there are two indexes of so-called P_{80} and K_{50} usually adopted. They are defined as the corresponding fragment size (mesh aperture) when the cumulative undersize content is of 80 and 50 %, respectively. Obviously, they are very similar as the first two indicators. But it must be noted that K_{50} is not the "mean size of fragments" index.

Another method for description of the degree of rock fragmentation by blasting is a general description method, namely the distribution of rock fragmentation by blasting (or the size composition of fragments).

In general, if the cumulative percentage of the fragments which pass the screen (or accumulated percentage remaining on the screen) is taken as the Y-axis and the mesh size of the screen (i.e., fragment size) as the X-axis, the rock fragmentation distribution after blasting can be described with a distribution curve (or distribution function expression):

$$Y = f(X)$$

There are the following advantages in describing the degree of rock fragmentation after blasting using the fragment distribution function:

(1) It fully reflects the degree of rock fragmentation in blasting;
(2) Usually only a few (usually two) distribution parameters can control the overall composition status of the rock fragmentation by blasting from the fine fraction to the coarse grade. It is not affected by the grading indicator. The cumulative percentage passing the screen and relative amounts of any size grade (including P_{80} and K_{50}) can be obtained from the distribution curve;
(3) Using this mathematical expression, it can further reveal the mechanism and rule of rock fragmentation in blasting. This is especially crucial for building relationships between blasting parameters and blasting results.

12.1.3.2 Distribution Functions of Rock Fragments

There are lots of functions to describe the rock fragment distribution in blasting in the published literatures by the researchers and scholars of different countries, but the most common are the following three functions:

(1) Rosin–Rammler function (R–R function)

$$Y = e - (d/d_0)^n \quad (12.1)$$

where

Y Cumulative percentage retained on screen (%);
d Size of fragment or screen;
d_0 Characteristic size, one of the parameter of distribution;
n Index of uniformity, one of the parameter of distribution.

(2) Gaudin–Meloy function (G–M function)

$$Y = \left(1 - \frac{d}{d_{\rm m}}\right)^n \quad (12.2)$$

where $d_{\rm m}$—maximum fragment size. The others are same as above.

(3) Gates–Gaudin–Scuhmann function (G–G–S function)

$$Y = \left(\frac{d}{d_{\rm m}}\right)^n \quad (12.3)$$

where all symbols have same meanings as above.

The above distribution functions are initially summarized from the experimental data of the particle size distribution of rock after mechanical crushing. Later it was discovered that the distribution of rock fragmentation by blasting can also be described using this function. For example, the former Soviet scholars, Baron [9], Kuznetsov [10], Japan's Prof. Kazuo Otsuka [11], Australia's Harries, just [12], described the size distribution of rock in their blasting experiments with R–R function. American's Dick [13] and Angola's Gama [14] used G–G–S function to describe the size distribution in rock blasting. Lovely [15] in Australia made a more detailed discussion on the application of G–M function in the distribution of rock fragmentation in blasting.

No matter if the rock is broken by mechanical crushing or blasting, the fact that the distribution of rock fragmentation can be described using the same kind of functions: the mechanism of rock breakage is mainly controlled by the characteristics of the rock material itself, not the form of external forces. Therefore, many people have attempted to get a reasonable explanation of the physical meaning of these functions from the mechanism of rock breakage. The results have shown that the random breakage theory of brittle materials can provide some explanation to these distribution functions of rock fragmentation. Some theoretical discussion on this issue was made by Kuznetsov et al., Austin, and Klimpel [16].

Because of the above-mentioned advantages of the distribution functions for description of rock fragmentation by blasting, they are increasingly used for blasting engineering calculations, especially in the blasting model for the prediction of rock fragmentation and computer-aided designs, such as the Kuz–Ram model, Gamma model, and SABREX design packages which will be discussed in the following sections of this chapter.

12.2 Some Typical Blasting Models

12.2.1 *Harries' Mathematical Model (Reproduced from Ref. [9, 17] with the Permission from the Australasian Institute of Mining and Metallurgy)*

In the 1970s, Dr. G. Harries (ICI Australia) proposed a mathematical model to simulate the progress of rock fragmentation by blasting.

In the model, it assumes that the rock is a continuous, homogeneous elastic medium; explosive energy mainly from the explosion pressure is applied on the wall of the blasthole after the detonation of explosive.

This model is actually a two-dimensional, quasi-static model.

When the explosive gas pressure acts on the borehole wall, it causes outward expansion of the borehole wall until equilibrium is reached. The tangential strain in the surrounding rock can be calculated according to the formula of a thick wall cylinder in elastic mechanics. At the hole wall, the tangential strain value at the borehole wall is given as:

$$\varepsilon = \frac{(1-\nu)P}{2(1-2\nu)\rho C_{\mathrm{P}}^2 + 3(1-\nu)K \times P} \tag{12.1}$$

where

P Explosive gas pressure acting on the borehole wall ($\mathrm{MP_a}$);
C_{P} P-wave velocity in the rock (m/s);
ρ Density of rock ($\mathrm{g/cm^3}$);

ν Poisson's ratio of rock; and
K Adiabatic index.

The strain generated by the explosion pressure in the rock decays by the negative exponential in the radial direction. The tangential strain value at the point which has a distance of r (cm) can be expressed as:

$$\varepsilon(r) = [\varepsilon/(r/b)] \times e^{-\alpha(r/b)} \tag{12.2}$$

where b Radius of the borehole, cm.

When the decay index $\alpha = 0$, the above formula becomes:

$$\varepsilon(r) = \varepsilon(b/r) \tag{12.3}$$

Under compressive stress, the radial displacement of rock causes tangential strains. When the tangential strain is larger than the dynamic tensile strength value T of the rock, radial cracks will be generated in the rock. The number of radial cracks generated in the rock at the distance r from the blasthole is as follows:

$$n = \varepsilon(r)/T \tag{12.4}$$

where

n The number of radial cracks generated in the rock at the distance r;
T The dynamic tensile strength of the rock.

As it is difficult to get the value of the dynamic tensile strength of the rock, Harries used back-analysis and trial-and-error method to reversely obtain the T values from the rock fragmentation and the crack size in small-scale experiments.

Harries' model also takes into account the role of explosive gas expanding the cracks and the role of reflecting tensile strain waves from the free surface. Under the dual role of high-pressure gas penetration and reflected tensile strain waves, the cracks which are toward the free surface and extended forward are double in length. Figure 12.1 shows the crack patterns generated by three blasthole firing. Figure 12.1b also shows that the growth of cracks which is generated by holes on a later delay is stopped by the previously formed cracks or existing fractures in the rock.

The rock fragmentation is calculated using the Monte Carlo stochastic method. A considerable number of random points are put into the blast area, and two mutually perpendicular straight lines are made through each of these points. The two lines intersect with the nearest cracks, the shorter one between the two segments of the two lines cut by the cracks represents of the size of the fragment blasted at that point in the rock mass. Figure 12.2 is an example of the comparison of the fragmentation distributions between the experimental result and the calculated result.

From 1980s, Harries tried to develop a 3D dynamic model to simulate the cylindrical explosive charge in bench blasting. In 1983, in his papers [17, 18], he

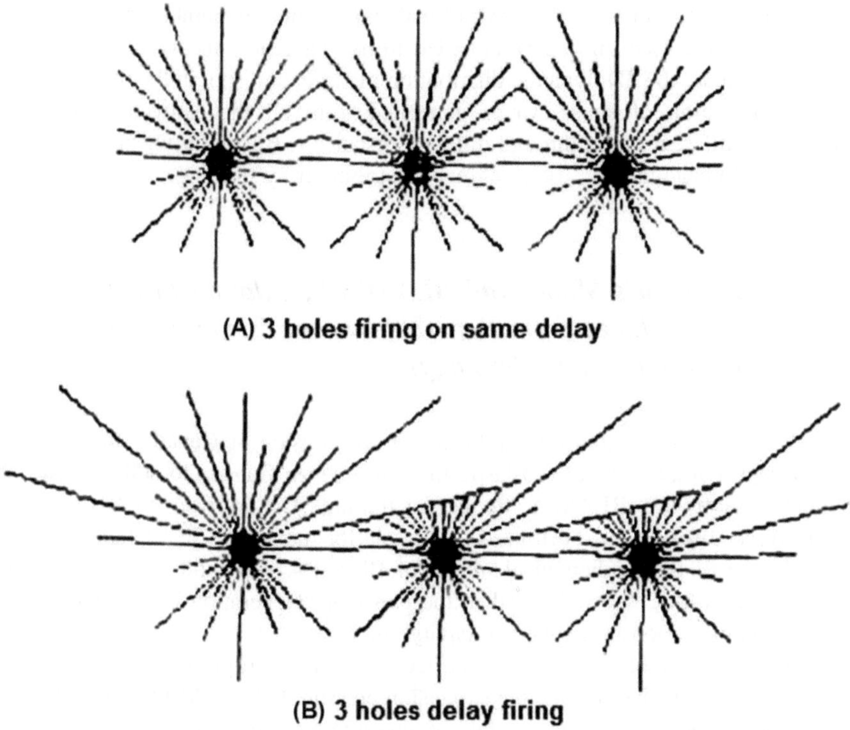

Fig. 12.1 Fracture pattern of three holes blasting (reproduced from Ref. [9] with the permission from The Australasian Institute of Mining and Metallurgy)

Fig. 12.2 Comparison of experimental results of ash of spacing/burden = 2.0 with calculated results assuming $K/T = 650$ (reproduced from Ref. [5] with the permission from The AusIMM)

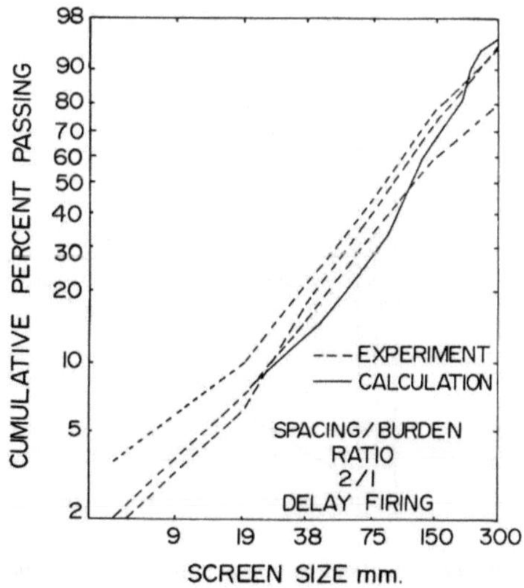

proposed to split up the long cylindrical charge into a number of equivalent spherical charge (see Fig. 12.3) and tried to use Favreau's theory of strain wave [19] to calculate the 3D dynamic distribution of strain at a spot in the rock mass by the superposition of stress waves generated from all the equivalent spherical charges in different time *t* through different distance *r*. But so far we have not seen the model to calculate the rock fragment distribution by blasting.

12.2.2 Favreau's Model and BLASPA Simulation Program (Reproduced from Ref. [20] by Permission of Luleå University of Technology)

Dr. R. F. Favreau used to be a Professor of Physics in Canada. In 1983, he established a small company, Blaspa Inc., in Quebec Canada to sell his blasting simulation software, BLASPA, for use in mining. The software of BLASPA as commercial software has been used in Canada for more than 30 years.

The core of the mathematical model of BLASPA was built on the theory of strain waves which was published in the paper of "Generation of strain waves in rock by an explosion in a spherical cavity" in 1969 [20].

In the theory, rock is assumed a perfectly elastic isotropic medium. The detonation of the explosives causes the expansion pressure to be suddenly applied to the cavity wall and the subsequent fall in pressure as the cavity expands can be described by a simple polytropic equation of state. It is also assumed that the

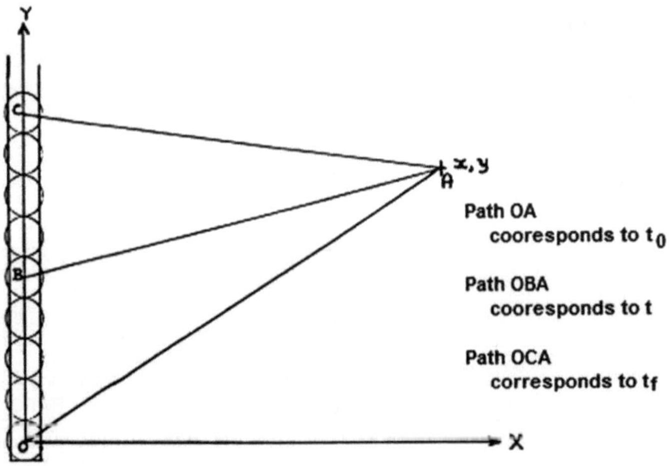

Fig. 12.3 Vertical section of simulation of cylindrical charge (reproduced from Ref. [18] with the permission from Luleå University of Technology)

expansion is small. The particle velocity (u) as a function of distance (r) and restarted time (τ) is given by:

$$u(r,t) = e^{-\frac{\alpha^2\tau}{\rho cb}}\left\{\left(\frac{Pb^2c}{\alpha\beta r^2} - \frac{\alpha\beta b}{\beta\rho cr}\right)\sin\frac{\alpha\beta\tau}{\rho cb} + \frac{Pb}{\rho cr}\cos\frac{\alpha\beta\tau}{\rho cb}\right\} \tag{12.5}$$

where

$$\alpha^2 = \frac{2(1-2\sigma)\rho c^2 - 3(1-\sigma)\gamma P}{2(1-\sigma)} \tag{12.6}$$

$$\beta^2 = \frac{2\rho c^2 - 3(1-\sigma)\gamma P}{2(1-\sigma)} \tag{12.7}$$

$$\tau = t - \frac{(r-b)}{c} \tag{12.8}$$

b Blasthole radius;
c Longitudinal wave velocity;
σ Poisson's ratio;
ρ Density of the rock;
γ Polytropic expansion coefficient of explosive product gas;
P Explosion pressure.

The above equation is only valid when

$$2\rho c^2 > 3(1-\sigma)\gamma P \tag{12.9}$$

As BLASPA is commercial software, more information about the mathematical-mechanical equations has not been seen so far.

The mechanisms of the blasting process that are adopted are shown in Fig. 12.4.

In the model, the creation of the primary cracks by both of the compressive wave and the reflected tension wave, C_R, which moves in the rock with a velocity of

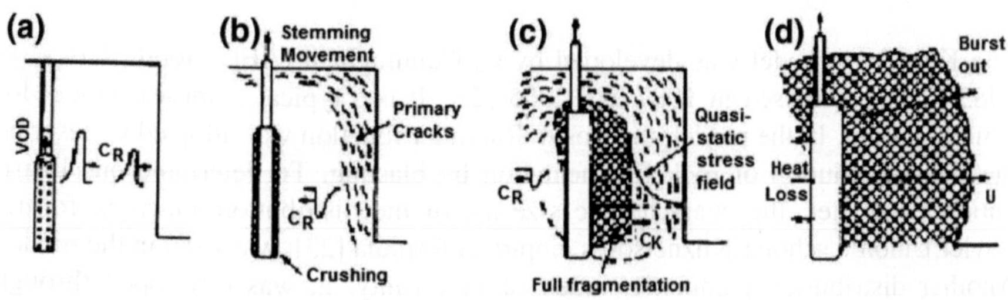

Fig. 12.4 Mechanism of the blasting process (reproduced from Ref. [20] with the permission from Luleå University of Technology)

about 5,00 m/s, is called "Brisance" action of the explosion (Fig. 12.4a, b). In the model, the creation of cracks in the rock mass is caused by the reflected tension wave as the rock's tensile strength is much lower than its compressive strength. So the criterion of rock failure is the dynamic tensile strength of the rock in the model. The rock weakened by the Brisance action is further fractured and converted to fully broken rock by the action of a quasi-static stress field generated by the high pressure of explosion gases. When the crack front with a velocity C_K of about 1,000 m/s reaches the free faces of the rock mass, Fig. 12.4c, d, the whole rock bench is fully broken and it bursts out with a burst velocity U of about 5–30 m/s, throwing the various fragments ahead into the muck-pile of broken rock. This outward movement of the rock mass is called the "Heave" action of the explosion, while the joint actions of Brisance, crack front development, plus comminution during the forward movement, all together lead to the final "fragmentation" of the rock.

For the whole simulation, the data needed to be input include the ingredients of the explosive, the physical and mechanical properties of the rock, the geometric parameters of the bench and blasting design, and the firing method and sequence. The output results include the effect of VOD, minimum delays, fragmentation size and distribution, the possibility of toe, the weakening of the rock at the back of the blast and how it promotes back-break, the displacement of the rock in various zones of the bench and how it relates to the ease of mucking, the shape of the muck pile, the ground vibration induced by the blasting, the possibility of flyrock in different positions of the bench and the maximum flyrock range. BLASPA can also calculate the costs of all procedures of drilling, blasting, loading, hauling, second breaking, and crushing to give the suggestion of the optimal parameters of the working chain to achieve a lowest overall cost.

12.2.3 Kuz–Ram Model (Reproduced from Ref. [21] with the Permission from Luleå University of Technology and Reproduced from Ref. [22] with the Permission of Copyright © 2005 European Federation of Explosives Engineers (EFEE))

The Kuz–Ram model was developed by C. Cunningham (AEL, South Africa) in 1983 [21] and revised in 1987 and 2005 [22]. It is a typical empirical model for bench blasting. In the model, the Rosin–Rammler function was adopted to describe the size distribution of rock fragmentation by blasting. To determine one distribution parameter, the characteristic size d_0, of the distribution function, former Soviet Union's scholar Kuznetsov's empirical formula [23], was used in the model. Another distribution parameter, index of uniformity, n, was developed through small- and large-scale experimental study and combination of crack analysis technique using Harries' mathematical model [5].

The revised equations in 2005 are given below:

(1) adopted Kuznetsov's empirical equation:

$$x_m = A \times A_{\mathrm{T}} \times K^{-0.8} \times Q^{1/6} \times \left(\frac{115}{\mathrm{RWS}}\right)^{19/20} \times C(A) \tag{12.10}$$

(2) adopted Rosin–Rommler equation:

$$R_x = \exp\left[-0.693\left(\frac{x}{x_m}\right)^n\right] \tag{12.11}$$

(3) The uniformity equation:

$$n = n_{\mathrm{s}}\sqrt{\left(2-\frac{30B}{d}\right)} \times \sqrt{\frac{1+S/B}{2}} \times \left(1-\frac{W}{B}\right)\left(\frac{L}{H}\right)^{0.3} C(n) \tag{12.12}$$

where

x_m	Particle size of K50 (it was mean particle size in Kuznetsov' equation), cm;
A	Rock factor, varying between 0.8 and 22, depending on hardness and structure, will be discussed later;
A_{T}	Timing factor as a multiplier of Eq. (12.10), will be discussed later;
K	Powder factor, explosive per cubic meter, kg/m^3;
Q	Mass of explosive in the hole, kg;
RWS	Weight strength relative to ANFO, 115 being the RWS of TNT;
$C(A)$	Correction factor to rock factor A, normally will be within the range 0.5–2;
R_x	Mass fraction retained on screen;
x	Particle size or screen opening;
x_m	Characteristic size;
n	Uniformity index, usually between 0.7 and 2;
n_{s}	Uniformity factor, will be discussed later;
B	Burden, m;
S	Spacing, m;
d	Hole diameter, mm;
W	Standard deviation of drilling precision, m;
L	Charge length, m;
H	Bench height, m; and
$C(n)$	Correction factor to uniformity index.

In the 2005 version of the Kuz–Ram model, some important changes had been made for improving the algorithms for mean fragmentation and uniformity, compared to the old versions of 1983 and 1987.

(1) Rock characterization: rock factor A

$$A = 0.06(\mathrm{RMD} + \mathrm{RDI} + \mathrm{HF}) \tag{12.13}$$

where RMD—the rock mass description. RMD is assigned according to the rock condition: powdery/friable = 10; massive formation (joint further apart than blastholes) = 50; vertically jointed—derive jointed rock factor (JF) (Note: JF may the rock factor A for the jointed rock—the Author) as follows:

$$\mathrm{JF} = (\mathrm{JCF}\,\mathrm{JPS}) + \mathrm{JPA} \tag{12.14}$$

where

JCF is the joint condition factor.
Tight joints JCF = 1;
Relaxed joints JCF = 1.5;
Gouge-filled joints JCF = 2.0;
JPS is the joint plane spacing factor.
Joint spacing <0.1 m, JPS = 10;
Joint spacing = 0.1–0.3 m, JPS = 20;
Joint spacing = 0.3 m to 95 % of $(B \times S)^{0.5}$, JPS = 80;
Joint spacing > $(B \times S)^{0.5}$, JPS = 80;
JPA is the joint plane angle factor.
Dip out of face JPA = 40
Strike out of face JPA = 30
Dip into face JPA = 20;
RDI—the density influence. Literature [22] missed to give the value range.
HF—the hardness factor:
If Y, 50, HF = Y/s;
If $Y > 50$, HF = UCS/5

where

Y Elastic modules, GPa;
UCS Unconfined compressive strength of rock, MPa.

(2) Inner-hole delay

Refer to other researchers' study: the optimum inner-hole delay times is of 3–6 ms per meter of burden; Conningham used a timing factor A_T to reflect the effect of the delay time between holes in a row on fragmentation adding to Eq. (12.10) as a multiplier:

$$A_{\mathrm{T}} = 0.66(T/T_{\mathrm{max}})^3 - 1.56(T/T_{\mathrm{max}}) + 2.1 \tag{12.15a}$$

And for higher values:

$$A_{\mathrm{T}} = 0.9 + 0.1(T/T_{\mathrm{max}} - 1) \tag{12.15b}$$

where

T The designed delay time between holes;

T_{max} The best delay time between holes in a row for maximum fragmentation, $T_{\mathrm{max}} = \frac{15.6}{C_x} B$,

where B is the hole burden and C_x is the longitudinal velocity in rock, km/s.

(3) Timing scatter

To address the adverse effect of timing scatter on uniformity of fragmentation, Cunningham introduced a uniformity factor to describe the crucial effect of precision on blasting:

$$n_{\mathrm{s}} = 0.206 + (1 - R_{\mathrm{s}}/4)^{0.8} \tag{12.16}$$

where

R_{s} Scatter ratio, $R_{\mathrm{s}} = 6\frac{\sigma_t}{T_x}$ and σ_t Standard deviation of initiation system, ms

T_x Desired delay between holes, ms. The uniformity factor n_{s} has the range 0.87–1.21.

(4) Effect of rock strength on uniformity

A correcting factor is used to reflect the effect of rock strength on Uniformity Index, n:

$$F(\Lambda) = (\Lambda/6)^{0.3} \tag{12.17}$$

(5) Rationalization of blasting geometry on uniformity

For the correct use of the Kuz–Ram model, some limitations (or so-called capping) were set to the geometric parameters for uniformity Eq. (12.12). The current uniformity index set and capping values are detailed in Table 12.1.

In Eq. (12.12), the case correction factor $C(n)$ is provided to overlay the above values $F(\alpha)$ and enable to reflect the effects of the geometric parameters of bench blasting.

Table 12.1 Geometric parameters for uniformity equation [22]

Parameter α	$F(\alpha)$	α range	$F(\alpha)$ range
S/B	$[(1 + \alpha)/2]^5$	0.7–1.5	0.92–1.12
30B/d	$(2 - \alpha)^5$	24–36	1.2–0.9
W/B	$1 - \alpha$	0–0.5B	1–0.5
L/H	$\alpha^{0.3}$	0.2–1	0.62–1
A	$(\alpha/6)^{0.3}$	0.8–21	0.5–1.45

As it is easy to apply, the Kuz–Ram model has been widely used in the world including incorporation in some computer-aided design software packages, such as ICI's SABREX Model.

12.2.4 BMMC Mathematical Model (Reproduced from Ref. [3] with the Permission of Copyright © 1987 SEM USA and [24, 25] with the Permission of AusIMM)

BMMC (**B**lasting **M**odel, **M**aanshan Institute of Mining Research, **C**hina) was developed by the author of this book, first in China in 1983 [24], then internationally in 1987 [3].

(A) Mechanism of rock fragmentation by blasting adopted in the model

- The fragmentation of the bench rock mass by blasting is the second fragmentation based on the situation that the rock mass has been cut into "natural blocks" which have a certain size distribution, by the weak faces formed by geological structure and previous blasts;
- In order to calculate quantitatively the fragmentation action of the explosion on rock, the rock mass is considered firstly as an isotropic, continuous, and elastic material. The fragmentation of the theoretical rock by blasting is calculated in the model. The final fragmentation is the probability sum of both the probability distributions of the natural blocks and the fragmentation probability distribution of theoretical rock by blasting;
- It is assumed that the breakage action of stress waves on the rock is the principal effect, which generates a great number of radial and tangential cracks in the rock around the blastholes. It is also assumed that the action of the explosion gases on the brittle rock only provides additional and expanding effects on the basis of the breakage action of stress waves. As the action processes and the effect to the size distribution of the explosion gases cannot be calculated in the model, a corrective coefficient K ($K > 1$) is introduced in the model in order to make up for the defect of neglecting the action of explosion gases.

(B) Three-dimensional distribution of the stress wave energy and rock fragmentation in the isotropic and elastic rock mass

The cylindrical explosive charge, of which the radius is b_c, is divided into numbers of segment charges. These segment charges can be considered as the spherical charges with equivalent radius b under the condition of height of segment charge not more than 3–4 times of charge diameter (Fig. 12.6). In the model, $b = 1.65b_c$ (the height of the segment charge is 3 times of the charge diameter). According to the superposition principle, the explosion stress field formed by the cylindrical charge in rock can be considered as the superposed result of all the stress fields produced by every equivalent spherical charge.

(1) Solution of the wave form function from solving the wave equation

The spherical compressive wave equation in the spherical coordinate (r, θ, φ) in which the original point is set at the center of the spherical charge is

$$\frac{\partial^2 u}{\partial r^2} + \frac{2}{r}\frac{\partial u}{\partial t} - 2\frac{u}{r} = \frac{1}{C_L^2}\frac{\partial^2 u}{\partial t^2} \tag{12.18}$$

where

u Particle displacement;
C_L Velocity of longitudinal wave;
T Time.

(a) The initial condition of the Eq. (12.18): $u = 0$ when $t = 0$;
(b) The boundary condition of the Eq. (12.18):

The exponential function used by Prof. I. Ito and K. Sassa is used in the model (Fig. 12.5):

$$P(t) = 4P_m\left(e^{-wt/\sqrt{2}} - e^{-\sqrt{2}wt}\right) \tag{12.19}$$

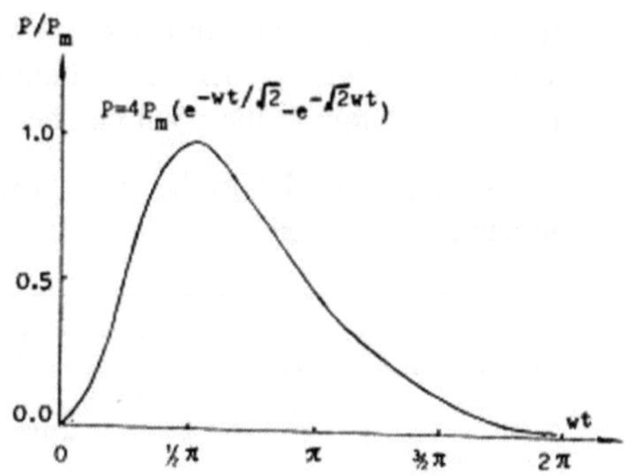

Fig. 12.5 Wave form function (reproduced from Ref. [3] with the permission from copyright © 1987 SEM USA

where

$P(t)$ Pressure on hole wall;
P_m Peak value of $P(t)$;
w Constant. According to Prof. Favreau's experiment results [17],

$$w = \frac{\sqrt{2}\ln 2}{3.64}\frac{C_L}{b} \tag{12.20}$$

Adopting Prof. Favreau's theory of strain wave and theory of elastic mechanics, when $r = b$,

$$u = u_0 = \frac{(1+v)b}{E}2\sigma_{m,r=b}\left(e^{-wt/\sqrt{2}} - e^{-\sqrt{2}wt}\right) \tag{12.21}$$

where

v Poisson's ratio of rock;
E Modulus of elasticity of rock.

(2) Stress state of a point in the bench

As the stress state of the rock mass derived from Eq. (12.18) only contain the geometrical attenuation and does not contain the physical attenuation, it is assumed in the model the displacement of any point in rock with a distance r from the original point can be expressed:

$$u(r,\tau) = u_p(r)u_w(\tau) \tag{12.22}$$

where

$$u_p(r) = \left(\frac{r}{b}\right)^{-n} \tag{12.23}$$

$$u_w(\tau) = u_0 = \frac{2(1+v)b}{E}\sigma_{m,r=b}\left(e^{-wt/\sqrt{2}} - e^{-\sqrt{2}wt}\right) \tag{12.24}$$

According to the theory of the stress wave in the spherical coordinates and the reflection laws of waves, each stress component of every stress wave initiating from each segment charge to the point can be calculated (see Fig. 12.6):

- Stress components DP of the incident wave from the center of a segment charge directly:

$$\sigma_{r,dp} = (\lambda + 2\mu)\left[u_0(\tau_{dp})\frac{du_p(r)}{dr} - \frac{u_p(r)v_0(\tau_{dp})}{C_L}\right] + 2\lambda\frac{u_r(r,\tau_{dp})}{r} \tag{12.25a}$$

Fig. 12.6 Cylindrical charge is divided into segment charges and the stress waves initiate from segment charges and reflect from free face (reproduced from Ref. [3] with the permission from copyright © 1987 SEM USA)

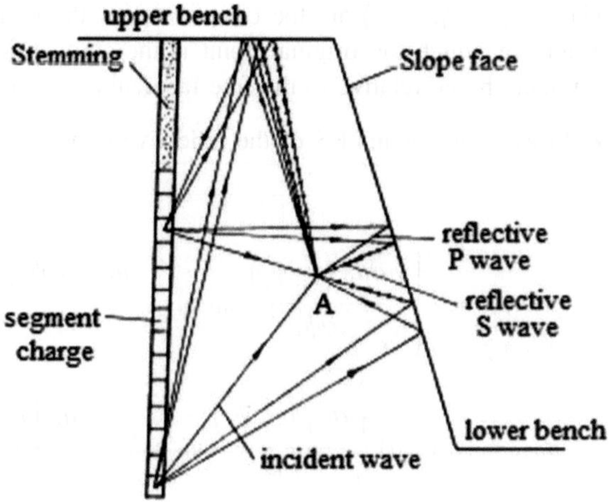

$$\sigma_{\theta,dp} = \sigma_{\varphi,dp} = 2(\lambda+\mu)\frac{u_r(r,\tau_{dp})}{r} + \lambda\left[u_0(\tau_{dp})\frac{\mathrm{d}u_p(r)}{\mathrm{d}r} - \frac{u_p(r)v_0(\tau_{dp})}{C_\mathrm{L}}\right] \quad (12.25\mathrm{b})$$

$$\tau_{r\theta,dp} = \tau_{\theta\phi,dp} = \tau_{r\phi,dp} = 0 \quad (12.25\mathrm{c})$$

where

$$\tau_{dp} = t - \frac{r}{C_\mathrm{L}}, \quad v_0(\tau_{dp}) = \frac{\partial u_r(r,\tau_{dp})}{\partial \tau_{dp}}$$

- Stress components RP of the reflective longitudinal wave from a free face:

$$\sigma_{r,rp} = (\lambda+2\mu)\left[u_0(\tau_{rp})\frac{\partial u_{pi}(r_i,\psi_i)}{\partial r_i} - \frac{u_{p_i}(r_i)v_{0_i}(\tau_{rp})}{C_\mathrm{L}}\right] + 2\lambda\frac{u_{rp}(r_i,\psi_i,\tau_{rp})}{r_i} \quad (12.26\mathrm{a})$$

$$\sigma_{\theta,rp} = \sigma_{\varphi,rp} = \lambda\left[u_0(\tau_{rp})\frac{\partial u_{p_i}(r_i,\psi_i)}{\partial r_i} - \frac{u_{p_i}(r_i)v_{0_i}(\tau_{rp})}{C_\mathrm{L}}\right] + 2(\lambda+\mu)\frac{u_{rp}(r_i,\psi_i,\tau_{rp})}{r_i} \quad (12.26\mathrm{b})$$

$$\tau_{r\psi,rp} = \frac{\mu}{r_i}\frac{\partial u_{p_i}(r_i,\psi_i)}{\partial \psi_i}u_0(\tau_{rp}); \quad \tau_{r\theta,rp} = \tau_{\theta\psi,rp} = 0 \quad (12.26\mathrm{c})$$

where $(r_i, \quad \theta_i, \quad \psi_i)$ are the coordinates of the point in the spherical coordinate system of which the original point is the symmetrical point of the center of the segment charge relative to the free face, and $\tau_{rp} = t - r_i/C_L$.

- Stress components RS of the reflective transverse wave from a free face:

$$\sigma_{r,rs} = \lambda\left[\frac{1}{r_i'}\frac{\partial u_{sp}\left(r_i', \psi_i'\right)}{\partial \psi_i'}u_0(\tau_{rs}) + \frac{u_{sp}\left(r_i', \psi_i'\right)}{r_i'}u_0(\tau_{rs})ctg\psi_i'\right] \tag{12.27a}$$

$$\sigma_{\theta,rs} = (\lambda+2\mu)\frac{1}{r_i'}\frac{\partial u_{sp}\left(r_i', \psi_i'\right)}{\partial \psi_i'}u_0(\tau_{rs}) + \lambda\frac{u_{sp}\left(r_i', \psi_i'\right)}{r_i'}u_0(\tau_{rs})ctg\psi_i' \tag{12.27b}$$

$$\sigma_{\psi,rs} = (\lambda+2\mu)\frac{u_{sp}\left(r_i', \psi_i'\right)}{r_i'}u_0(\tau_{rs})ctg\psi_i' + \lambda\frac{1}{r_i'}\frac{\partial u_{sp}\left(r_i', \psi_i'\right)}{\partial \psi_i'}u_0(\tau_{rs}) \tag{12.27c}$$

$$\tau_{r\psi,rs} = \mu\left[\frac{\partial u_{sp}\left(r_i', \psi_i'\right)}{\partial r_i'}u_0(\tau_{rs}) - \frac{u_{sp}\left(r_i', \psi_i'\right)v_{0_i}(\tau_{rs})}{C_L} - \frac{u_{sp}\left(r_i', \psi_i'\right)}{r_i'}u_0(\tau_{rs})\right] \tag{12.27d}$$

$$\tau_{\theta\psi,rs} = \tau_{r\theta,rs} = 0 \tag{12.27e}$$

where $\left(r_i', \theta_r', \psi_r'\right)$ are the coordinates of the point in the spherical coordinate system of which the original point is at the normal line of the free face, and the position of the original point can be obtained according to the reflection law of the transverse wave, and $\tau = t - \frac{r_{is}}{C_L} - \left(r_i' - r_s\right)/C_L$.

According to the theory above, all the stress components of every stress wave can be calculated not only under the conditions of single cylindrical hole charge and single free face, but also under various conditions of multiple cylindrical hole charges (in single row or in multiple rows), multiple free faces, firing simultaneously or in millisecond delays while taking the corresponding coordinate transformation.

(3) Energy density of stress waves at a point and the three-dimensional energy field in bench rock mass

Although all stress components are the dynamic vector quantities with different phases and different directions, the average energy densities of every stress wave arriving at a point are the scalar quantities irrelative to time and direction.

At the time τ, the specific stress energy of a stress wave arriving at a point is that:

$$U_0 = \frac{1}{2E}\left[\left(\sigma_\tau^2 + \sigma_\theta^2 + \sigma_\psi^2\right) - 2v\left(\sigma_r\sigma_\theta + \sigma_\theta\sigma_\psi + \sigma_\psi\sigma_r\right) + 2(1+v)\left(\tau_{r\theta}^2 + \tau_{\theta\psi}^2 + \tau_{\psi r}^2\right)\right] \quad (12.28)$$

Taking the average value during a period of $t_P (t_P = 2\pi/w)$, the average energy density of all the stress waves arriving at the point can be obtained:

$$\begin{aligned} U_P &= \sum_{j=1}^{m(n+1)} \frac{K}{t_P}\int_0^{t_P} U_{0,j}(\tau)\mathrm{d}\tau \\ &= \sum_{j=1}^{m(n+1)} \frac{K}{t_P E}\int_0^{t_P} \left[\begin{array}{l}\left(\sigma_{r,j}^2 + \sigma_{\theta,j}^2 + \sigma_{\psi,j}^2\right) - 2v\left(\sigma_{r,j}\sigma_{\theta,j} + \sigma_{\theta,j}\sigma_{\psi,j} + \sigma_{\psi,j}\sigma_{r,j}\right) \\ \qquad + 2(1+v)\left(\tau_{r\theta,j}^2 + \tau_{\theta\psi,j}^2 + \tau_{\psi r,j}^2\right)\end{array}\right]\mathrm{d}\tau \end{aligned} \quad (12.29)$$

where m—the number of segment charges; n—the number of free faces; and K—the corrective coefficient for making up the defect of neglecting the effect of explosion gases.

So, the three-dimensional distribution of the stress energy in bench rock mass can be obtained while numerous points are placed uniformly in the bench rock mass, and the average energy densities of all stress waves including incident and reflective waves arriving at each point are calculated. An example shown in Fig. 12.7 is the distribution of average energy densities of stress waves of a three-hole bench blast on both horizontal and vertical sections through the bench rock mass.

(4) Calculation of the blasting fragmentation distribution of the isotropic and elastic rock mass

Dividing the values of average energy densities of all the points into a series of grades is expressed as e_i (i = 1, 2, …, n): $e_1 > e_2 > \ldots e_j > e_{j+1} > \ldots e_n$.

Taking the ratio of the number of points between every two energy density grades to the total number of all points as the volume distribution of rock in the bench which have different grades of energy density and expressed as $V_i(e_i)$.

According to Griffith's fracture theory of the brittle material, the fracture of a material which has undergone multiple stress actions is the result of the cumulative effect of all the stresses. So, all stresses have contributed to the rock fragmentation although their directions and time phases are different.

It is assumed that all stress energy in the rock transfer to the surface energy of the new surfaces produced when rock is broken. Assuming the specific surface energy (the energy needed for producing a unit of new surface area) is q, the distribution density of new surface can be obtained:

$$s_i = e_i V_i(e_i)/q \quad (i = 1, 2, \ldots, n) \quad (12.30)$$

Assuming that the rock which has the volume of $V_i(e_i)$ has been broken into m pieces with size d_i, the size d_i can be calculated by the equation:

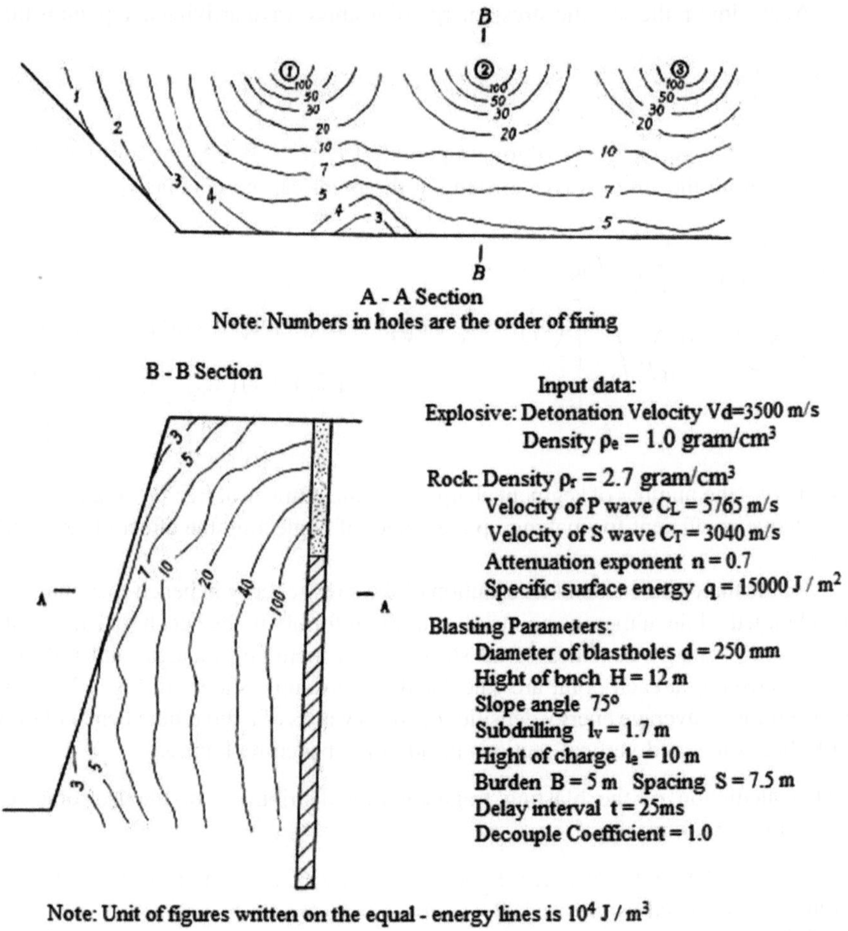

Fig. 12.7 Distribution of average energy densities of stress wave on both horizontal and vertical sections of a 3-hole bench blasting (reproduced from Ref. [3] with the permission from copyright © 1987 SEM USA)

$$d_i = \frac{c_s}{c_v s_i} V_i(e_i) = \frac{c_s q}{c_v e_i} \tag{12.31}$$

where

c_v Volume coefficient, $c_v = 0.6 - 0.7$;
c_s Area coefficient, $c_s = 2 - 3$.

It can be seen that n grades of energy densities e_i determine the n grades of d_i of rock fragments.

The probability distribution density $\Phi(d_i)$ of fragments in the size grade d_i in isotropic and elastic rock by blasting can be calculated:

$$\Phi(d_i) = V_i(e_i)/\sum_{i=1}^{n} V_i(e_i) \tag{12.32}$$

And the cumulative ratio of rock passing screen mesh of size d_i can also be calculated:

$$F_i(d_j \leq d_i) = \sum_{j=1}^{i} V_j(e_j)/\sum_{j=1}^{m} V_j(e_j) \tag{12.33}$$

(C) The fragmentation distribution of the practical rock bench by blasting

As stated in the beginning of this section, the weak planes formed by geological structures and previous blasts play an important role in rock fragmentation distribution by blasting. The practical bench rock fragmentation by blasting is the second fragmentation on the basis of that the rock mass has been broken by weak planes into "natural blocks" which have a certain distribution.

- Assume the probability distribution of the "natural blocks" of size $d < d_i$ in the bench rock mass before blasting as:

$$F_0(d < d_i) \tag{12.34}$$

- As the probability of the "natural blocks" of size $d > d_i$ is $[1 - F_0(d < d_i)]$, the probability of those blocks being broken into small fragments of size $d < d_i$ is:

$$F_1(d < d_i)[1 - F_0(d < d_i)] \tag{12.35}$$

Here, the fragmentation probability distribution $F_1(d < d_l)$ can be calculated using the fragmentation theory described above, in which the rock is regarded as isotropic and elastic.

- So, the final probability distribution of the fragmentation of size $d < d_i$ after blasting can be obtained:

$$F(d < d_i) = F_0(d < d_i) + F_1(d < d_i)[1 - F_0(d < d_i)] \tag{12.36}$$

(D) Method of obtaining the probability distribution of "natural blocks" in bench rock formed by weak planes

There are three methods that can be used to obtain the probability distribution of "natural blocks."

(1) Measuring size and keeping count of "natural blocks" which have more than two old faces on the pile after blasting. The size distribution of the counted blocks can be approximated as the distribution of "natural blocks" in the rock mass if the number of measured blocks is great enough.
(2) Back calculation from known blasting result.

When the size distribution of fragments after blasting is known as $F(d<d_i)$ and the fragmentation distribution of the isotropic and elastic rock mass by blasting has been calculated using computer program "BMMC" as $F_1(d<d_i)$, the probability distribution of "natural blocks" can be obtained by the back calculation using the formula below:

$$F_0(d<d_i) = \frac{F(d<d_i) - F_1(d<d_i)}{1 - F_1(d<d_i)} \tag{12.37}$$

(3) Computer simulation of spatial distribution of weak planes in rock mass.

This method was illustrated in detail in the author's previous published paper [10]. A Monte Carlo simulation technique was used in this method.

Set up three scan lines along three orthogonal orientations on the exposed bench face which will be blasted and measure the spacing of weak planes with reference to the distance, d, between the point where weak planes intersect the scan lines, three one-dimensional probability density distributions along the three orientations, $f_j(d), j = 1, 2, 3$, which can be considered as independent each other, can be obtained. The common distribution of the weak plane spacing is a negative exponential function (Fig. 12.8 shows two examples):

$$f_j(d) = \exp(-\lambda_j d) \quad j = 1, 2, 3 \tag{12.38}$$

The three-dimensional distribution of "natural blocks" as the joint distribution of the three independent distributions can be determined by the product of the three marginal distributions:

$$f_0(d) = f_1(d) f_2(d) f_3(d) \tag{12.39}$$

As a computer usually provides a standard procedure which can generate random numbers from a uniform distribution over (0, 1) interval, generation of a random number of the negative exponential distribution needs to use an inverse transformation function:

$$d = -(\ln u)/\lambda_j \tag{12.40}$$

where u—random number of the uniform distribution between (0, 1) interval.

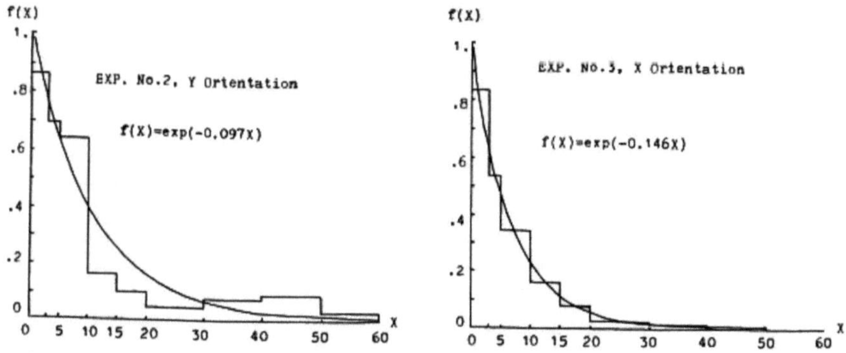

Fig. 12.8 Frequency distribution of weak planes along X- and Y-orientations of two blasts before blasting (reproduced from Ref. [3] with the permission from copyright © 1987 SEM USA)

The probability distribution of the natural blocks with the size $d < d_i$ is:

$$F_0(d < d_i) = \frac{1}{\delta} \sum\nolimits_{k=1}^{i} d_k f_0(d_k) \tag{12.41}$$

where

δ the average value of "natural blocks" distribution $f_0(d)$;
d_k the screen size of the k grade.

The theoretical fragmentation distribution calculated by the BMMC model with the parameters of bench blasting experiments of both small scale and production scale, respectively, are in accordance with the practical fragmentation distribution. Figure 12.9 is the comparison of the calculated data with the data of experiments conducted by Dick et al. [13]. Table 12.2 is the comparison of the results of the computer simulation and the data measured from the experiments in production scale (reproduced from Ref. [25] with the permission from AusIMM).

12.2.5 *Blasting Models for Jointed Rock Mass*

12.2.5.1 Gama Model (Reproduced from Ref. [26] with the Permission from Luleå University of Technology)

Dr. C. Dinis da Gama (Brazil) in 1971, according to Bond's third theory of comminution—the concept of the Bond's work index, through a series of tunnel blasting and bench blasting experiments, obtained an empirical formula for

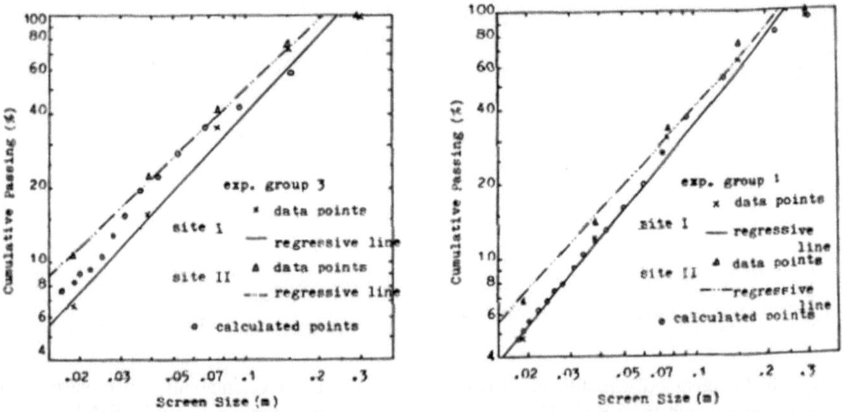

Fig. 12.9 Comparison of calculated distributions by computer against experimental distributions of Dick [13] (reproduced from Ref. [3] with the permission from copyright © 1987 SEM USA)

computing the blasting fragmentation distribution of the continuous homogeneous rock mass:

$$P_S = aW^b \left(\frac{S}{B}\right)^c \tag{12.42}$$

where

P_S The percent cumulative undersize weight of the fraction size, S;
W The total explosive energy input per unit weight of blasted rock, it can be calculated from the detonation heat of the explosive, kWh/ton;
S Size of fragment;
B Burden of the charges; and
a, b, c Empirical factors depending on the explosive type, rock properties, and blasting pattern

In 1983, Dr. Gama revised the formula (12.42) using the multiple regression analysis according to the data collected from a field study of blasting in jointed rock masses performed in several quarries in Brazil:

$$P_S = aW^b \left(\frac{S}{B}\right)^c F_{50}^{-d} \tag{12.43}$$

where

F_{50} The average size of the natural blocks which were cut by the joints in the bench rock mass before blasting, and
d Empirical factor.

Table 12.2 Comparison of the results of the computer simulation and the experiments in production scale (reproduced from Ref. [25] with the permission from Australasian Institute of Mining and Metallurgy)

Pattern Size (m)	Data source	Cumulative passing % under fragment size, x (m)										Parameters of R–R function		Relative errors (%)		K_{50} (m)	P_{80} (m)
		<0.1	<0.2	<0.3	<0.4	<0.5	<0.6	<0.7	<0.8	<0.9	<1.0	*A*	*B*	Aver.	Max.		
5 × 5	Expe.	21.5	48.6	66.0	74.7	78.3	80.0	80.1				1.0581	0.9841	15.0	28.6	0.24	0.55
	Calcu.	27.6	45.1	56.9	68.5	79.1	88.6	93.5	96.6	98.4	99.3	1.2002	1.0111			0.21	0.49
6 × 6	Expe.	18.6	43.7	61.5	71.7	76.7	78.9	79.8				1.0625	1.0663	10.3	21.3	0.26	0.58
	Calcu.	20.4	36.7	48.4	60.2	71.4	81.8	88.6	93.8	97.0	98.8	1.0816	1.1115			0.27	0.58
7 × 7	Expe.	12.5	24.6	36.9	57.4	76.4	78.9	79.9				1.1163	1.2514	12.1	22.7	0.31	0.60
	Calcu.	15.0	30.2	42.0	54.5	66.5	77.2	86.0	92.1	96.0	98.1	1.0294	1.2371			0.32	0.64
8 × 8	Expe.	12.6	30.8	46.4	58.2	66.4	71.9	75.4				0.8917	1.2131	11.3	19.5	0.35	0.71
	Calcu.	13.5	24.8	37.6	50.7	63.4	75.0	84.3	91.1	96.0	97.9	1.0916	1.4071			0.35	0.65

Rosin–Rammler (R–R) function: $Y = 1 - (x/x_0)^a$, $A = a \times \ln x_0$; $B = a$

The size distribution of natural blocks cut by different geological weakness faces was completed by a computer program, named COMPART, and developed by Dr. Gama in 1977, based on the data of geological survey of all important weakness faces in the rock bench.

12.2.5.2 Bezmatlih Model [27]

Bezmatlih et al (B. X. Безматлих, scholars of the former Soviet Union) developed a mathematical model in 1971 for calculation of the fragmentation distribution of jointed rock mass by blasting. His point of view was basically the same as other scholars of the former Soviet Union who studied the influence of joints and fissures on the distribution of rock fragmentation by blasting, so it has some representation. The basic idea is:

- The probability of the blocks, which have size not greater than X_k cut by the joints and fissures in the rock mass before blasting, is $P_0(X \le X_k)$;
- The probability of the blocks, which have size not greater than X_k produced by explosive energy in an intact rock mass during blasting, is $P_1(X \le X_k)$;
- Then the probability of the blocks, which have size not greater than X_k produced by explosive energy in joined rock mass after blasting should be the sum of the above two items;
- According to the theory of random breakage, P_0 and P_1 shall be subject to the Poisson distribution. So that, Bezmatlih derived the distribution function of Rosin–Rammler for fragmentation of jointed rock mass after blasting:

$$P(X \le X_k) = 1 - \exp\left[-\left(X_0^{-1} + \beta_0 J\right)X\right] \quad (12.44)$$

where X_0—average size of the natural blocks cut by the joints and fissures before blasting. If the number of joints and fissures within unit length measured in field is α_0, then $X_0 = 1/\alpha_0$;

J—specific impulse of explosives, $J = qD$ where q is the powder factor and D is the velocity of detonation of the explosive used;

β_0—constant related to the blasting conditions.

- If it is specified that the blocks whose size are greater than X_h are the unqualified blocks, the rate of unqualified blocks can be calculated from the Eq. (12.42):

$$X_h = \exp\left[-\left(X_0^{-1} + \beta_0 J\right)X_h\right]$$

or:

$$X_h = \exp\left(-\frac{X_h}{X_0}\right)\exp(-\beta_0 J X_h) = V_e \exp(-\beta_0 J X_h) \tag{12.45}$$

where $V_e = \exp\left(-\frac{X_h}{X_0}\right)$ is the rate of natural unqualified blocks in the rock mass before blasting.

The data measured in the Glafulei open pit verified that the calculated rate of unqualified blocks is close to the measured results.

12.2.6 SABREX Model (Reproduce from Ref. [28] with Permission from the Canada Institution of Mining, Metallurgy and Petroleum)

SABREX is a modular computer program which is used for surface and underground blast simulation developed by ICI Explosive group (this group had been sold to Orica) Ref. [28]. The program can be run on a microcomputer and it provides predicted blast results as dynamic color graphics and tables. Figure 12.10 shows a block diagram of the module and associated technologies related to the program.

Variables input:

- Explosives: BLEND code and CPEX code. The former is the ideal detonation code, and the latter is the non-ideal detonation code which requires the necessary data from extensive field and laboratory testing.
- Rock properties: These are obtained by laboratory testing, and the data that are normally used as input to SABREX are density, Poisson's ratio, Young's Modulus, unconfined compressive strength, tensile strength, dynamic tensile strength, and shock attenuation. In situ measurements of the rock to be blasted may also be required and have generally consisted of seismic determination and structural orientation mapping.

Once the explosives and rock properties have been assembled and inputted, the following controllable parameters may be varied and tested on SABREX: blasthole diameter, pattern (burden-spacing-special face condition), blasthole inclination, sub-grade drilling, collar distance, initiation pattern, delay timing, coal seam depth, and spoil pile location (for cast blasting in coal operation) and costs for explosives accessories and drilling.

The operational modules of SABREX program mainly consist of the following:

- Fragmentation modules: Kuz–Ram module (see Sect. 12.2.3) and CRACK module (see Sect. 12.2.1) offer independent absolute predictions of fragmentation resulting from a given blast.

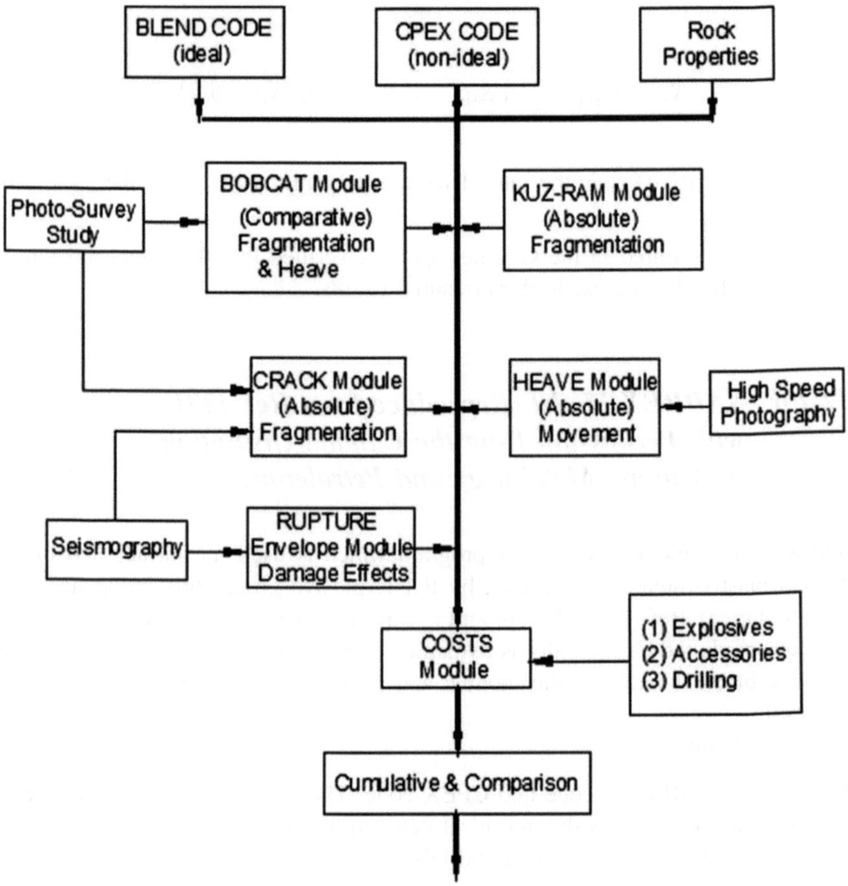

Fig. 12.10 Block diagram of the modular and associated technologies of SABREX (reproduced from Ref. [28] with the permission from the Canada Institution of Mining, Metallurgy and Petroleum)

- The HEAVE module: This module uses calculations based on gas expansion to determine the velocity and time of initial movement of blocks within the burden and then tracks the motion of the blocks until they come to rest. Factors in the calculation include limitations of subsequent rows by the first row movement, swell factor, and angle of repose.
- Damage module: The RUPTURE envelope module—The ability to model damage effects from blasting is an important tool in design and maintenance of mine structures. In this study, the rock properties are determined and then damage envelopes run for spherical charges of several explosives types. The radius of damage behind the charge gives the resultant crater radius and the subgrade crack penetration.

- COSTS module: At the lowest possible costs, obtaining the desired blasting results of qualified fragmentation, uplift, and permissible extension of damage and flyrock, constitutes blasting optimization.

The field experiences using SABREX model have proven that useful results can be obtained in line with the actual mining practice.

12.3 Some Typical Program of Computer-Aided Design

12.3.1 Blast Maker [33]

Blast Maker (www.blastmaker.kg) is a symbiosis of the "Blast Maker" industrial enterprise and the research Institute of Communications and Information Technologies (ICIT). Blast Maker, in conjunction with the Institute of Communications and Information Technologies of Kyrgyz-Russian Slavic University, offers a system of computer-aided design of mass blasts in open pits and is developed under the support of International Science and Technology Center (www.istc.ru) in collaboration with MEPhI and Mines Paris (Ecole des Mines de Paris).

Computer-aided design package for drilling and blasting operations (DBO CAD) is an integral part of Blast Maker® HSC and is designed to perform procedures concerned with the preparation of technical documentation for production drilling and blasting operations at open pit mines. The structure of the CAD package consists of a set of separate modules that can operate together or independently. This approach provides opportunity to vary the package configurations according to the amount of tasks of the mining enterprise. Data exchange between the individual modules is performed through the single database containing all the necessary information for the CAD package.

The main modules of CAD package are as follows:

- Deposit digital module: The module is designed for collection, analysis, and storage of geological, physical, and mechanical data and to process 3D information about the deposit;
- Open pit surface digital mode: The module provides digital mapping of the open pit surface and mining-geometric analysis;
- DBO design module: The module is designed to calculate the parameters of single blasting and placement of boreholes on the block exploded.
- Single blasting simulation module: This module is designed to assess the quality of the blasting operation and predict the parameters of the broken rock;
- Data import and export module: The module provides interfacing of the CAD package with the DBO software of third parties that will be used in mining;

- Documentation output module: The module is designed for graphics and analytical works related to the preparation and production of design documentation for a single blasting operation.

Figure 12.11 shows the flowchart of Blast Maker system. The principal software and hardware include **Blast Maker, KOBUS controller, Stress module, and Split Analyzer**.

- **Blast Maker**: The Blast Maker program-technical complex has many housekeeping functions. It can predict the results of explosion in accordance with selected project parameters. The forecast is visually represented in 3D image form. The system also includes economic calculations that allow choosing the type of explosives and rational charge structure. The designer can review various blast design variants and choose the most optimal one. The output documents module generates all necessary views and formats (Fig. 12.11).
- **KOBUS** controller: One of the effective measures to obtain the information relevant to properties of the rock mass exploded is the information obtained directly from the drilling rig during blastholes drilling, including parameters such

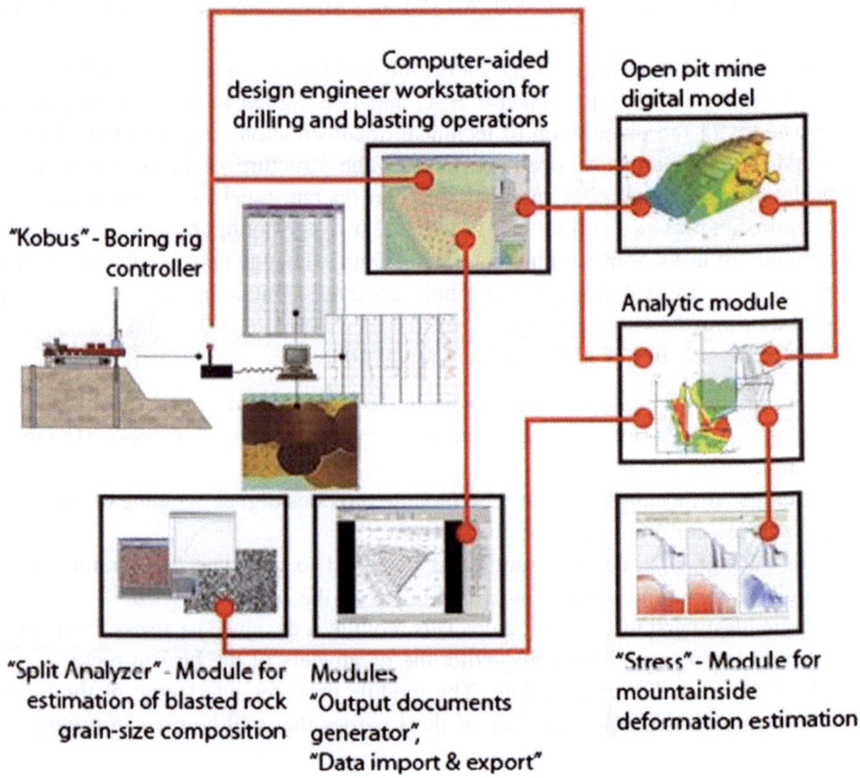

Fig. 12.11 Flowchart of Blast Maker system, [33]

Fig. 12.12 KOBUS—hardware for collection and record the drilling information [33]

as specific energy of drilling. KOBUS controller (Fig. 12.12) is the hardware component of the CAD of drilling-and-blasting in open pits. The system allows the continuous acquiring of data about the parameters of the drilling of blastholes online creating the unique system of drilling modes monitoring. The effective performance of the controller is ensured by using the original software. The program performs the filtering and processing of the acquired data, estimates the momentary energy intensity of drilling in each blasthole, graphically represents incoming information enabling it to accumulate and analyze drilling parameters. Figure 12.13 shows the KOBUS controller. The KOBUS drilling rig controller is intended for the acquiring, preprocessing, indicating, and transmitting of data to a control station from regular or additionally installed monitoring transducers. The controller is installed in the booth of the drilling rig.

- **Stress** module: Stress software module is intended for the evaluation of the slope stability of pit walls, dumps, and other mine's technical structures. The package is one of the elements of the Blast Maker. The forecasting of the stability of mine openings is carried out based on the numerical simulation of the deflected mode of the rock mass and the continuous monitoring of deformations. The kit of geomechanical models sufficiently represents the various

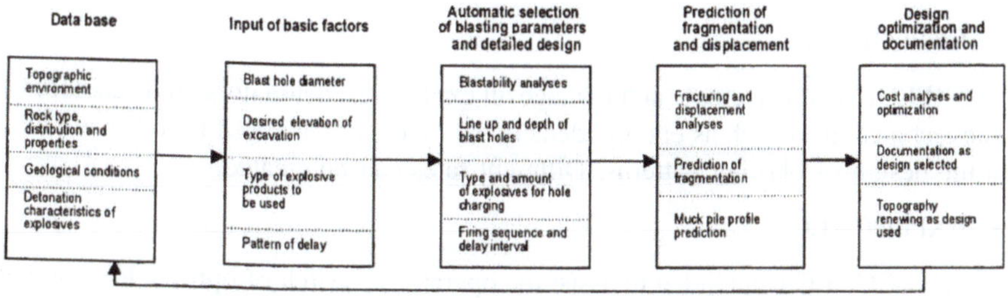

Fig. 12.13 Schematic description of the Blast-Code model (reproduced from Ref. [29] by permission of author Qu et al.)

mechanisms of rock destruction. The interaction of the Blast Maker system and the Stress package makes it possible to choose the optimal parameters of drilling design ensuring the high quality of blasting and the stability of mine technical structures.

- Split Analyzer: The Split Analyzer system carries out the evaluation of the grain-size composition of blasted rock mass. Split Analyzer software module performs objective, human-factor-independent estimation of blast quality. It is intended for the determination of the grain-size composition of blasted rock mass based on the processing of digital photographs. It allows distribution grid of particular fractions to be obtained and analysis of the acquired data. The system has the feature of printing images and statistical data acquired as a result of analysis. It can work with several photographs simultaneously.

The Blast Maker has been use in some surface mines of Russia, Kazakhstan, and Kyrgyzstan.

12.3.2 EXPERTIR [34]

EXPERTIR (www.epc-france.com), designed by the company EPC-FRANCE, is designed to handle the blasting in mines and open pits. It includes modules that have features and different levels of complexity.

EXPLOBASE is a module managing an explosives database and their rates.
EXPLOCALC is a computer program to optimize the theoretical drilling pattern depending on the actual energy explosives, while optimizing the cost of blasting.
EDITIR is a module to help in the design of firing sequences software. It works on the basis of a theoretical cutting face.
FRONTIR is an extension of EDITIR to real fronts, incorporating 3D measurement of the front.
EXPERTIR 3D is an expert software combining the functionality of FRONTIR and optimization of explosive energies.
The software is available in French and English.

- EXPLOBASE

EXPLOBASE can create and manage all explosives, detonators, and accessories. The technical data such as energy, density, and the costs are used by other modules in the design of blasting patterns. Different rates can be created.

- EXPLOCALC

EXPLOCALC is intended to assist the operator in search of optimal loading and drilling geometry. This blasting plan guarantees the desired result while minimizing the cost per cubic meter shot, integrating explosive costs, priming, and drilling.

EXPLOCALC will quickly compare all possible configurations and shooting plans based on explosives available on the market and will adopt the most economical solution for the site. Its organization around a database allows a blasting plan management with feedback and an adaptation of the software to the actual site conditions.

- Plan interactive loading
  After setting the initial parameters of the shot to achieve:
  - Rock type,
  - Height of the bench, and
  - Blasthole diameter…
    this option allows for graphically interactive loading plans.
    The results calculation option will then determine the optimal geometric parameters:

  - Burden and
  - Spacing…

depending on the available explosive energy and will display the specific costs incurred.

The windowing allows the user to design simultaneous multiple loading plans for comparison.

- Automatic Loading and Geometry

By specifying only the explosives list specified, this option automatically calculates within seconds the loading plan and optimal geometric parameters that guarantee sufficient energy is imported to the rock in question to minimize the cost of the shot per meter cube. The desired loading plan may be either with or without intermediate stemmings.

- EDITIR–FRONTIR–EXPERTIR 3D

These three modules are three different versions of the same complex blasting software. They integrate the expertise of EPC FRANCE and its partners into opencast mining. The main features are summarized in the following Table 12.3.

12.3.3 Blast-Code (Reproduced from Ref. [8, 29] by Permission of the Author, Prof. Qu S. J. et al.)

Blast-Code was developed by the University of Science and Technology Beijing in 2001. It was developed for Shuichang surface mine of Beijing first, then used in some large surface mines in China [29, 30].

Figure 12.14 shows the schematic description of the Blast-Code model.

Table 12.3 Main features of EDITIR–FRONTIR–EXPERTIR 3D [34]

	EDITIR	FRONTIR	EXPERTIR 3D
Creating a type front with preestablished views or by manual drawing of the front	X		
Creation of the drilling pattern, the number of holes and rows	X	X	X
Work on each hole differentially	X	X	X
Implantation drilling according to homogeneous plane and markers		X	X
Rapid alignment functions (bottom altitudes, positions, holes, heads, constant benches, etc.)	X	X	X
Importing a real forehead from 3D topographic measurement		X	X
Photographic mapping of the front face		X	X
Importing the actual positions of the holes		X	X
Importing hole deviation measurements probe-type distance meter (Tiara)	X	X	X
Importing hole deviation measurements inclinometer sensor types (Pulsar, Boretrack)		X	X
Integration and storage hole by hole drilling reports		X	X
Importing and viewing continuous signals drilling parameters		X	X
Determination of minimum profiles per hole for the prevention of risk of projection		X	X
Integration of the positions of discontinuities or abnormalities visible on the front face		X	X
Determination of connection diagrams and sequences for non-electrical firing	X	X	X
Semi-automatic sequence determination by electronic detonators	X	X	X
Electrically sequential-type sequences design	X	X	X
Propagation simulation of seismic waves	X	X	X
Loading holes with explosives and detonators found in the database	X	X	X
Calculation of explosives hole by hole consumption (g/m^3)	X	X	X
Calculation of the specific energy per hole (MJ/m^3), and foot column			X
Automation of loading for mining constant energy. Editing blast design, printing			X
Editing blast design, printing	X	X	X
Export of contact information and technical data (text, Excel, XML, etc.)	X	X	X

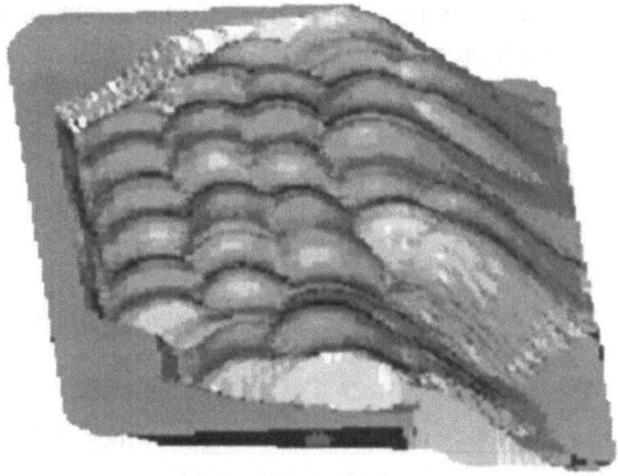

Fig. 12.14 Predicted profile of muck pile of a blast in 3D mode (reproduced from Ref. [29] by permission of author Qu et al.)

The Blast-Code model consists of a database representing geological and topographical conditions of the mine and the modules Frag + and Disp + for blast design and prediction of resultant fragmentation and displacement of rock mass.

12.3.3.1 Database for the Blast-Code Model

The database stored the following data and can be automatically loaded and used in blast design and simulation with the Blast-Code model.

- Topography of the area to be blasted including the free face geometry from the survey results;
- Geological conditions, mainly the average spacing s and spacing distribution coefficient n of geological discontinuities.
- Rock properties, mainly the compressive strength σ and density γ of each type of the rock to be blasted;
- Explosive's detonation characteristics: detonation velocity D and bulk density ρ_e of the explosives to be used in blasting.

12.3.3.2 Module FRAG$^+$ for Blast Design

Fragmentation and displacement are the two main issues involved in bench blasting. As no theoretical equations can directly be used in the design of bench blasting operations, both qualitatively theoretical analyses and engineering statistics are used for construction of the FRAG$^+$ module. Based on the theory of crater blasting, a series of qualitative equations representing the relationship of resultant fragmentation and displacement of rock mass with free face conditions, geology, rock properties, explosive characteristics, and the blasting parameters considered are established. With modification of the established equations by multielement regression on basis of the data collected from 85 field production blasts in the 3 submines of Shuichang surface mine. The following relationships are established:

- Explosive strength η and blastability index ξ of the rock to be blasted:

$$\eta = a_1 \rho_e D^2 \quad (12.46)$$

$$\xi = a_2(\gamma\sigma^{\alpha}) \log(a_3 s)^{\beta} \quad (12.47)$$

where

σ	Compressure strength of rock, kPa;
γ	Density of rock, g/cm^3;
s	Average spacing between geological discontinuities, cm; and
$a_1, a_2, a_3, \alpha, \beta$	Constants determined from regression analyses.

- Blasting parameters: B—burden, S—spacing, L—blasthole depth, and q—powder factor:

$$A = c_1\left(\eta^{\delta}\xi^{\theta}\right)\phi^{\kappa} \quad (12.47)$$

$$B = (A/m)^{1/2},\ \text{m} \quad (12.48)$$

$$S = A/B,\ \text{m} \quad (12.49)$$

$$h = c_2\left(A\eta^{\delta}\xi^{\theta}\right),\ \text{m} \quad (12.50)$$

$$L = \left(H_{\text{top}} - H\right) + h,\ \text{m} \quad (12.51)$$

$$q = c_3\left(\xi\eta^{\lambda}\right)\gamma,\ \text{kg/t} \quad (12.52)$$

where

A	Top surface area of crater from a blasthole, m^2;
ϕ	Diameter of the blasthole, mm;
m	Ratio of spacing to burden;
H_{top}	Elevation of the blasthole, m;
H	Desired elevation of excavation after loading, m;
h	Depth of subdrilling, m;
γ	Rock density, g/cm^3;
$c_1, c_2, c_3, \delta, \theta, \kappa, \lambda$	Constants determined from regression analyses

- Hole charge amount Q and collar stemming length l:

$$Q = BS(L - h)\gamma q, \text{ kg} \quad (12.53)$$

$$l = L - 4Q/\left(\pi\phi^2\rho_e\right), \text{ m} \quad (12.54)$$

- Delay interval

In Blast-Code model, the delay interval (millisecond delay between rows is used in Shuichang surface mine) is determined empirically and is expressed as:

$$u = k_t B, \text{ ms} \quad (12.55)$$

where

k_t An empirical constant, k_t = 4–8 ms/m, normally k_t = 6.6 ms/m in the model.

12.3.3.3 Module DISP$^+$ for the Prediction of Muck Pile Profile and Fragmentation

- Prediction of Profile of Muck Pile after blasting

The module DISP$^+$ is a software package, in conjunction with the module FRAG$^+$, allows prediction of the scope of fragmentation and muck pile in 2D and/or 3D model.

The module DISP$^+$ is established based upon the assumption that the profile of muck pile at rest after blasting follows the Weibull distribution of statistics. After collection of the data of muck pile surface from production blasts, geological data, rock properties in situ and blasting parameters, multielement regression analyses method is used to get the coefficients of the Weibull distribution structure. Figure 12.15 is an example of the predicted muck pile profile of a blast with 5 rows of blastholes in 3D mode. Figure 12.16 shows the 2D profile at a vertical section of the same muck pile in Fig. 12.16.

- Prediction of Fragment Size and Distribution

Fig. 12.15 Vertical section of the profile of muck pile in 2D mode (reproduced from Ref. [29] by permission of author Qu et al.)

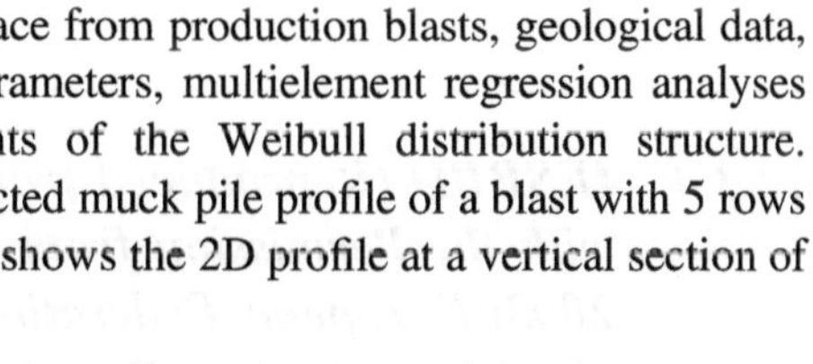

The fragmentation distribution of blasted rock is expressed by Rosin–Ramler equation. The characteristic size x_c and distribution coefficient n are greatly controlled by blast design, rock mass geological condition, rock properties explosive charge, firing sequence, and timing of delay. The Rosin–Ramler function and its parameters x_c and n can be expressed as:

$$y = \left\{1 - \exp\left[-\left(\frac{x}{x_c}\right)^n\right]\right\} \times 100\,\% \tag{12.56}$$

$$x_c = K_1 \gamma^{\alpha_1} f^{\alpha_2} m^{\alpha_3} s(q\rho_e D^2)^{\alpha_4} T_u \tag{12.57}$$

$$n = n_n\left\{1 + K_2\left[C_1 - \frac{B}{\phi}\right]\left(\frac{1+m}{C_2}\right)^{\frac{1}{2}} \frac{l_{\exp}}{L}\right\} \tag{12.58}$$

where

y	Cumulative percentage by weight or volume of rock fragments which size are no bigger than x, %;
x	Size of rock fragments, cm;
T_u	A coefficient as a function of u;
n_n	Distribution coefficient of rock mass before blasting;
$l_{\exp}$	Charge length of the blasthole, m; and
$K_1, K_2, C_1, C_2, \alpha_1, \alpha_2, \alpha_3, \alpha_4$	Coefficients determined empirically.

The above equations are established based on qualitative analyses, and all coefficients are determined with results of photographic fragmentation analyses of 36 muck piles.

The Blast-Code model also permits interactive parameter selection based on the comparison of the predicted fragmentation and displacement as well as the cost for drilling, explosives, and accessories until the most effective option can be selected.

12.3.4 IESBBD (Reproduced from Ref. [31] with the Permission from Copyright © 2003nEuropean Federation of Explosives Engineers (EFEE) & [32] by Permission of the Author, Prof. Zhang J.C. et al.)

IESBBD (**I**ntelligence **E**xpert **S**ystem for **B**ench **B**lasting **D**esign) was developed by Prof. Zhang et al in China [31, 32] and published in 2003. It includes choosing blasting parameters, the automation of blasting design, the standardization of design charts, and the systemization of blasting management, and improves the quality and speed of production blast design. The IESBBD system has been initially applied to production blasting. Improved blast results were experienced while achieving the production requirements.

12.3.4.1 The Composition and Structure of IESBBD

The IESBBD system consists of five parts: database, intelligent study, design, prediction, and analysis. The composition and structure of IESBBD are shown in Fig. 12.16.

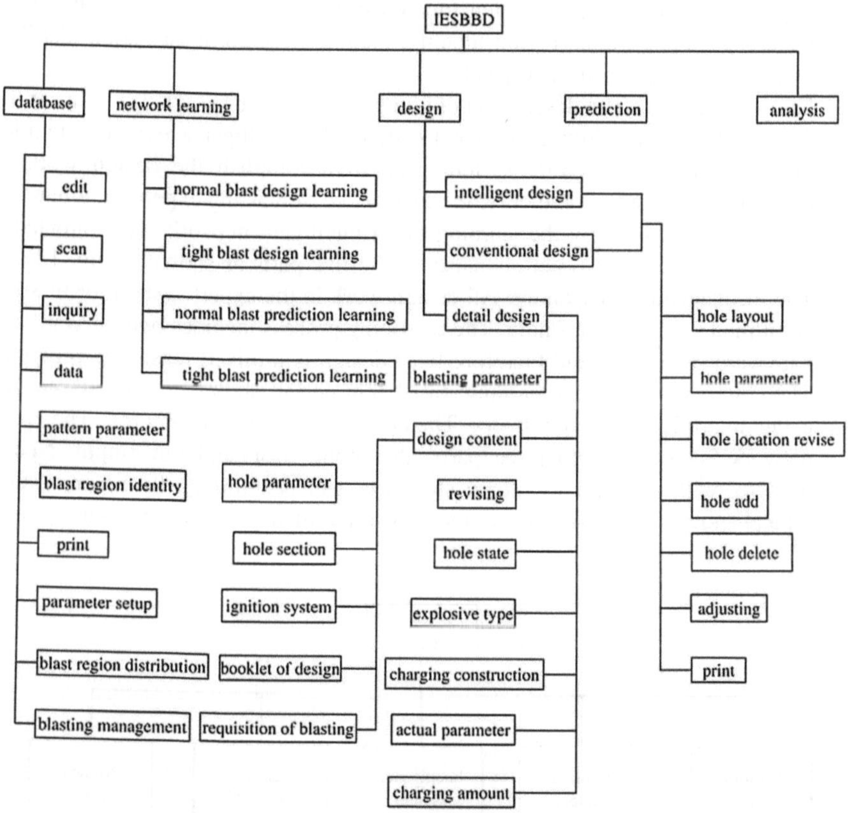

Fig. 12.16 Composition and structure of IESBBD (reproduced from Ref. [31] with the permission from Copyright © 2003 European Federation Of Explosives Engineers (EFEE))

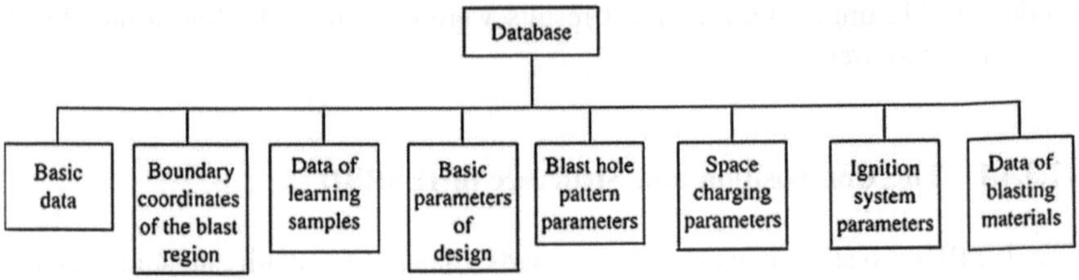

Fig. 12.17 Basic content of the database [reproduced from Ref. [31] with the permission from Copyright © 2003nEuropean Federation of Explosives Engineers (EFEE])

1. Database: The database of the IESBBD is composed of eight parts as shown in Fig. 12.17. The database has ten functions: edit, scan, inquiry, data, blasthole pattern parameter, blast region identity, parameter setup, blast region distribution, blast management, and print.
2. Knowledge Base: The description of knowledge and the knowledge base are the foundations for developing an expert system. The intelligence expert system for blast design is a software system used to accomplish the design work of blasting. It is based upon theory, the experience of blasting experts, and blasting knowledge acquired by the learning of an intelligent network. The knowledge base of the IESBBD system consists of the five parts as shown in Fig. 12.18.
3. The mechanism of reasoning: All design work in the expert system for blasting is fulfilled based on reasoning. The reasoning process includes both forward and backward reasoning, and therefore blast design is a mixed reasoning process.
4. The system of intelligent learning: Intelligent design and prediction are realized using an artificial neural network. The elementary unit is the neuron. The entire network is divided into three parts: the input, deal, and the output. Neural network with different topologic structures can be built up by neurons. One typical model in engineering is the feed-forward network model with reverse

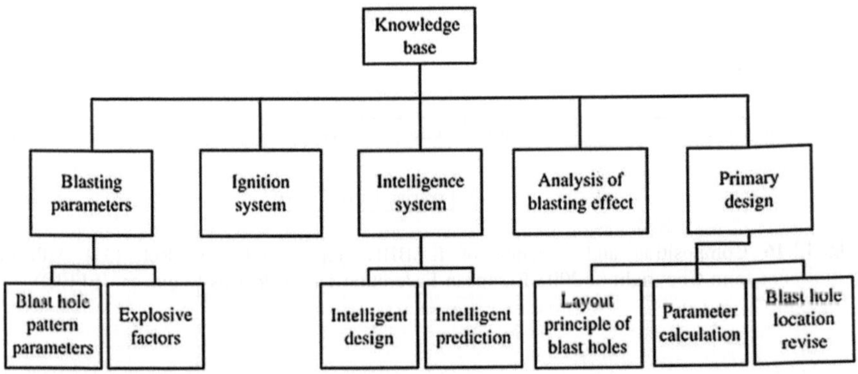

Fig. 12.18 Composition of knowledge base [reproduced from Ref. [31] with the permission from Copyright © 2003nEuropean Federation of Explosives Engineers (EFEE)]

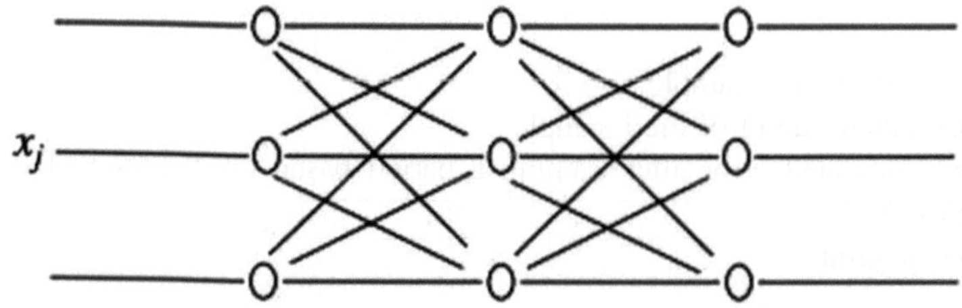

Fig. 12.19 Structure of feed-forward network model [reproduced from Ref. [31] with the permission from Copyright © 2003nEuropean Federation of Explosives Engineers (EFEE)]

propagation. Figure 12.19 shows the structure of the feed-forward network model. It consists of one input layer, one output layer, and one or more hidden layers. All elements are completely joined. The input layer receives the outside samples. The network partially adjusts the weighting parameter through training the samples, and the output layer creates an output. At the same time, the expected output corresponding to the input acts as a teacher signal. If there is a difference (error) between the actual output and the expected output, the network system automatically regulates the corresponding weighting toward decreasing the error. The error gradually becomes smaller by repetitive training, and finally, the actual output coincides with that expected.

The action function can be defined as:

$$p_i = \sum\nolimits_j \omega_{ij} x_j \tag{12.59}$$

$$y_i = f(p_i, v_i) \tag{12.60}$$

where

- x_j The inputs for the k layer element j;
- p_i The inputs for the $k + 1$ layer element i;
- ω_{ij} The associated weighting values for the k layer element j and the $k + 1$ layer element i;
- y_i The outputs of the $k + 1$ layer element i;
- $f(p_i, v_i)$ The action function; and
- v_i The bias factors.

The weighting values are changed toward the negative gradient of the error function by using the steepest descent method from nonlinear programming. The error function E can be defined as:

$$E = \sum\nolimits_k \sum\nolimits_i \varepsilon_i = \sum\nolimits_k \sum\nolimits_i (y_i - \hat{y}_i)^2/2 \tag{12.61}$$

where

y_i The output of the i sample;
$\hat{y}_i$ The expected output of the i sample;
E The accumulated error after training a period based on N group of training samples; and
ε The error limit.

when $E \leq \varepsilon$, the training process is terminated.
The neural network learning process is shown in Fig. 12.20.

5. Analysis system: The blasting of a rock mass is considered to be a grey system; hence the principal factors affecting the quality of rock blasting are determined by the grey correlation analysis. There are two kinds of parameters in rock blasting. These are the target parameters and the affecting factor parameters. The former judges the blasting quality and the latter controls the blasting quality. In a grey analysis, the target parameters are defined as system characteristic variables and the affecting factor parameters as the relative factor variables.

The analysis targets of the IESBBD system are the ratio of large-size fragments, the maximum height of the muck pile, and the extent (width) of back break. The

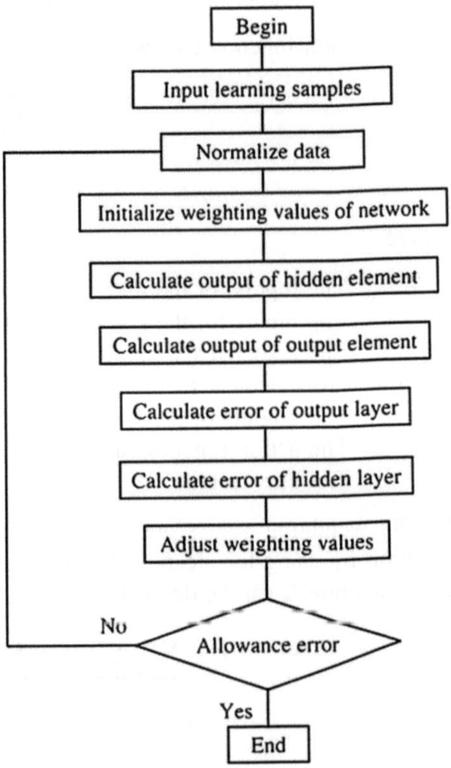

Fig. 12.20 Learning process for a neural network [reproduced from Ref. [31] with the permission from Copyright © 2003nEuropean Federation of Explosives Engineers (EFEE)]

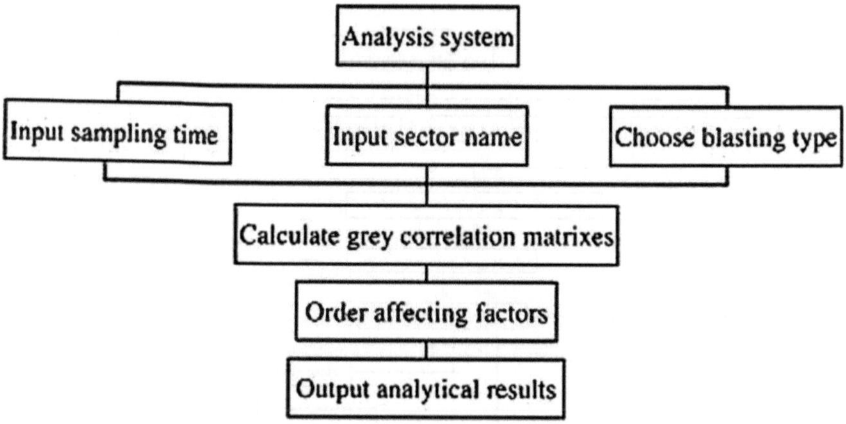

Fig. 12.21 Calculating process of grey correlation analysis in IESBBD [reproduced from Ref. [31] with the permission from Copyright © 2003nEuropean Federation of Explosives Engineers (EFEE)]

affecting factors are the extent of joint growth, the hole spacing in the intermediate row, the toe burden, the explosive factor, the hole spacing in the back row, the row width, the maximum charge of a single hole in the back row, and the width and porosity of a tight muck pile. The flowchart used in the grey correlation analysis is shown in Fig. 12.21.

When the analysis is completed, the system outputs the names of the five affecting factors which have the greatest influence on the targets and sorts their order in light of the domination ability.

12.3.4.2 Design Flowchart of IESBBD

The entire blast design is divided into two parts in IESBBD: the primary design and the detailed design. The primary design is further divided into conventional design and intelligent design depending on the source of blasting parameters.

Conventional primary design is carried out according to the data contained in the knowledge database of the expert system. The experts determine the blasthole pattern parameters and the explosive factors. The intelligent primary design is carried out according to the data from the knowledge base acquired after learning from the blasting parameter samples by neural network. The primary design has many functions. These include things such as revising the control points of the back boundary of the blast region, modifying blasthole locations, adding or deleting blastholes, adjusting blasthole locations after the blasthole layout, drawing the map showing the plan layout of the blastholes, and printing the instruction booklet regarding the distribution of holes. The detailed design is made after checking the blasthole pattern. Its main function is to input or choose the blasthole state (wet hole or dry hole). The explosive type and the actual charge amount, the actual charge

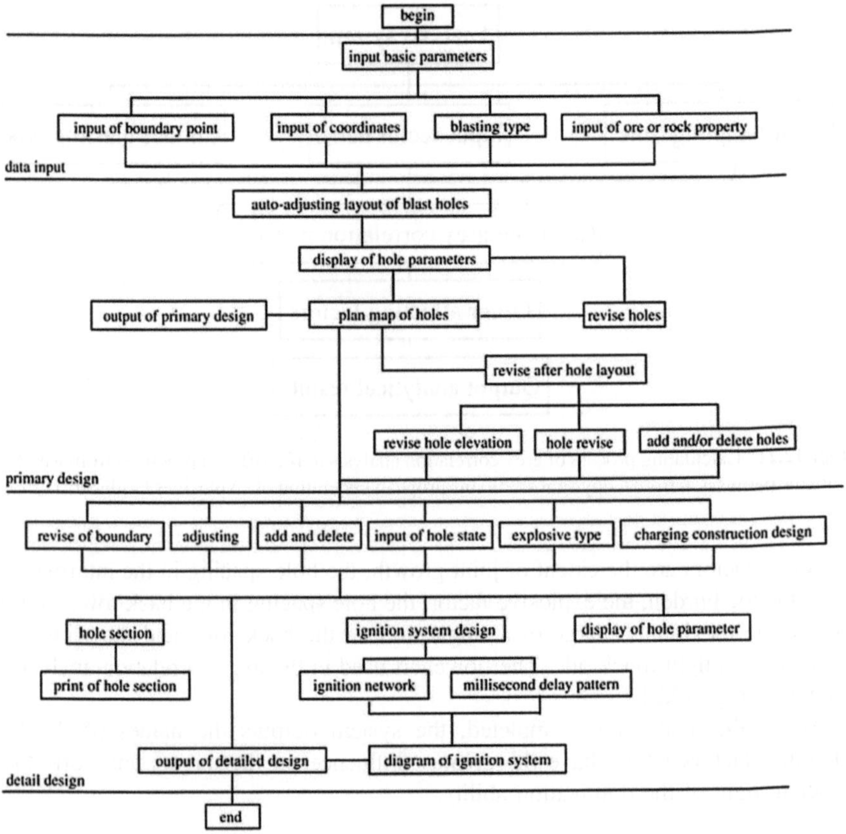

Fig. 12.22 Design flowchart of IESBBD [reproduced from Ref. [31] with the permission from Copyright © 2003nEuropean Federation of Explosives Engineers (EFEE)]

configuration, the ignition system, as well as the millisecond delay pattern draw the section chart showing the charge configuration for single or multiple holes, output the instruction booklet for the blast design, the requisition of the blasting order, and prepare the section drawings of the blastholes. The design flowchart of the IESBBD system is shown in Fig. 12.22.

The IESBBD system has been tested and applied to production blasting at the Lanjian iron mine of Panzhihua Steel Group of China.

References

1. Yang J et al (1999) Theoretical models and numerical calculation for rock blasting. Science Press (in Chinese) 杨军 等, 岩石爆破理论模型及数值计算, 科学出版社, 1999
2. Alabamian Z et al (2013) Simulation of dynamic fracturing of continuum rock in open pit mining. Geomaterials 3:82–89
3. Zou D (1987) A three dimensional mathematical model in calculating the rock fragmentation of bench blasting in the open pit. In: 2nd international symposium on rock fragmentation by blasting, Keystone, Colorado, Aug 1987, OSTI Identifier: 6726464
4. Favreau RF (1980) "Blaspa" a practical blasting optimization system, 6th conference on explosives and blasting technique 2/5–8, Tempa
5. Harries G et al (1977) The use of a computer to describe blasting. In: 15th APCOM symposium, Brisbane, Australia, pp 317–324. Libraries Australia ID: 9325500
6. A rigid Block Mechanics Code, University of Minesota and Dames and Moorre Ltd. 1987
7. Yang R, Scovian DS (2011) A model for near and far-field blast vibration based on multiple seed waveforms and amplitude attenuation. Blasting Fragm 4(2):91–116
8. Trivedi R et al (2014) Prediction of blast-induced flyrock in india limestone mines using neural networks. J Rock Mech Geotech Eng 6(2014):447–454
9. И. Барон и др (1967) "Тест Розин-Раммлера Уравнение Заявка на диаметр расчета Взрывных рок".// Санкт-П. "Взрывные работы", Недра, 1967. # 62/19
10. Zou D et al (1986) Computer simulation of spatial distribution of weak planes and its influence on rock blasting. In: Proceedings of the international symposium on intense dynamic loading and its effects, June 1986, Beijing, China
11. Kazuo Otsuka et al (1981) Particle size distribution characteristics of the crushing products—basic research of rock blasting. Nippon Min Assoc Mag 97(1121):521–526 (in Japanese) 大塚一雄等; 破砕產物の粒度分布特性 - 岩石爆破の基礎的研究, 《日本鉱业会志》97 (1121):521–526
12. Just GD (1973) The application of size distribution equations to rock breakage by explosives. In: National symposium on rock fragmentation. Astralia Geomechanics Society, pp 18–23
13. Dick RA et al (1973) A study of fragmentation from bench blasting in limestone at a reduced scale. USBM report investment no. 7704
14. Gama CD (1971) Size distribution general law of fragments resulting from rock blasting. Trans Soc Min Eng AIME 250(4):314–316
15. Lovely BG (1973) A study of the sizing analysis of rock particles fragmented by a small explosives blast. In: National symposium on rock fragmentation. Astralia Geomechanics Society, pp 24–34
16. Austin LG et al (1968) Statistics of random fracture. Trans Soc Min Eng AIME 241(2):219–224
17. Harries G (1990) Development of a dynamic blasting simulation. In: 3rd international symposium on rock fragmentation by blasting, Brisbane, Australia, Aug 1990. ISBN: 0 949 106 55 0
18. Harries G (1983) The modeling of long cylindrical charge of explosives. In: 1st international symposium on rock fragmentation by blasting, Lulea, Sweden, Aug 1983, Lulea University of Technology. ISBN: 91-7260-851-x
19. Favreau RF (1969) Generation of stress waves in rock by an explosion in a spherical cavity. J Geophys Res 74:4267–4280
20. Favreau RF (1983) Rock displacement velocity during a bench blast. In: 1st international symposium on rock fragmentation by blasting, Lulea, Sweden, Aug 1983. Lulea University of Technology. ISBN: 91-7260-851-x
21. Cunningham C (1983) The KUZ-RAM model for prediction of fragmentation from blasting. 1st international symposium on rock fragmentation by blasting, Lulea, Sweden, Aug 1983. Lulea University of Technology. ISBN: 91-7260-851-x

22. Cunningham C (2005) The KUZ-RAM fragmentation model-20 years on. In: Brighton conference proceedings, 2005. EFEE 2005. ISBN 0 = 9550290-0-7
23. Kuznetsov VM (1973) The mean diameter of the fragments formed by blasting rock. Sov Min Sci 9(2):144–148
24. Zou D (1983) A three dimensional mathematical model on the rock/mineral fragmentation process in bench blasting in the open pit (in Chinese). Selected papers on opencast blasting technology. Fuzhou 75–82. 邹定祥(1983), 露天矿台阶爆破矿岩破碎过程的三维数学模型,"露天矿爆破技术论文选编", 1983, 福州,75–82
25. Zou D (1990) Experimental study of computer simulation of bench blasting in production scale. In: 3rd international symposium on rock fragmentation by blasting, Brisbane, Australia. Aug 1990. ISBN: 0 949 106 55 0
26. Da Gama CD (1983) Use of comminution theory to predict fragmentation of jointed rock mass subjected to blasting. In: 1st international symposium on rock fragmentation by blasting, Lulea, Sweden, Aug 1983. Lulea University of Technology. ISBN: 91-7260-851-x
27. Zou D (1985) The laws of rock fragmentation distribution by blasting and their application in blasting engineering, < Blasting > no. 2 (in Chinese) 邹定祥, 矿岩爆破块度分布规律及其在爆破工程中的应用, < 爆破 > No. 2 (1985)
28. Jorgenson GK, Chung SH (1987) Blasting simulation—surface and underground with SABREX model. CIM Bull 80(904):68–71
29. Qu SJ et al (2002) The blast-code model—a computer-aided bench blast design and simulation system. Fragblast 6(1):85–103
30. Qu SJ et al (2001) Blast-code model for design aad simulation of open-pit blasting and its application in Shuichang Iron Mine (in Chinese. Eng Blast 7(2):18–24. 璩世杰等 (2001)露天矿爆破设计与模拟 Biast-code 模型及其在水厂铁矿的应用,《工程爆破》第7卷,第2期,2001年六月。第18-24页。
31. Zhang JC et al (2003) An intelligence-expert system for bench blasting design and its application at the Lanjian iron mine. Explosives and Blasting Technique. In: Proceedings of the EFEE 2nd world conference, Prague, Czech Republic, Sep 2003. CRC Press, pp 617–624. ISBN 9789058096050 - CAT# RU41361
32. Chang C et al (2004) Structures and theory of intelligence expert system for bench blasting design (in Chinese). J Liaoning Tech Univ 23(2):194–196. 常春等(2004), 台阶爆破设计专家系统结构与原理研究, 《辽宁工程技术大学学报》,第23卷,第2期,第194–196页。
33. Blast Maker (2011) The computer-aided-design of drilling-and-blasting in open pits, mining world Russia 2011: mining equipment, mining and milling of ores and minerals. In: 15th anniversary international exhibition and conference, Moscow, April 2011. http://www.blastmaker.kg/en/
34. EPC France - 4 rue de St Martin - 13310 SAINT MARTIN DE CRAU. www.epc-france.com

Part III
Underground Excavation

Chapter 13
Introduction

13.1 Type and Features of Underground Excavation

There is a layer of very thick lithosphere below the surface of the earth. The rock surface has been weathered to be soil and forms soil layers of different thicknesses, covering most of the land. Rock and soil in their natural state are entities in which the landscape is formed by the external actions. In Karst formations, many underground spaces can form due to the action of water, to form underground caves or underground rivers. Early humans started to use these natural spaces, such as the "cavemen", and also began artificial underground excavation in order to obtain the living space, such as the "caves" in northwest China's Loess Layer. More widespread human action is the underground mining for obtaining required mineral resources. It was not until modern times, due to the rapid development of science and technology, in addition to large-scale underground mining carried out due to industrial development, the human race has been carrying out large-scale underground excavation to expand living space, to improve living conditions.

Based on their function and utilities, underground excavations (tunnels and large openings) can be classified as follows [2, 4]:

Category	Function and utilities
1. Transportation/conveyance	Peoples and goods
	–Pedestrian and cycle subway
	–Railways and metros
	–Highways
	–Water and sewage
	–Tunnels and canals
	–Hydroelectric power stations

(continued)

D. Zou, *Theory and Technology of Rock Excavation for Civil Engineering*,
DOI 10.1007/978-981-10-1989-0_13

(continued)

Category	Function and utilities
2. Storage and plants	Civilian function
	–Car parks
	–U/g commercial centers
	–Oil and gas storage caverns
	–U/g nuclear power stations
	–Disposal of radioactive waste
	–Military stocks and plants
3. Protection openings	Air raid shelters
	U/G hospitals
	Military shelters
4. Underground mining	

From the geological engineer's point of view, the most meaningful classification of underground excavations is one which is related to the degree of stability or security which is required of the rock surrounding the excavation. This, in turn, is dependent upon the use for which the excavation is intended. Barton, Lien, and Lunde suggest the following categories of underground excavations (an extract from [1]):

A. Temporary mine openings.
B. Vertical shafts.
C. Permanent mine openings, water tunnels for hydroelectric project (excluding high pressure penstocks), pilot tunnels, drifts, and headings for large excavations.
D. Storage rooms, water treat plants, minor road and railway tunnels, surge chambers, and access tunnels in hydroelectric projects.
E. Underground power station caverns, major road and railway tunnels, civil defence chambers, tunnel portals, and intersections.
F. Underground nuclear power stations, railway stations, sports and public facilities, and underground factories.

It is evident that the stability requirements, as a consequence the cost of the support (including geological investigations, support design, and installation), increase from category A to category F as one progresses through the list given above.

The features of underground excavation or the differences from the surface excavation are mainly as follows:

- The ground condition surrounding the excavation space, including rock characteristics, geological structures, and stability, is vital to the excavation project and much more important than the surface excavation. The ground condition determines the choice of excavation method, excavation design, and equipment, safety measures during the progress of excavation, temporary, and permanent supporting method and design and the overall costs of the excavation project.

- For most underground excavation, groundwater is usually a serious problem. Effective control of the ground water must be carried out before and during the excavation process. Sealing or channeling ground water should be also included in the permanent support design.
- Underground excavation is a more complicated and difficult operation than surface excavation because the only free face that initial rock breakage can take place toward is the excavating face. When drilling and blasting method is adopted, because of the high degree of constraints or fixation, larger charges will be required, leading to a considerably larger specific charge than in bench blasting.
- All operations of ground excavation are carried out in a confined space. Although the risk of flyrock to the public is eliminated due to the surrounding rock mass and blasting door at the tunnel portal, other risks, such as toxic fumes generated in blasting, ground settlement during or after excavation, rock fall from the roof and walls of the opening, water flood when an unexpected water-bearing stratum is broken during excavation, may occur during the operation. So special safety measures shall be carried out in underground excavation when compared to surface excavation.

13.2 Relationship Between Underground Excavation and Environment

13.2.1 Effects of Underground Excavation to the Environment

(1) May cause ground settlement (Fig. 13.1);
(2) May cause surface or ground water loss;
(3) Cause ground vibration when drill and blast method is used;
(4) Produce air overpressure and blast noise, especially in the early stage of blasting;
(5) Produce toxic fumes during blasting; and
(6) Delivery and storage of explosives may bring some potential risk to adjacent public area when blasting method is used.

Fig. 13.1 Ground collapse caused by underground construction

13.2.2 Environment Restrictions to the Underground Excavation

(1) PPV restriction to the underground blasting works

- Nearby Buildings, Structures, Utilities and Services Structures, Residential Houses, Historical Buildings, and Temples;
- Slopes (existing and newly formed), retaining walls, natural terrain, and boulders.

(2) Underground and surface water sources of which the water may ingress to the excavation space during and after construction of the underground project, so that some special measures may be required to be taken to protect the water source.

(3) For reducing the effect to adjacent public or traffic, the working time of delivery of explosives and firing the blast may be restricted in a special time window.
So, a very detailed assessment report for the underground excavation works is very important.

13.3 Methods of Underground Excavation in Rock

According to the geological condition or surrounding environment, the following methods for underground excavation in rock are usually adopted:

1. Mechanical excavation.
   - Drilling and breaking.
   - Excavation by boring machines.
     a. Shaft excavation by raise boring.
     b. Excavation by roadheaders.
     c. Tunnel boring machine (TBM).
2. Excavation by drilling and blasting.
3. Other methods of underground excavation
   - Cut and cover tunnel.
   - Pipe jacking

All the methods will be discussed in the following chapters.

In modern tunnel and underground cavern excavation, it is possible to select from many different methods. The following factors should be taken into consideration when selecting the method:

- Tunnel dimensions
- Tunnel geometry
- Length of tunnel, total volume to be excavated
- Geological and rock mechanical conditions
- Ground water level and expected water inflow
- Vibration restrictions
- Environment conditions and allowed ground settlements.

Figures 13.2 and 13.3 (refer to [3]) show the methods of rock excavation in underground, and their application depending on the rock strength (UCS).

The drill and blast method is still the most typical method for medium to hard rock conditions. It can be applied to a wide range of rock conditions. Some of its features include versatile equipment, fast start-up, and relatively low capital cost tied to the equipment. On the other hand, the cyclic nature of the drill and blast method requires good work site organization. Blast vibrations and noise also restrict the use of drill and blast in urban areas.

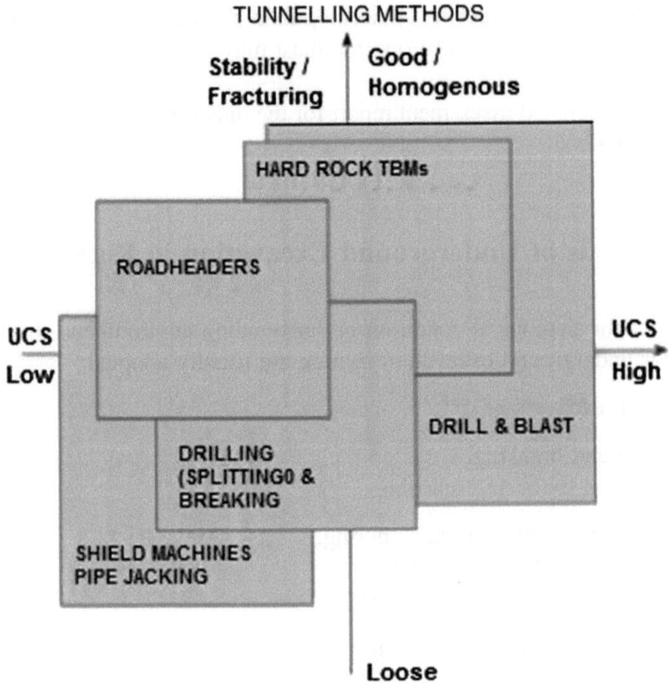

Fig. 13.2 Tunnelling methods in different rock/soil conditions (reproduce from Ref. [3] by permission of Sandvik)

Fig. 13.3 Range of methods compared to uniaxal compressive strength (UCS) (reproduce from Ref. [3] by permission of Sandvik)

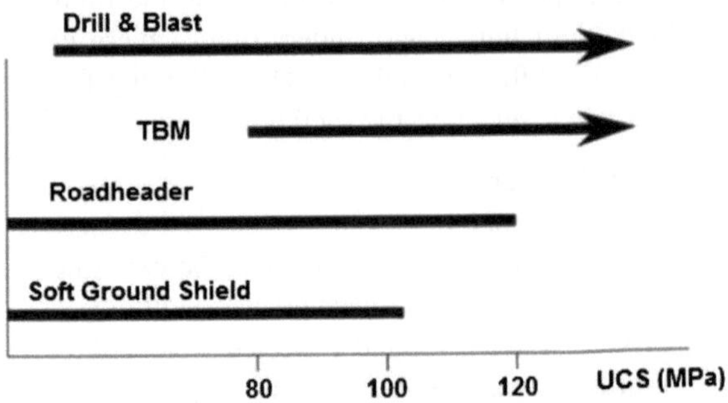

Hard-rock TBMs can be used in relatively soft to hard rock conditions and are best when rock conditions are predictable and have little variation. The TBM is generally most economical method for longer tunnel lengths, in which its high investment cost and long preparation time can be offset by the high advance rate of excavation. TBM excavation produces a smooth tunnel with low rock reinforcement cost and is optimal in terms of flow resistance in long ventilation or water tunnels.

Shielded TBMs or shield machines are used in loose soil and mixed ground, and in conditions, where high water ingress is expected. The mechanical and/or pressurized shield prevents ground settlement and ground water inflow. Because of continuous ground control and no blast vibrations, this method is commonly used in urban tunneling. Pipe-jacking is a special application, in which the tunnel lining is continuously pushed by heavy hydraulic jacks as the tunnel advances. Microtunneling is a special application of pipe-jacking in no-man-entry sized tunnels.

Roadheaders can be used for tunneling in stable rock conditions of low-to-medium hardness. Where it is applied, the roadheader combines the versatility of drill and blast for producing various tunnel geometries, and the continuity of full-face mechanical excavation. As it lacks blast vibration, this method can be used in sensitive urban areas. In harder rock conditions, use of roadheaders is limited by a shorter lifetime of tools and increasing cutting tool cost.

Drill and break tunneling has gained popularity mainly in the Mediterranean countries, Hong Kong, and Japan due to its low equipment costs and relative less disturbance to the surrounding environment and ground especially when the tunnel passes beneath the sea/river bed. The tunnel face geometry is unlimited, and the method is effective in rocks of low-to-medium compressive strength or relatively fractured. In hard and compact ground, application is limited by low production rate.

References

1. Hoek E, Broun ET (1980) Underground excavation in rock. Institution of Mining and Metallurgy, London. ISBN: 9780415461658
2. Tatiya RR (2013) Surface and underground excavation, methods, techniques and equipment, 2nd edn. CRC Press. ISBN: 978-0-415-62119-9
3. Heiniö M (Editor in Chief) (1999) Rock excavation handbook for civil engineering. Sandvik & Tamrock
4. Persson PA et al (1993) Rock blasting and explosives engineering. CRC Press, Boca Raton. ISBN: 0-8493-8978-X

Chapter 14
Mechanical Underground Excavation in Rock

14.1 Drilling and Breaking Method

In some places, where blasting is prohibited due to the effects and safety to the public, nearby buildings and structures or some particularly sensitive objects, or if the tunnel is underneath the bed of river (lake or sea) and blasting may cause the water inflow, the drilling and breaking method has to be used for underground excavation.

Usually, this kind of project has a relative small scale and quantity.

Drilling and breaking is also used in the initial stage of the tunnel excavation prior to commencing drill and blast method.

If the rock mass is highly jointed, the excavation can be performed only by the hydraulic breaker (hydraulic hammer), that is called "hammer tunneling."

But for hard rock, as well as the drill rig and hydraulic breaker, hydraulic splitters are used as well for cracking the rock.

The working procedure is drilling–splitting or cracking–breaking.

The measurements are as follows: Drilling—hole diameter: 76–110 mm depending to the diameter of the hydraulic splitter, hole depth: 2–3 m, and hole spacing: 400 × 400 mm to 600 × 700 mm according to rock strength and the capacity of the splitter.

Splitting—There are different splitters that can be applied, refer to 7.2 of Chap 7.

Breaking—Heavy hydraulic breaker is usually adopted.

Excavation usually starts at the center of the tunnel at a height of 1.0–1.5 m. Sometimes, a big hole (110 mm in diameter) is used for creating a free face for splitting. Splitting and breaking then continue from the sides of the large hole as close as possible to the final sides of the tunnel. Once this stage is reached, works continue in the same way from the floor up until the roof of the tunnel has been formed (Figs. 14.1 and 14.2).

D. Zou, *Theory and Technology of Rock Excavation for Civil Engineering*,
DOI 10.1007/978-981-10-1989-0_14

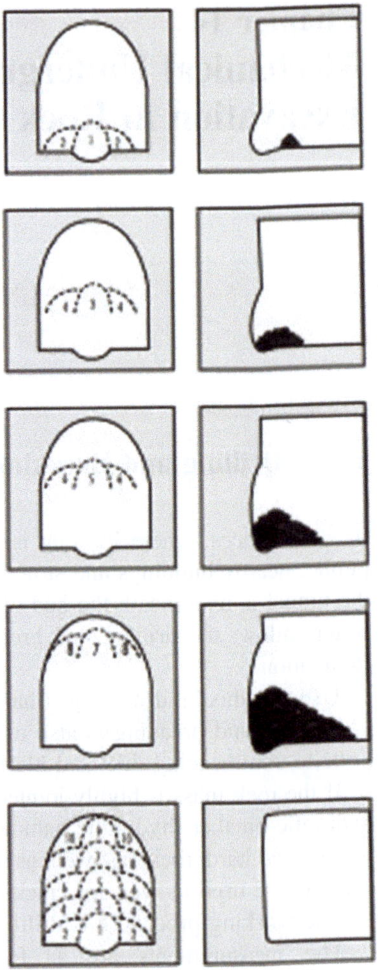

Fig. 14.1 Working cycle of D and B tunneling (Reproduced from Ref. [3] by permission of Sandvik)

Fig. 14.2 Breaking tunnel rock with wedge hydraulic splitter

14.2 Roadheader Excavation

The roadheader is also called Alpine boom miner or boom miner. It is a partial face boring machine which works on discrete areas of the face rather than excavating the whole face in one go. The roadheader was originally developed for use in coal mines; however, they are now used in a broad range of applications including but not limited to tunneling. In a similar manner to a full-face machine, the cutterhead uses a rotating motion and is pressed against the rock face, so it is the best for softer rocks [1, 2].

The roadheader consists of a boom-mounted cutting head, a loading device usually involving a conveyor, and a crawler traveling track to move the entire machine forward into the rock face.

Depending on the direction of the rotation of the cutterhead, it is divided two types: axial (Fig. 14.3) and transverse (Fig. 14.4).

A roadheader with an axial cutterhead will be heavier for the same power of the head motor in order to absorb the reaction forces of the forward drive compared to one with a transverse cutterhead. The axial type can be referred to as milling, radial auger, or inline. With the most common cutting method used, the inline, the roadheader rotates in line with the cutter axis and, applying the cutter force sideways, reduce the utilization of the machine's weight in the cutting process.

The transverse type of roadheader is often referred to as "ripping." Its best cutting performance is in weak rocks. The slewing force is at a right angle to the circumferential force. Cutting in the direction of the face makes it more stable. It is more adaptable to a wider range of conditions. It is less affected by changing rock conditions, including hard rock bands, and it is highly maneuverable.

Fig. 14.3 Mitsui Miike S300A Roadheader, Gathering Ver., Mitsui-S300, Supplied by RTM Equipment

Fig. 14.4 SANDVIK ATM 105 for Alpine Tunnel

Table 14.1 Classification of roadheaders (Reproduced from Ref. [3] by permission of Sandvik)

Roadheader class	Range of weight (ton)	Range of cutterhead power (kW)	Range of operation			
			Roadheaders with standard cutting range		Roadheaders with extended cutting range	
			Max. section (m^2)	Max. U.C.S. (MPa)	Max. section (m^2)	Max. U.C.S. (MPa)
Light	8–40	50–170	~25	60–80	~40	20–40
Medium	40–70	160–230	~30	80–100	~60	40–60
Heavy	70–110	250–300	~40	100–120	~70	50–70
Extra heavy	>100	350–400	~45	120–140	~80	80–110

The roadheaders based on the machine weight and cutterhead power could be classified as light-to-heavy-duty units. Table 14.1 indicates the range of defined classes with regard to their main features and limit of operation.

Roadheaders are now offered with an operating weight of 13 tons up to 135 tons and with a cutting power of up to 500 kW, making it possible to cut relatively hard rock with the largest units. The Sandvik model MT720 roadheaders, for example, can cut rock up to an unconfined compressive strength of almost 30,000 psi (206.8 MPa). The transverse roadheader can cut large cross sections, a large machine can cut 100 m^2 (1076 ft^2) [5, 6].

As the rock is cut from the face by the tools on the rotating head, it falls on to the apron. The muck is then collected by the gathering arms to the conveyor. The conveyor transports the muck to the rear of the machine and drops it onto another conveyor or into a muck car, train, or excavator bucket. Figure 14.5 shows the most common loading devices.

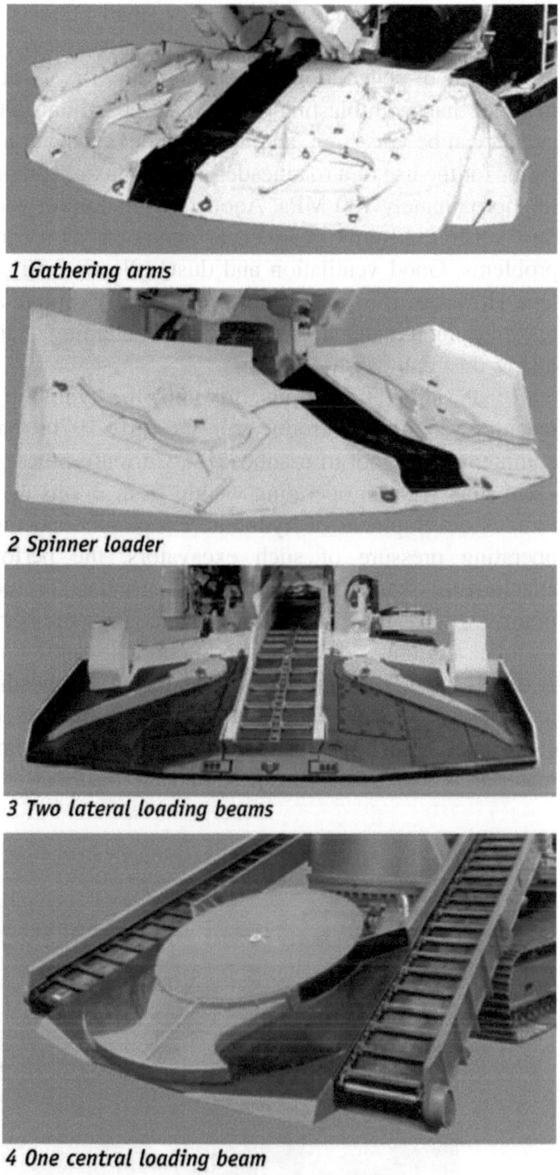

Fig. 14.5 Main type of loading assemblies on roadheaders (Reproduced from Ref. [3] by permission of Sandvik)

The roadheader offers several advantages. It has versatility and mobility. That is, it can mine varied cross sections; there is easy access to the heading, providing good access for ground support, and large faces can be subdivided simultaneously. The cost is relatively low, approximately 20 %, and can vary by 10 %, depending on the size of the roadheader compared to a TBM. Roadheaders are often available for renting, which makes it ideal for small projects. The roadheader is relatively easily mobilized.

The rate of penetration of the roadheader is very dependent on the rock strength, more so than with blasting or TBMs. Depending on the rock hardness, the cost of cutters can be very high, adding to overall excavation costs. Generally, the limiting factor for the use of a roadheader is the uniaxial compression strength of the ground of approximately 100 MPa. Another limitation for widely using a roadheader is the dust generated by the roadheader operation, as it can cause severe environmental problems. Good ventilation and dust collection can effectively alleviate the problem. However, it is impossible to completely capture all the dust produced. For this reason using a roadheader in a ground containing quartz is generally not possible as quartz dust can cause cancer.

Roadheading attachments, driven by the hydraulic system of excavators or other machinery, initially introduced in the early 1970s, have become a more and more complementary tool to roadheaders. Currently, these attachments are available for excavators with an operating weight from 2 tons up to 150 tons. Due to the continuous development of hydraulic excavators and the possibilities to increase the operating pressure of such excavators, the performance of the roadheading attachments—sometimes called rock grinders or rock cutters—has substantially improved. (see Fig. 14.6). Attachments are extremely flexible in large tunnels for scaling, cutting manholes and corners, and removing excessive grouting. A disadvantage of using an excavator with roadheading attachments is its inability to remove muck simultaneous to cutting; it must be stopped and moved in order to clear muck, which results in a reduced production rate, which includes cutting the material as well as moving it behind the machine.

Fig. 14.6 Roadheader Attachment AQ-4 (Andraquip) is mounted on an excavator

14.3 Raise Boring

In the past, all shafts and raises were excavated by drilling and blasting method. However, during the last decades, full-face raise boring methods have by and large surpassed drill and blast methods for making raises both in mining and civil contracting.

Raise boring is a non-blasting, mechanized full-face raise excavation method. It can be used for raise with any inclination from horizontal to vertical (0–90°).

The raise boring rig is equipped with specially designed cutters that are fitted with special carbide inserts. The cutters are attached to a reamer which is connected through guide rods to the rotating boring unit.

14.3.1 *Methods of Raise Boring*

There are different methods to apply the raise boring technologies:

- Conventional Raise Boring

The raise borer is set up on the upper level of the two levels to be connected, on an evenly laid platform (typically a concrete pad). A small-diameter hole (pilot hole) is drilled to the level required; the diameter of this hole is typically 230–445 mm (9″–17.5″), large enough to accommodate the drill string. Once the drill has broken into the opening on the target level, the bit is removed and a reamer head, of the required diameter of the excavation, is attached to the drill string and then raised back toward the machine while it is rotated. The drill cuttings from the reamer head fall to the floor of the lower level. The finished raise has smooth walls and may not require rock bolting or other forms of ground support. Figure 14.7

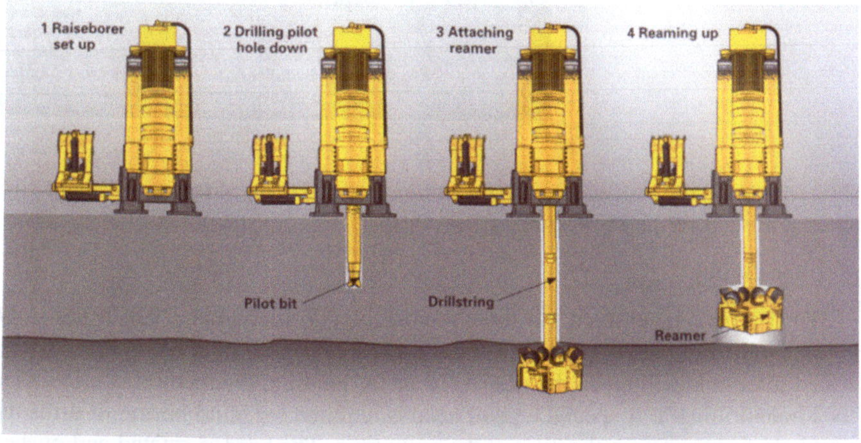

Fig. 14.7 Working procedure of conventional raise boring (Reproduced from Ref. [18] by Permission of Atlas Copco)

Fig. 14.8 The Robbins 34RH C raise boring machine (Reproduced from Ref. [18] by Permission of Atlas Copco)

Fig. 14.9 A reamer is raising for reaming (Reproduced from Ref. [3] by permission of Sandvik)

shows the working procedures of conventional raise boring. Figures 14.8 and 14.9 show a raise borer and a reamer.

- Blind Boring

When a raise is required but there is no access to the upper level, it has to be bored blind from below, usually without a predrilled pilot hole (sometimes a pre-drilled pilot hole will reach the desired lever for guiding, the method is also called "box hole drilling"). A special type of head is required for blind boring. It drills the pilot hole and reams out the raise at the same time. The head is rotated and pushed upward. The cuttings fall out of the hole by gravity.

Normal blind raise diameters are from 0.6 to 1.8 m. Since the drill string is under compression during blind boring, special large-diameter stabilizers are needed to support the drill string. The blind boring method is used to produce so-called slot raises, ore passes, and manways. Figure 14.10 is a sketch to show how blind boring works (from Tamrock).

- Down boring with a predrilled pilot hole

There are two kinds of arrangement of the location of the boring rig for down boring methods.

The one arrangement locates the boring rig at the upper level (see Fig. 14.11, from TAMROCK), Another one locates the boring rig at the lower lever (see

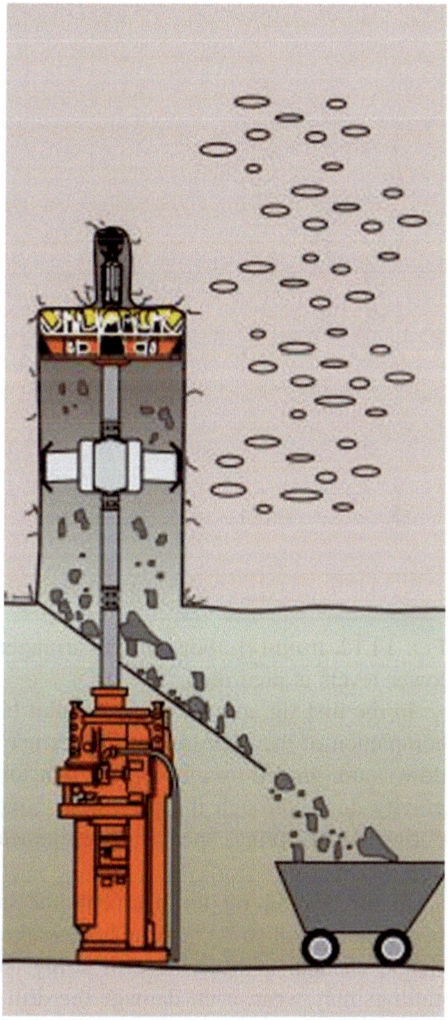

Fig. 14.10 Blind boring (Reproduced from Ref. [3] by permission of Sandvik)

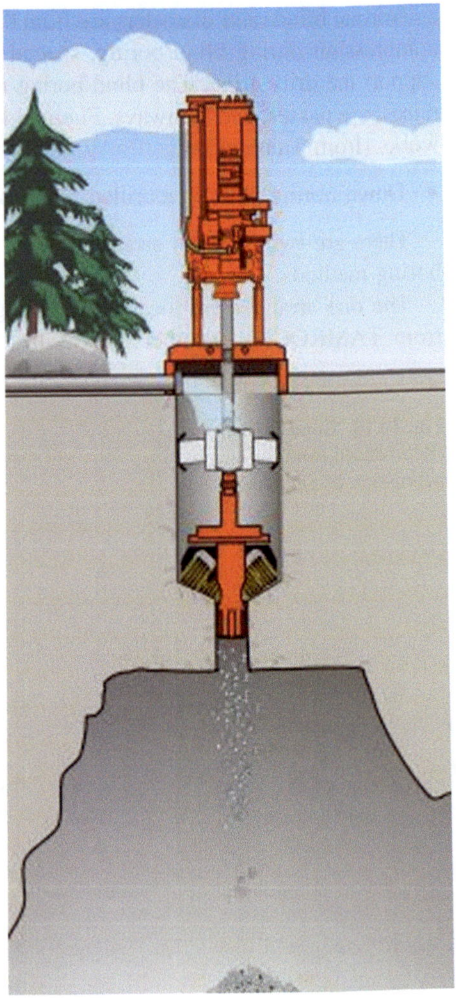

Fig. 14.11 Down boring with the boring rig at the upper level (Reproduced from Ref. [3] by permission of Sandvik)

Fig. 14.12, from [7]). For both rig arrangements, a pilot hole connecting upper and lower levels is predrilled.

In the first rig arrangement, the pilot bit and drill string are then removed after completion of the pilot hole and a reamer fitted. The reamer is pushed and rotated downward, guided by a nosepiece that follows the pilot hole. The cuttings fall by gravity down through the pilot hole. Since the drill string is under compression during down boring, special large-diameter stabilizers are needed to support the drill string.

In the second rig arrangement, the cuttings fall down through the annulus between the drill string and the hole wall to the lower level. As the pilot hole is the shared channel of both the drill string and the cuttings, the falling down of the cuttings may wear, even damage the drill string.

Fig. 14.12 Down boring with the boring rig at the lower level (Reproduce from Ref. [7] by permission of JRMGE)

- Horizontal boring

Horizontal boring is an excellent method in urban construction projects where drilling and blasting is restricted or forbidden and tunnel boring machines (TBMs) are too bulky. First, a horizontal pilot hole is drilled, with the aid of a directional drilling system if necessary.

When the pilot bit breaks through, it is removed and replaced with a reaming head. Because the hole is horizontal, the reamer must be equipped with a special cuttings removal system (see Fig. 14.13, from TAMROCK).

Typical diameters for horizontal reaming are from 0.6 to 4.5 m. The method is used to drill tunnels for cables, escape routes, sewage, etc., without disturbing the environment unduly.

Horizontal boring requires good rock stability.

14.3.2 Main Uses of Raise Boring in Civil Construction [3]

The main uses of the raise boring in civil construction are as follows:

- ventilation holes for road and railway tunnels;
- various holes and raises for hydropower station and underground storage halls;
- holes used as pilot holes for big diameter shaft sinking;
- raises in areas where environmental restrictions (noise, vibrations, etc.) limit use of other methods,for example, urban areas, nuclear power plant, or nuclear waste storage vicinities;

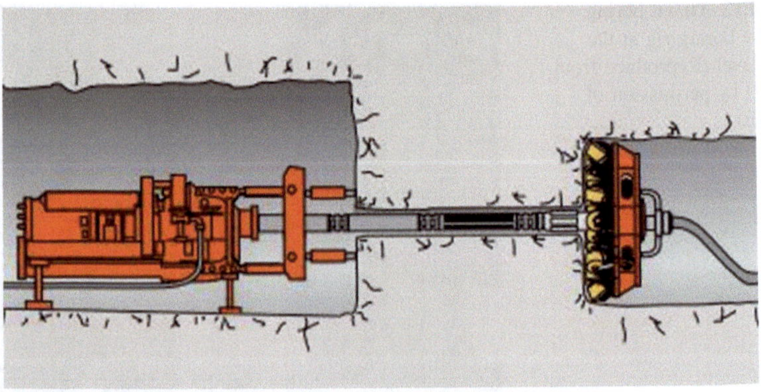

Fig. 14.13 Horizontal boring (Reproduced by permission of TAMROCK)

- Raise boring can only be used in ground that will be stable and self-supporting, with no temporary support. While reaming is taking place, it is not possible to access the shaft, and any instability may collapse the shaft or jam the reaming head, and the shaft may need to be abandoned.

14.3.3 Main Benefits of Raise Boring [3]

The main benefits of the raise boring methods are as follows:

Safety:

- Always working in a safe area; no working under newly blasted roof;
- Clean environment: no dust, blasting fumes, exhaust gases, or oil mist;
- Low noise level and minimum vibration (compared to blasting);
- Optimal shape of the raise is strong against rock pressure.

Speed, Efficiency:

- Raise boring can be typically 2–3 times faster than other methods;
- Only one operator required in a modern raise boring machine.

Quality:

- Round cross section and smooth walls are optimal in terms of flow characteristics (ventilation, water flow) and require a minimum amount of additional support;
- A regular, round cross section makes it easy to assemble prefabricated equipment in the hole;
- It does not cause any fractured zones or cracking to surrounding rock.

14.4 Tunnel Boring Machine (TBM)

14.4.1 Introduction [9–11, 15]

A tunnel boring machine (TBM) is a machine used to excavate tunnels full face with a circular cross section through a variety of soil and rock strata. The first tunnel boring machine appeared in the world in the mid-1950s, the TBM has come to the point where mechanical excavation can now be considered for virtually any tunneling conditions. They can bore through anything from hard rock (U.C.S. > 400 MPa) to soil and sand. Tunnel diameters can range from a meter (done with micro-TBMs) to 17.4 m to date. Tunnel boring machines are used as an alternative to drilling and blasting (DandB) methods in rock and conventional "hand mining" in soil.

Comparing TBM with the conventional tunnel excavation method (mainly drill and blast method), TBM offers certain advantages:

- Much higher advance rate. According to the database of 631 TBM projects mainly in North America and Europe, assembled by the University of Texas in 1963 to 1964, the average advance rate was 375 m/month and the highest was 2084 m/month (from [8]). Half a century has passed, following the quick development of TBM technology, all indexes of TBM performance have increased greatly, especially on the advance rate in hard rocks. Nonetheless, the above figures are still encouraging.
- Exact excavation profile with smooth tunnel wall. This significantly reduces the cost of lining the tunnel and makes them suitable to use with no permanent lining in some circumstances.
- Very few environmental hazards such as noise, explosive gases, and vibration. They do not disturb the surrounding public and residents when the site is situated in an urban area.
- Very light damage to the surround rock. It is a very important advantage when the tunnel was excavated beneath the water (river, lake, or sea bed) for prevention of water penetrating to the tunnel.
- Better working conditions and safety.
- Less operators are needed.
- Automated and continual operation.
- Low cost if the tunnel is sufficiently long.

The disadvantages of the TBM are as follows:

- More geological investigation and information is necessary than for drilling and blasting.
- High investment resulting in longer tunnel stretches being necessary.
- Longer lead time for design, manufacture, and mobilization of the machine.
- Fixed profile of the tunnel (mostly circular).
- Limited flexibility in response to extremes of geologic conditions.
- Limitations on curve radii and enlargements.

- Long learning phase of the crew.
- Transport of the machine with trailers to the tunnel portal.

A TBM is a system that provides thrust, torque, rotational stability, muck transport, steering, ventilation, and ground support.

In most cases, these functions can be accomplished continuously during each mining cycle.

The TBM cutterhead is rotated and thrust into the rock surface, causing the cutting disk tools to penetrate and break the rock at the tunnel face. Reaction to applied thrust and torque forces may be developed by anchoring with braces (grippers) extended to the tunnel wall, or bracing against support installed behind the TBM.

So there are the following four system groups (Fig. 14.14).

- Boring system.
- Thrust and gripping system.
- Muck removal system.
- Support system.

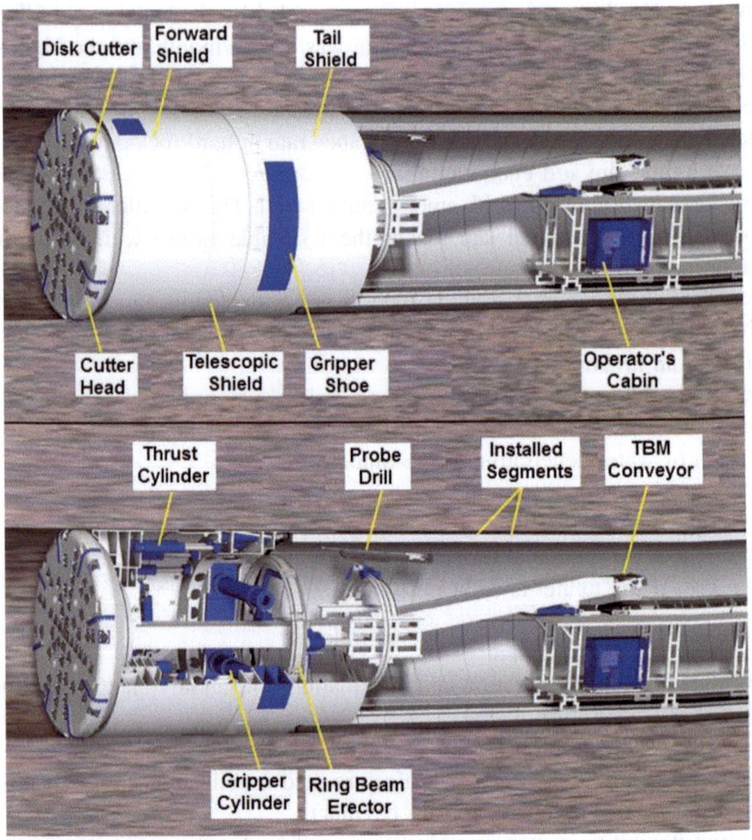

Fig. 14.14 Basic elements of a double-shield TBM (Courtesy of Robbins)

14.4.2 Classification of TBM

There are different ways to classify the full-face TBM. The below classification is mainly based on the type of face support:

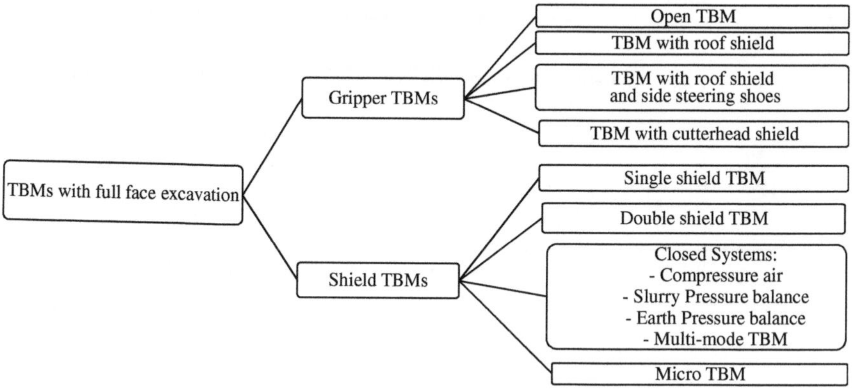

14.4.2.1 Gripper TBM

The gripper TBMs, often widely described as open TBMs and also called "the main beam TBM," are utilized in stable rock conditions with low water ingress. The gripper machine locks itself against the tunnel wall laterally using hydraulically moved "gripper shoes" to establish the required face pressure.

Gripper TBMs are further categorized into open TBMs, TBMs with roofs, with partial shield and with cutter shield (see Fig. 14.15).

According to the number of grippers to be used, there are two types of gripper TBMs: single-gripper machines and double-gripper machines (see Figs. 14.16, 14.17). Single-gripper machine is used more frequently in standard tunnel projects.

14.4.2.2 Shield TBM

In contrast to gripper TBMs, shield TBMs have an extended shield over the front part of the machine. The shield has the function of supporting the ground and protecting the personnel, thus allowing safe erection of the tunnel lining. There are two basic types of shield TBMs for hard rock available: the single shield and double shield.

The single-shield TBMs (Fig. 14.16) are primarily for use in hard rock with short stand-up time and in fractured rock where there is a risk of ground collapse. With these machines, the pushing forces are maintained axially against the installed lining segments. One of the advantages of a single-shield machine is that it can be converted to a closed mode if high groundwater ingress is likely to be encountered.

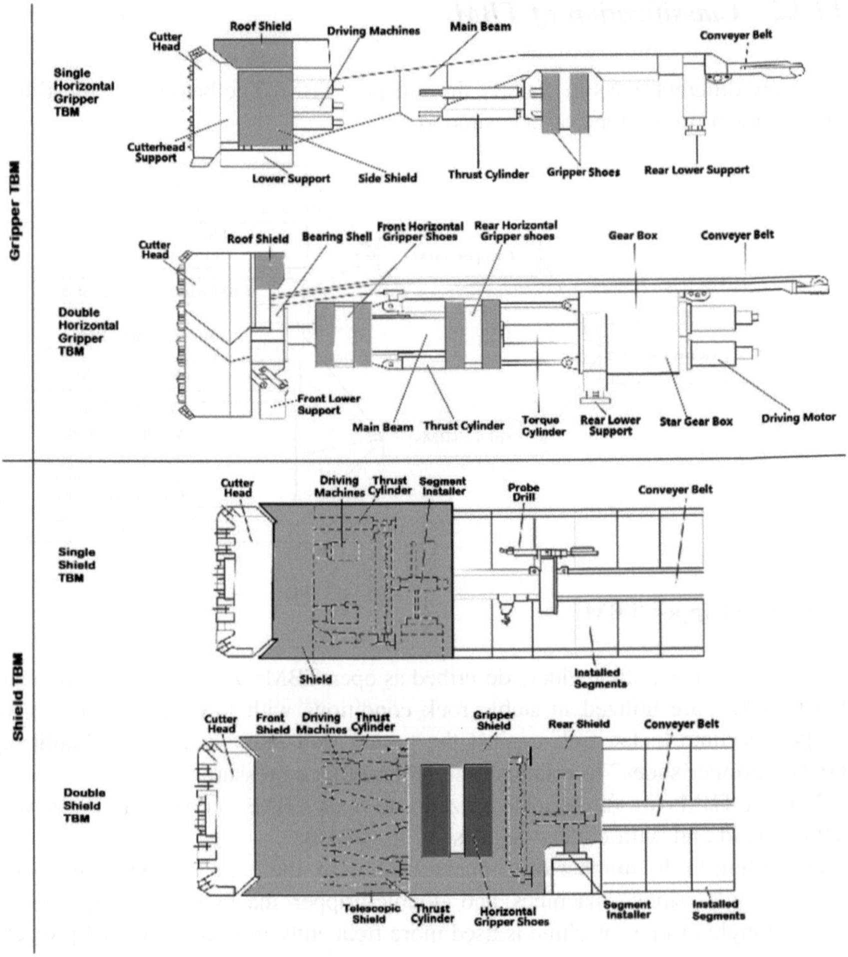

Fig. 14.15 Various types of full-face rock TBM (Courtesy of Central South University, China)

The double-shield TBMs, or telescopic shield TBMs, combine the ideas of the gripper and single-shield techniques and can therefore be applied to a variety of geological conditions. The double-shield TBM (Fig. 14.17) consists of two main components, the front shield with cutterhead, and a gripper section with gripper shoes, tail shield, and auxiliary thrust jacks. Both shield parts are connected to each other with a telescopic shield. The double-shield TBM has an essential disadvantage compared with single-shield TBM. When used in fractured rock with high strength, the rear shield can be blocked due to the rock fragments getting into the telescopic joint. In such conditions, it is possible to lock the two shields together, effectively operation in single-shield mode.

Fig. 14.16 Single-gripper TBM for hard rock (Reproduced by permission of Copyright © 2005, Tunnelseis)

Fig. 14.17 Double-gripper TBM for hard rock, TB880E, Dia. 8.8 m, developed by China Railway Group (Courtesy of Central South University, China)

14.4.2.3 Closed Systems

Closed shield TBM systems are also called pressured support systems. The principle is that pressure is created at the front of the shield, and this supports and stabilizes the tunnel face. In addition, this can be used to control water flow into the tunnel. There are three types of pressurized closed-face tunneling systems, compressed air, slurry tunneling machines (STMs), and earth pressure balance machines (EPBMs). Closed systems mainly are used for soft, or highly fractured ground as well as under water table conditions.

- Earth Pressure Balance Machines (EPBMs)

EPBMs use the excavated material to support the tunnel face during excavation of the ground. The excavated material enters and fills the excavation chamber which is under pressure in a fluid or plasticized state after having been mixed with a

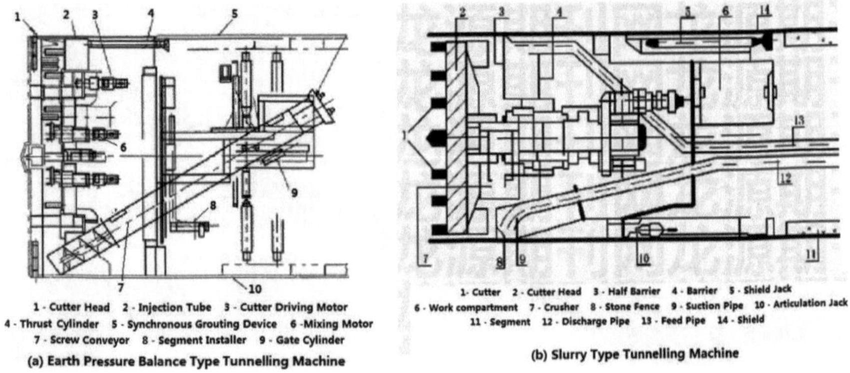

Fig. 14.18 Section views of earth pressure balance-type tunneling machine and slurry-type tunneling machine

conditioning agent (Fig. 14.18a). The plasticized spoil is removed from the chamber by using a screw conveyor. The screw conveyor is used to remove the excavated material in a very controlled manner so that pressure is maintained in the chamber. The pressure in the chamber should be high enough to maintain the ground stability and is controlled by a combination of thrust on the cutterhead and the rate of removal of material from the chamber via the screw conveyor.

- Slurry Tunneling Machines (STMs)

Slurry tunneling machines use a pressurized fluid to stabilize the face during excavation of the ground. There are two systems in use today for maintaining a balanced force pressure. One simply uses the fluid in the pressurized chamber behind the cutterhead to provide the pressure, and the other uses an air bubble system (Fig. 14.18b). The air bubble system has a bubble of compressed air introduced into the chamber in the roof behind the bulkhead and this acts as an accumulator and thereby ensuring that a constant pressure is maintained at the face. The slurry does not only stabilize the face, but also mixes with the excavated material to allow it to be transported out of the machine. The fluid is pumped to the face, where it mixes with the excavated material. The mixture is then pumped out of the excavation chamber through a slurry line where it is conveyed to a separation plant. To prevent the slurry line being blocked by large pieces of material it passes through a crushing unit prior to entering the slurry line. In the separation plant, the excavated material is removed from the transportation fluid using screens and cyclones.

- Mixshield TBM

A shield TBM which is operated with two or more modes is called mixshield TBM. The machine type was conceived for use in very changeable geology. The original idea incorporated some of the technologies used in slurry shield, EPB shield, and the pressure shield. Experience has also shown that when using this type of shield, some features of hard rock TBM have to be integrated to the system.

14.4.3 Mechanism of Rock Breakage by Cutting Tools

The cutting tools are the essential components in the process of cutting the rocks. The type of cutter that is to be used with a TBM depends upon the type of rock for which this equipment is to be used. The very soft rock requires very high torque and low thrust. A soft to medium hard rock needs very high thrust and medium torque. For hard rocks, high thrust and torque are required. Depending on the type of cutter design, there are toothed, disk or button disk cutters. Table 14.2 gives the different type of cutting tools (cutters) used in different type of rocks.

In TBMs, the cutting head configuration takes three district zones, namely center, the face, and the outer gauge cutters (see Figs. 14.19 and 14.20). Disk or roller cutters, depending upon the hardness of rock (and including drag cutters for the soft rocks, see Fig. 14.21), usually excavate the main face area. The gauge cutters are located at the outside edge of the cutting head to excavate the opening of the desired size which are most prone to wear (abrasion) as they cover the longest distance for each rotation of the cutterhead.

Table 14.2 Reproduced from Ref. [4] p. 360 Table 11.2 Cutting tools used in conjunction with various tunnel machines (by permission of Taylor and Francis Group)

Rock type	UCS (MPa)	Cutting tools	Type of attack
Very soft to soft	0–124	Toothed (drag, chisel, picks)	Point, i.e., applying force parallel to rock surface
Soft to hard	140–180	Disk cutter	Small surface area of contact, cutting force normal to rock surface (indenture tools)
Hard	>180	Button disk cutter	Large surface area of contact, cutting force normal to rock surface (indenture tools)

UCS unconfined compressive strength

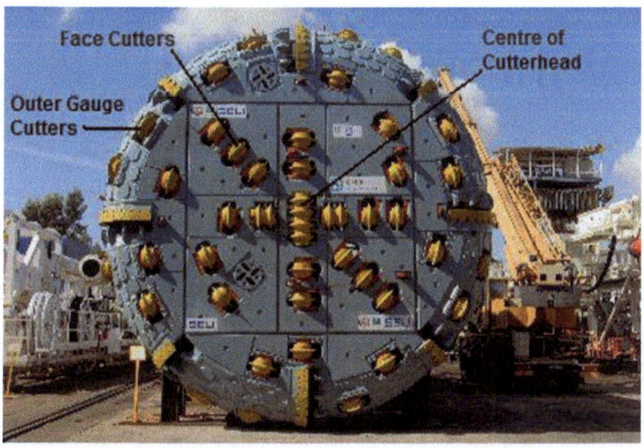

Fig. 14.19 Cutterhead of 7.27-m-dia TBM manufactured by SELI for the excavation of storm drainage tunnel in Tsuen Wan, Hong Kong, in 2008

Fig. 14.20 Cutters of three district zones of TBM cutterhead, made by JUN Engineering Co. Ltd

Fig. 14.21 An example of a drag tooth for soft rock TBM

In Sect. 2.1.1. of Chap. 2, the mechanism of rock breakage by tool penetration has been discussed. Figure 14.22 shows the cutting process of the disk cutters. The cutters are pressed against the rock face. The contact pressure between the disk and rock pulverizes the rock on contact and induces lateral cracking toward the neighboring kerf—and rock chipping. To achieve the best performance, kerf spacing (distance between two adjacent tracking cutters) and cutter load must have suitable values for each rock type. The determination of the spacing of the roll tracks was

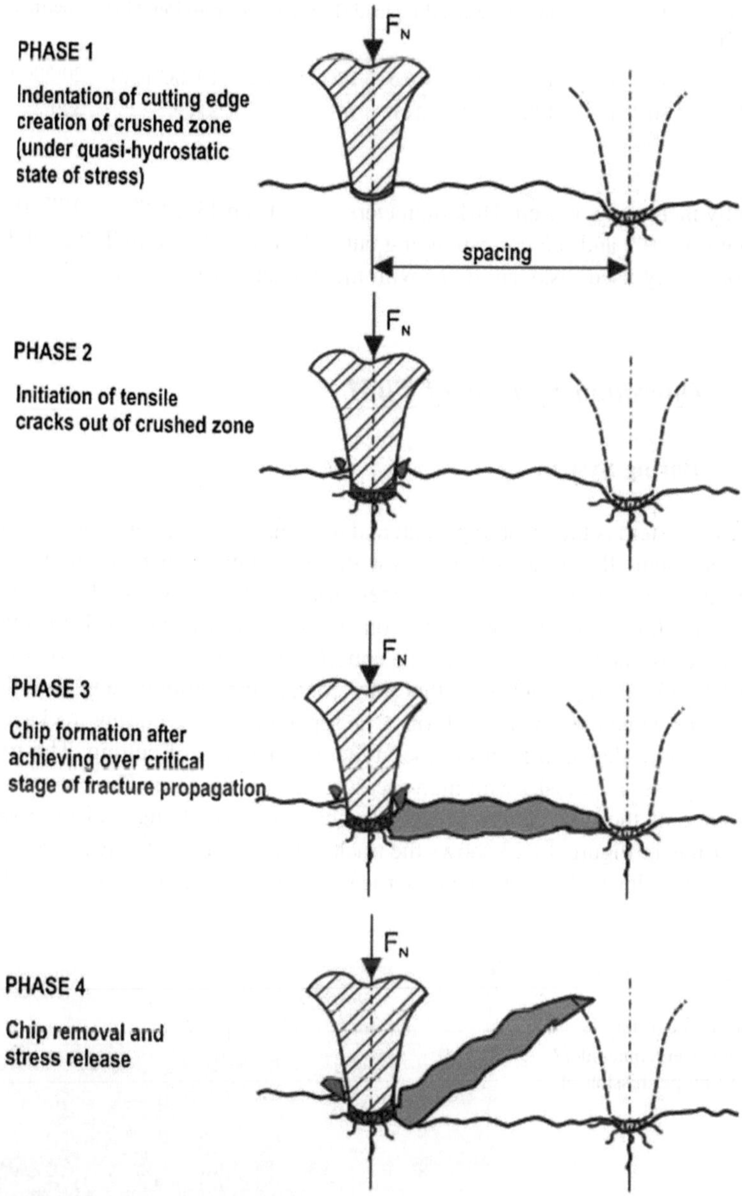

Fig. 14.22 Cutting process of disk cutters (Reproduced from Ref. [3] by permission of Sandvik)

done empirically. This spacing was, according to manufacturer, between 65 and 95 mm. With the use of large cutters and the resulting greatly increased disk loading, spacing of 80–95 mm is now used. With the continuous development of technology and the use of more high-strength materials, disk cutter diameter and cutter load are

Table 14.3 Disk diameters and corresponding load (Reproduced from Ref. [11] by permission of Science Press, China)

Disk dia. in	11	12	14	15.5	16.5	17(432 mm)	19(483 mm)	20(500 mm)
Load (kN)	80	133	178	222	245	267	311	320

constantly increasing as well. Disk diameters are of from 11″, 12″, …, 17″, 19″ to a maximum of 20″, and the corresponding cutter load are shown in Table 14.3. The most commonly used disks are those with the diameters of 17″ and 19″.

14.4.4 Operation Systems of TBM

14.4.4.1 Boring System

The boring system is the most important and determines the performance of a TBM. It consists essentially of cutter housings with disk cutters, which are mounted on a cutterhead. The cutterhead is a rigid steel structure that supports the cutters and loads the muck onto a belt conveyor or other spoil removal system. Depending on the machine size and site conditions, the cutterhead can be one piece or of sectional design. The disks are so arranged that they contact the entire cutting face in concentric tracks when the cutterhead turns. The separation of the cutting tracks and the disks are chosen depending on the rock type and the ease of cutting. The rotating cutterhead presses the disks with high pressure against the face. The pressure at the cutting edge of the disk cutters exceeds the compressive strength of the rock and locally grinds it. Figure 14.23 shows the tracks of the cutter on the rock face. Cutter drive system is located behind the cutterhead and mainly refers to a series of drive

Fig. 14.23 Track of the cutter on the face (Reproduced from http://www.bartz.es/en/contenido/?idsec=517 by permission of Bartz)

motors arranged in a ring-shape. There are two kinds of drives, electric drive and hydraulic drive, wherein the electric drive has a mature technology, power, small size, low cost, it has a wider application.

14.4.4.2 Thrust and Gripping System

The thrust and clamping system is responsible for the advance thrust and the boring progress. The cutterhead with its drive unit is thrust forward with the required pressure by hydraulic cylinders (jacks). The length of the thrust cylinder restricts the maximum stroke which can be achieved with a typical maximum stroke length of 2 m. In stable rock condition with low water ingress, the gripper technique can be utilized. Figure 14.24 shows the principle of the gripper machine technique. The operating principle of a double shield is based on the gripper shoes pressure against the tunnel wall while excavation and segment installation are performed at the same time (the segment installation takes place at the rear of the whole machine). Figure 14.25 shows the advance model of a double-shield TBM.

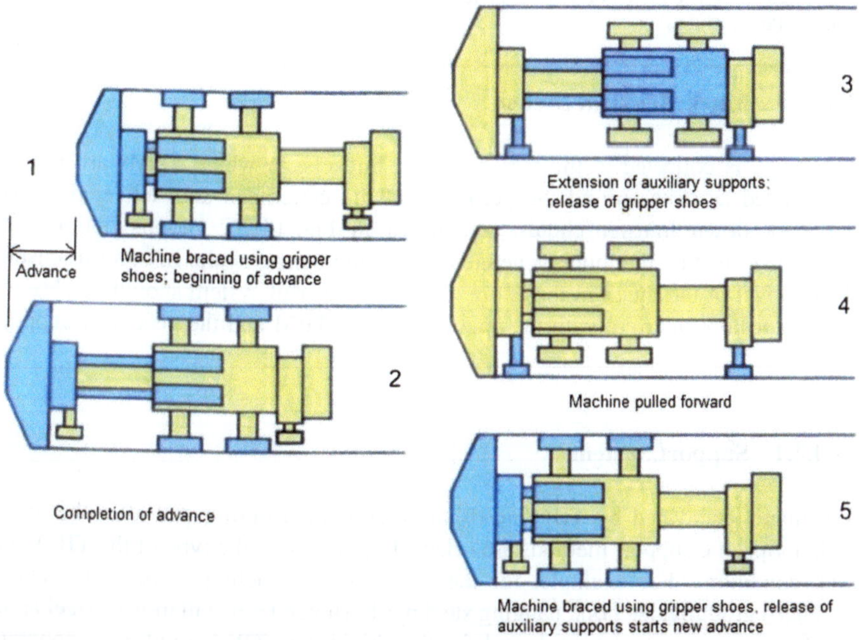

Fig. 14.24 Principle of advancing of the double-gripper TBM

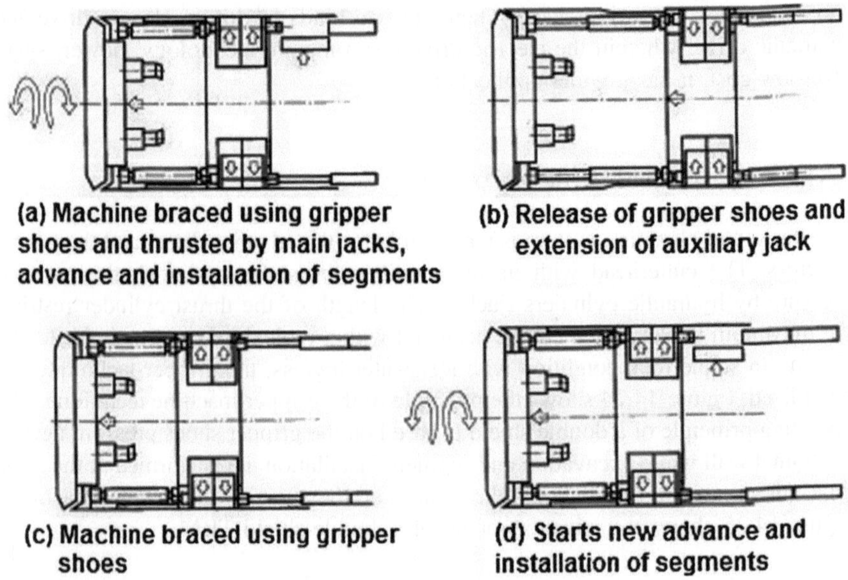

Fig. 14.25 Advance model of double-shield TBM (Reproduced from Ref. [11] by permission of Science Press, China)

14.4.4.3 Muck Removal System

The muck is collected at the face by cutter buckets, in which TBMs are mostly constructed as slots around the perimeter of the cutterhead and delivered to the conveyor down transfer chutes. As shown in Fig. 14.26, various options are available to handle the muck generated during the boring cycle. In order to ensure the removal of the muck in the entire tunnel, a powerful system should be chosen, which does not interfere with the operation of the TBM and the necessary support measures.

14.4.4.4 Support System

The tunnel excavated by TBM needs to be supported during and after the excavation. Specific support methods are selected according to the type of the TBM and rock conditions. For example, for the open TBM tunneling in poor formation, advanced grouting, anchors, hanging steel mesh, shotcrete, installation of steel arch, and lagging, etc., can be used, and for the shield-type TBM, reinforced concrete segments are mostly used for the support. The TBM support system can be divided into three regions, namely L1, L2, and L3 according to its different position in the TBM.

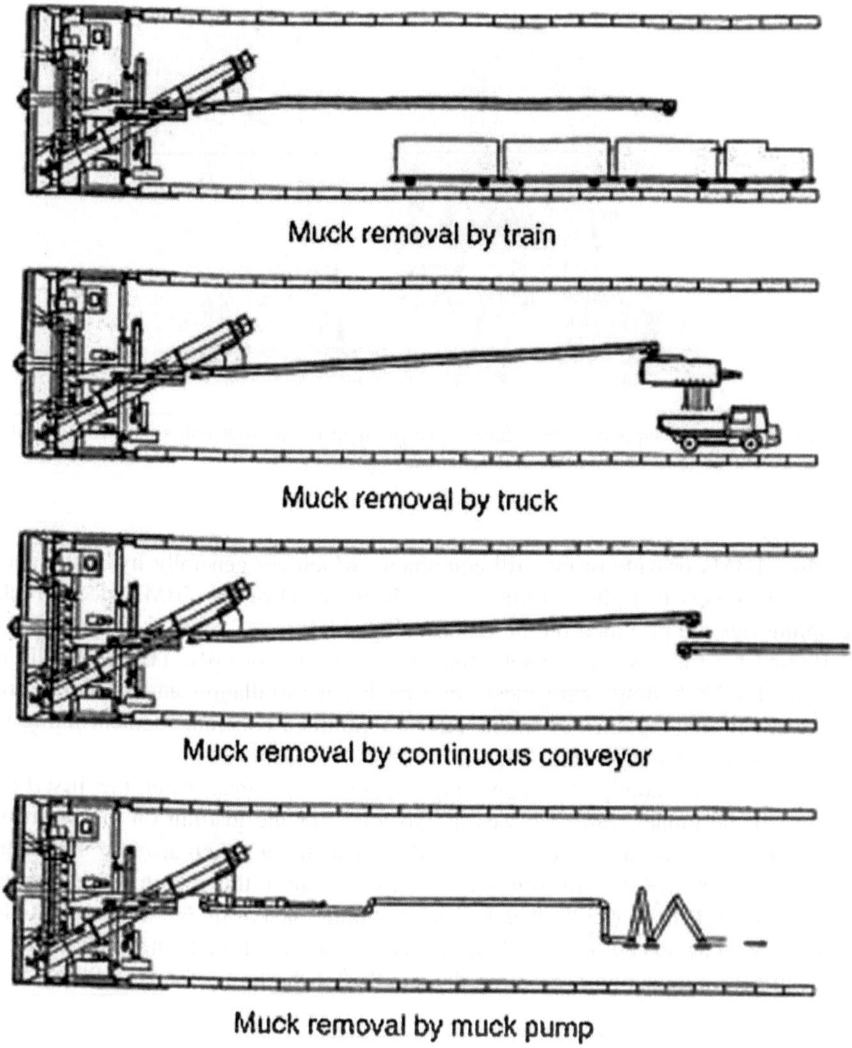

Fig. 14.26 Reproduced from Ref. [4] p. 366 Fig. 11.2 (a) Muck disposal arrangement and schemes (by permission of Taylor & Francis Group)

Region L1 support system is located directly behind the TBM cutterhead support. Its purpose is supporting the excavated rock face immediately ensuring the safety of the machine and personnel. For dealing with weak rock conditions, provisions can be made by installing a pair of roof bolters just behind the shield (Fig. 14.27). This allows a primary roof support in poor ground conditions. In some TBMs, some equipment, such as work platform, shotcrete robot, ring beam erector, mesh erector, and mesh transport system are equipped, especially in the large-diameter TBMs.

Fig. 14.27 The rock support anchors have to be installed by the roof bolting systems directly behind the shield. (*Source* www.herrenknecht.com/)

Most TBMs provide probe drill equipment, which are generally hydraulic percussion hammers that allow drilling up to 50 m ahead of the TBM. Special rock sampling rigs are provided on the request (Fig. 14.28).

Region L2 follows immediately behind region L1. For open TBMs, usually a system of rock bolting, wire mesh, or ring beam installation and shotcrete are performed in this region. For shield TBMs, reinforced concrete segment lining is carried out in this region (Fig. 14.29).

Region L3 is mainly the supporting reinforcement area. After the first two regions, if the tunnel wall collapses, is too loose, or the loading on the segment lining appears excessive, there is ground settlement or other adverse situation, appropriate reinforcement measures will be carried out in this region: for example, for the open TBM, after bolting and shotcrete lining, increasing the thickness of the shotcrete wall again can be carried out to ensure stability of the tunnel wall; and for the segment lining support, grouting the annulus of the lining for void filling.

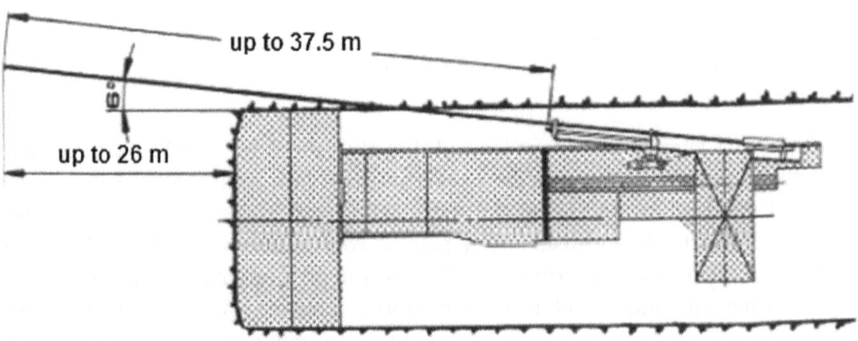

Fig. 14.28 Probe drill arrangement. (Reproduced from Ref. [3] by Permission of Sandvik)

Fig. 14.29 Installation of segment lining for TBM tunnel

14.4.4.5 Operation System

The operation system controls all the work of the TBM through the programmable logic controller (PLC) which is installed in the PLC box and consists of a computer module, multiple input/output modules, and some communication modules (Fig. 14.30). Under normal circumstances, except for a few hardwired safety electrical components, all electrical devices of the TBM are controlled via PLC.

The PLC controls the process of TBM's excavation. According to the ground conditions, the control models which can be selected for TBM include automatic

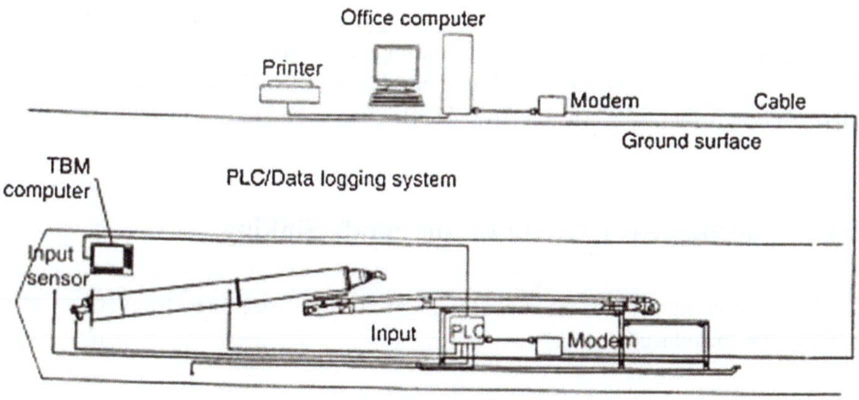

Fig. 14.30 Reproduced from Ref. [4] p. 366 Fig. 11.12 TBM's monitoring through PLC/data logging system (courtesy of Lovat) (by permission of Taylor & Francis Group)

torque mode, automatic thrust mode and manual mode. Automatic torque mode is suitable for soft rock, automatic thrust mode is suitable for homogeneous hard rock, and manual mode, which is operated manually according to the experiences of the operator, is suitable for different ground conditions.

14.4.4.6 Other Backup Systems

Other than the muck removal system, support system and operation system, other backup systems include:

- Ventilation system: Dust suppression (wet or dry), fresh air flexible ventilation duct cassettes, and booster fans for fresh air;
- Electrical system: High voltage cable reels, Emergency generator and emergency lighting;
- Water and compressed air: Backup mounted compressors (or hose reels), Compressed air distribution system and water systems (including recirculation system);
- Signal lighting system for trains and other equipment;
- Communication system for TBM and backups and TBM to portal, and
- Direction (centerline and grade-line laser devices) survey system;
- Safety refuge chamber and worker facilities.

14.4.5 TBM Type Selection

Full-face TBMs mainly consist of three types: open, double shield, and single shield, and each is suitable for different geological conditions. Selection of the TBM type should be based on a comprehensive analysis of all aspects of the factors, such as engineering geology and hydrogeology, construction environment, schedule requirements, economical, and technological conditions. Table 14.4 gives a comparison of three types of TBMs on different conditions (translated from "Description Table 6.13" of reference [12]).

14.5 Mechanical Excavation for Shaft Sinking

In this section, two kinds of mechanical excavation machines for vertical shaft sinking are introduced.

Table 14.4 Comparison of open and shield TBMs (Refer to: [12])

Item	Open TBM	Double-shield TBM	Single shield TBM
Geological adaptability	Generally used in good geological condition. Best adaptability in hard rock. Ground prereinforcement is needed for weak rock mass. Suitable for the tunnels mainly in Grade II and III rock mass	Same as open TBM in hard rock. Better to use single-shield TBM than double-shield TBM in soft weak rock mass. More suitable for the tunnels mainly in Grade III rock mass	Used in the geological relatively poor conditions (but the excavating face can self-stabilize). More suitable for the tunnels mainly in Grade III and IV rock mass
Driving performance	Under the premise of playing driving speed, mainly applicable to hard rock (50–150 MPa) with medium intact to intact formations and good self-stability. After effective supporting, it can be used in soft rock tunnel as well, but the driving speed will be limited	Under the premise of playing driving speed, mainly applicable to soft–hard rock (30–90 MPa) with medium intact formations and fair self-stability	Applicable to medium–long tunnel in the soft rock with certain self-stability
Construction speed	Under good geological conditions, it has a high speed due to less support with bolts-mesh-shotcrete. Under the worse geological condition, it has a slow speed due to the supporting workload	Under good geological conditions, it has a fair speed as the driving and segment installation can be carried out at the same time using the gripper to brace against the tunnel wall to establish the required thrust force. In weak ground, only single shield can be used, driving and segment installation cannot be carried out at the same time, the construction speed is limited	Driving and segment installation cannot be carried out at the same time, the construction speed is limited
Safety	As equipment and people are exposed in the surrounding rock, need to strengthen protection	Under the protection of the shield, people have good safety. Under ground with high stress, there is a danger of being stuck	Under the protection of the shield, people have good safety. Under ground with high stress, there is a danger of being stuck

(continued)

Table 14.4 (continued)

Item	Open TBM	Double-shield TBM	Single-shield TBM
Driving speed	Greatly affected by the geological conditions	Less influence by the geological conditions than the open TBM	Less influence by the geological conditions than the open TBM
Lining method	Depends on the actual situation, the secondary concrete lining may be carried out	Segment lining	Segment lining
Geological Mapping during construction	It is convenient for geological mapping during TBM driving. There is less risk when there is a lack of detailed geological exploration data	It is impossible to do the geological mapping and also difficult to carry out convergence deformation measurement. There is more risk when there is a lack of detailed geological exploration data	It is impossible to do the geological mapping and also difficult to carry out convergence deformation measurement. There is more risk when there is a lack of detailed geological exploration data

14.5.1 *Roadheader Excavation: Vertical Shaft Machine (VSM)* [14]

The VSM has been developed for the mechanized excavation of shallow shafts in water-bearing soils (Suhm 2006). This technology is based on a roadheader boom with a cutter drum equipped with cutter bits or soft ground chisels. A typical setup is shown in Fig. 14.33.

The roadheader boom is attached to a mainframe which can be equipped with gripper pads to stabilize and support the machine. The whole machine is designed to operate in submerged conditions, remotely controlled from the surface (Figs. 14.31 and 14.32).

Two options are available for rock support:

- precast concrete segments (segmental lining), and
- rock bolts, wire mesh, and sprayed concrete.

The segments are erected on the surface and the whole lining is lowered as the shaft is sunk—the usual option for soft ground conditions. In stable to poor soft rock conditions, rock bolting, installation of wire mesh, and shotcreting can be performed from special designed working platforms with permanently installed rock drills and shotcreting equipment.

VSM technology has been used on a number of projects in soil and soft rock conditions with rock strengths of up to 120 MPa. As a conveying system, a slurry circuit is used for soft soil below ground water level and a pilot hole for dry conditions.

Fig. 14.31 Core component of vertical shaft machine. (*Source* [14])

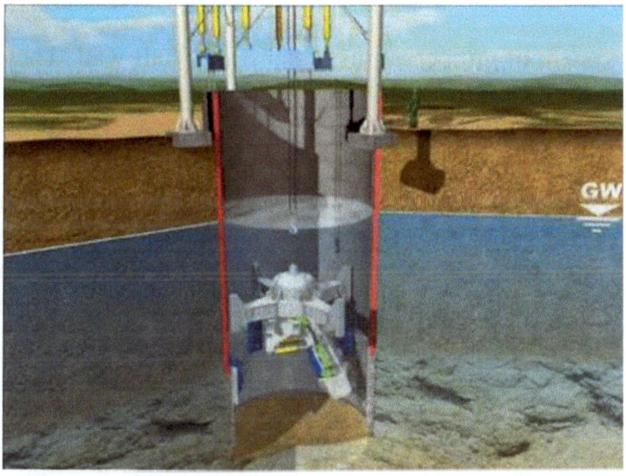

Fig. 14.32 A VSM machine is working below the table of groundwater. (Reproduced from Ref. [17] by permission of the author, Garcia, J. V.)

14.5.2 Shaft Boring Machine (SBM)

14.5.2.1 Reproduce from Refs. [13] and [14] by permission of Herrenknecht AG

The shaft boring machine (SBM) is a vertical application of proven rock boring technology (Fig. 14.33). The machine applies a rotating cutter wheel with disk cutters for shaft excavation. The wheel is thrust against the rock by hydraulic cylinders and slews about the shaft bottom as it rotates. Cuttings are removed by a clam shell device similar to conventional shaft mucking, and the muck is hoisted by buckets. The entire machine moves down (and up) the shaft through the use of a

Fig. 14.33 Overview of a SBM and its main functional parts. (Reproduced by permission of Herrenknecht AG)

system of grippers thrust against the shaft wall. These grippers and their associated cylinders also provide the means to maintain verticality and stability of the machine. The machine applies the same principles as tunnel boring machines but in a vertical mode. Other shaft construction activities such as rock bolting, utility installation, and shaft concrete lining can be accomplished concurrent with shaft boring. The method is comparable in cost to conventional sinking to a depth of about 460 m (1500 ft) beyond which the SBM has a clear cost advantage. The SBM has a greater advantage in productivity in that it can excavate significantly faster than drill and blast methods.

The excavation process is divided into two steps:

In the first step, the cutting wheel penetrates the rock like a circular saw, thus creating a slit with a depth of 1.5 m. In the second step, it rotates around the vertical axis of the machine to cut out the entire shaft profile. In doing so, the cutting wheel not only loosens the rock but also serves as a paddle wheel which transports the muck via integrated channels to the center. There the material is transferred to a vertical belt conveyor, which transports it to the transfer point for shaft conveyance. Up to 3 gripper systems brace against the shaft wall and thus stabilizing the entire system during the tunneling procedure.

The SBM can be separated into the main functional areas as shown in Fig. 14.35.

In the development of the SBM, a strong focus was placed on increasing occupational safety. As is the case with a hard rock TBM, shotcrete can be applied by remote control directly behind the cutting wheel. The disk cutters are replaced in a specially secured working area which is easily accessible and protected against falling rock. Thus, no personnel have to remain in dangerous areas during normal operation (Fig. 14.34).

Fig. 14.34 SBM's cutting wheel penetrates the rock like a circular saw. Disk cutters can be replaced from a safe working position. (Reproduced by permission of Herrenknecht AG)

14.6 Bored Piles in Rock

Bored piling is commonly used in construction as a foundation, especially for bridge work and tall buildings as well.

A bored pile is a non-displacement form of foundation that is cast in situ and provides economical load bearing and walling solutions suited to a wide range of ground conditions and applications. Key advantages include the following:

- Able to carry very high load/shear/moment capacity,
- Low noise and vibration,
- Ground is "seen" during construction allowing validation of design assumptions,

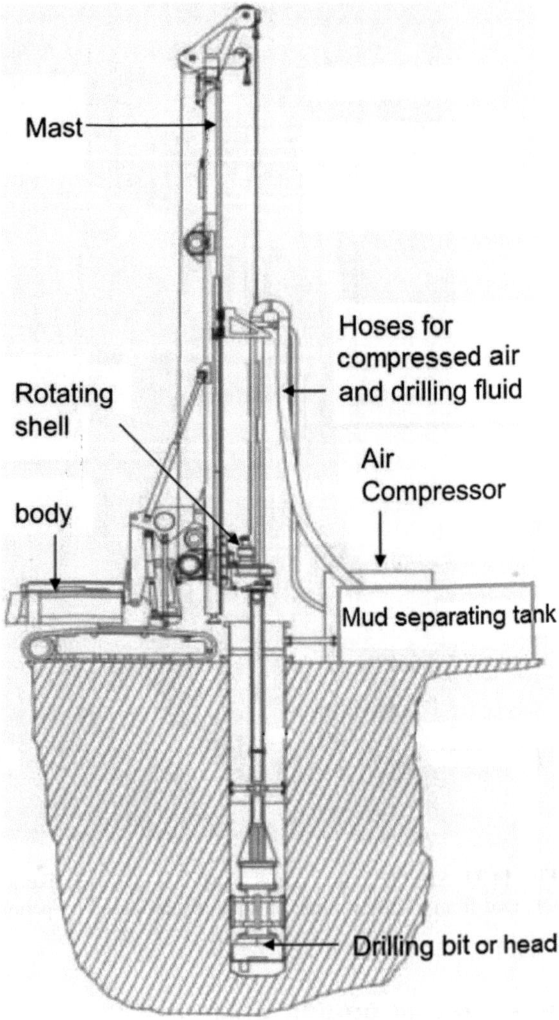

Fig. 14.35 Operation of the drill rig. (Reproduced from Ref. [16] by permission of City University of Hong Kong)

- Can overcome adverse ground including natural or manmade obstructions,
- Can drill into hard rock, and
- Can be constructed to tight tolerances.

Bored piles can be constructed using crane-mounted or track-mounted hydraulic drill rigs. A wide range of drill rigs is available (rotary, crane mount, excavator mount, low headroom) for drilling borehole.

14.6.1 *Small- to Medium-Sized Bored Pile (Generally Refers to Piles Ranging in Size from 300 to 900 Mm in Diameter)* [16]

- The use of drilling rigs of an appropriate capacity is required. Due to the possible collapse of subsoil during drilling, the forming process usually requires temporary protection by the use of steel casings and some kind of drilling fluid such as bentonite slurry.
- Due to the rapid development of a wide range of highly effective mechanical drilling equipment, this foundation method is becoming quite popular for the construction of medium- to high-rise buildings around the world.

Figure 14.35 shows the operation of a drill rig for small- to medium-sized bored pile (quoted from [16].

14.6.2 *Large-Diameter Concrete Bored Pile [16]*

- The boring process can be done manually or mechanically. In general, piles ranging in size from 1 to 3 m diameter can be formed by mechanical methods while piles of 3 m diameter and above are dug using manual methods (Fig. 14.36).
- Mechanical boring can be done by the use of grab-and-chisel or reverse circulation drilling (RCD), both of which require the use of a steel casing to

Fig. 14.36 Borehole formed by chisel and grab and support with casing. (Reproduced from Ref. [16] by permission of City University of Hong Kong)

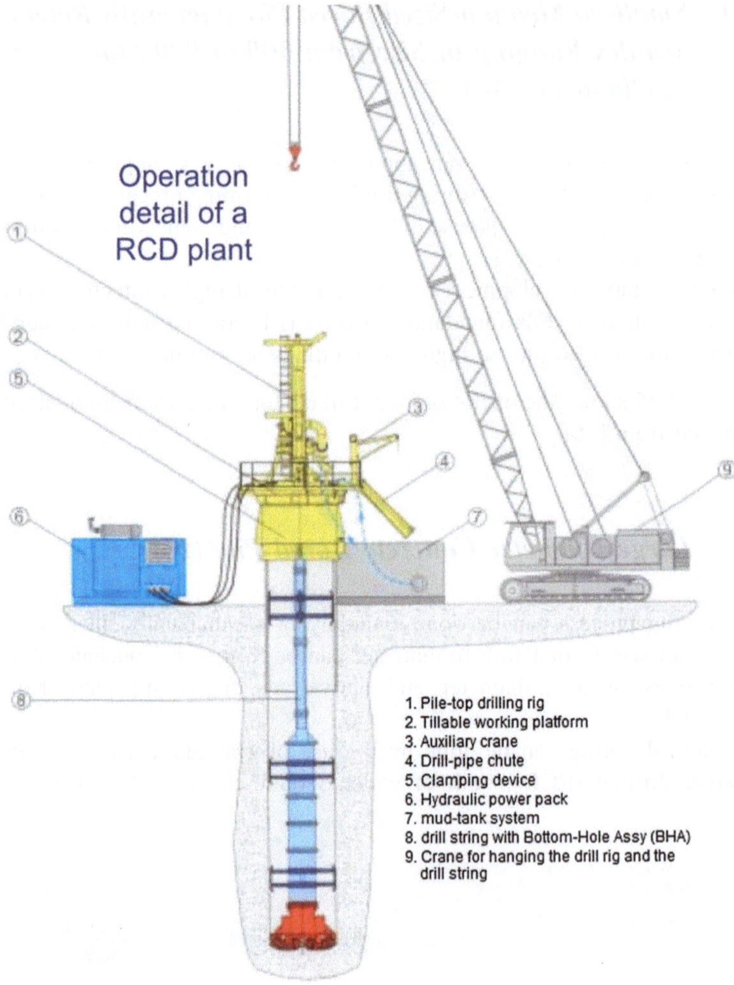

Fig. 14.37 Bored pile formed by reverse circulation drilling (RCD) method (Reproduced from Ref. [16] by permission of City University of Hong Kong)

stabilize the bore during excavation through soft ground in the upper portion of the bored hole.

- Sometimes, super large-sized piles of up to 6 to 8 m diameter can be constructed. In this case, a cofferdam formed by sheet piles, soldier piles, or in situ concrete piles is provided for soil retaining purpose (Fig. 14.37).

Figure 14.38 shows a bored hole formed by chisel and grab and support with casing. Figure 14.37 shows the operation detail of a RCD plant, and Fig. 14.38 shows the principle of removal of spoil from the borehole.

Fig. 14.38 Removal of spoil from the borehole in RCD method (Reproduced from Ref. [16] by permission of City University of Hong Kong)

References

1. Chapman D et al (2010) Introduction to tunnel construction, Spon Press, London, ISBN10: 0-415-46841-8
2. Hemphill GB (2013) Practical tunnel construction, Wiley, Hoboken, ISBN: 978-0-470-64197-2
3. Heiniö M (Ed) (1999) Rock excavation handbook for civil engineering, Sandvik, Tamrock
4. Tatiya RR (2013) Surface and underground excavation, methods, techniques and equipment, 2nd edn. CRC Press, Baco Raton
5. Copur H et al (1997) Roadheader application in mining and tunneling industries. Earth Mechanics Institute, Colorado School of Mines, Golden Colorado
6. Puhakka T (1997) Underground drilling and loading handbook, Tamrock
7. Liu Z, Meng Y (2015) Key technologies of drilling process with raise boring method. J Rock Mech Geotech Eng 7(2015):385–394. doi:10.1016/j.jrmge.2014.12.006
8. Williams O (1997) Engineering and design-tunnels and shaft in rock. Department of the Army, U.S., Army Corps of Engineers, 20 May 1997
9. Maidl B et al (2008) Hardrock tunnel boring machines. Erst and Sohn, Berlin, Germany. ISBN: 978-3-433-01676-3
10. 白云 等 (2013), 隧道掘进机施工技术(第二版),中国建筑工业出版社, 2013,北京。Bai Y et al (2013), Operation technology of tunnel boring machines, 2nd edn. (in Chinese), China Construction Industry Press, Beijing

11. 龚秋明 (2014),掘进机隧道掘进概论,科学出版社,2014, 北京。 Gong Q (2014) Introduction to Tunnel Excavation Using TBM (in Chinese). Science Press, Beijing. ISBN: 978-7-03-039970-0
12. 中华人民共和国行业标准 (2007): 铁路隧道全断面岩石掘进机法技术指南, 铁建设[2007] 106号,2007-05-22发布及实施,中国铁道出版社,2007,北京。 People's Republic of China Industry Standard (2007) Technical guide of the full face tbm method for railway tunnels. Publication and Implementation on 22 May 2007, China Railway Press, Beijing
13. Rennkamp P (2015) Shaft boring machine (SBM) safe and quick construction of bblind shafts down to depths of 2,000 meters. Herrenknecht AG, 23 Aug 2015
14. Frenzel C et al (2010) Shaft boring systems for mechanical excavation of deep safts. Newsletter, Australian Centre For Geomechanics, Vol 34, May 2010
15. ITA Working Group No. 14 ("Mechanized Tunneling") (2000) Guide lines for selecting TBMs for soft ground. Japan and Norway, Published in "Recommendations and Guidelines for Tunnel Boring Machines (TBMs)" pp. 1–118, 2000, by ITA–AITES, www.ita-aites.org, Cedex, France. ISSN 1267-8422
16. Raymond W (2012) Foundation, (PowerPoint). City University of Hong Kong. http://www6.cityu.edu.hk/construction_archive/features%5Cjune%5CConst_Tech_series_Foundation_Technology.pdf. Accessed 15 April 2016
17. Garcia JV (2010) Vertical Shaft Sinking Machines (VSM-007). Geotechnical Engineering. http://johnny-villavicencio-garcia.blogspot.hk/2010/11/vertical-shaft-sinking-machines-vsm-007.html, Accessed 12 May 2016
18. Atlas Copco (2015) Underground Construction—a global review of tunneling and subsurface installations, 1st edn, Atlas Copco, Rock Drills AB

Chapter 15
Other Underground Excavation Methods

15.1 Cut-and-Cover Method

The cut-and-cover method for constructing tunnels offers an alternative approach to underground construction techniques. This method, in which a trench is excavated from the surface and then recovered, is usually more economical and more practical than mined or bored tunneling. Pipelines, such as sewers, vehicular tunnels, and metro tunnels, are often constructed using this technique. It is especially practical at shallow depths (10–15 m). However, the depth of 30 m is quite common for metro tunnels.

15.1.1 Construction Methods of Cut-and-Cover

Two basic forms of cut-and-cover tunneling are available:

- ***Bottom-up method***: This technique involves stepped excavation and implementation of support either by means of temporary walls and bracing systems in order to support the sides of the excavation. In cases of extremely adverse geotechnical conditions, prestrengthening might be necessary in order to minimize or avoid stability problems during the excavation phase. Therefore, bolting–mesh–shotcrete for hard rock and anchored retaining wall, sheet piles, or diaphragm wall for soft ground have become common practice in "cut-and-cover" construction. Once the foundation level has been reached, concreting of the tunnel commences starting with the base slab, followed by the walls and roof, then finally followed by waterproofing and placement of backfill. Figures 15.1 and 15.2 show examples.
- ***Top-down method***: This method was originally developed for construction of shallow underground structures in congested urban areas, where open excavation techniques would cause significant disruption to traffic. The fundamental

D. Zou, *Theory and Technology of Rock Excavation for Civil Engineering*,
DOI 10.1007/978-981-10-1989-0_15

Excavation is mostly by excavator and hydraulic breakers with a section of drill+blast needed through a zone of volcanic rock for a distance of less than 1km from 29th to 35th Streets. Support of the trench consists of shotcrete and wire mesh and rockbolts in the rock section. Behind the invert base slab, the reinforced walls, center divide and top deck of the box structure are cast as one composite pour using 20m long forms supplied by Peri.

Fig. 15.1. Construction of Vancouver's rapid transit Canada Line by cut-and-cover (Reprinted from Ref. [3] with the permission from TunnelTalk)

Cut and Cover Tunnel for the Airport Railway at the Central Reclamation in Hong Kong in Late 1996. Section of the tunnel pit formed by diaphragm wall and strutted by the use of steel tubes. The construction of the tunnel sections using in-situ method can be seen inside the pit.

Fig. 15.2. Cut-and-cover tunneling in Hong Kong

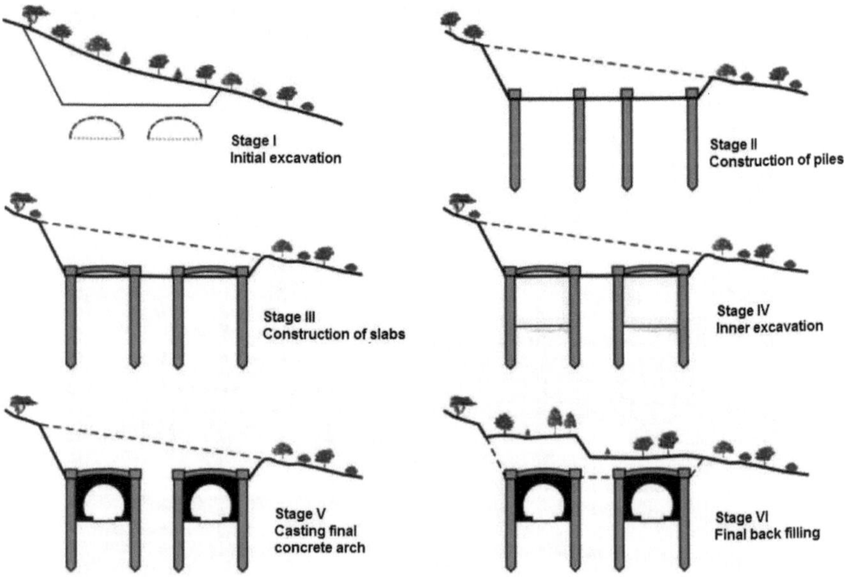

Fig. 15.3. Construction stages for "cover-and-cut" method. (Reproduced from Ref. [1] with the permission from EJGE)

concept of the method comprises of several stages. Side support walls and capping beams are constructed from ground level by such methods as slurry walling, or contiguous bored piling. Then, a shallow excavation allows making the tunnel roof of precast beams or in situ concrete. The surface is then reinstated except for access openings. This allows early reinstatement of roadways, services, and other surface features. Excavation then takes place under the permanent tunnel roof, and the base slab is constructed.

Top-down method is also called "cover-and-cut" method. Figure 15.3 shows the six distinct stages of the construction process of this method (from [1]).

15.1.2 Support Methods for the Sidewalls of the Excavation

There are many methods that can be used to support excavation side walls for the cut-and-cover method. The design of the excavation support system will depend on many factors, which include the following (Fig. 15.4):

- The geological conditions of the ground, including the characteristics of the soil/rock and the groundwater level where the excavation will take place;
- The size (width, length, and depth) of the excavation trench;
- The proximity of the excavation to adjacent structures and buildings;
- The number, size, and type of utilities crossing the proposed excavation or adjacent to the excavation;

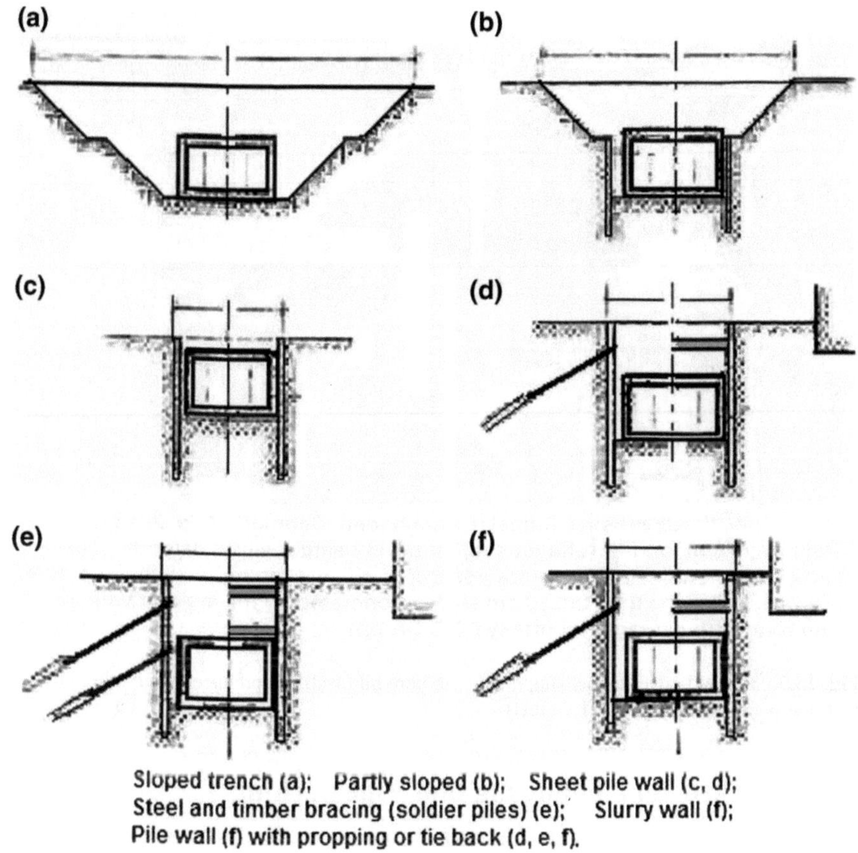

Fig. 15.4. Ground support for cut-and-cover method

- The surcharge loading adjacent to the excavation from traffic or construction equipment; and
- The influence on the surrounding environment and the restriction of noise and dust in urban area.

According to the site conditions, the following support methods are usually adapted for cut and cover:

- Shotcrete, typically together with rock bolts and wire mesh, is usually used for stable rock walls (see Fig. 15.1);
- Sheet pile with walings and struts, or ground anchors;
- Ground anchors;
- Soldier piles and horizontal poling boards (lagging) (Fig. 15.5);
- Slurry walls;
- Large-diameter bored piles, contiguous, or overlapping.

Nordhavnsvej Tunnel, Copenhagen, Denmark, Aug. 2013
The new 620 m long Nordhavensvej tunnel is being realized using the cut-and-cover method. The walls are concreted against the up to 25 m deep bore pile wall. Following this, the 80 cm slab is concreted by means of a Variokit slab formwork carriage. (courtesy PERI Group)

Fig. 15.5. Cut-and-cover tunnel supported with bore pile wall. (Reproduced from http://peri.com with the permission from PERI GmbH)

15.2 Jacked Box Tunneling and Pipe Jacking

As both the methods of jacked box tunneling and pipe jacking are mainly used in soft ground such as soil, sand, or mud (pipe jacking can be used for rocks when a micro-TBM is used), only a simply introduction is described in this section.

15.2.1 Jacked Box Tunneling

Jacked box tunneling or tunnel jacking is a non-intrusive method of constructing railway or subway tunnels, large oversized culverts, or an underbridge by pushing a huge concrete box through the soil using specialized jacking equipment and hydraulic jacks [2]. The obvious benefit of the tunnel jacking method is that construction of shallow tunnels can resume under existing infrastructure or structures above with minimal disturbance [5].

The technique is outlined in Fig. 15.6 which shows the stages of construction for a new vehicular underbridge beneath a four track railway. A reinforced concrete box is cast on a jacking base adjacent to the railway or highway embankment (see Fig. 15.6a). A purpose designed tunnel shield is provided at its leading end, and

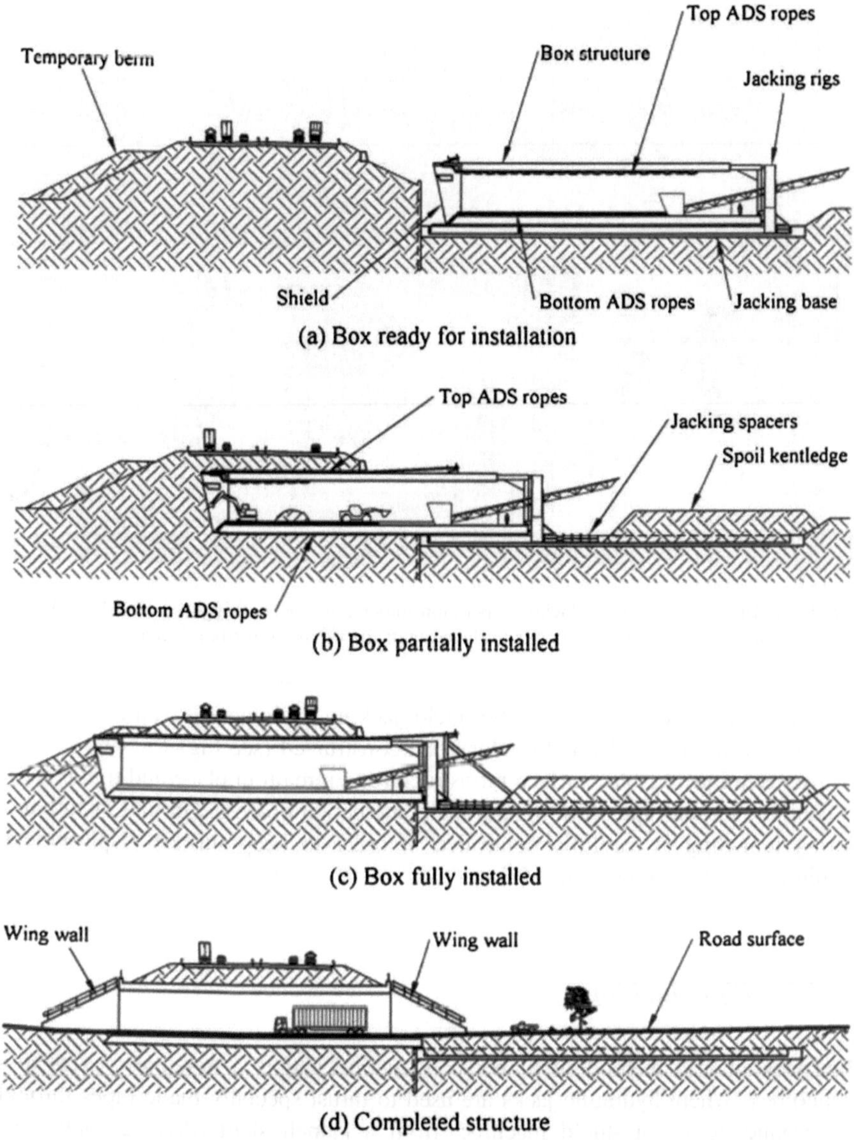

Fig. 15.6. Jacked box tunnel installation (Reproduced from Ref. [4] with the permission from © Institution of Mechanical Engineers 2007)

thrust jacks are provided at its rear end reacting against the jacking base. Anti-drag systems (ADSs) are installed, and the box is jacked forward to the embankment. The box is then tunneled through the ground by carefully excavating 150 mm off the face and jacking the box forward by a corresponding increment, and this sequence is being repeated many times (see Fig. 15.6b). When the box has reached

Fig. 15.7. Box culvert during Jacking operation under rail line at Wiggins Island Coal Export Terminal Gladstone, Qld Australia (Reproduced with the permission from Tunnelcorp Pty Ltd.)

its final position (see Fig. 15.6c), the shield, jacking equipment, etc., are dismantled and the portal wing walls and roadway are constructed (see Fig. 15.6d).

The technique enables surface infrastructure to remain in place and traffic usage to continue uninterrupted throughout the construction process. Ground disturbance during tunneling is carefully controlled in order to keep surface displacements within acceptable limits (Fig. 15.7).

15.2.2 Pipe Jacking

Pipe jacking is a method of tunnel construction and is a trenchless excavation technology, where hydraulic jacks are used to thrust specially made pipes through the ground behind a shield machine, from a launch shaft (driving shaft) to a reception shaft (see Fig. 15.8). The term microtunneling is also often used to describe this method of pipe installation.

Pipe jacking is used to install conduits below ground for a variety of applications including [6]:

- sewerage pipelines,
- storm water pipelines,
- road and rail culverts,
- pressure pipelines,

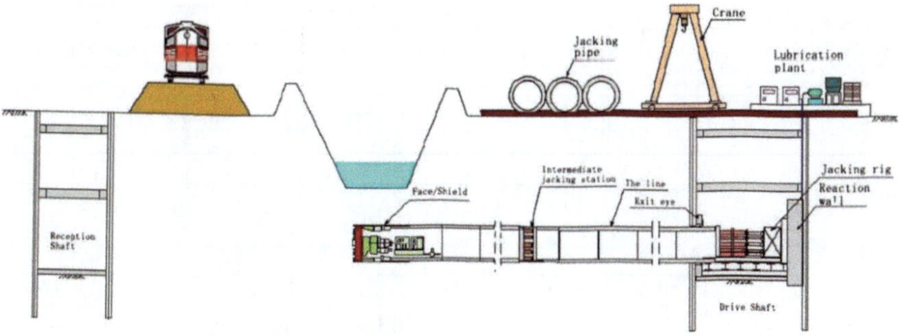

Fig. 15.8. Basic elements of pipe jacking (microtunneling) technology

- as a sleeve pipe for other utility pipelines (water, sewage, and electricity and communication cables), and
- pipe replacement and relining.

The pipes are usually made of high-strength concrete to withstand the high jacking forces. They form driving elements and permanent supports at the same time. Intermediate jacking stations are used to reduce the jacking pressures and facilitate curved drives. With longer tunnels, there are intermediate shafts which may be crossed by the pipe section.

Except for manual (or mechanical means) excavation for some larger pipes (minimum of 1 m in diameter) where workers need to enter the pipe for excavation and spoil removal process during the jacking operation, the micro-TBM (MTBM) is usually used for pipe jacking. According to the American Society of Civil Engineers Standards, MTBM is defined as a remotely controlled and guided pipe jacking technique that provides continuous support to the excavation face and does not require personnel entry into the tunnel. The system simultaneously installs pipe as spoil is excavated and removed. Figure 15.9 shows MTBM for hard rock. Figure 15.10 shows the pipe jacking operation in the driving shaft.

Fig. 15.9. YS-1800 Hard Rock Slurry Balance Pipe Jacking Machine by Shanghai Huayu Machinery Manufacture Co., Ltd

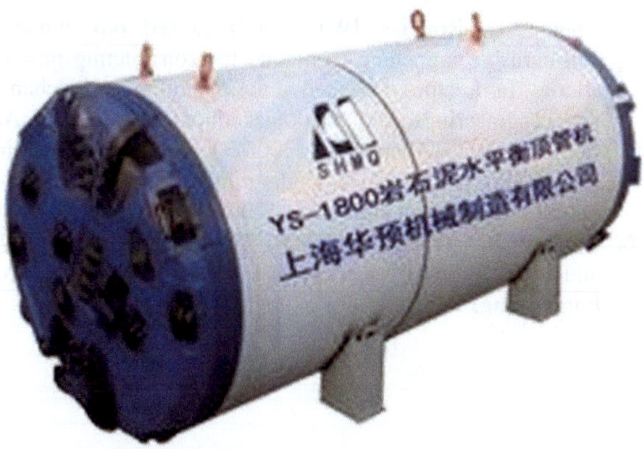

Fig. 15.10. Pipe jacking operation in the driving shaft (reproduced with the permission from Ed. Züblin AG—Singapore Branch)

References

1. Mouratidis A (2008) The "cut-and-cover" and "cover-and-cut" techniques in highway engineering, EJGE, vol 13, Bund. F. http://www.ejge.com/2008/Ppr0864/Ppr0864.pdf. Accessed 12 May 2016
2. Chapman D et al (2010) Introduction to tunnel construction. Spon Press. ISBN: 10.0-415-46841-8
3. Wallis S (ed) (2007) Canada Line on track to meet Olympic deadline. TunnelTalk Direct by Design, Mar 2007. http://www.tunneltalk.com/Canada-Line-March07-New-LRT-on-track.php. Accessed 12 May 2016
4. Allenby D, Ropkins JWT (2007) Jacked box tunnelling using Ropkins System™, a non-intrusive tunnelling technique for constructing new underbridge beneath existing traffic arteries. In: Lecture presented at the Institution of Mechanical Engineers, London 2007. http://www.engineeringsociety.co.uk/jackedboxtunnelling.pdf. Accessed 12 May 2016
5. Ropkins JWT, Allenby D (2005) Jacked box tunnelling, a non-intrusive technique for constructing underbridges. Underground Space Use: Analysis of the Past and Lessons for the Future—Erdem & So/ak (eds), ©2005 Taylor & Francis Group, London, ISBN: 04 1537 452 9
6. Dinescu S, Andras A (2008) Environmental friendly equipment and technology for underground civil excavations. Annals of the University of Petrosani, Mechanical Engineering, 10, pp 47–52

Chapter 16
Introduction to Underground Excavation by Drilling and Blasting

16.1 Introduction

16.1.1 Working Cycle of Excavation by D & B

Drilling and blasting are the most common methods for tunneling and underground excavation. The process begins with holes of predetermined diameter, depth, and spacing being drilled into the rock. Explosives are placed in the holes, and when detonated, the energy generated by the explosive reaction fractures the rock.

The drilling and blasting cycle consist of:

- Drilling—drilling holes in the rock for the placement of explosives,
- Charging/loading—insert the explosives in the holes drilled,
- Blasting and ventilation—detonation of the explosives and waiting for the dust and gases generated by the blast to clear,
- Scaling—removing the loose rock in the crown and sidewalls,
- Installing temporary support, and
- Loading and transporting blasted rock (muck).

The order of the last two may be reversed. Figure 16.1 shows the D & B cycle.

16.1.2 Working Condition

Drilling and blasting for underground excavation can be used in soft rocks with low strength, e.g., marl, loam, clay, gypsum, and chalk, to the hardest rocks, such as granite, gneiss, basalt, or quartz. Due to this large range of possible usage, drill and

D. Zou, *Theory and Technology of Rock Excavation for Civil Engineering*,
DOI 10.1007/978-981-10-1989-0_16

Fig. 16.1 Drilling and blasting cycle for underground excavation. Reproduced from Ref. [5] with the permission from Sandvik & Tamrock

blast can be advantageous for very changeable ground conditions [2, 4]. In addition, tunneling by using drill and blast is often preferable to TBM or roadheader tunneling if, for example, the tunnel is relatively short so that the high investment costs needed for a tunneling machine are not economic, or when the rock hardness is very high so that a high wear of the cutter tools leads to an uneconomic application of the machine. Drilling and blasting are also suitable for these cases that the cross section of the tunnel or cavern changes along the alignment or the cross section has an irregular shape [3].

16.1.3 Drilling Equipment, Explosives, and Blasting Design

The drilling equipment and explosive products and initiation accessories are described in Chaps. 2, 3, and 4.

The blasting design of underground excavation of D & B will be discussed in detail in the later chapters.

16.2 Excavation Methods for Tunnels and Caverns

Figure 16.2 shows the drilling equipment for tunnel and cavern blasting. According to the tunnel (Cavern) size, drilling jumbos with one to four booms can be chosen.

16.2.1 Full-Face Excavation

The following conditions affect if full-face excavation is carried out:

(a) The reach of the drilling jumbos and the platforms used for support works, which limits the height and width to 7 × 11 to 11 × 18 m depending on the plant used (Figs. 2.29, 2.30 for reference);
(b) Rock quality that means the area of unsupported roof that can be exposed at any one time; and
(c) Limitation on the quantity of explosives discharged in a round given by vibration acceptance criteria.

It is normally more economical to excavate full face if conditions permit.

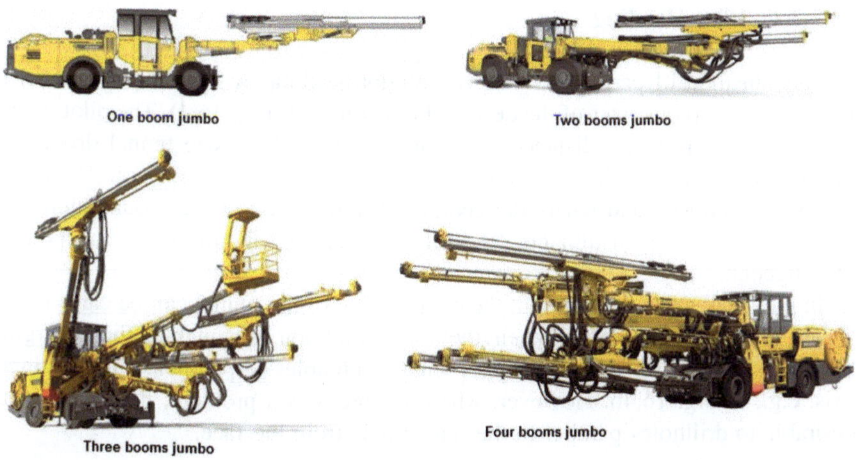

Fig. 16.2 Drilling equipment for underground excavation. Reproduced with the permission from Atlas Copco

16.2.2 Partial Face Excavation

In the case that above conditions cannot be satisfied, the methods of partial face excavation can be used. The partial excavation methods include top heading and benching, pilot heading, and sidewall drift.

16.2.2.1 Top Heading and Benching

The most common advance heading method is top heading. The crown is excavated before the bench (Fig. 16.3). The temporary support of the crown with rock bolts and shotcrete gives safe working conditions for the excavation of the lower levels of the tunnel or cavern.

Bench excavation is cheap because the large free surfaces allow the use of quarrying principles rather than tunneling technology. Support costs for benching are low because the roof already has been supported, leaving only some wall support to be done.

Bench excavation may be done with vertical drillholes as in a quarrying operation, or with horizontal drillholes for using the same drilling jumbo as the top heading.

16.2.2.2 Pilot Heading

Another method of partial advance is the pilot heading. A smaller pilot tunnel is driven in the top or center of the designed larger tunnel (Fig. 16.4). The pilot tunnel can be driven a certain distance in advance of the whole size tunnel driving or through the entire length of the designed tunnel first before enlarging it. Under the complex conditions and where the geological information of the ground along the tunnel alignment is insufficient, the pilot tunnel has an important function of investigation.

In the weak ground condition, the enlargement of the tunnel can be carry out in stages, e.g., in Fig. 16.4a. When the rock condition is suitable, the enlarging blastholes can be drilled from the pilot tunnel with holes perpendicular to the tunnel axis, e.g., in Fig. 16.4b. However, when overbreak is a problem, it may be more desirable to drillholes parallel to the tunnel axis from the face.

16.2.2.3 Sidewall Drift

Under the condition of weak ground with low strength, sidewall drift method is usually used. The side galleries are excavated and supported first. They serve as abutment for the support of the crown, which is subsequently excavated (Figs. 16.5,

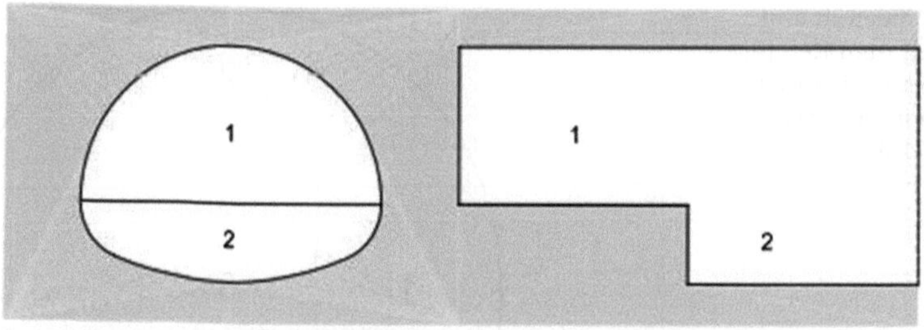

(a) Top heading, cross and longitudinal sections. 1: calotte, 2: bench

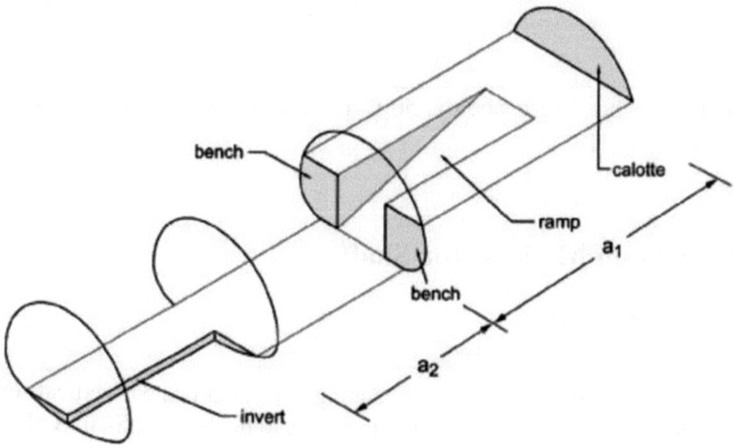

(b) Schematic representation of top heading

(c) Bench blasting with Horizontal or vertical drill holes

Fig. 16.3 Top heading and benching excavation method for tunneling. **a**, **b** Reproduced from Ref. [6] with the permission from ©2008 Springer-Verlag Berlin Heidelberg)

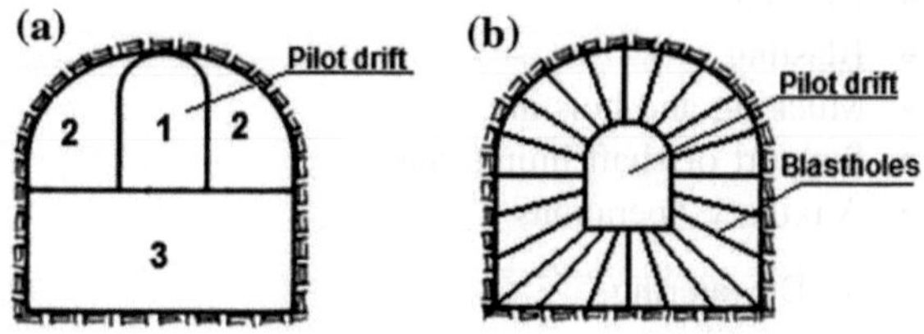

Fig. 16.4 Pilot tunnel method: **a** top method, the figure is the excavation order; **b** central pilot

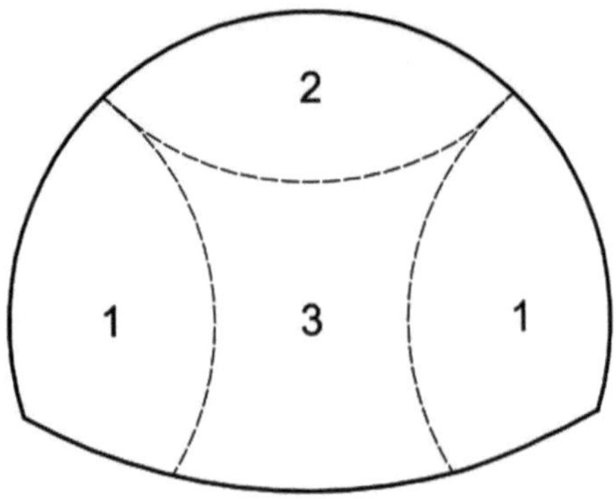

Fig. 16.5 Sidewall drift (reproduced from Ref. [6] with the permission from ©2008 Springer-Verlag Berlin Heidelberg)

16.6). This type of heading is approx. 50 % more expensive and slower than top heading excavation [1].

16.3 Excavation Methods for Shaft

16.3.1 Shaft Sinking

Shaft sinking refers to the method of excavating a vertical or near-vertical tunnel from the top-down, where there is initially no access to the bottom. Shaft Sinking is one of the most difficult of all development methods: Restricted space, gravity, groundwater, and specialized procedures make the task quite formidable.

Shaft dimensions are determined by shaft purpose and geological and rock mechanical conditions. Most shafts have a diameter of 5–8 m, with only a few reaching 10 m or more in diameter. Shafts are usually round in shape, but some times, there are rectangular shafts for special purposes.

After exploring the geology and groundwater conditions, overburden is removed. If the overburden requires stabilizing, it is typically lined with concrete rings. Once the rock surface has been exposed, it is reinforced and grouted. The collar for the head frame is installed after excavation has progressed a short distance. The head frame includes the hoisting system for the shaft sinking equipment. At this point, the actual shaft sinking begins.

A sinking cycle includes the following operations:

- Drilling,
- Blasting,
- Mucking and hoisting,
- Support or shaft lining, and
- Auxiliary operations
  - Dewatering,
  - Ventilation,

- Lighting or illumination, and
- Shaft centering.

Rock drilling can be carried out either by handheld or machine-mounted rock drills, but in small shafts, there is very limited space for machinery. In modern shaft sinking, the drilling rig is a two to 6 or more-boom drilling jumbos designed specifically for the dimensions of the shaft and sinking platform. Figs. 16.7 and 16.8 show two kinds of shaft drilling equipment. The drill pattern for shaft sinking blast will be discussed in late chapter.

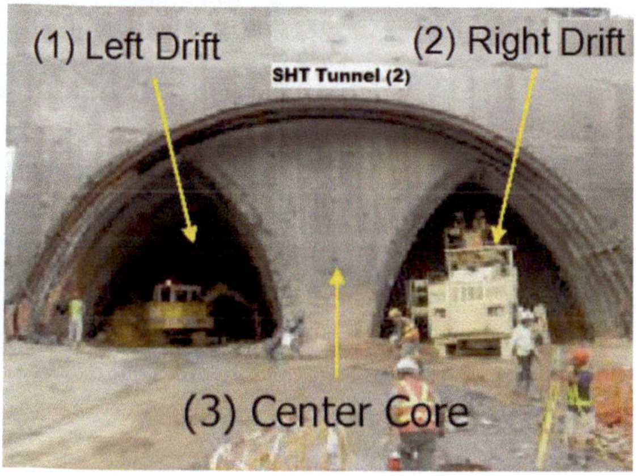

Fig. 16.6 Sha Tin Height tunnel, sidewall drift

Fig. 16.7 A small two boom jumbos are drilling blastholes in shaft

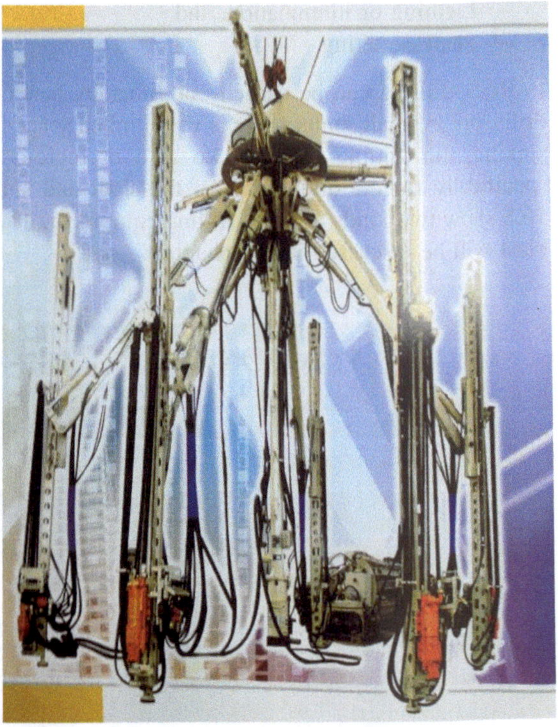

Fig. 16.8 Vertical shaft drill made by Qingtao Grandplan Technology Co. Ltd., China

The methods for shaft sinking can be divided into three groups:

- The full-face sinking (full-bottom) method.
  The full-face sinking method is used frequently in shaft sinking as it suits either rectangular or round section shafts.
- The benching method (half-bottom method).

Benching can be used as an alternative to full-bottom shaft sinking when rock conditions do not allow full-face excavation. Benching is an older method that is suitable for square-shaped shafts. Benching is done in halves. While one-half of the cross section is being drilled with a fan and blasted, the lower half serves as a water sump and spoil dump. Work continues downward in alternately lowering benches (Fig. 16.9).

- The spiral method.

Spiraling is a variation of the benching method. Excavation spirals downward. This method is suitable for fairly large round- or oval-shaped shafts, or when the full bottom is not otherwise possible. Drilling and blasting progress with half of the face at a time. The holes in each half are drilled parallel in the same length, as there always has to be a free face which descends with each position (Fig. 16.10).

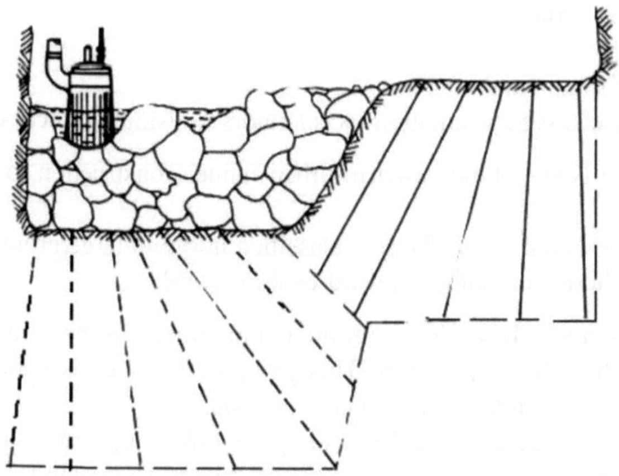

Fig. 16.9 Benching method (reproduce from Ref. [5] with the permission from Tamrock)

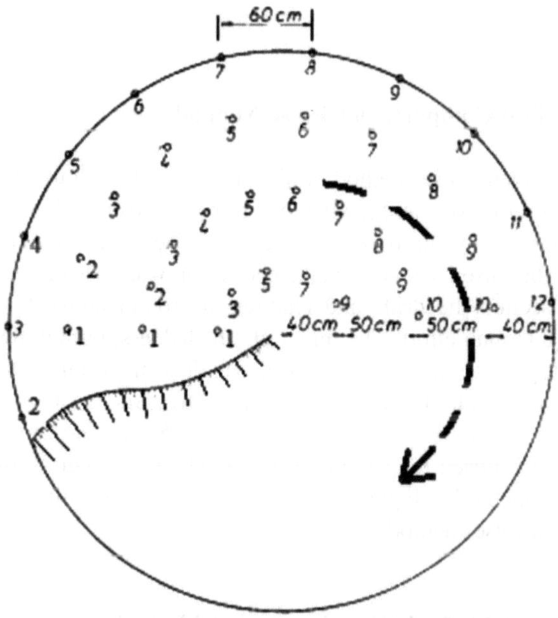

Fig. 16.10 The spiral method (reproduce from Ref. [5] with the permission from Tamrock)

16.3.2 Raise Driving

A raise is a vertical excavation proceeding from a lower elevation to a higher elevation, perhaps from one tunnel to another, or from a tunnel to the ground surface.

Raises are one of the important structures in many civil and construction projects:

- Hydroelectric projects
- Water supply,
- Wastewater shafts, and
- Tunnel projects, e.g., ventilation, accelerators housings, or access ways.

Raise driving is sometimes used in urban tunnel construction to minimize surface disruption.

Raise driving with drill and blast is classified into two large groups according to the drilling method used, either upward or downward:

- Upward drilling by hand drill with the compartment method (raise building), a Jora lift or the Alimak platform. This group of methods is manual raise excavation method and a tough and dangerous job.
- Downward drilling. Long holes with pilot hole cut, crater cut, "VCR" cut (vertical crater retreat), and the full-face method. These kinds of methods are much safer as the workers do not work under the excavation space directly, but they require high drilling accuracy and good rock condition (relatively integral and stable).

16.3.2.1 The Two-Compartment Raise Method

The two-compartment raise method, which is also called "raise building," is the oldest method of raise excavation, where the miner builds a timber wall dividing the raise into one open and one rock-filled section. The open section is used by the miner to climb the raise by a ladder attached to the wall. Then, standing on top of the rock-filled section, he drills and charges the round above his head. Manual raising is 100 % human effort. The miner climbs ladders, pulling his rock drill and material to the top of the raise, tied to a rope. He drills the round, charges blastholes lower his equipment, climbs down, and finally triggers the blast. The height of manually excavated raised are normally limited to 50 m, due to the heavy strains which miners are exposed to. Therefore, this method has been mostly replaced by more advanced methods. Figure 16.11 shows the conceptual diagram of the two-compartment raise method

16.3.2.2 Jora Lift Method (Hanging Basket Method)

This method can be used for vertical or inclined raises (Fig. 16.12). The main character of this method is the drilling of a pilot hole with a diameter between 75 and 100 mm through which the cable which holds the lift is lowered. The main components are the work platform, the lift basket, the hoisting mechanism, and, in inclined raises, the guide rail.

During drilling, the platform can be fixed against the raise sides with some jacks. While drilling around, the parallel holes are drilled around the central hole, which

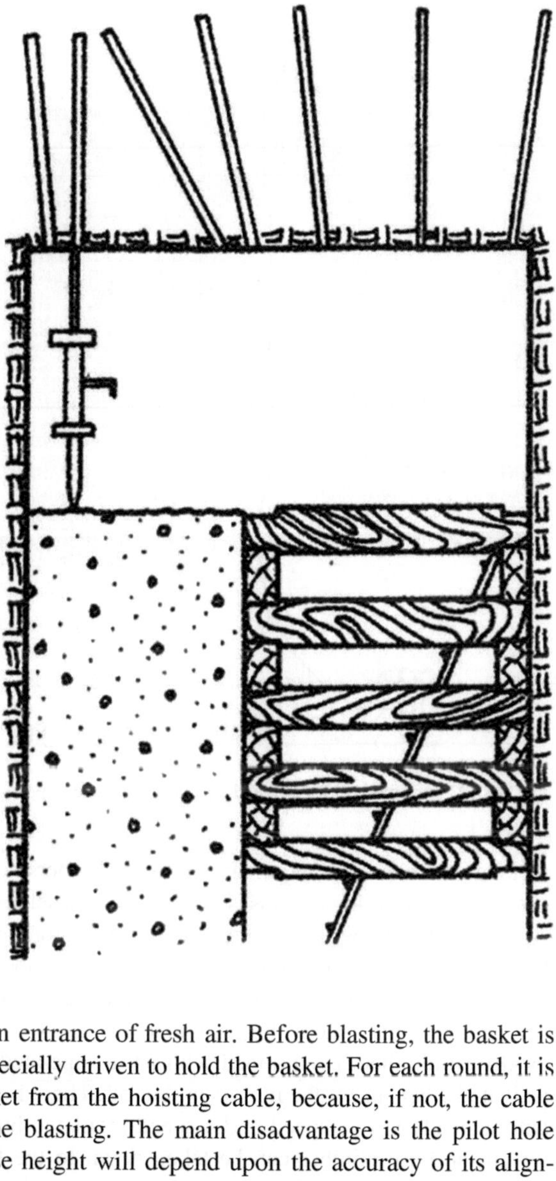

Fig. 16.11 The two-compartment raise method (reproduce from Ref. [7] with the permission from Taylor and Francis Book, UK

acts as a free face and also an entrance of fresh air. Before blasting, the basket is lowered down in an access specially driven to hold the basket. For each round, it is necessary to remove the basket from the hoisting cable, because, if not, the cable would be damaged during the blasting. The main disadvantage is the pilot hole drilling, as the maximum raise height will depend upon the accuracy of its alignment. Its practical and economical field of application is between 30 and 100 m.

16.3.2.3 Alimak Raise Climber Method

Alimak Company (Sweden) introduced this technique in 1957, and even today, it is often used in driving blind raises which have long lengths. It makes it possible to drive very long raises, vertical or inclined, straight or curved, and mostly

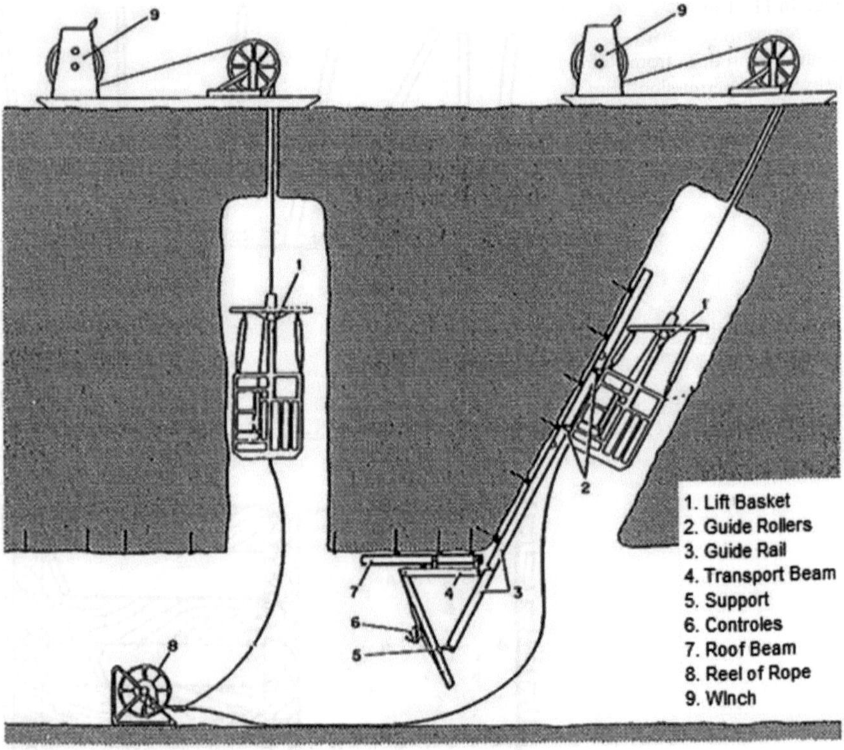

Fig. 16.12 Jora method for vertical and inclined raises, reproduced by permission of Atlas Copco

rectangular shape. Air motor-driven raise climbers normally are used in mines with short raises. For longer raises, electric or diesel hydraulic drives are used. The Alimak raise climber is designed entirely in accordance with the instructions of the Swedish Board of Industrial Safety. There are good margins of safety concerning the breaking strength of the material, and the raise climber comprises safety devices and features that practically eliminate the risks of accidents. The drive gear of air-driven climbers is equipped with an air-operated brake which is automatically activated when air for the motors is shut off. There is a safety device which activates automatically at overspeed and a speed-regulated brake for descent by gravity, by which the raise climber can be taken down to the bottom of the raise in case of cutoff of the air supply. The raise climber climbs along a pin rack welded to a guide rail which also comprises pipes for air and water. The guide rail can be extended by using 1- or 2-m (3.2 or 6.4-ft) sections as the driving progresses. Each guide rail section is bolted to the rock wall by special expansion bolts of Alimak design. Men ride up to the face in the cage, traveling comfortably and easily while material is transported on the platform. There is no danger in climbing up ladders. The men are well protected during ascent and arrive in good condition at the face. The greatest advantages of raise climber systems are flexibility and versatility. The equipment provides a comfortable working position in vertical as well as inclined raises of any area and length (Fig. 16.13).

The Alimak raising method consists of five steps, which together make up a cycle (Fig. 16.14). The five steps are all dependent on the raise climber, which serves both as a working platform and a means of transport up to the work site. The five steps are as follows:

1. Drilling—drilling is done from a platform on the raise climber. The platform is adapted to fit the size and shape of the raise required.
2. Loading—loading is also done from the platform. Explosive charges are placed in the holes by hand.
3. Blasting—before blasting, the raise climber is moved down to the nest for protection against falling rocks. Then, the blast can be triggered from a well-protected spot close by.
4. Ventilation—noxious gases and dust created by the blast are cleared by spraying a mixture of water and air from the top of the guide rail.

Fig. 16.13 Drilling on Alimak platform (reproduced by permission of Atlas Copco.)

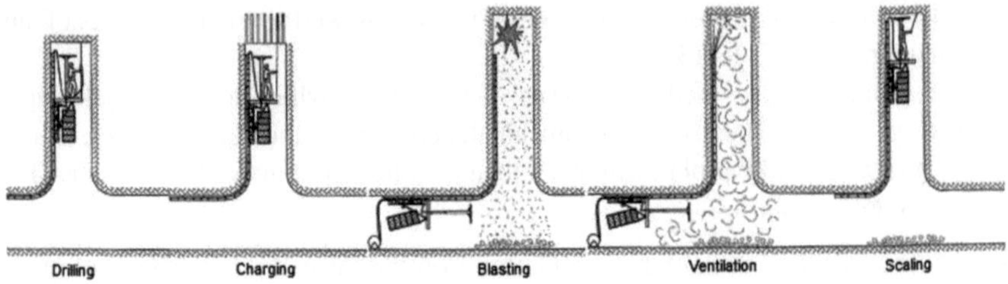

Fig. 16.14 Working cycle of Alimak raise climber method (reproduced by permission of Atlas Copco.)

5. Scaling—scaling is also done from the platform under the protective canopy.

The Alimak raising method can be used for any length and inclination raises in a safe working condition. The Alimak method is a relatively inexpensive alternative for construction sites that have a few variable length raises.

16.3.2.4 Long Hole Blasting Method

The long hole method is suitable for raises with more than a 45° inclination (sufficient for rock removal). Maximum raise length, normally from 10 to 60 meters, depends on drilling accuracy, hole alignment, and geology. For successful blasting, maximum hole deviation should not exceed 0.25 m (10″).

Excavation via the long hole method starts by drilling the blastholes to the total whole length of the raise from the top level and later firing the blasts in stages from the bottom-up with hanging charges. Usually, a relief hole with larger diameter (100–200 mm) is favorable (Fig. 16.15).

16.4 Explosive Charging

16.4.1 Manual Charging

When cartridge explosive is used for underground blasting, manual charging is the most common method. Workers use a wooden or plastic stick to push the explosive cartridge, primer cartridge first, into the blasthole one by one and tamp them with the stick (Fig. 16.16).

For blasthole stemming, wet newspaper (or cardboard), Hessian bags, and soil are usually used. For better stemming effect, crushed rock stemming (circa 10 mm) in thin plastic sausages is recommended as crushed rock stemming will 'lock' in the blasthole due to its angular shape. For the holes which are inclined upward or

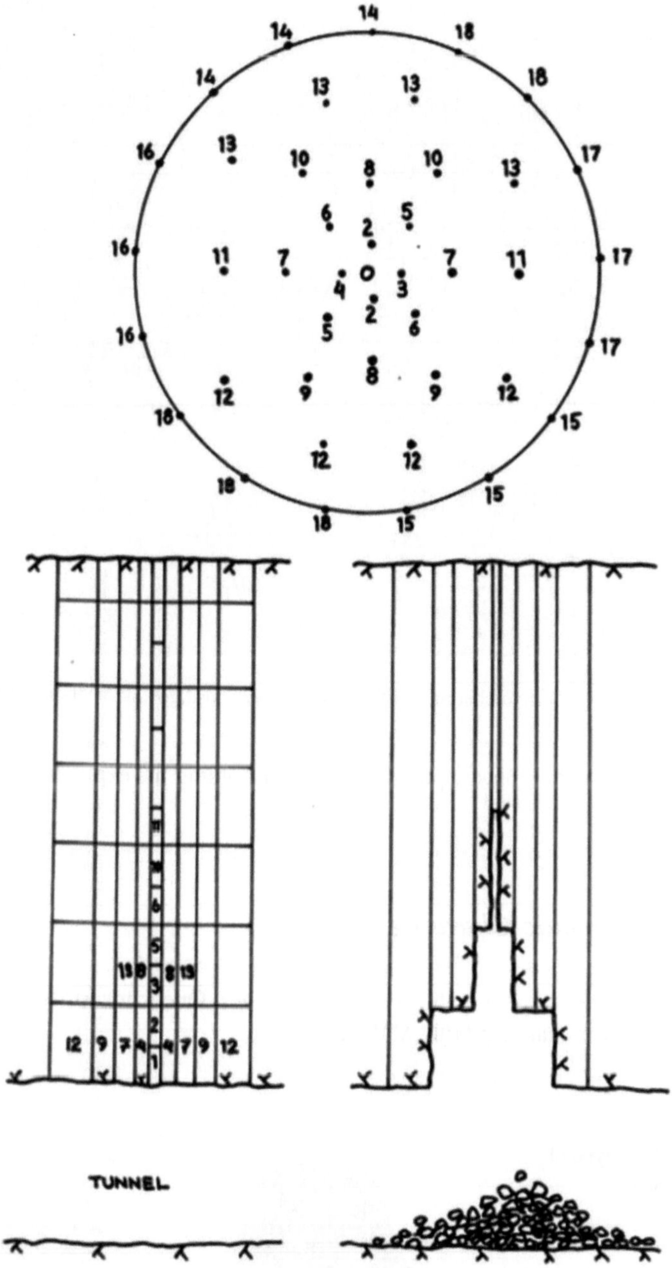

Fig. 16.15 Drilling pattern and working sequence of long hole blasting method in raise excavation. Reproduced from Ref. [5] by permission of Sandvik

Fig. 16.16 Manual explosive charging in face. *Source* www.dsd.gov.hk/others/HKWDT/

vertical from the collar, plastic plugs are used for holding the explosives in position (see Fig. 16.17).

16.4.2 Pneumatic Charging

There are two kinds of pneumatic charging machines: One is for cartridge charging, and another is for charging bulk ANFO.

Fig. 16.17 Two type of blasthole plugs

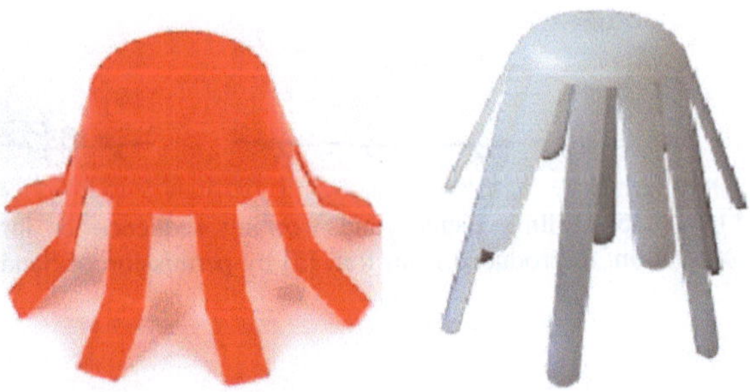

For preventing the generation of static electricity which may accumulate in sufficient amount to cause premature detonation of the priming charge, according to AS 2187.2006, where pneumatic charging devices are used, they shall be efficiently electrically earthed, and semiconductive (antistatic) loading tubes shall be used. Such tube shall have a resistance of not less than 15×10^3 Ω/m and shall be such length that its resistance is not more than 2 MΩ. The resistance shall be measured as described in BS 2050.

- Pneumatic cartridge loader

Pneumatic cartridge charging machine was developed in Sweden during the 1950s. These machines allow the charging of blastholes with diameters between 35 and 100 mm, obtaining a 15–20 % increase in loading densities comparing with manual tamping. Figure 16.18 shows the pneumatic cartridge loader manufactured by Can-Blast Inc. But in recent years, it is seldom used in tunnel blasting due to the increasing use of pumpable emulsion explosives.

- Pneumatic ANFO charging machine

Pneumatic ANFO charging machine is commonly used in tunneling. For horizontal and upward blastholes, the principal method of charging is via pneumatic charging devices. The principle of the charging machine is that the ANFO is transported from the container through a plastic hose, into the blasthole by pneumatic pressure. Figure 16.19 shows a pneumatic ANFO charging unit. Prilled ANFO can be charged in upward blastholes with an inclination of up to 35° without running out. The flow of ANFO is remotely controlled via charger. As ANFO is highly corrosive, all machine parts that come in contact with ANFO are made of stainless steel. The charging machine is a combined pressure/ejector unit for charging of prilled ANFO in blastholes with diameters between 32 and 51 mm and a depth of up to 45 m. The charging hose is antistatic as the ANFO is transported through the hose at high velocity causing a risk of static electricity

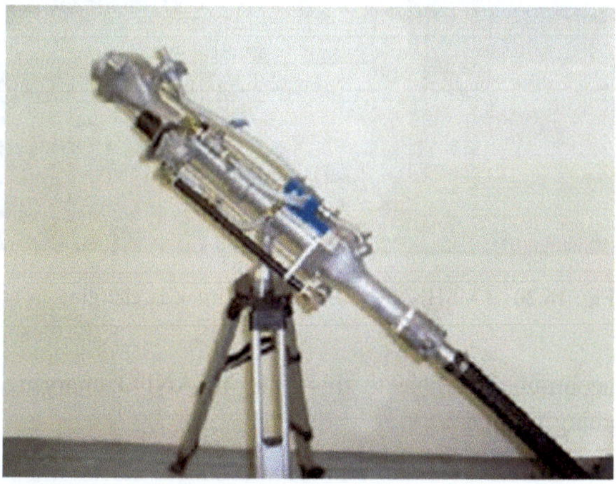

Fig. 16.18 Pneumatic cartridge loaders made by Can-Blast Inc. *Source* www.can.blast.com/

Fig. 16.19 Pneumatic ANFO face charging unit

Fig. 16.20 AMSHK's emulsion pump truck is charging in tunnel face (courtesy of AMSHK)

accumulation. Due to this risk, all ANFO charging unit must be earthed during charging operation.

16.4.2.1 Pumpable Emulsion Charging

Following the progress of explosive technique, bulk emulsion explosives are more widely used in tunnel blasting due to its outstanding advantages of water resistance, safety, and low cost. Similar to the bulk emulsion pump truck used in surface blasting, the sensitization process of emulsion matrix, which is contained in different tanks separate from the sensitizer in the truck, is performed during the charging. As the emulsion matrix and sensitizer are not explosive materials, the storage and transportation in site are safer than other explosives and blasting agents (Fig. 16.20). As the pumpable emulsion explosive is a product that is not detonator sensitive, mini boosters are needed to be used in blasting (Fig. 16.21).

Care must be taken when charging holes contains water. The charging hoes must be introduced to the bottom and is pulled out gradually during charging at the same rate that the hole is filled to avoid separation of the explosive column by water pockets.

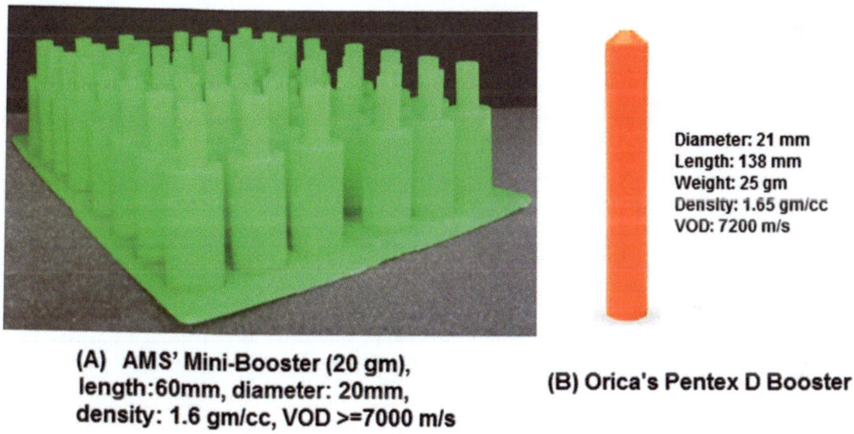

Fig. 16.21 Two types of mini boosters for tunnel blasting

References

1. Egger K et al (2004) Santiago's metro expands, NAT 2004, Atlanta GA. http://www.dr-sauer.com/resources/presentations-lectures/378. Accessed 25 May 2016
2. Chapman D et al (2010) Introduction to tunnel construction. Spon Press, London. ISBN 10: 0-415-46841-8
3. Berthelsen OJ (1992) Geoguide 4: guide to Cavern engineering. GEO, CEDD, Hong Kong
4. Hemphill GB (2013) Practical tunnel construction. Wiley, Hoboken. ISBN 978-0-470-64197-2
5. Matti H (1999) Rock excavation handbook for civil engineering. Sandvik & Tamrock, Danbury
6. Kolymbas D (2008) Tunnel and tunnel mechanics. Springer, Heidelberg. ISBN 978-3-540-25196-5
7. Jimeno CL et al (1995) Drilling and blasting in rock. A. A. Balkema, Rotterdam (Copyright ©1995 A. A. Balkema). ISBN 90 5410 199 7

Chapter 17
Contour Blasting for Underground Excavation

17.1 The Characters of Contour Blasting for Underground Excavation

In practice, over-break along the planned outline of the excavation occurs after blasting a round. This results in higher cost for supports, more mucking, and generation of cracks in the surrounding rock and exceptionally into roof falls.

For reducing the damage to surrounding rock beyond the excavation profile and also achieving a relative smooth surface of the roof and walls, smooth blasting as the main contour blasting technique is usually used in underground excavations, and for large tunnels or caverns, presplitting may be adopted.

The mechanism of smooth blasting and presplitting has been illustrated in Chap. 10, and only their application in underground excavation will be emphasized in this chapter.

Smooth blasting involves drilling a number of closely spaced parallel boreholes along the final excavation surface, placing low charge density-decoupled charges in these boreholes and detonating all of these charges together after detonation of the remainder of the blastholes in the face. The final face of rock is peeled off the undamaged final excavation surface by the smooth blasting (see Photo 17.1).

In most cases, the smooth blasting holes are drilled, charged, and fired in the same tunneling cycle as the cut and main blastholes. But in some cases, the cut and main blasts are carried out during one cycle and the smooth blasting is carried out during a later cycle, and especially, a pilot tunnel method is utilized for tunnel excavation.

In spite of lots of advantages and cost savings associated with utilizing smooth blasting techniques, if a rock mass is highly fractured or weathered, or if the joint orientation is unfavorable, the geology alone will control the result, and the smooth blasting is of limited or no value. The truth is that perimeter control procedures are even more important in poor quality rock and that good results can be obtained.

D. Zou, *Theory and Technology of Rock Excavation for Civil Engineering*,
DOI 10.1007/978-981-10-1989-0_17

Fig. 17.1. Tunnel rock face formed by smooth blasting

However, poor rock requires even closer spacing of perimeter and lighter loading [2].

17.2 Charge Calculation for Smooth Blasting

In smooth blasting, the spacing between the boreholes is usually 15–16 times the hole diameter and the burden (the distance between the boreholes and the free face created by the previous detonated blastholes (see Fig. 17.2)) is 1.25 times the spacing.

According to Svanholm et al. [1], the minimum linear charge concentration [charge weight (kg) per 1 m length of the borehole] for smooth blasting is given by

$$w = 90 * d^2 \tag{17.1}$$

Fig. 17.2. Drillhole pattern for smooth blasting

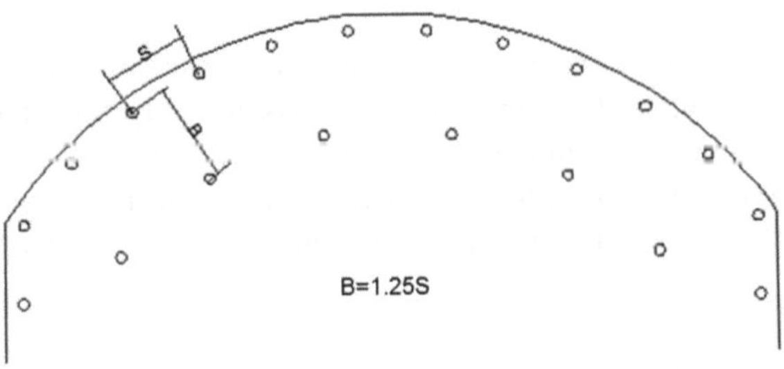

Table 17.1 Parameters of smooth blasting recommended by Svanholm et al.

Drillhole diameter (mm)	Charge diameter (mm)	Charge concentration kg (ANFO)/m	Burden m	Spacing (m)
25–32	11	0.08	0.30–0.45	0.25–0.35
25–48	17	0.20	0.70–0.90	0.50–0.70
51–64	22	0.44	1.00–1.10	0.08–0.90

Table 17.2 Suggested initial perimeter hole spacing, burden, and loading for perimeter control blasting in tunnel (reproduced from Ref. [2] Table 29.3, courtesy of ISEE)

Drillhole diameter (mm)	Rock quality	Burden feet (m)	Spacing feet (m)	Approximate linear charge concentration Ibs/ft (kg/m)
1.0–1.5 (25–38)	Good	2.00 (0.61)	1.5 (0.46)	0.10 (0.15)
	Poor	1.3 (0.40)	1.0 (0.31)	0.5 (0.07)
1.5–2.0 (38–50)	Good	2.75 (0.83)	2.0 (0.61)	0.15 (0.22)
	Poor	2.25 (0.69)	1.5 (0.46)	0.08 (0.12)
2.0–3.0 (50–70)	Good	3.5 (1.02)	3.0 (0.91)	0.30 (0.45)
	Poor	3.0 (0.91)	2.0 (0.61)	0.15 (0.22)

where

w Linear charge concentration of ANFO-equivalent explosive in kg/m;
d Borehole diameter in meters.

Svanholm et al. gave the following recommendations for smooth blasting in underground excavation:

As stated in previous sections, rock quality seriously affects the result of smooth blasting. Table 17.2 provides the recommendations for perimeter hole spacing and burden based on rock quality [2]. These recommendations are generalized conditions and should be considered ranges of spacing and loadings which should be adjusted based on preliminary results, as rock conditions change. Using the light loading recommended, good results, with half cast factors of 50–90 %, should be achieved, even in relatively fractured bedrock (Table 17.1).

17.3 Blasthole Charging for Smooth Blasting

There are several methods to charge the smooth blastholes:

- Small-diameter cartridges separated from each other and connected with a light detonation cord. For easy operation, the charges often are tied to a bamboo piece (Fig. 17.3).
- Two heavy-loaded detonating cords with a cartridge in the hole bottom. Usually, according to the quality of the rock, two lines of detonating cords with strengths

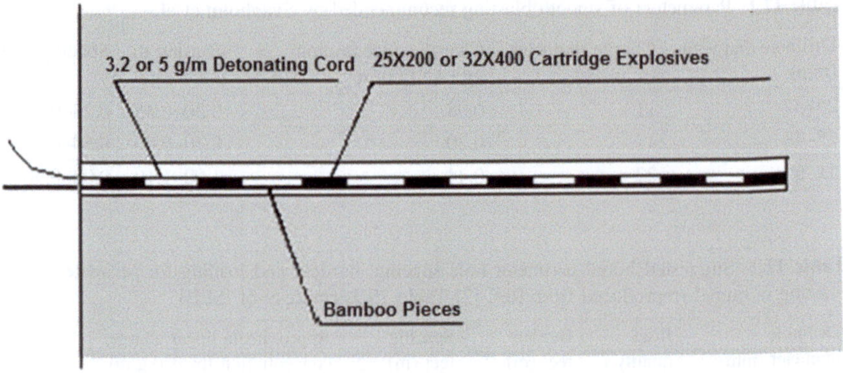

Fig. 17.3. Loading smooth blastholes with small-diameter cartridge explosives

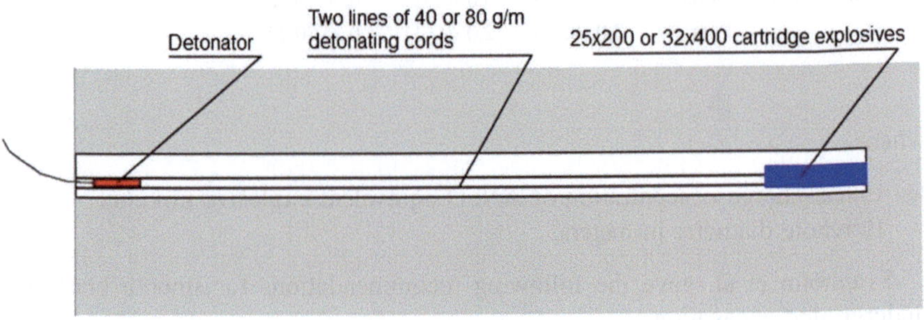

Fig. 17.4. Detonating cords used as the charge for smooth blasting

of 40 or 80 g/m together with one cartridge of 25 × 200 or 32 × 400 emulsion explosives placed at the bottom of the holes are used for smooth blasting for underground excavation (Fig. 17.4).

- Special explosives for smooth blasting. To charge the smooth blastholes, special explosives are manufactured for this purpose by explosive suppliers. Figures 17.5 and 17.6 are examples.
- Charging with lightly loaded bulk explosives. When bulk emulsion explosives are used for the perimeter holes for smooth blasting, the charging tube is dragged out faster than for the main blastholes, using the charging machine or manually, to ensure the emulsion explosive remaining in the hole only occupies around 1/4 of the hole volume (Fig. 17.7).

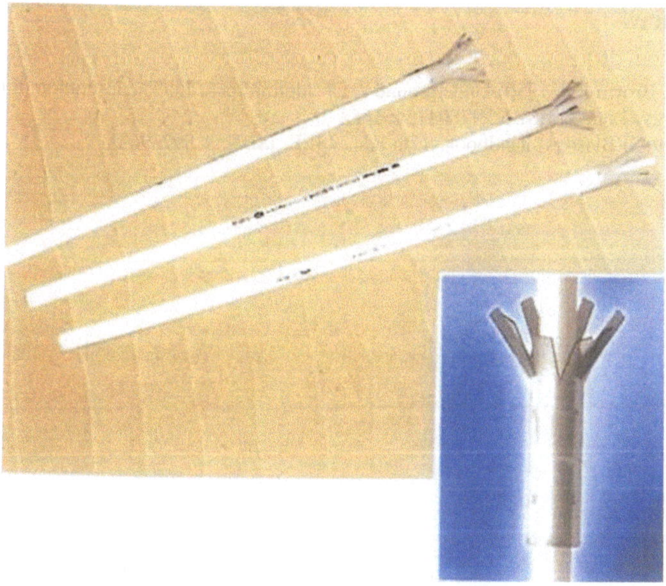

Fig. 17.5. Explosive in small-diameter rigid plastic pipes for smooth blasting

Fig. 17.6. Orica's Senatel ™ Powersplit ™ detonator sensitive, cord traced emulsion explosive for decoupled charging

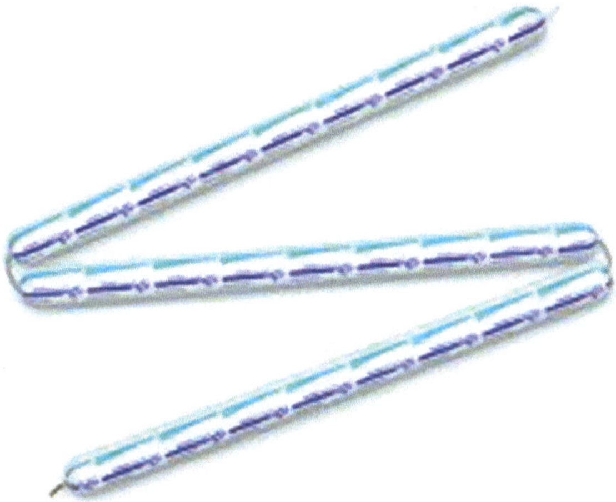

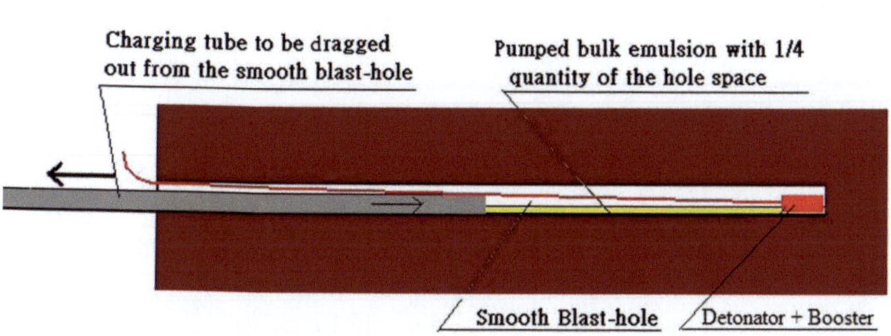

Fig. 17.7. Charging smooth blastholes with bulk emulsion explosives

References

1. Hoek E, Brown ET (1980) Underground excavation in rock. In: The institution of mining and metallurgy. London. ISBN: 9780415461658
2. ISEE (1998) Blasters' handbook, 17th edn., ISEE. ISBN: 1-892396-00-9

Chapter 18
Blasting Design for Underground Excavation

18.1 Blasting Design for Tunnel (Cavern)

18.1.1 Hole Layout and Firing Sequence

The blasts in tunnels are characterized by the initial lack of an available free surface toward which breakage can occur, only the tunnel heading itself. The principle behind tunnel blasting is to create an opening by means of a cut, and then, stoping (reliever) is carried toward the opening. For easy illustration of the blasting procedure of tunneling, we divide the tunnel face into five separate zones, A–E (see Fig. 18.1):

A the cut section,
B the stoping holes breaking horizontally,
C the stoping holes breaking downward,
D contour holes, and
E the lifter holes.

The general firing sequence must be A > B > C > D > E.

The most important operation in the blasting procedure of tunneling is to create an opening in the face in order to develop another free face in the rock. This is the function of the cut holes. If this stage fails, the round can definitely not be considered a success.

18.1.2 Types of Cut-Hole Pattern

The cut-hole pattern can be classified in two large groups:

- angled hole cuts and
- parallel-hole cuts.

D. Zou, *Theory and Technology of Rock Excavation for Civil Engineering*,
DOI 10.1007/978-981-10-1989-0_18

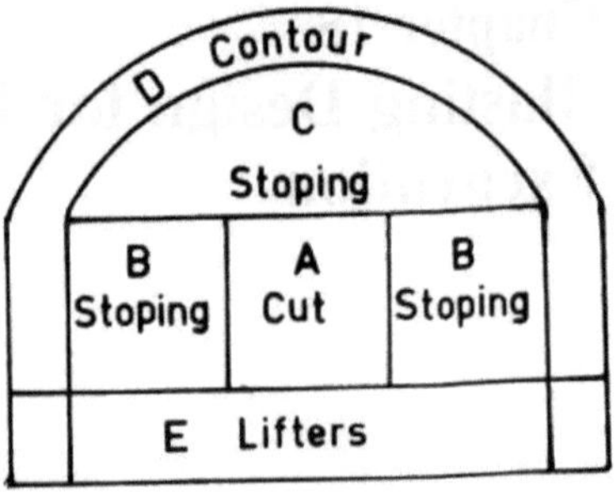

Fig. 18.1 Zones in tunnel blasting

The most common type of cuts today are the parallel-hole cuts and angled hole cuts. The angled hole cut is the old type and is still occasionally used in construction. It is quite an effective type of cut for tunnels with a fairly large cross section, and it requires fewer holes than a parallel cut.

The parallel-hole cut was introduced when the first mechanized drilling machines came to the market and made accurate parallel drilling possible.

18.1.2.1 Angled Hole Cuts

The angled cut is a traditional cut type based on the symmetrically drilled, angled holes. It has lost some of its popularity with the widespread adoption of the parallel cut and longer rounds. However, it is still commonly used in wide tunnels where the tunnel width sets no limitations on drilling. The advantage is a lower drilling length and explosive consumption than the parallel cut because there is better utilization of the free face surface and the possibility of orientation toward the visible discontinuities in the section. But its biggest disadvantage is that it ejects rock violently, the rock is thrown a considerable distance resulting in services being destroyed, e.g., ventilation, power, air, and communications, and its use is banned in some places.

The following explains the most common angled cuts.

- V-cut

V-cut is also called wedge cut. With this angular cut, a wedge is detonated out of the center of the face and after that, the remaining part of the advance length is detonated. The V-cut can be positioned vertically or horizontally (or at an angle depending on the layering of the rock mass) as a single or staged wedge. Figure 18.2 shows a horizontal V-cut.

The angle at the bottom of the cut holes should not be less than 60°. Maintaining the right angle is the main difficulty in V-cut drilling; in addition, the correct drilling angle limits the round length in narrow tunnels (Fig. 18.3). Tunnel width limits the use of the V-cut. In narrow tunnels, the advance per round can be less than one-third of the tunnel width.

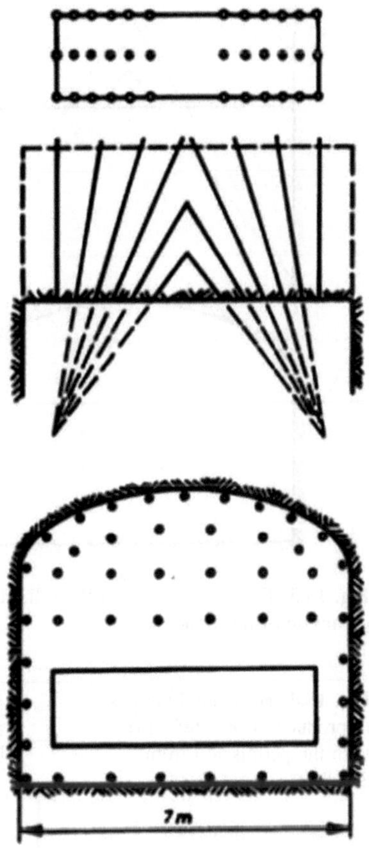

Fig. 18.2 Horizontal "V"-cut blast (reproduced from Ref. [14] with the permission from Sandvik)

- Fan Cut

The fan cut is also an angled cut. For this arrangement, several drillhole rows are placed in a fan shape. They have different lengths and angles. Figure 18.4 shows a fan cut. This type of cut was widely used before, but it is not favored nowadays because of the complicated drilling.

18.1.2.2 Parallel-Hole Cut

Parallel-hole cuts are also called cylindrical cuts. The characteristic for this cut is that the drillholes are the same length and obviously parallel to each other. At the moment, this type of cut is the most frequently used in tunneling and cavern blasting, regardless of their dimensions. This type of cut consists of one or several uncharged or relief blastholes toward which the charged holes break at intervals.

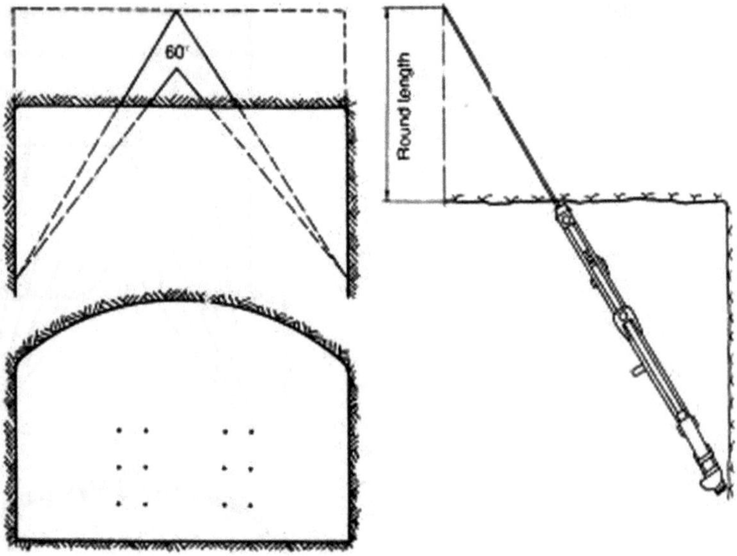

Fig. 18.3 Feed setup and drilling limitation in V-cut (reproduced from Ref. [14] with the permission from Sandvik)

Fig. 18.4 Horizontal fan cut (reproduced from Ref. [14] with the permission from Sandvik)

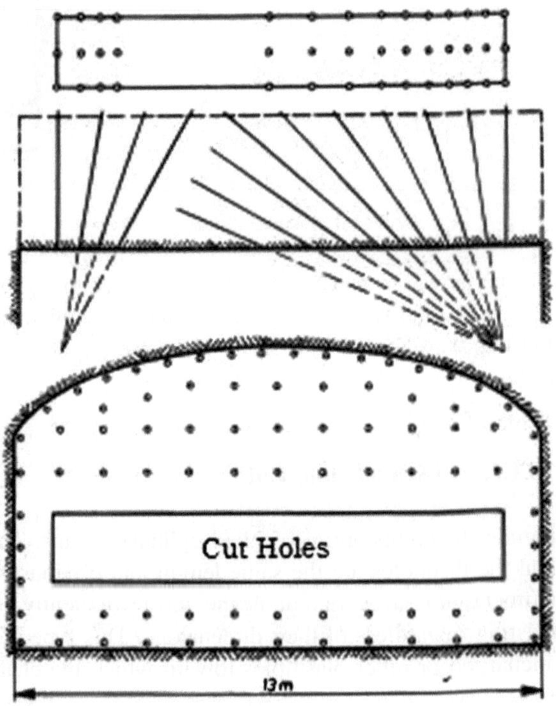

The principal advantages of the parallel-hole cuts are as follows:

1. The depth of the round is not dependent on the working space available for drilling holes at an angle.
2. The cut allows a deep pull even in tough rock formations.
3. It is relatively simple to drill, because all holes are parallel.
4. There is generally less throw with better fragmentation.
5. The resultant muckpile is higher, so it provides a better platform for scaling and bolting work.
6. Round length may be shortened or lengthened without any difficulties.

Principal disadvantages of the parallel-hole cuts are as follows:

1. If large relief holes are used, it requires reaming or larger drilling equipment.
2. Drilling and explosives requirements (powder factor) are higher.
3. Drilling must be accurate; otherwise, the results will be unfavorable.

There are two kinds of parallel-hole cuts: burn cuts and parallel-hole cuts with large empty hole(s). They will be discussed in separate sections later.

Burn Cuts

Burn cuts are also called Swedish cuts as they were first used in Sweden. In this cut, all the blastholes are drilled parallel and with the same diameter. Some are charged with a large quantity of explosive, while others are left empty. The empty holes provide a free face for reflection of shock waves. It is important that these holes are accurately drilled and parallel to each other in order to achieve a good blasting result. Figure 18.5 shows some hole patterns of burn cuts.

As the charge concentration is high, the fragmented rock is centralized in the far end of the cut and is difficult to breakout, so that the advance is reduced and does not surpass 2.5 m per round.

Parallel-Hole Cuts with Large Empty Hole(S): Cylinder Cuts

The difference with the burn cut is that the uncharged or relief hole(s) is (are) larger than the charged holes. The large diameter blastholes (76–175 mm) are drilled with reamer bits which are adapted to the same drill steel which is used to drill the rest of blastholes.

All the blastholes in the cut are placed with little spacing, in line and parallel. Figure 18.6 shows some parallel-hole cuts with large relief holes.

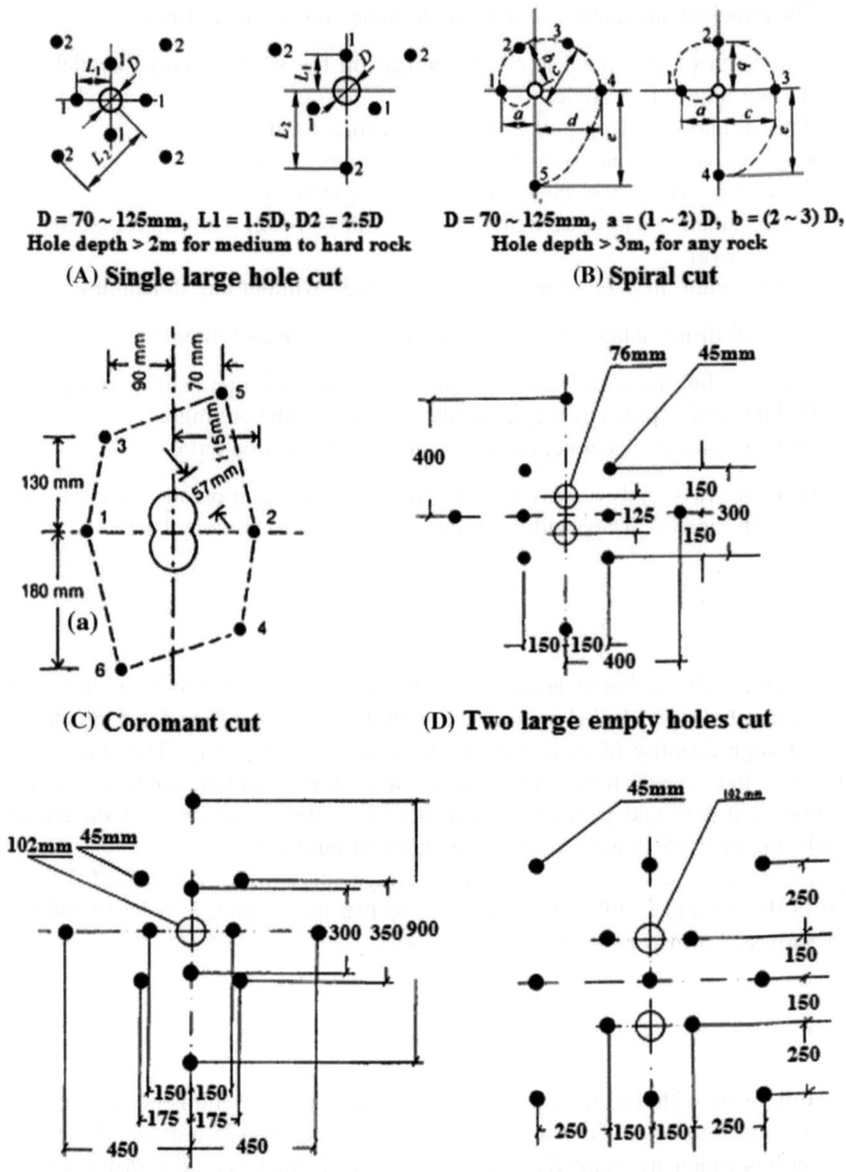

Fig. 18.5 Examples of parallel-hole cuts with large relief hole(s)

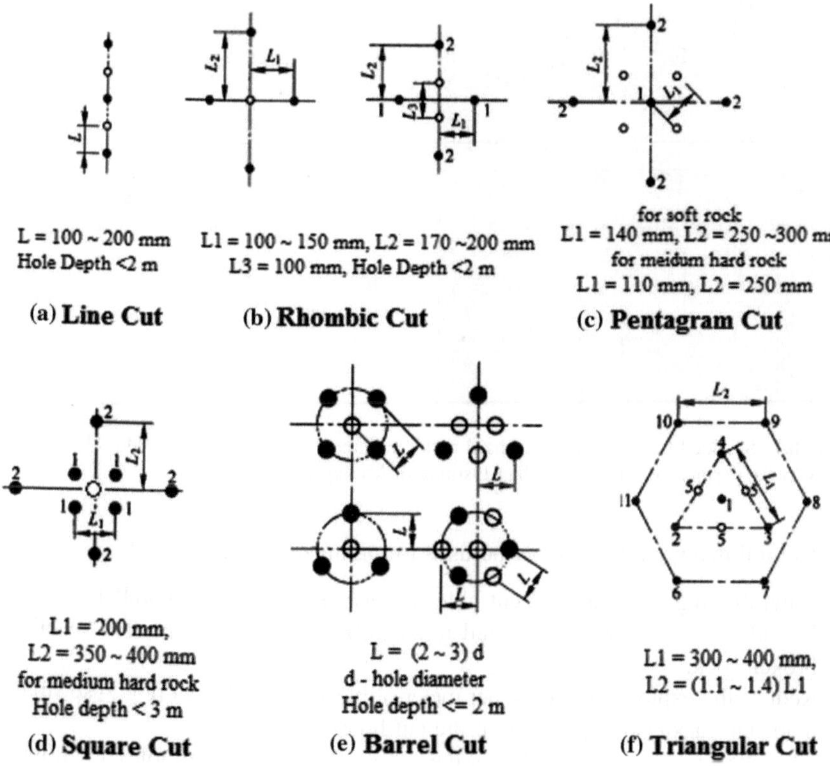

Fig. 18.6 Examples of burn cut

18.1.3 Some Important Issues on Cut Holes and Tunnel Blasting

18.1.3.1 The Concept of Cut "Freezing"

As the cut holes are tightly spaced and overloaded, some of the problems that can rise in blasting with parallel-hole cuts are sympathetic detonation and dynamic pressure desensitization. Any one of these two problems can cause the cut to "freeze" and fail. The first phenomenon can appear in a hole that is adjacent to the detonating hole when the explosive used has a high degree of sensitivity, such as those with nitroglycerine in their composition. On the other hand, the dynamic pressure desensitization takes place in many explosives, and water-based emulsion or water gel explosives are most susceptible to dead-pressing failure when they are used in closely spaced holes because the shock wave of a charge can elevate the density of the adjacent charge above the critical or death density (refer to Chap. 3, Sect. 3.7.8).

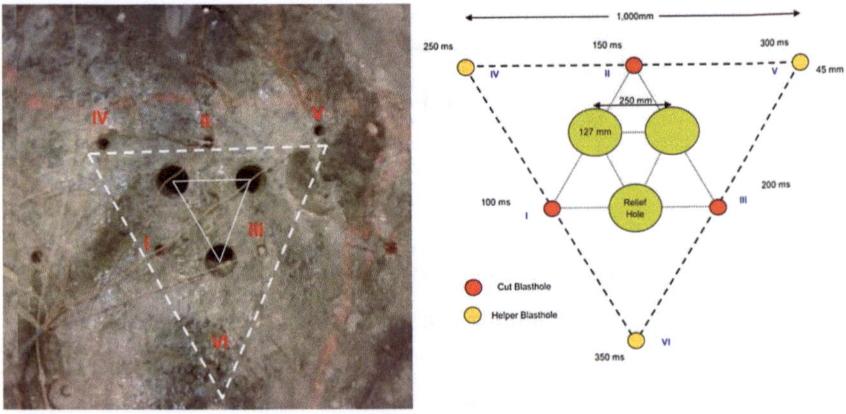

Fig. 18.7 Shielded burn-cut layout in hard rock recommended by Hagan (reproduced from Ref. [1] with the permission from Australian Institute of Mining and Metallurgy)

The following guidelines can be used to minimize cut "freezing":

1. Carefully align and drill all cut holes to ensure that they are parallel.
2. Provide more or large uncharged relief holes to accommodate the "swell" of broken rock.
3. Reduce the explosives energy per meter of blasthole in the cut (e.g., use smaller diameter packaged explosives in the cut holes).
4. Alter the geometry and spacing of the cut blastholes and relief holes, to allow for changes in ground conditions.
5. Ensure that blastholes in the cut area fire in a controlled sequence, with adequate time between successive detonations.

Figure 18.7 is suggested by Dr. T. N. Hagan [1] for eliminating sympathetic detonation and dynamic pressure desensitization.

If the cut fails due to precompression, try spreading the loaded holes farther apart. Adding more tightly spaced holes will aggravate the problem and unnecessarily increase the costs. In soft or seamy rock, adjacent loaded holes that are separately delayed should be at least 30 cm apart.

To pull rounds deeper than 2.5 m, use parallel-hole cut designs with an adequate volume of uncharged relief holes. In rounds exceeding 2.5 m, the relief-hole area should be at least 25 % of the total area in the immediate cut.

To further aid rock ejection from the burn, a small "kicker" or "booster" charge can be placed at the bottom of the normally empty void holes. These charges are delayed to fire just after the other fully loaded holes in the immediate cut have fired.

Generally, long-period delays are used to ensure that there is sufficient time for the rock from each hole, or group of holes, to break and be ejected from the cut, before subsequent holes fire. Some amount of bootleg normally occurs in the cut area. When this occurs, a similar or greater length of advance is usually lost in the rest of the face area. To minimize this lost advance, when drill steel length allows it, the cut holes should be overdrilled by 15–30 cm (6–12 in.). If this extra drilling for a few holes returns an equal length of advance for the whole face, it is well worth the investment in extra drilling.

18.1.3.2 Stemming

Stemming refers to an inert material that is placed in the borehole between the top of the explosive column and the collar of the hole.

There are two completely opposite views on the roles of blasthole stemming.

One view is that it is unnecessary to use stemming for blastholes in tunnel blasting. They said: "It has been discovered that the explosive used in tunneling the stemming does not have an improved effect for a long charged column. The inertia of the molecules within the air column in the drillhole (in relation to the very high detonation speed) is sufficient to act as the stemming." "Therefore the decision is often taken not to use stemming, in order to save cost and time" (refer to 5.6.4 Stemming on p. 169 of [2]).

But most practice has shown that it is necessary to confine the charge to localize the effect of the gases produced by the explosive reaction. When a stemmed blasthole detonates, the stemming momentarily retains the energy within the blasthole, only a few milliseconds but that is sufficient. When a blasthole is improperly stemmed or not stemmed at all, the resulting action is called "rifling." Rifling gets its name from the action of blowing the stemming material, much like a bullet from a rifle. The action of rifling can be flyrock, increased air overpressure, poor fragmentation, and boulders.

Stemming material can consist of sand, drill fines, gravel, or pea stone; sandy clay is usually used. Sometimes, wet cardboard, wet paper, or hessian bags are used as the stemming material, but practice shows they are not effective.

Practice has shown that using stemming of crushed rock granular with a blasthole-to-size ratio of 17:1 (refer to [3]) in thin plastic sausages can get the best result as the granular rock can "lock" in the blasthole due to its angular shape. Table 18.1 (refer to Table 5.5, p. 151 in [3]) gives the idea crushed rock sizes based on the blasthole diameter.

Table 18.1 Common stemming sizes based on hole diameter (reproduced from Ref. [3] with the permission from John Wiley & Sons Ltd)

Hole diameter (mm)	Hole diameter (in)	Size of stemming	Size of stemming
38	1.5	10 mm minus chips	3/8 in minus chips
50–90	2–3.5	10–13 mm chips	3/8–1/2 in chips
100–127	4–5	16 mm chips	5/8 in chips
127 and above	5 and above	19 mm minus chips	3/4 in minus chips

Source: From Atlas Powder Company (1987)

Generally, the amount of stemming material required will range from 0.7B to 1B, where B represents the burden or 10 times of the blasthole diameter. The rock characteristics affect the amount of stemming material, and more stemming amount is required in a highly fractured rock mass.

18.1.3.3 Lookout Angle

To maintain the designed cross section from one round to another, the contour holes have to be angled outside the designed cross section. This is to make sure that when drilling the next round, there is required space for the rock drill. If the contour holes were drilled parallel to the designed line of the tunnel, the tunnel face would get smaller and smaller after each round. The lookout is depending on the equipment used, but normally amounts to not less than 0.1–0.2 m (see Fig. 18.8). Modern drilling rigs have electronic or automatic lookout angle indicators that enable correct adjustment of the lookout angle relative to standard alignment. Computerized drilling jumbos make setting, adjustment, and monitoring of the lookout angle even easier.

18.1.4 Parallel-Hole Cut Design: Cylinder Cuts

The parallel-hole cut has a large number of minor variations, but the basic layout always involves the drilling of one or several uncharged large diameter holes at or very near the center of the cut. These holes give empty space for the adjacent blasted holes to swell into. As the drilling equipment has become more and more powerful, this type of cut has become more and more used. The large diameter holes (65–175 mm) are drilled with reamer bits which are adapted to the same drill steel which is used to drill the rest of blastholes.

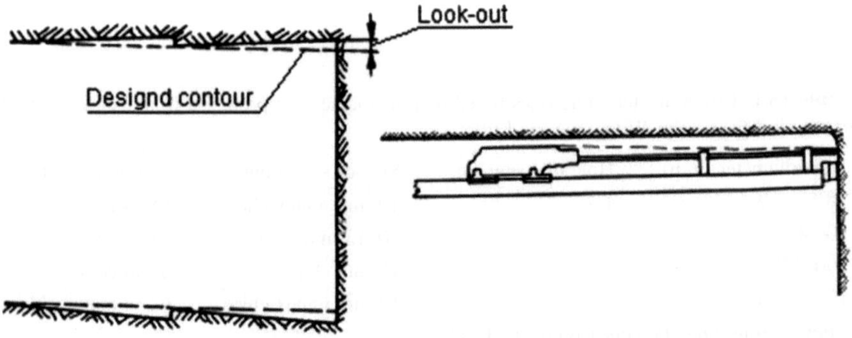

Fig. 18.8 The lookout angle (reproduced from Ref. [14] with the permission from Sandvik)

18.1.4.1 Large Hole Diameter and Number

It is proved that the diameter of the large hole should be a function of the blasthole depth used. In order to reach an acceptable advance/round 95 % of the blasthole depth, an "equivalent" hole diameter, d_f, can be calculated:

$$d_f \approx (3.2 \times l)^2 \tag{18.1}$$

where

d_f equivalent hole diameter, in mm;
l blasthole depth, in m.

Individual hole diameter now can be calculated as follows:

$$d_l = d_f / \sqrt{n} \tag{18.2}$$

where

n number of large holes.

18.1.4.2 Spacing of Cut Holes

The charged holes closest to the large hole(s) are named "cut holes."

Generally, the center-to-center distance between the empty large hole and the cut hole should be about 1.5 times the diameter of the large empty hole. That means the burden of the cut holes, v, should be as follows:

$$v = 1.5 \times d_f = 1.5 \times d_l \times \sqrt{n} \tag{18.3}$$

For two or more large holes, v is calculated for the holes marked in black in Fig. 18.9 and the remaining holes are added in order to achieve a square system.

U. Langefors and B. Kilhström indicated that v should not be more than 1.7 d_f to obtain fragmentation and a satisfactory movement of the rock [4]. The conditions of fragmentation vary greatly depending upon the type of explosive, rock properties,

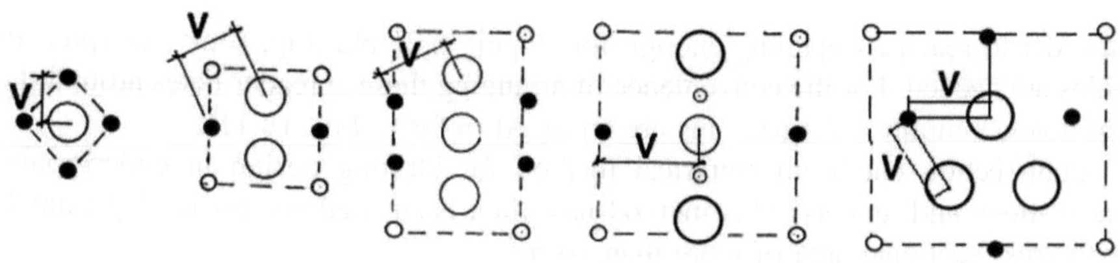

Fig. 18.9 Burden of cut holes

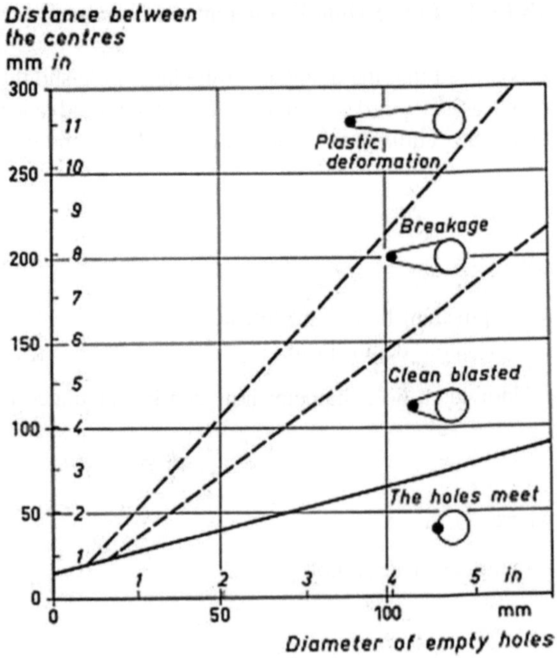

Fig. 18.10 Result when blasting toward an empty hole at different distance and dimension of the empty hole (reproduced from Ref. [4] with the permission from John Wiley & Sons Ltd

and the distance between the charged blastholes and the large empty hole. As reflected in Fig. 18.10, for burdens larger than 2 d_f, the break angle is too small and a plastic deformation of the rock between the two holes is produced.

It is quite obvious that the accuracy when drilling these holes is extremely important. When drilling deviation is more than 1 %, the practical burden is calculated from:

$$v_1 = 1.7d_f - E_p = 1.7d_f - (\alpha \times L + e') \tag{18.4}$$

where E_p is the drilling error (m), α is the angular deviation (m/m), L is the blasthole depth (m), and e' is the collaring error (m) (from [5]).

18.1.4.3 Four-Section Design for Parallel-Hole Cut

In order to reach an opening enough for stoping hole blasting, some cut spreader holes are needed. For the convenience of arranging these spreader holes around the cut holes, a number of quadrants are arranged (refer to Fig. 18.11).

Four-section cut is an empirical method for blasting design in underground excavations and tunnels. This method has often been used for excavating tunnels with cross-sectional area of more than 10 m^2.

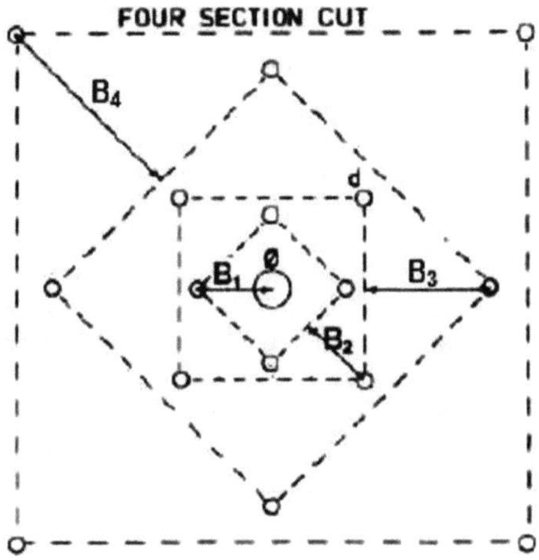

Fig. 18.11 Four-section cut method

Table 18.2 Equation for blasting pattern design of four-section cut model (refer to [15])

Section	Burden (B)	Spacing (S, X)	Stemming (S_t)
First square cut	$B_1 = 1.5\emptyset_{e2}$	$X_1 = B_1\sqrt{2}$	$S_{t1} = B_1$
Second square	$B_2 = B_1\sqrt{2}$	$X_2 = 1.5B_2\sqrt{2}$	$S_{t2} = B_1\frac{\sqrt{2}}{2}$
Third square	$B_3 = 1.5B_2\sqrt{2}$	$X_3 = 1.5B_3\sqrt{2}$	$S_{t3} = \frac{\sqrt{2}}{2}(B_1\frac{\sqrt{2}}{2} + B_2)$
Fourth square	$B_4 = 1.5B_3\sqrt{2}$	$X_4 = 1.5B_4\sqrt{2}$	$S_{t4} = \frac{\sqrt{2}}{2}(\frac{\sqrt{2}}{2}\left(B_1\frac{\sqrt{2}}{2} + B_2\right) + B_3)$

The four-section cut is based on the parallel-hole cut. This model started with Lagnefors and Kihlström in 1963 [5] and has been further developed afterward.

The method suggests the experimental equations listed in Table 18.2. In this table, X is the length of each quadrangle side (see Fig. 18.11).

Four-section cut method includes an empty large hole in the center. If the number of empty holes is more than one, equivalent diameter is calculated by the Eq. (18.2), i.e., $d_f = d_l\sqrt{n}$. In Table 18.2, $\emptyset_{e2} - d_f$ is the diameter of the equivalent hole.

18.1.5 Blasthole Pattern for Stoping

The object of the cut is to create a free surface toward which the rest of the blasting can be carried out. The purpose of stoping holes is to attain just as large an advance with the remainder of the round as created by the cut holes, to get a satisfactory fragmentation, and to get a suitable disposal of the broken rock. At the same time, the remaining rock face should also be left undamaged.

Table 18.3 Concentration of the charge in kg/m for various burden (*V*) and extension (*B*) of the free face (preliminary table; reproduced from Ref. [4] with the permission from John Wiley & Sons Ltd)

Max. burden (V)		Extension (*B*)											
		m	0.10	0.15	0.20	0.25	0.30	0.35	0.40	0.50	0.60	0.80	1.4
		ft	0.3	0.5	0.7	0.8	1	1.2	1.3	1.7	2	2.7	4.7
m	ft	Concentration of the charge, kg/m (×2/3 for lb/ft)											
0.10	0.3		0.12	0.08	0.06								
0.15	0.5		0.30	0.18	0.13	0.11	0.09						
0.20	0.7		0.60	0.35	0.24	0.20	0.16	0.14	0.12				
0.25	0.8		1.0	0.60	0.35	0.30	0.26	0.22	0.18				
0.30	1		1.3	0.9	0.60	0.50	0.35	0.31	0.26	0.22	0.18		
0.35	1.2			1.2	0.9	0.65	0.45	0.40	0.35	0.30	0.25		
0.40	1.3			1.6	1.2	0.9	0.7	0.6	0.50	0.40	0.30	0.24	
0.50	1.7				2.0	1.6	1.3	1.0	0.7	0.60	0.50	0.36	
0.60	2					2.2	1.9	1.6	1.3	1.0	0.7	0.52	
0.70	2.3						2.5	2.2	1.8	1.3	0.9	0.7	
0.80	2.7							3.2	2.4	1.8	1.4	1.0	0.6
1.00	3.3								4.0	3.0	2.4	1.4	0.9
1.20	4									4.4	3.8	2.5	1.2
1.40	4.7										5.0	3.6	1.6
1.60	5.3											4.8	2.4
2.00	6.7												4.0

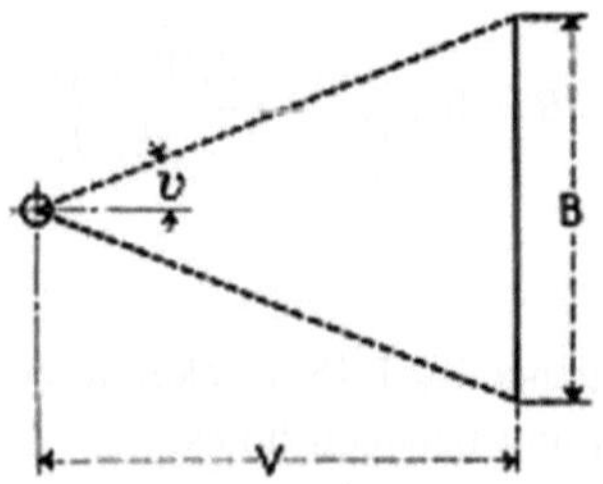

Fig. 18.12 Blasting toward a narrow opening

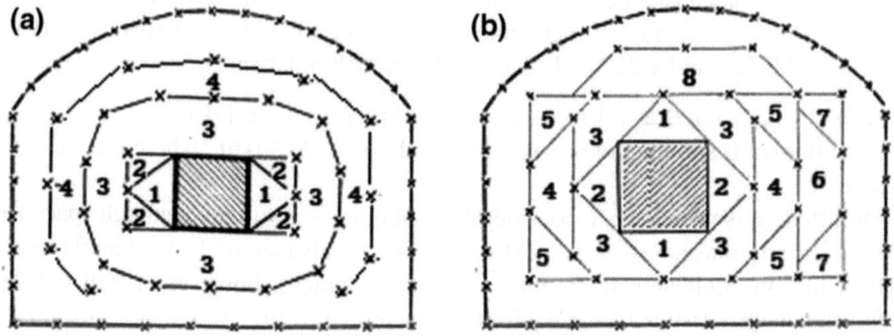

Fig. 18.13 Construction of stope pattern should be made as in the right-hand figure and not as in the left-hand one

The maximum burden of the stoping holes recommended by Langefors and Kihlström is given in Table 18.3 (reproduced from Table 7.2 in p. 187 of [4]) below and refer to Fig. 18.12.

The burden of the stoping holes should be not greater than the figure in the column of maximum burden (V). The best way to decide the place of stoping holes is using the principle of rectangularity. The right-hand stope pattern in Fig. 18.13 (from [4]) shows that in this way, the sequence of ignition to be used is clearly defined and tearing in the surrounding rock is reduced to a minimum. The left-hand stope pattern in this figure should be avoided.

18.1.6 Lifter Holes

Similar as the bench blasting, just taking the advance of the tunnel as the height of the bench height, the following equation can be used to calculate the burden of the lifter holes (from [6]):

$$B = 0.9\sqrt{\frac{q_e \times \mathrm{PRP_{ANFO}}}{\bar{c} \times f \times (S/B)}} \quad (18.5)$$

where

f	Fixation factor, generally 1.45, is taken to consider the gravitation effect and the delay timing between holes.
q_e	Explosive loading density (kg/dm^3).
$\mathrm{PRP_{ANFO}}$	Weight strength of explosive relative to ANFO (1–1.4).
S/B	Ratio of spacing/burden is usually considered equal 1.
$\bar{c}$	Rock constant calculated from c. Constant c is the quantity of explosive necessary to fragment one cubic meter of rock, normally in surface blasts and with hard rock c = 0.4 which is taken.

$\bar{c} = c + 0.05$ for $B \geq 1.4$ m and $\bar{c} = c + 0.07/B$ for $B < 1.4$ m.

The burden B should comply with the condition of $B \leq 0.6L$, where L is the hole length.

In lifters, it is necessary to consider the lookout angle γ to give enough space for the rig drilling blastholes of the next round. For an advance of 3 m, an angle of 3°, which has an equivalent of 6 cm/m, is usually enough; however, it will depend upon the characteristics of the equipment, as shown in Fig. 18.14.

The number of lifter holes can be given by:

$$\mathrm{NB} = \text{integer of}\left[\frac{B + 2L \times \sin\gamma}{B} + 2\right] \quad (18.6)$$

The stemming is fixed in T = 10 × d_1, where d_1 is the drillhole diameter.

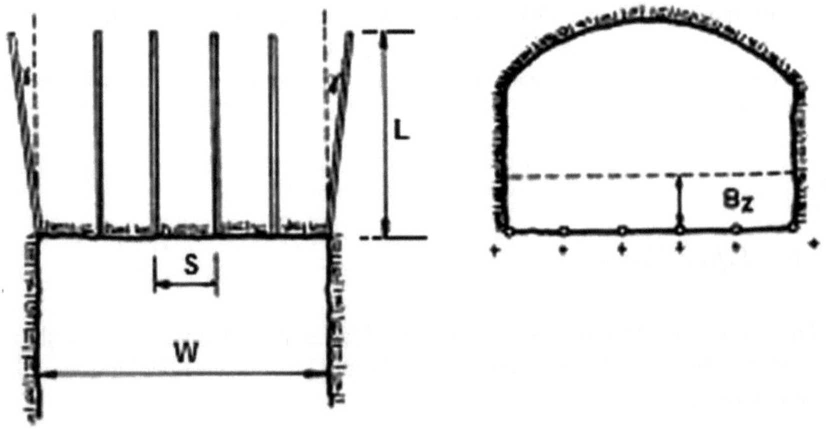

Fig. 18.14 Geometry of the lifters (reproduced from Ref. [6] with the permission from Taylor & Francis Book UK)

18.1.7 Contour Holes

The contour holes, also called perimeter holes, of tunnel blasting, especially the roof holes, are usually blasted using the smooth blasting method. The technique and parameters of smooth blasting have been discussed in Chap. 17.

If the blast does not need contour or smooth blasting, the parameters are calculated as for the lifters with the following values [6]: fixation factor, $f = 1.2$; S/B = 1.25; column charge concentration, $q_c = 0.5q_f$, where q_f is the bottom charge concentration.

18.1.8 Lineal Charge Concentration of Blasthole

18.1.8.1 Cut Holes

Langefors and Kihlström gave the guidelines of the charge concentration of the cut holes closest to the empty large holes in Table 18.4.

18.1.8.2 Stoping Holes

The recommended charge concentration of stoping holes by Langefors and Kihlström has been given in Table 18.3.

18.1.8.3 Lifter Holes

The lineal charge concentration is same as the stoping holes in practice. In fact, considering the gravitational and fixated effects, the burden and spacing are smaller than the stoping holes and the specific charge (kg/m^3 rock) is obviously higher than stoping holes.

Table 18.4 Concentration of charge (l) in kg/m for cylinder cuts and greatest distance (a) when blasting toward empty holes with diameter between $\varphi = 2 \times 57$ and 200 mm (d indicates the diameter of the loaded hole)

φ (mm)		50	2 × 57	75	83	100	2 × 75	110	125	150	200
D (mm)	32	0.2	0.3	0.3	0.35	0.4	0.45	0.45	0.5	0.6	0.8
	37	0.25	0.35	0.35	0.4	0.45	0.53	0.53	0.6	0.7	0.95
	45	0.30	0.42	0.42	0.50	0.55	0.65	0.65	0.7	0.85	1.10
a (mm)		90	150	130	145	175	200	190	220	250	330

The weight strength of the explosive is s = 1.0 (reproduced from Ref. [4] with the permission from John Wiley & Sons Ltd)

18.1.8.4 Contour Holes

If the tunnel blasting does not need smooth blasting, the charge concentration of the contour holes can use the following equation:

$$q_{lc} = 90 \times d_1^2 \quad (18.7)$$

where

q_{lc} lineal charge concentration of contour holes other than smooth blasting, in kg/m;

d_1 drillhole diameter, in m.

18.1.9 General Information for Tunnel Blasting Design

The following information can be used as a reference for the estimation of the tunnel blasting in the initial stage of a project and should be adjusted according to the geological condition and blasting practice along with the project progress.

18.1.9.1 Simplified Calculation for Designing Drilling and Blasting Pattern in Tunnels

The following table can be used as a quick initial design method for tunnel blasting with parallel-hole cuts (Table 18.5).

Table 18.5 Quick design of drilling and blasting pattern for tunnel blasting with parallel-hole cuts (reproduced from Ref. [6] with the permission from Taylor & Francis Book UK)

Part of round	Burden (m)	Spacing (m)	Length of bottom charge (m)	Charge concentration (kg/m)		Stemming (m)
				Bottom	Column	
Lifters	B	$1.1B$	$L/3$	q_f	q_f	$0.2B$
Wall*	$0.9B$	$1.1B$	$L/6$	q_f	$0.4q_f$	$0.5B$
Roof*	$0.9B$	$1.1B$	$L/6$	q_f	$0.36q_f$	$0.5B$
Stoping						
Upwards	B	$1.1B$	$L/3$	q_f	$0.5q_f$	$0.5B$
Horizontal	B	$1.1B$	$L/3$	q_f	$0.5q_f$	$0.5B$
Downwards	B	$1.2B$	$L/3$	q_f	$0.5q_f$	$0.5B$

q_f—charge concentration in bottom of hole = $7.85 \times 10^{-4}\, xd^2\rho$; d—cartridge diameter (mm); ρ—explosive density (gm/cc); B—burden in stoping area, $B = 0.88\, q_f^{0.35}$; L—hole depth in the round

*In some cases, smooth blasting is essential and these relationships are not applicable

18.1.9.2 Relationship between Drillhole Number and Tunnel Cross-Sectional Area

- Figure of ICI's Handbook of Blasting Table [7] (Fig. 18.15)
- C. L. Jimeno et al. Figure [6] (Fig. 18.16)

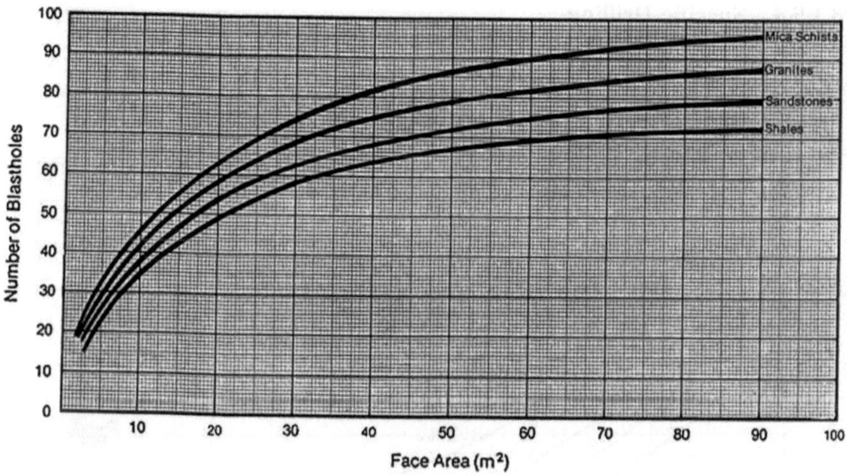

Fig. 18.15 Relationship of number of blastholes to face area (refer to [7])

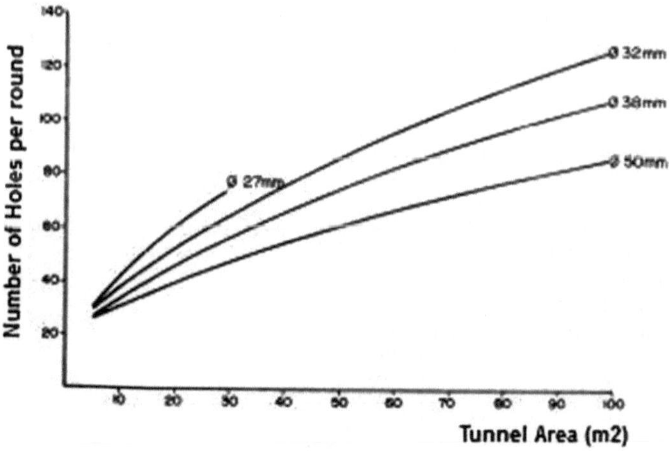

Fig. 18.16 Number of blastholes per round in function of tunnel's area (reproduced from Ref. [6] with the permission from Taylor & Francis Book UK)

18.1.9.3 Relationship between Explosives Consumption and Tunnel Cross-Sectional Area

- Figure of ICI's Handbook of Blasting Table [7] (Fig. 18.17)
- TAMROCK's Figure [8] (Fig. 18.18)

18.1.9.4 Specific Drilling

The following Fig. 18.19 is reproduced from Fig. 22.19 on p. 225 of [6]:

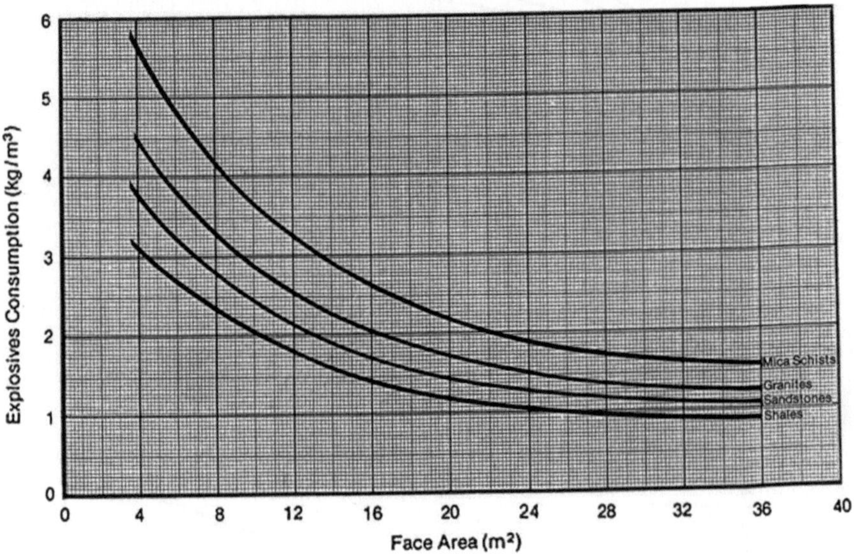

Fig. 18.17 Relationship of explosives consumption to face area (refer to [7])

Fig. 18.18 Powder factor as a function of tunnel area and blasthole diameter (reproduced from Ref. [8] with the permission from Sandvik)

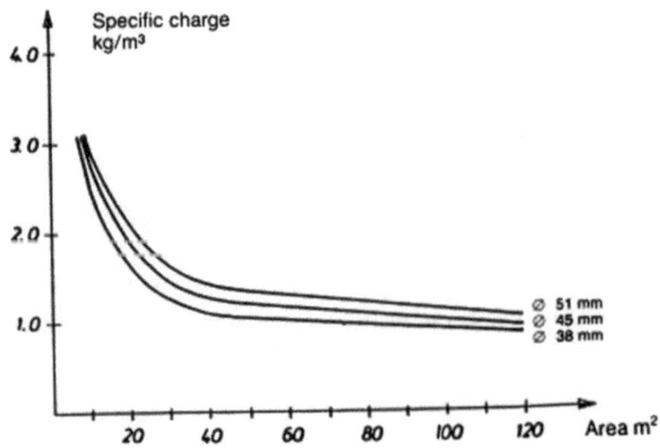

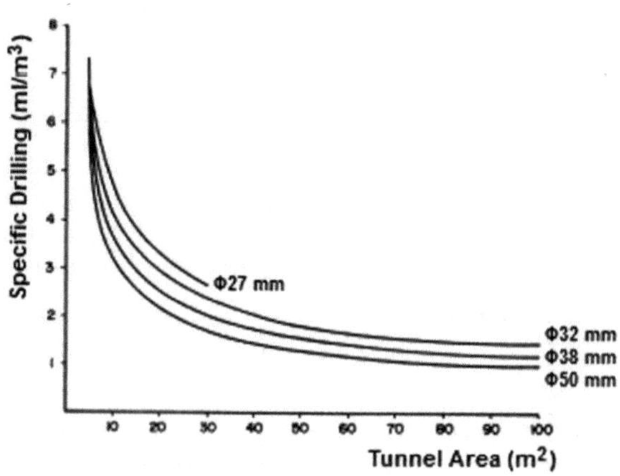

Fig. 18.19 Specific drilling as a function of the tunnel area and drillhole diameter (reproduced from Ref. [6] with the permission from Taylor & Francis Book UK)

18.2 Blasting Design for Shaft: Full Face Sinking

The full face method is used frequently in shaft sinking as it suits either rectangular- or round-section shafts. The principles described earlier for tunnel excavation may be applied with some modifications for special circumstances. As with tunnel blasting, the "cut" is critically important. There are various techniques for cut-hole design to create a free face with a few blastholes, V-cuts, or cone cuts, and parallel-hole cuts with relief holes.

18.2.1 Types of Cut-Hole Pattern

18.2.1.1 "V"-Cut

"V" cuts are used in rectangular-section shafts. The planes of the dihedrals formed by the blastholes that are inclined between 50° and 75° should be parallel to the discontinuities, in order to use them to advantage during breakage. Figure 18.20 is a sample of "V" cut.

18.2.1.2 Cone Cut

Cone cuts are used most in round-section shafts. It is easy to drill blastholes with a shaft jumbo. The holes are placed so as to form several inverted cone areas in the central part, as shown in Fig. 18.21.

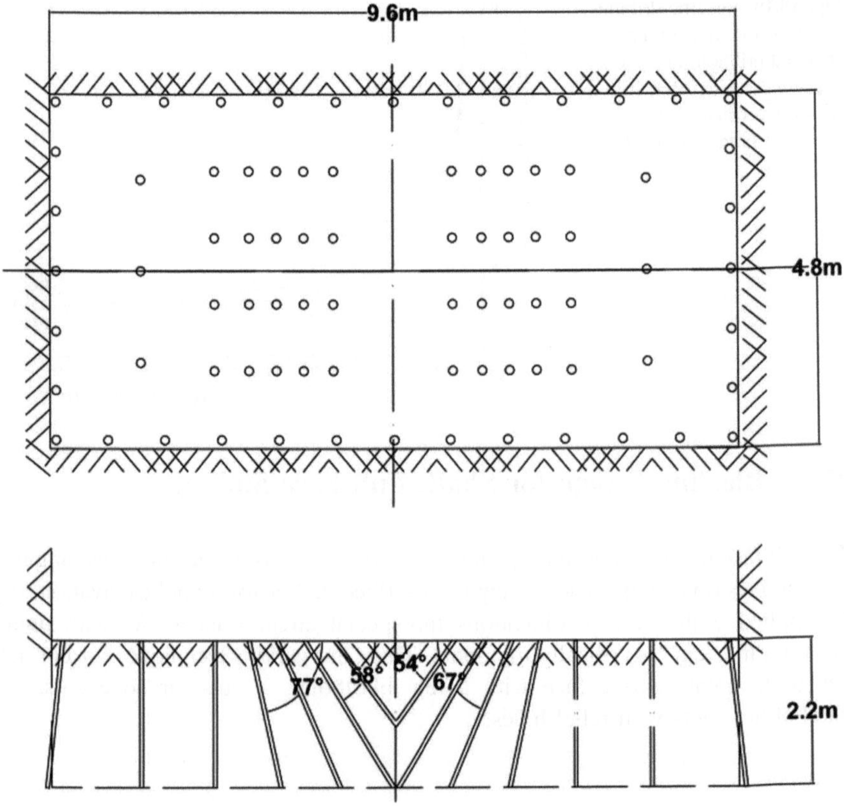

Fig. 18.20 Drilling pattern in a rectangular-section shaft

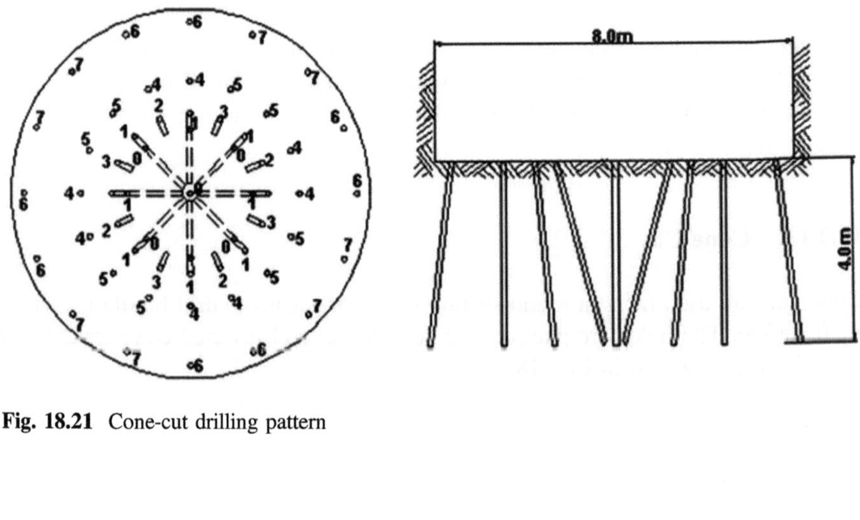

Fig. 18.21 Cone-cut drilling pattern

18.2.1.3 Parallel-Hole Cut with Relief Hole(S)

Similar to tunnel blasting, parallel-hole cut with large empty hole(s) is widely used in shaft blasting, especially for relatively small shafts. Figure 18.22 is a sample.

18.2.2 Blasting Parameters for Shaft Blasting

The pull of the rounds, as well as the number of blastholes, depends upon many factors such as the type of rock mass, the diameter of explosive charge, the blasting pattern, type of cut, the shaft size to be excavated, and the restriction of the surrounding environment (i.e., the charge weight per delay as the vibration limitation).

All blasting parameters should be adjusted time to time during the blasting practice to suit the above factors.

At the early stage, the following formula can be used to estimate the number of blastholes when 32-mm-diameter explosive charge is used:

$$\mathrm{NB} = 2D_p^2 + 20 \tag{18.8}$$

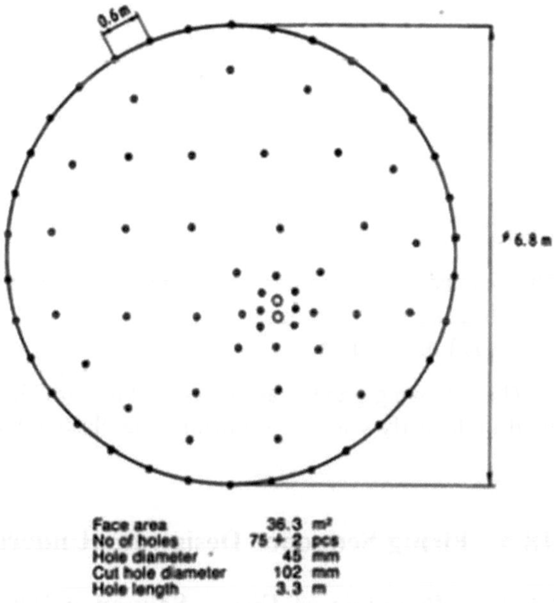

Fig. 18.22 Drillhole pattern using parallel-hole cut with an empty large hole (reproduced from Ref. [14] with the permission of Sandvik)

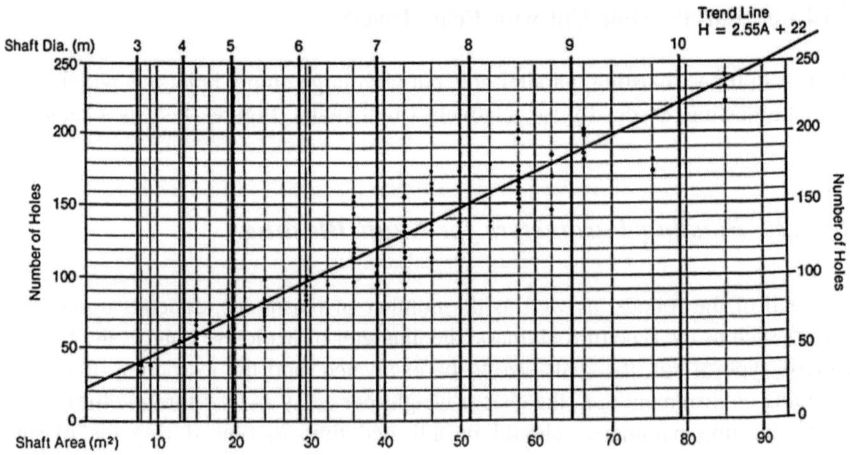

Fig. 18.23 Graph of number of holes in shaft of various size (*Source* [7])

Fig. 18.24 Round advance as function of shaft area (reproduced from Ref. [6] with the permission from Taylor & Francis Book UK)

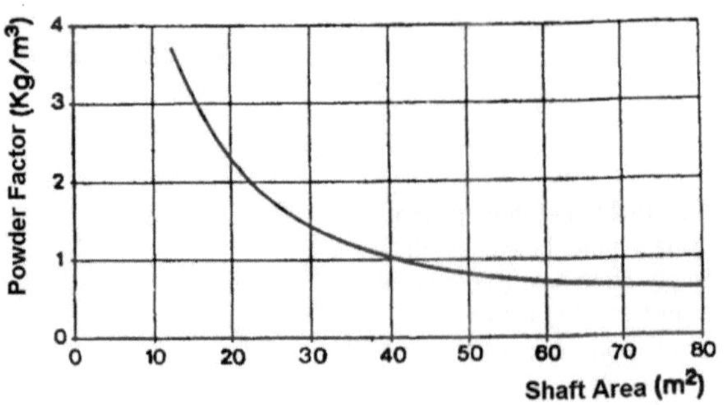

where

NB number of blastholes excluding the perimeter holes if contour (smooth) blasting is carried out;

D_p shaft diameter (m).

The following graphs, as shown in Figs. 18.23, 18.24 and 18.25, can also be used in the early stage for estimating the blasting parameters of shafts.

18.3 Firing Sequence Design for Underground Blasting

18.3.1 Principle of Firing Sequence Design

As there is only one free face in underground blasting (excluding benching), the blastholes should be fired in a certain sequence. The firing pattern must be designed

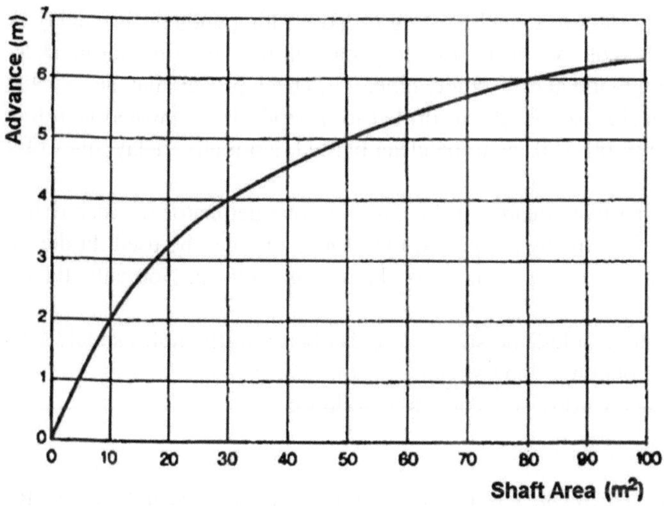

Fig. 18.25 Powder factor as function of shaft area (reproduced from Ref. [6] with the permission from Taylor & Francis Book UK)

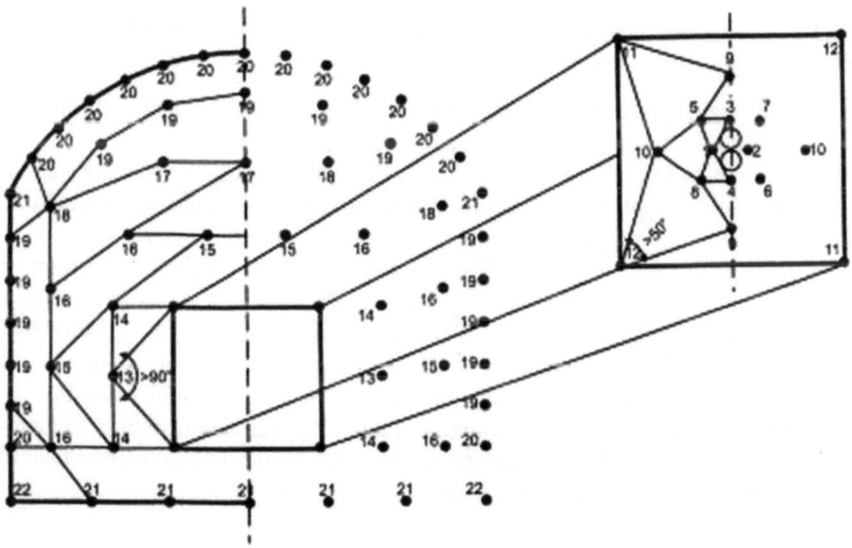

Fig. 18.26 Firing sequence for small tunnel in numerical order (reproduced from Ref. [14] with the permission from Sandvik)

so that each hole has free breakage. The angle of breakage is smallest in the cut area where it is around 50°. In the stoping area, the firing pattern should be designed so that the angle of breakage does not fall below 90° (see Fig. 18.26).

It is important in tunnel blasting to have a long enough time delay between the holes. In the cut area, it must be long enough to allow time for breakage and rock throw through the narrow empty hole. It has been proven that the rock moves with a velocity of 40–70 m/s. A cut drilled to a depth 4–5 m would therefore require a delay time of 60–100 ms to be clean blasted. Normally, delay times of 75–100 ms are used.

In the first two squares of the cut, only one detonator for each delay should be used. In the following 2 squares, two detonators may be used. In the stoping area, the delay must be long enough for the rock movement. Normally, the delay time is 100–500 ms.

For contour holes, the scatter in delay between the holes should be as small as possible to obtain a good smooth blasting effect.

The general firing sequence is as follows:

For tunnel (cavern):

Cut holes → Cut spreader holes → Stoping holes → Wall holes → Roof contour holes → Lifter holes.

For shaft:

Cut holes → Cut spreader holes → Stoping holes → Wall contour holes.

There are some important principles that must be kept in mind when designing the firing sequence:

(a) The firing sequence must start from the central cut holes and then progress outward to the tunnel contour gradually.
(b) The minimal time interval of two adjacent holes must be no less than 25 ms, especially in shaft blasting with wet holes to avoid the "water hammer effect."
(c) For the large tunnel or cavern, the blastholes on the working face can be divided into several groups. The shock tubes of all holes in each group are bunched together and connected with a surface delay connector (refer to Sect. 18.3.3 below). The delay time of the first firing in-hole detonator must longer than the longest delay time among all surface detonators; otherwise, the first fired blasthole will damage the initiation net for the tunnel blasting.
(d) If there is a restriction of ground vibration, the maximum explosive charge weight per delay of any simultaneously fired holes must be no greater than the allowable charge weight per delay.

18.3.2 Small Tunnel

If it is possible to have enough numbers of required delay intervals for a blasting round, all blastholes of the whole tunnel face can be fired in one time. Fig. 18.26 is a sample.

18.3.3 *Large Tunnel*

For the large tunnel or cavern, the working face can be divided into several sectors. There are two kinds of connection methods: bunch connector method and detonating cord ring connection method.

18.3.3.1 Bunch Connector Method

The shock tubes of all holes in each sector are bunched together and connected with a surface delay connector (see Fig. 18.27).

The following figure (Fig. 18.28) is a good example of using bunch connectors to fire a large cross-sectional tunnel.

18.3.3.2 Detonating Cord Ring Connection Method

In this method, all shock tubes have a "J" hook and all detonators within a delay sector are connected to a ring (circle) of two strands of 5 gm/m detonating cord (see Fig. 18.29).

- The shock tube from the detonators has "J" hooks at their end which clip onto the detonating cord (see Fig. 18.30).
- The "ring" of detonating cord should be flush with the face and should not touch any other shock tube tail hanging from an adjacent delay sector.
- Delay sectors are linked by the appropriate non-electric surface delay connector.
- The surface connectors are attached to a 0-ms delay bunch block which initiates the detonating cord.

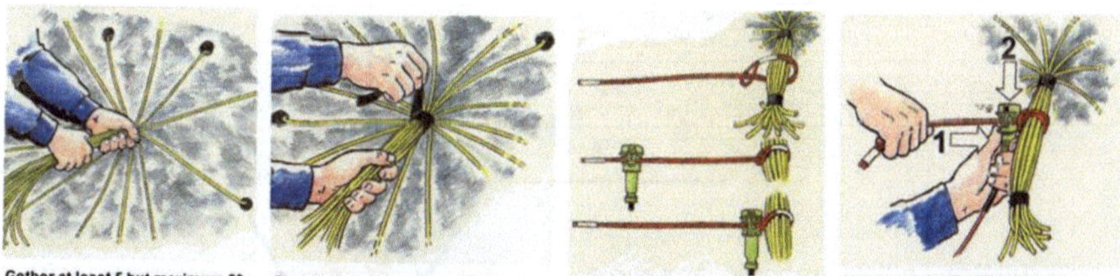

Fig. 18.27 Initiation by means of bunch connectors for a large section of tunnel (courtesy of "Nonel User's Guide" of Dyno Nobel)

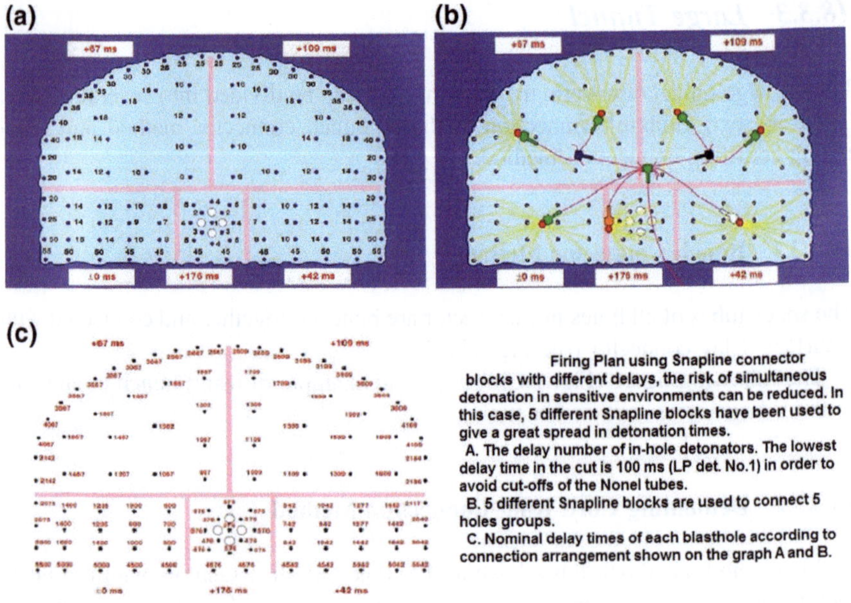

Fig. 18.28 Firing plan using snapline connector blocks with different delays (courtesy of Dyno Nobel)

Fig. 18.29 Detonating cord ring connection method (courtesy of Orica)

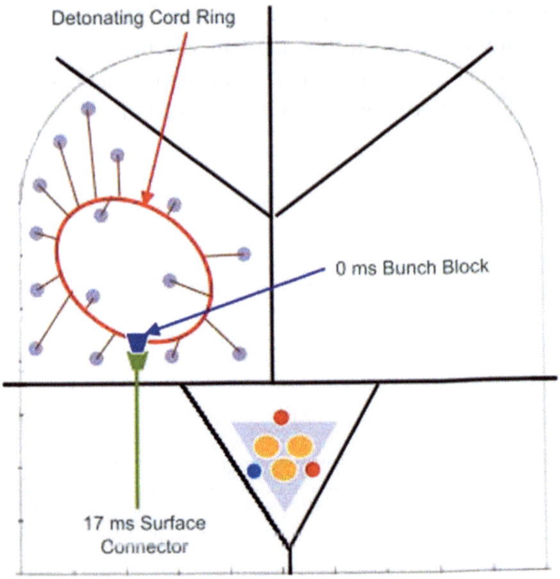

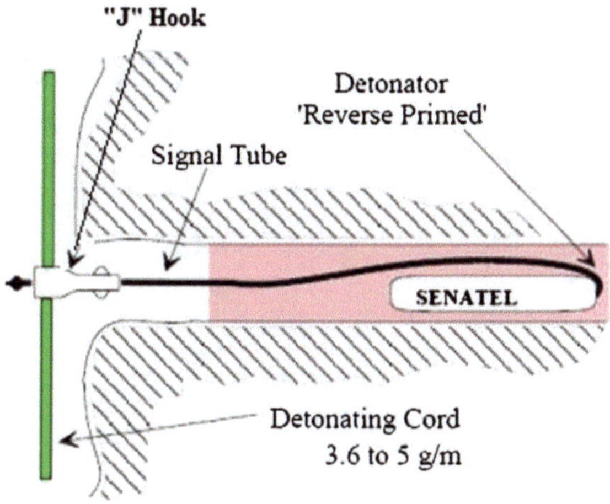

Fig. 18.30 Detonator's "J" hook clips the detonating cord (courtesy of Orica)

18.3.4 Tunnel Blasting with Electronic Detonators

Electronic detonators possess the advantages of high precision in delay time so that the blasting is more guaranteed for safety, particularly for ground vibration control. It is more and more used for construction blasting, including underground blasting, in sensitive environments, especially in urban areas. The following examples briefly show how electronic detonators are used in tunnel blasting.

18.3.4.1 Example 1: SmartShot Electronic Detonators' Tunnel Blasting Design

Referring to Fig. 18.31, all blastholes are divided into 5 subsectors.

The initiation sequence is as follows: SS3 > SS4 > SS1 > SS5 > SS2. Their igniting times are as follows: 0, 2151, 2181, 2593 and 2691 ms, respectively, with different delay intervals between holes in each sector.

18.3.4.2 Double-Deck Tunnel Blasting Using Electronic Detonators [9]

Fig. 18.33 is a tunnel blasting design using Orica's eDev™ electronic detonators. This tunnel excavation was carried out in the busy urban area of Hong Kong in 2013. Due to the complex environment of the tunnel project, the surrounding highly sensitive receivers made the allowable maximum instantaneous charge (MIC) for tunnel blasting very low, only 0.2–0.4 kg/delay in some tunnel sections.

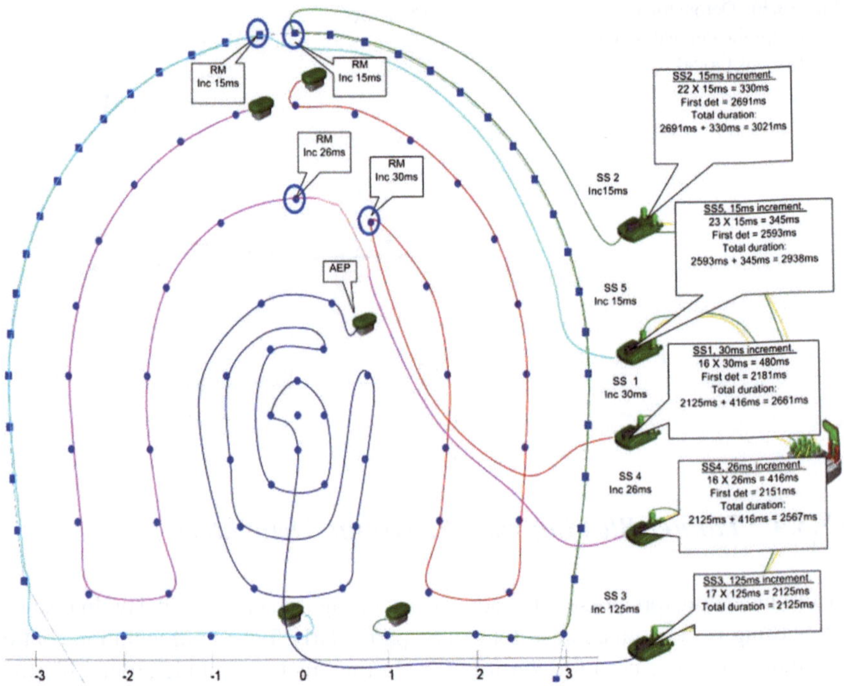

Fig. 18.31 Tunnel firing sequence design using SmartShot Electronic System (Courtesy of AMS)

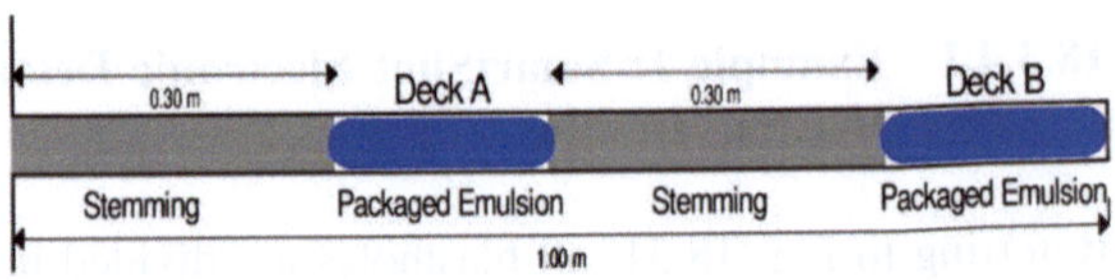

Fig. 18.32 Blasthole charging structure (reproduced from Ref. [9], courtesy of ISEE)

In this design, double-deck blasting technique was applied to increase the advance of each round. The charging structure is shown in Fig. 18.32. The allowable MIC is 0.2 kg per delay. The drillhole and firing sequence pattern are shown in Fig. 18.33.

Holes surrounding the cut were moved closer, reducing burdens, and the initiation sequence was changed substantially. Commencing with hole by hole firing, the angle of initiation was gradually opened, as shown in the figure. In three dimensions, this results in expanding conical angles of initiation.

Loading and tie in time for this blast, also consisting of 358 decks, was 185 min; a small amount of experience with the previous blast offered immediate improvement in productivity. There were no notable safety issues or increase in risk. All measured ground vibration was below 1 mm/s, or monitor thresholds were not triggered. The majority of this blast pulled 100 % or more than the drilled 1.0 m length. There were very few small butts (sockets) remaining in the face (refer to [9]).

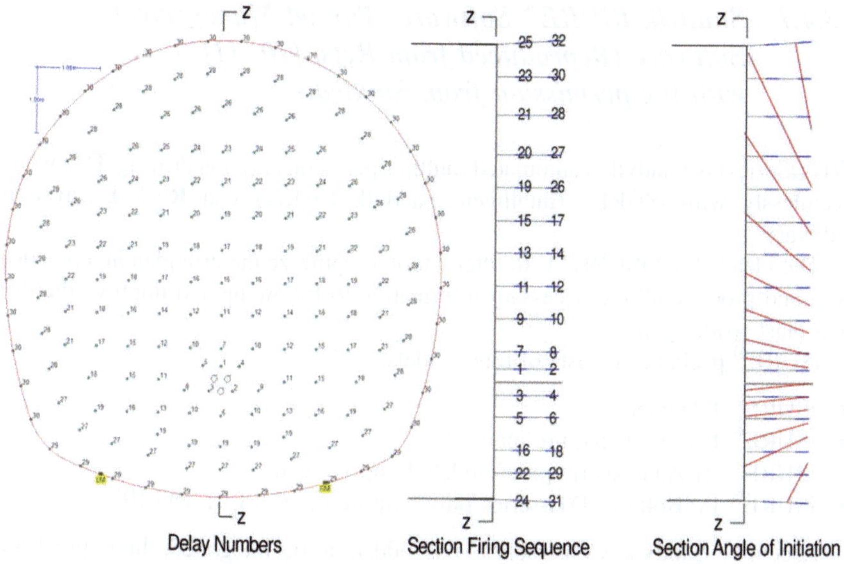

Fig. 18.33 Tunnel blasting design using double charging technique and electronic detonators (reproduced from Ref. [9], courtesy of ISEE)

18.4 Computer-Aided Tunnel Design and Management

Some drill manufacturers together with some universities have developed some software to help the engineer to do the works of tunnel blasting design and management. The manufacturers supply the software together with the drill equipment and train the engineer, drillers, and shotfirers.

Among them, Atlas Copco's "Underground Manager" and Sandvik's "iSURE®" are the most popular software. Their prominent feature is that the software is integrated with the drill operation system, and it not only guides the drilling machine to drill blastholes in accordance with the designed pattern accurately, but also collects all real-time available drilling information during the drilling process, using the MWD (Measure While Drilling) data acquisition systems, to analyze and assess the rock properties and geological features of the rock mass for guiding tunnel blasting design.

iSURE® and Underground Manager are based on Windows so the user who are familiar with windows system will not have difficult to master them.

18.4.1 *Sandvik ISURE® Software: Tunnel Management Software (Reproduced from Refs. [10, 11] with the permission from Sandvik)*

DTi (iSeries) is Sandvik's automated multipurpose construction drill rig. DTi works seamlessly with iSURE® (**i**ntelligent **S**andvik **U**nderground **R**ock **E**xcavation) software.

The idea behind iSURE® is to offer a tool to optimize the drill plan in a practical way and produce all the necessary information to follow up and improve the drill and blast work cycle.

iSURE® package consists of four models:

- iSURE® I Tunnel,
- iSURE® II Report (require model I),
- iSURE® III Analysis (require models I and II), and
- iSURE® IV Bolting (DTi-series only, require models I, II, and III).

Recently, Sandvik Construction has added a rig-integrated, high precision, online rock mass analysis and visualization system to its existing offering for tunneling process optimization—geoSURE. This new option is fully integrated to the iSURE® tunneling project management software, providing rock mass information and a view inside the drilled rock. Its unique features improve the overall tunneling process in terms of efficiency and quality.

- iSURE® I Tunnel: Drill and Blast Design

Tunnel is the basic module of iSURE®; it will always come with the package. It includes project files management, tunnel profiles, tunnel location, drill and blast design, and drilling and blasting patterns. This module offers one of the most revolutionary features in the iSURE®: pattern design in the end of the round, providing hole burden calculation and optimization of hole location (Fig. 18.34).

The design of the theoretical profile can be drawn manually or chosen from the standard profiles provided in iSURE®. It is also possible to import a profile in .dxf format from AutoCad. As part of the drill plan design, the tunnel module also includes detonator design into which surface delay design can be incorporated. It significantly speeds up the design work and decreases the number of errors. The charger can be supplied with the reports on the explosives and detonator demand. This allows explosives planning to start immediately.

In the drill pattern, iSURE® offers a possibility to define a range of different drilling types such as contour holes, field holes, and grouting holes.

ISURE® pullout analysis software (DTi-series only) can compare the start positions of holes to end positions of the previous round. This can be used to modify the charging or hole distances in program areas and improve the productivity.

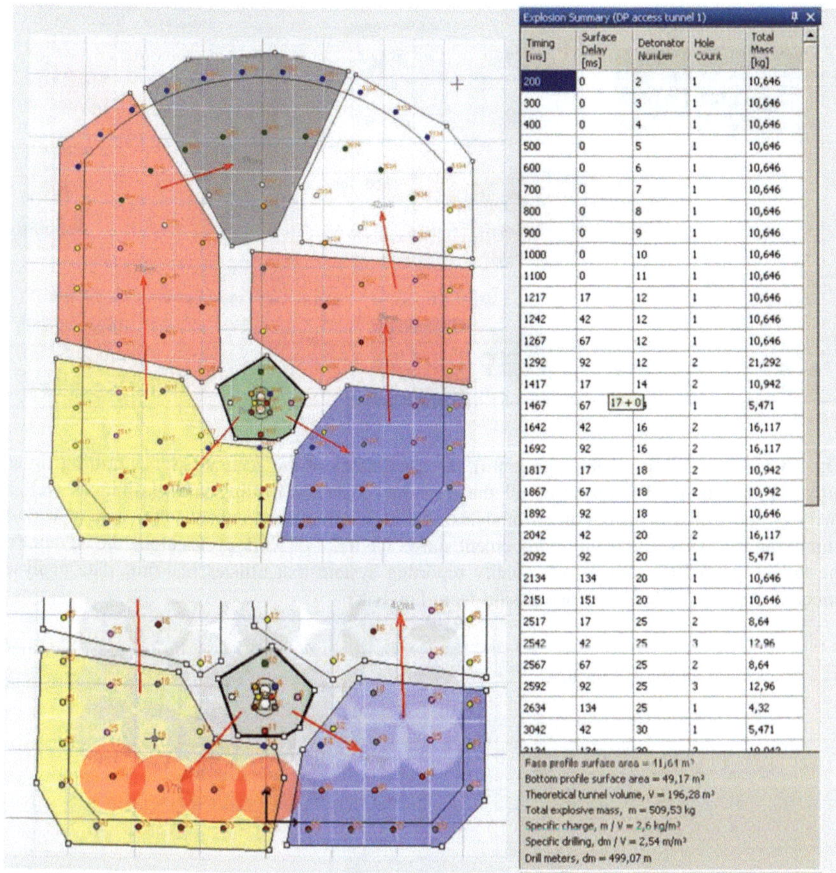

Fig. 18.34 Use iSure tunnel module to do drilling and blasting design-firing sequence design (reproduced fro Ref. [11] with the permission from SANDVIK)

- iSURE® Report: Process Control and Reporting

The report module supplies clear reports that can be used directly for reporting. Information on the actual drilling process can be received on four levels: per round, per user (instance of use), per service (maintenance interval), and per lifetime. The reports include data on the time used for different phases in the drilling process and rock tools' consumption per set intervals (Fig. 18.35).

- iSURE® Vibration Feedback (Fig. 18.36)

iSURE® introduces a practical approach to vibration control, as it knows the charges, detonation timing (momentary situation), and planned and drilled drill plan (burden). Blasting vibration data from third-party systems can be imported into

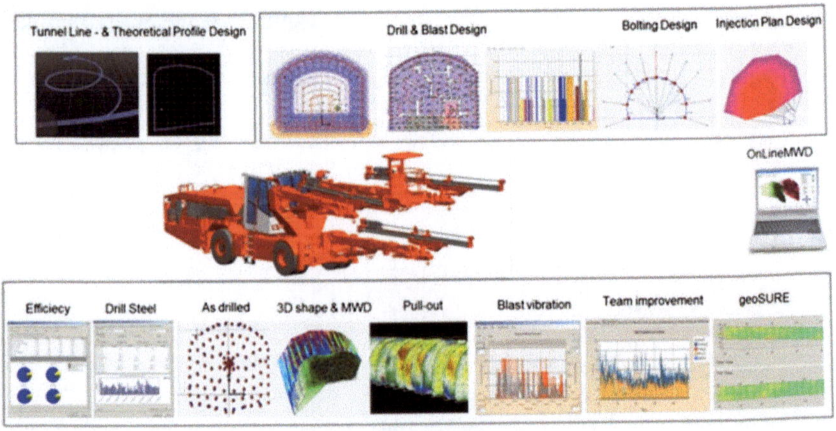

Fig. 18.35 With iSURE (Office program), the capabilities of the accurate iSeries drilling rig are fully capitalized. iSURE produces all the necessary data for drilling and blasting as well as analyzes the drill rig data. Pullout analysis and blast vibration feedback offer a new way of improving the D&B, and team improvement shows the trend of work cycles along the advance of the tunnel. geoSURE adds a rock quality reporting system that utilizes real-time data analyses onboard (reproduced with the permission from Sandvik)

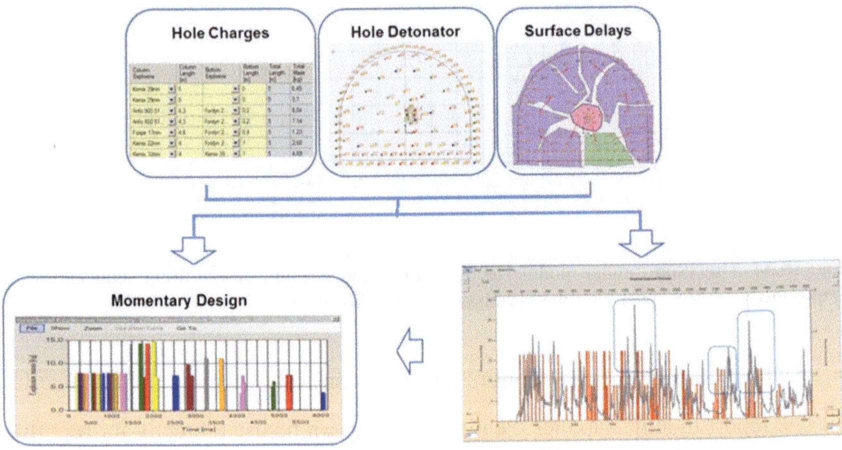

Fig. 18.36 Deviation in kg versus PPV can be pinpointed into the drill and blast plan. This connection can be used in charging work quality control, adjustment of the blast/detonation and tuning the round length while approaching a sensitive area (reproduced with the permission from Sandvik)

iSURE in order to compare blasting-generated vibration against momentary mass of explosives. This comparison of information is then used to reveal the problem areas in the drill and blast plan.

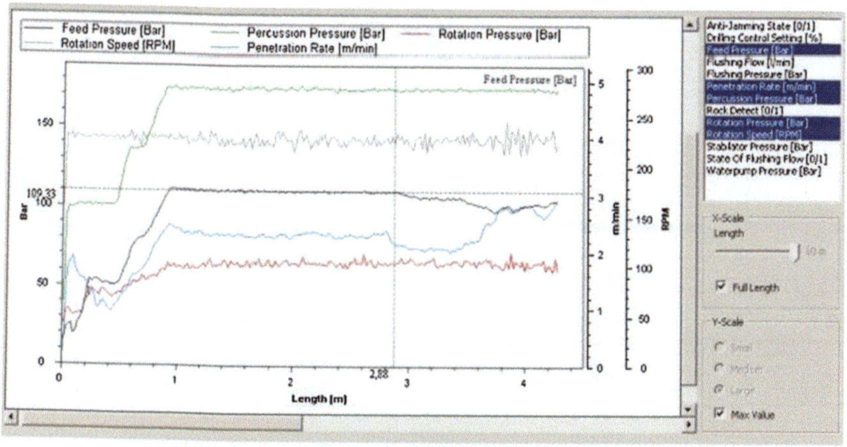

(A) Graph of parts of measured drilling parameters

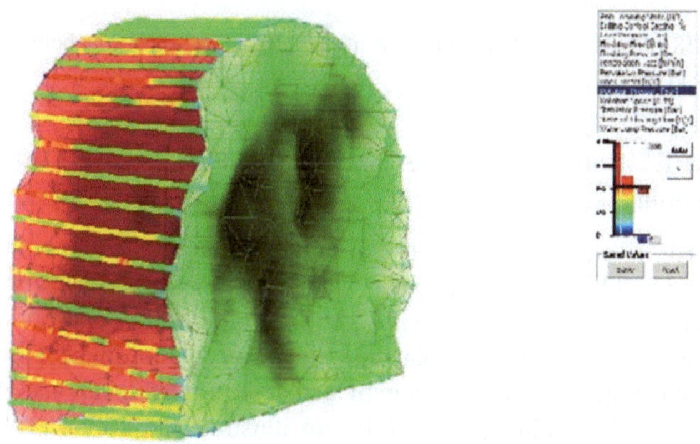

(B) MWD can be visualized as colour freely rotated 3D-images of rock conditions during drilling

Fig. 18.37 iSURE use MWD to collect drilling data and analyze rock characters (reproduced with the permission from SANDVIK)

- iSURE® Analysis: MWD Data Collection and Reporting

iSURE® Analysis offers Measuring-While-Drilling (MWD) data collection and reporting for analyzing rock structure and characteristics. The module collects data on 19 parameters—more than any other software in the market—among those being parameters such as airflow, feed pressure setting, rotation speed setting, anti-jamming state, and drilling control setting. The MWD data can be studied and analyzed after drilling (Fig. 18.37).

As an option, an online MWD is available. It offers MWD data and visualization of the rock surface in real time in 3D. The data can be exported to external programs from iSURE® to do further analysis in other analysis environment.

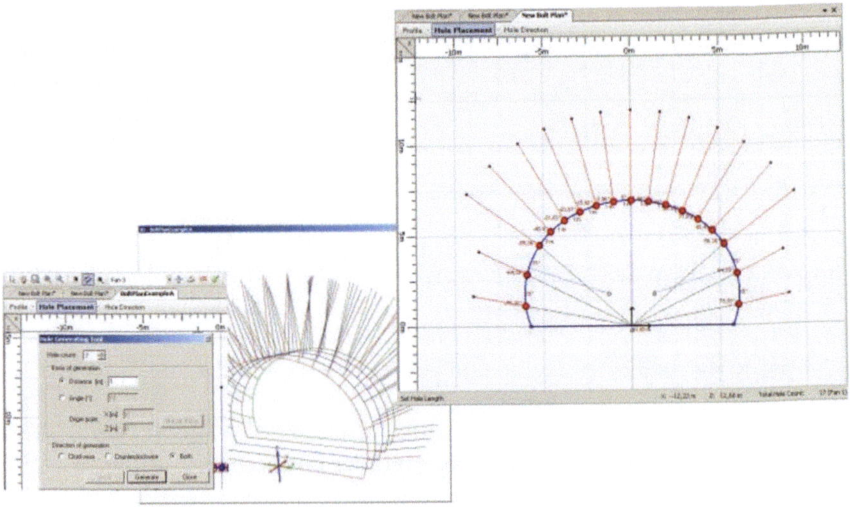

Fig. 18.38 iSURE IV Bolting module is used to design and report the bolting works (reproduced with the permission from Sandvik)

- iSURE® Bolting (DTi-series only)—Bolt Plan Design, Data Collection, and Reporting

iSURE® Bolting makes the use of one machine possible in both drilling the face holes and bolt holes simultaneously. The program enables utilization of the drill and bolt plans at the same time, with one navigation when appropriate. Reporting is possible on both the drill plan and the project coordinate systems, promoting systematic bolting and reporting of actual bolt-hole locations. Additionally, the iSURE® Analysis offers a possibility to illustrate the tunnel profile based on the actual bolt locations.

The bolting module allows the design of up to 5 bolting fans in the same plan. The plan includes hole placement and direction, hole generating and fan management tools, and 3D visualization (Fig. 18.38).

- geoSURE

Based on the MWD, Sandvik Construction recently launched geoSURE. The new system provides accurate geological information for companies or individuals involved in tunneling or underground operations, through a completely new, rig-integrated onboard rock mass analysis and visualization system. Designed to be used with Sandvik underground construction drill rigs, it delivers real-time onboard analysis of the rock mass which includes features such as fracture, rock strength, and water indication. The extended analysis of the data allows the evaluation of the rock strength class, rock quality class, and rock quality index. These features can then be further visualized using the iSURE® tunneling project management software. The 2D planar view provides an overall outlook of the tunnel section,

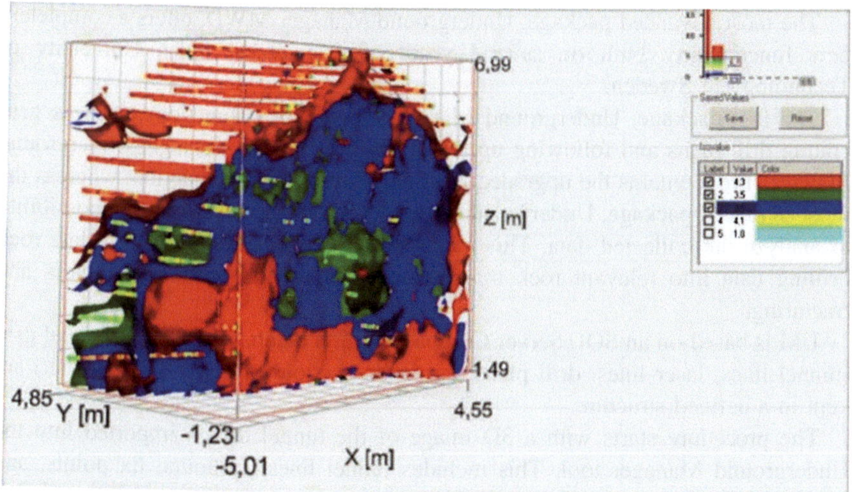

Fig. 18.39 geoSURE—rock mass analysis and visualization system (reproduced with the permission from Sandvik)

including 2D interpolation, side view, top view, and unrolled view. For more detailed inspection, the 3D structural view of the tunnel section can be used. This features 3D interpolations, plane intersections, and iso-surfaces, and curves. The iSURE® system also provides the former 3D view and one hole MWD diagram with the new geoSURE variables.

Not only is it an easy way to fulfill the most advanced reporting requirements in the industry, but it also acts as an important tool for the assessment of rock reinforcement or injection requirements. Additionally, it serves as an assisting tool for charging and blasting control as well as a complementary tool for geological mapping (Fig. 18.39).

18.4.2 Atlas Copco: Underground Manager MWD (Reproduced from Refs. [12, 13] with the permission from Atlas Copco)

Underground Manager (Tunnel Manager) is a support software for planning, administration, and evaluation of the drilling operation in mining and tunneling projects.

The Windows-based program Underground Manager (UM) is suited for Atlas Copco Boomer rigs equipped with Rig Control System (RCS) and the Advanced Boom Control function (ABC Regular or ABC Total). The upgraded Tunnel Manager software is available in three different packages: Tunnel Manager, Tunnel Manager Pro, and Tunnel Manager MWD.

The most advanced package, Underground Manager MWD, offers a completely new functionality built on several years of research at Luleå University of Technology in Sweden.

The first package, Underground Manager, offers basic functions, such as generating drill plans and following up the result. The second package, Underground Manager Pro, contains the upgraded Measure-While-Drilling function, whereas the most advanced package, Underground Manager MWD, also contains the possibility to analyze the collected data. This means that the user can swiftly translate rock drilling data into relevant rock mass characteristics such as rock hardness and fracturing.

UM is based on an SQL Server Compact Edition database, where all tunnel data (tunnel lines, laser lines, drill plans, contours, section logs, MWD logs, etc.) are kept in a defined structure.

The procedure starts with a 3D image of the tunnel that is imported into the Underground Manager tool. This includes tunnel lines, contours, fix points, and laser points.

- Drill Plan Generator and Charging and Firing Pattern

The drill plan designing is a key function of UM where all lines, shapes, and holes (length, lookout, type, and diameter) are allocated. Each defined section of a tunnel can also be given an individual contour design for generating drill plans as shown in Fig. 18.43. Any given number of holes can be allocated that may vary between the defined or interpolated contours. Once the drill plan is complete, the UM can be linked up with Atlas Copco's Rig Remote Access System (Fig. 18.40).

Following proper analysis of the geometry and rock conditions, a lot of preparation goes into selecting the right explosives and firing sequences in order to achieve good contours and also make sure that vibrations do not exceed stipulated limits and regulations for the project. All these can be designed and simulated using the UM, which offers dynamic drawing tools for both charging and blasting. By selecting designated sections and using a drawing tool, the UM enables various blasting scenarios to be tried and tested. Sections of blastholes can be given individual delay times according to a chosen sequence, including block and surface delays, as shown in Fig. 18.41. This is very useful in ensuring that the right ignition sequence is employed. Any changes to the drill plan will be synchronized and shown in the firing and charging pattern.

- MWD and reporting

The technique is to extract rock mass properties while drilling is called MWD, which stands for "Measurement While Drilling."

The parameters recorded are as follows: penetration rate, feed force, percussive pressure, rotation pressure, rotation speed, damp pressure, water pressure, and water flow.

The MWD technology consists of two separate processes: registration of data and presentation/evaluation/interpretation of data.

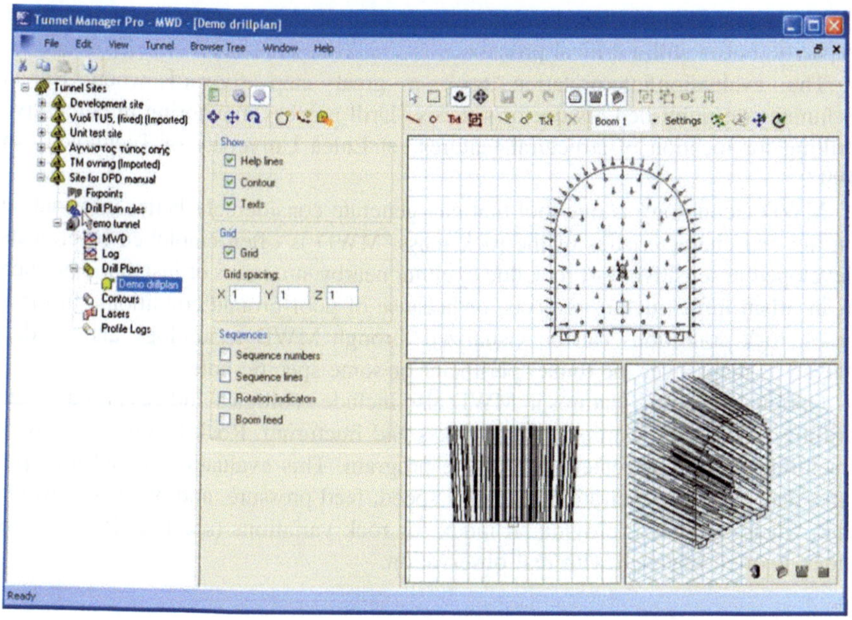

Fig. 18.40 Edit a drill plan using tunnel manager MWD (reproduced with the permission from Atlas Copco)

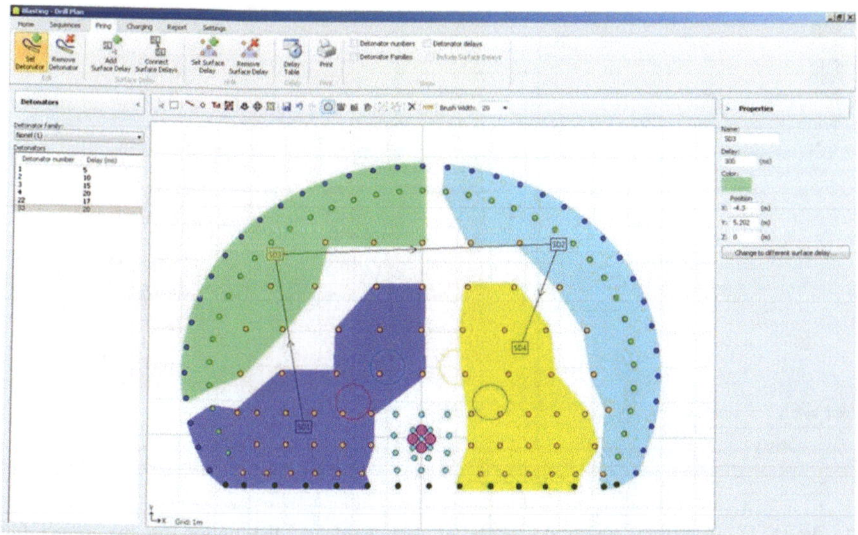

Fig. 18.41 Firing sequence design (reproduced with the permission from Atlas Copco)

The registration part has developed as the drill rigs get more and more computer capacity, but is still a critical process.

The evaluation/interpretation took a great step forward when Håkan Schunnesson published his doctor's thesis "Drill process monitoring in percussive drilling for location of structural features" at Luleå University of Technology in 1997.

MWD technology is one tool that can generate considerably better information on the real characteristics of the rock mass. MWD is often employed in projects developed in sensitive host rock due to either nearby structures or installations, such as in urban infrastructure projects, or because of poor ground conditions. In these cases, rock conditions can be visualized through MWD technology and be displayed as maps over the tunnel perimeter or some specific holes.

Optional reporting features in MWD also include geological indices based on the drilling process and estimates of hardness and fracturing. Rock hardness and rock fracturing indices are calculated by the program. This evaluation model incorporates data such as penetration rate, drill speed, feed pressure, and other parameters that, when combined, provide an index for rock variations (see Fig. 18.42). This index is often matched with real observation.

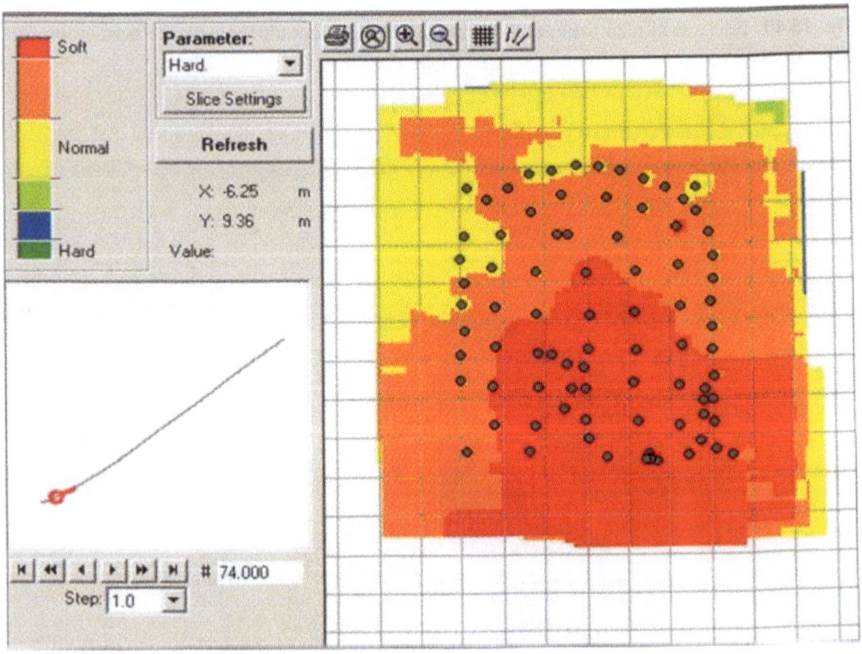

Fig. 18.42 Using MWD output the index of rock hardness distribution on the tunnel face (reproduced from Ref. [13] with the permission from Atlas Copco)

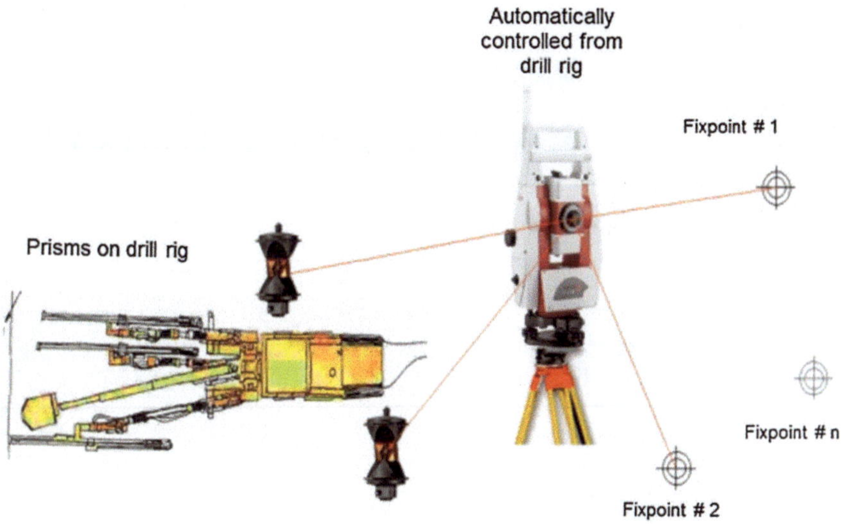

Fig. 18.43 Total navigation system with a tripod-mounted (reproduced with the permission from Atlas Copco)

- Total navigation system

That is an input method of surveying with a total station which starts with a station setup where the instrument is placed on a tripod and either centered over a known point, or set up as a free station.

As shown in Fig. 18.43, the instrument is centered over an existing point with known coordinates and the orientation of the instrument is obtained by reference direction measured against another known point. The second case is free station which means that the instrument is set up in any location, preferably next to the item to be measured in order to obtain a higher accuracy. Determination of the total station's locations is carried out by measuring against other known objects. Known point is the most common method of station establishment, but the method of free station is preferred when measuring is made within a small geographic area.

In recent years, the total station has evolved to become increasingly automated, thereby simplifying the surveying work. An example of this is the so-called automatic prism lock, or automatic target recognition (ATR), which enables the instrument to be directed toward the prism automatically using a CCD camera (Jodahl and Larsson 2007). The technology requires only a rough orientation of the instrument, while the fine tuning is handled automatically. ATR technology along with the development of the automatic prism called LOCK (or automatic target tracking) has also enabled more effective surveying.

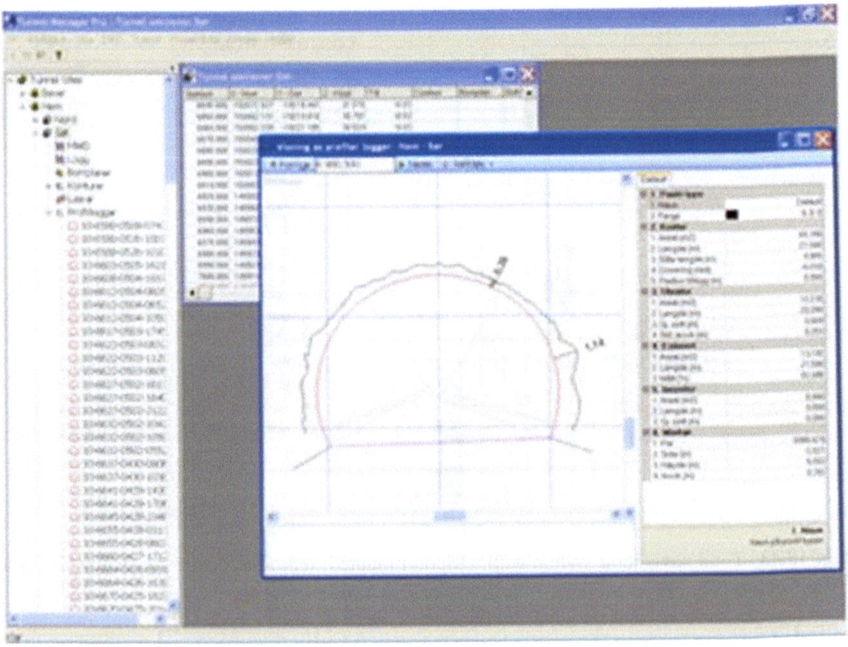

Fig. 18.44 Using tunnel profiler to measure overbreak (reproduced with the permission from Atlas Copco)

By attaching the total station control panel directly on the prism pole and through radio connection to communicate with the instrument, the meter could move from the instrument directly to the object to be measured. This has completely changed the method of data entry. Today, in many cases, only one person is required to carry out field work. Another technique that is still evolving toward a better accuracy is reflectorless measurement, where neither prism nor reflectors are used. Instead, the distance is measured directly at the object via the reflection of the laser beam. The technology allows inaccessible objects to be measured, however, with a slightly reduced accuracy than when the prism is used.

- Tunnel profiler (Fig. 18.44)

Tunnel profiler is a fully integrated system for measuring the excavation profile and is reliant on network communications. The accuracy of the scanned surface is 3–5 cm, by saving the overbreak scanning of the profile and making adjustments on the drilling pattern may save some 5 cm on overbreak. By limiting the overbreak, it can reduce the cost of concrete for secondary lining, reduce the mucking cost of extra rock material, and shorten the construction time.

References

1. Hagan TN (1981) Large diameter blastholes—a proposal means of increasing advance rate. In: 4th Australian tunnelling conference, Melbourne. ISBN: 0909520593
2. Chapman D et al (2010) Introduction to tunnel construction. Spon Press, London. ISBN 10: 0-415-46841-8
3. Hemphill GB (2013) Practical tunnel construction. Wiley, Hobokon. ISBN 978-0-470-64197-2
4. Langefors U, KihilstrÖm B (1978) The modern technique of rock blasting, 3rd edn. Wiley, New York. ISBN 0-470-99282-4
5. Persson P et al (1993) Rock blasting and explosives engineering. CRC Press, Boca Raton. ISBN 0-8493-8978-X
6. Jimeno CL et al (1995) Drilling and blasting in rock. A. A, Balkema, Rotterdam. ISBN 90 5410 199 7
7. ICI Australia Operations Pty Ltd (1980) Handbook of blasting table, Metric edition
8. Puhakka T (1997) Underground drilling and loading handbook. Tamrock Corp., Tampere, Finland
9. Iwata O et al (2013) Double deck blasting for rapid tunnel advance in Hong Kong. J Explos Eng 30(4):6–14, ISEE, Cleveland, Ohio, USA
10. Sandvik (2012) Sandvik iSURE®—Tunnel Management Software, Technical Specification, 5-9700-D, 2012-11-02, Sandvik Mining and Construction
11. Sandvik (2013) iSURE® Brochure 2013, Sandvik iSURE®—a revolution in precision, 7–10
12. Atlas Copco (2015) Specifications of underground manager, Version 1.0, Mar 05 2015
13. Atlas Copco (2015) Underground constrction—talking technically, 1st edn
14. Heiniö M (1999) Rock excavation handbook for civil engineering. Sandvic & Tamrock
15. Sharma PD (2009) Four-section parallel hole cut model (Swedish Method) for tunnel blasting design. https://miningandblasting.files.wordpress.com/2009/09/four-section-parallel-hole-cut-model-swedish-method-for-tunnel-blast-design.pdf, Email: sharmapd1@gmail.com, Website: http://miningandblasting.wordpress.com/. Accessed 7 Nov 2016

Chapter 19
Loading and Transportation for Underground Excavation

19.1 Loading and Haulage in Tunnel

As stated in Sect. 16.1.1 of Chap. 16, removing (loading and hauling) excavated material from the working face is one of the most important phases during the working cycle of tunneling.

According to the conditions of the project, there are different loading (mucking) and transportation (hauling) methods and different combinations of them.

- Loading Methods:
  - Overhead shovel loader. This machine usually is used in small tunnels with narrow rails (Fig. 19.1).
  - Continuous mucking machine.
  - Crawler or rubber tire excavator.
  - Load-haul-dump (LHD).
- Hauling Methods:
  - Rail transport—locomotive (electric, battery, diesel driving, or combination).
  - Dumping truck.
  - Shuttle car.
  - Belt conveyor.

The most common combinations of loading-hauling are as follows:

Wheel loader and dump truck.
Continuous mucking machine and shuttle car train/belt conveyor.
Crawler and rubber tire excavator/wheel loader and shuttle car.
Crawler and rubber tire excavator/overhead shovel loader and rail transportation.

D. Zou, *Theory and Technology of Rock Excavation for Civil Engineering*,
DOI 10.1007/978-981-10-1989-0_19

Fig. 19.1 Overhead shovel loader, rail-mounted

The principal factors when determining what type of mucking transport systems to use are as follows [1, 2]:

Size of tunnel;
Length of tunnel;
Type of material;

Time of the loading phase according to one excavation cycle required by the project working process;

Availability of equipment, and
Capital cost.

The size of the tunnel will have a direct effect on the size of equipment used for mucking. Obviously, a large tunnel cross section permits larger equipment. The length of the tunnel does not necessarily affect the choice of mucking equipment, but it does affect the choice of muck transport. That is, if it is a very long distance from the heading to the portal, a scooptram would not be efficient for hauling. It could still be used for mucking and dumping the muck into a muck car/shuttle car or a conveyor hopper. The type of material also affects the mucking and transport choice. Soft and light rock is suitable to use a continuous mucking machine or very abrasive material may cause undue damage to a conveyor belt.

A rule of thumb is that the rate of advance is totally dependent on the time required for each critical phase in the working cycle of the excavation process, which explains why contractors strive to minimize these phases as far as possible. For instance, a simple change in the choice of mucking out equipment will have a profound effect on the rate of advance.

Is the equipment available? This can be paired with capital costs. In a short tunnel, contractors may use equipment they already have or rent/buy some secondhand equipment rather than purchasing new equipment that may be more

Fig. 19.2 Wheel loader and dump truck combination (*left*) and Minetruck MT 5020 (*right*) (Reproduced by permission of Atlas Copco)

suitable for the task. The newer or more modern equipment may be more efficient, but is the additional cost worth the increasing capital cost? Whatever equipment is chosen, it is probably going to have to provide the lowest project cost.

19.1.1 Wheel Loader and Dump Truck

When the cross section of tunnel is large enough and the rubber tire wheel loader can move laterally, wheel loader and dump truck or shuttle car combination is usually used for tunnel excavation.

To suit the limited space, special articulated mine trucks with lower body are manufactured for underground excavation. The right picture of Fig. 19.2 is an example.

Load-Haul-Dump (LHD) Machine
Load-haul-dump machines are capable, as the name indicates, of loading, hauling, and dumping the muck; they function similar to front-end loaders and trucks combined. The machine is not as tall as a front-end loader and generally longer (Fig. 19.3).

Fig. 19.3 Sandvik LHD-LH203E

The LHD was originally specially designed for the pillarless sublevel caving method in Sweden's underground mines, but it has been widely used in tunnel excavation nowadays as LHD offers fast and effective face cleaning and greatly reduces the advance cycle time.

The basic aim of LHD is to clear the face of blasted material and haul it to waiting trucks (see Fig. 19.4a). In large tunnel, the loading on trucks can be done in the tunnel itself, a tunnel niche is not required. If trucks are not available, the LHD will dump the material onto a secondary muck pile to keep the excavation area clean (see Fig. 19.4b, c).

19.1.2 Continuous Mucking Machine and Shuttle Car/Belt Conveyor

Continuous loading is a method used in tunnel operation. It employs one or two digging arms that load the rock onto a conveyor, which fills the transportation vehicle (dump truck or shuttle car) at a constant flow. Figure 19.5 is an example. Figure 19.6 is another type of continuous loading equipment that is used in soft rock, such as coal. Figure 19.7 is a shuttle car. Shuttle cars are usually electric-powered, rubber-tired vehicles—essentially a large trough with a scraper conveyor running along the floor, a driver's position mounted to one side and a jib section at one end of the conveyor which can be raised for discharging the load. They are four wheel-steered, four wheel-driven vehicles with power supplied by a trailing cable reeled onto and off a cable reel mounted opposite the driver's position.

Shuttle cars are also can be rail-mounted for long haulage distance and where the tunnel floor is difficult to maintain (Fig. 19.8). Rail-mounted shuttle car can be used separately, or several cars can be connected to form a shuttle train. The scraper conveyor moves the loaded muck from the loading end to the last car to fill the complete train. Due to the large capacity of the trough, all the rock volume blasted in a round can be loaded at once. This method minimizes the need for time-consuming car switching and tramming throughout the tunnel, so that it greatly reduces the mucking and hauling time, especially when the continuous loading machine is used for soft rock.

Conveyors are often the most economical transportation mode. Belt conveyors in tunneling application have been limited due to its maximum practical particle size of approximately 300 mm. So the conveyor belt application in tunneling always works together with the mobile or semi-mobile crushing equipment which enables crushing close to the face (see Fig. 19.9). This has made it possible to change from hauling by trucks to the more economical method of continuous transportation by belt conveyors. Belt conveyors can be adopted with any loading method. Another advantage is that conveyor belt can be used over long distances without any loss of efficiency (see Fig. 19.10).

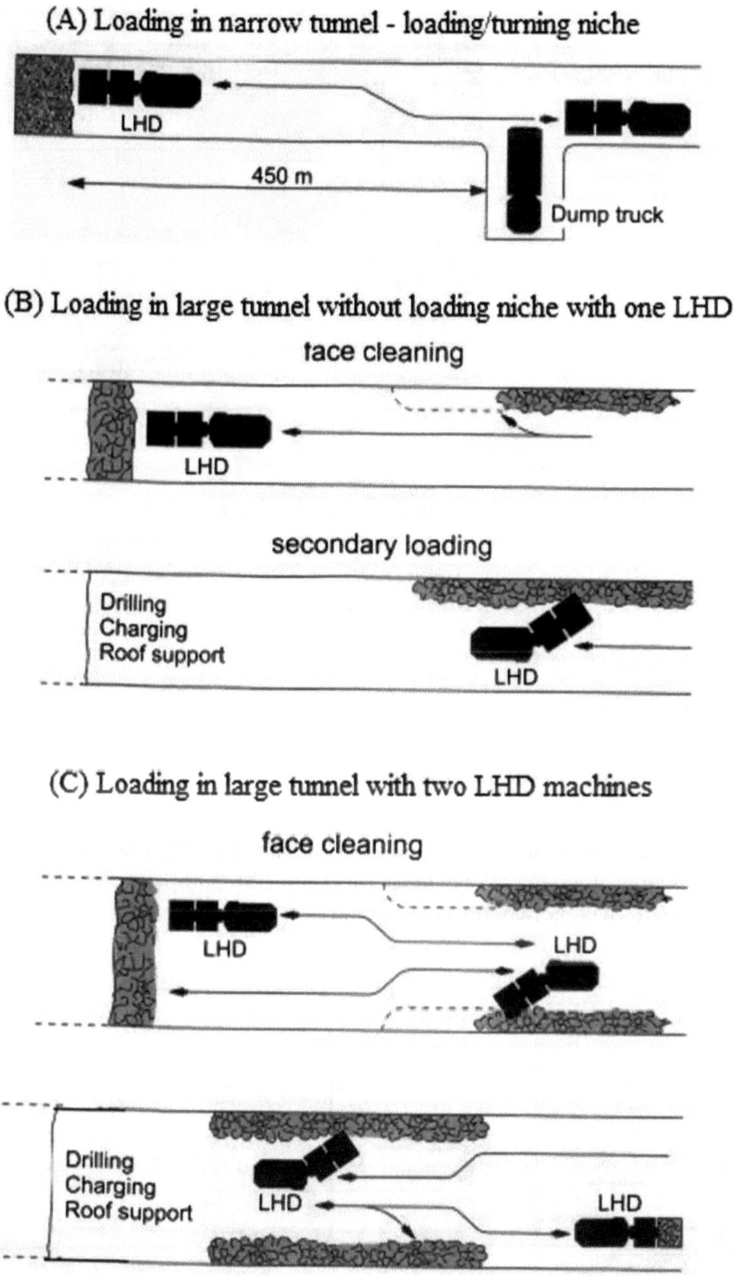

Fig. 19.4 LHD machines can be used in narrow and large tunnel for rapidly cleaning work face (Reproduced by permission of Sandvik [3])

Fig. 19.5 Atlas Copco's continuous loader: Häggloader 10HR/10HR-B

Fig. 19.6 Gathering arm-conveyor loader:14BU27 (*Source* www.joyglobal.com)

19.1.3 Railway Transportation

Railway haulage is best suited as a long distance transportation of excavated materials with a gradient in the range of 1 in 200–300. The system is flexible compared to a belt conveyor system. Good road, efficient maintenance, large

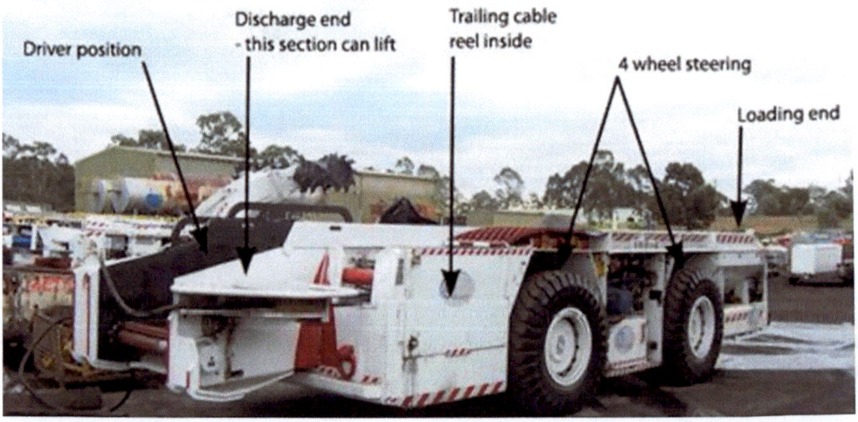

Fig. 19.7 Typical shuttle car viewed from discharge end (*Source* www.undergroundcoal.com.au)

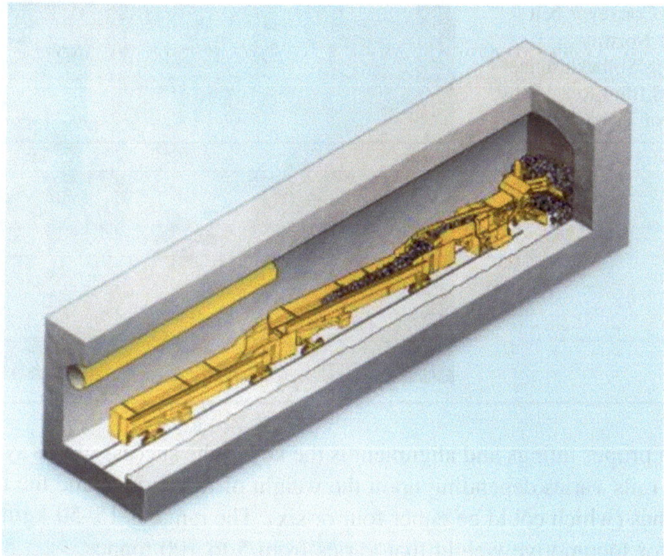

Fig. 19.8 Digging arm-conveyor loader is combined with a rail-mounted shuttle car string (Reproduced from Ref. [2] by permission of Atlas Copco)

Fig. 19.9 Sandvik underground mobile crusher (Reproduced from Ref. [4] by permission of Sandvik)

output, and adequate ventilation are the basic requirements for the success of this system. Well-drained, properly graded, and minimum turning with smooth curves constitutes a good road. Laying the rails of suitable size (i.e., weight per meter or

Fig. 19.10 Conveyor belt was used in Norway 14.3-km-long Solbakk tunnel (Reproduced from Ref. [5] by permission of TannelTalk)

yard) with proper fittings and alignment is the key to the success of this system. The weight of rails varies depending upon the weight of locomotive and the number of wheels it has (which could be either four or six). The range is 15–50 kg/m (30–100 ib./yard) for locomotive weight that varies from 5 to 100 tonnes.

Locomotives for underground use are either diesel or electric power driven. The electric power systems include battery, trolley wire, and combined trolley battery. When the diesel locomotive is used, good ventilation is required. Figure 19.11 shows an electric-powered locomotive.

Fig. 19.11 Electric locomotive for tunnel transportation

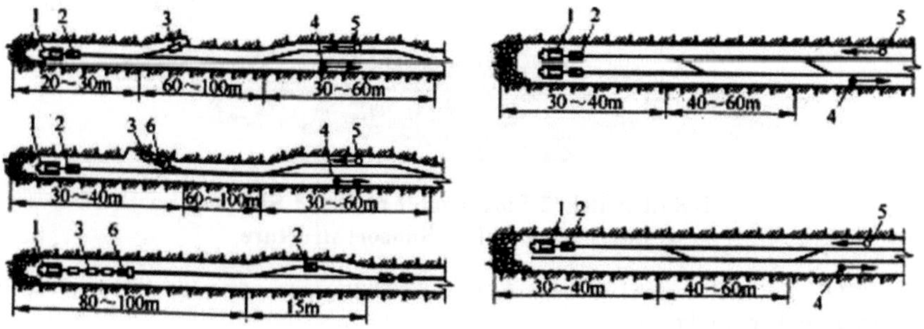

1- Loader; 2-Muck Car; 3-Empty car; 4-Track for loaded cars; 5- track for emplty cars; 6-Locomotive

Fig. 19.12 Fixed switch tracks (Reproduce from Ref. [6] by permission of Metallurgy Industry Press)

There are various muck cars for railway transportation: bottom-dump muck cars, side-dump cars, dumped by the scraper conveyor running along the car's floor such as shuttle car's and dumped by a specialized turning machine when the car passes though the machine.

Some tunnels are large enough to have two parallel tracks in the tunnel to provide transportation of men and materials. However, some tunnels are too narrow and more than one train cannot operate in the tunnel at the same time. A switch track can be fabricated to allow the trains to pass each other. Figures 19.12 and 19.13 show two types of switch tracks and Fig. 19.14 is a float double-line switch for the single-track layout in the narrow tunnel shown in Fig. 19.13c.

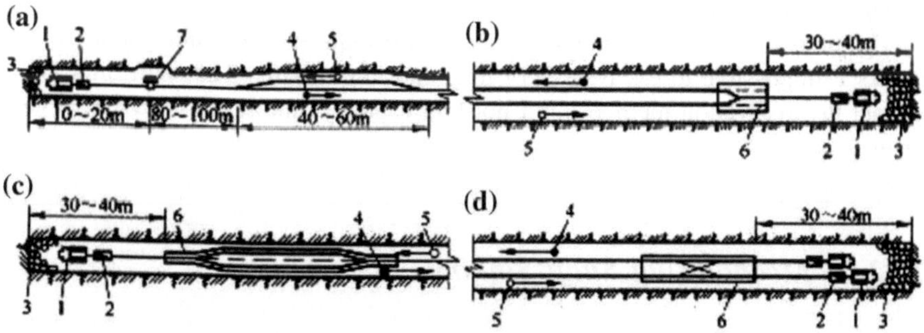

1-load; 2-muck car; 3- Mulk Pile; 4-Loaded car moving direction; 5- Empty car moving direction; 6-Float trck switch; 7-Translational shunting device

Fig. 19.13 Movable shunting devices (Reproduce from Ref. [6] by permission of Metallurgy Industry Press)

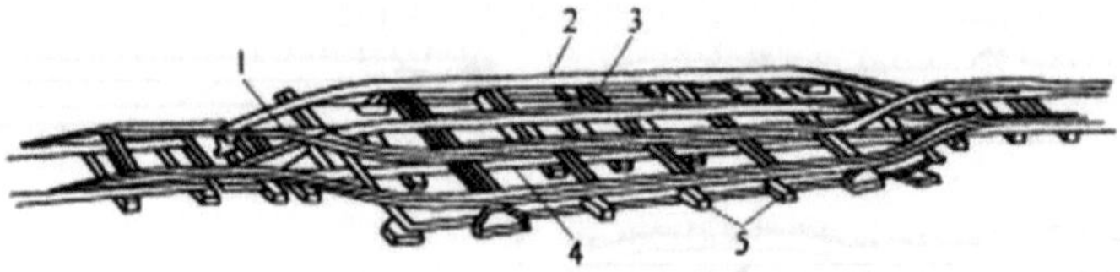

1-Rail switch; 2-Float double tracks; 3-Sleeper; 4-Single track rail: 5- Support structure

Fig. 19.14 Float double-track switch for single-track rail (Reproduce from Ref. [6] by permission of Metallurgy Industry Press)

19.2 Loading and Hoisting in Shaft Excavation

After exploration of the geology and groundwater conditions, overburden is removed. If the overburden requires stabilizing, it is typically lined with concrete rings. Once the rock surface has been exposed, it is reinforced and grouted. The collar for the head frame is installed after excavation has progressed a short distance. The head frame includes the hoisting system for the shaft sinking equipment. At this point, the actual shaft sinking begins.

19.2.1 Manual Shaft Sinking

Manual shaft sinking is hard work and time-consuming with slow progress. Usually, simple equipment is used, such as handheld rock drills and shoveling the rock manually into small buckets. A crane is used to transport the buckets up and down to transport rock materials.

The number of workers and amount of effort required for manual shaft sinking makes it impossible to excavate very long and large shafts. The shaft dimensions restrict the excavation method, limit hoisting capacity, and restrict the use of large plant.

19.2.2 Mechanizing Shaft Sinking

19.2.2.1 Simple Mucking and Hoisting Equipment

If the shaft is not deep (<100 m) or it is only for temporary use and with relative small size, some simple equipment for mucking, e.g., small loader and skip, and hoisting, e.g., crawler crane or gantry traveling crane, are usually used. Figure 19.15 shows this simple equipment for shallow shaft sinking operation. The skips should be only 75 % loaded to avoid the risk of falling rocks.

Fig. 19.15 Simple equipment for shallow shaft sinking operation

19.2.2.2 Specialized Mucking and Hoisting System

For deep (50 m or deeper) and large shafts, a specialized mucking and hoisting system is necessary for a safe and efficient shaft excavation. The system consists of a loading machine, a loading container (skips) and hoisting and dump system.

Mucking

Several types of shaft muckers are available; the prominent among them are as follows:

- Cactus grab slung below the sinking stage;
- Arm loaders such as Cryderman grab—wall mounted;
- Rocker shovel such as Eimco 630.

Grab is suspended from a 360°-rotation turret boom or lashing unit, for rapid rates of shaft sinking. An alternative may be high-performance Cryderman grab, wall-mounted muckers.

Fig. 19.16 Cactus grab manufactured by DH Mining Systems (*Source* www.dhms.com.en)

Fig. 19.17 Eimco 630 air mucking machine (*Source* www.savonaequipment.com/en)

A Cryderman mucker is a loader that operates by means of pneumatic cylinders and a telescopic boom. This is suspended from independent hoisting system mostly located at the surface and used in rectangular, circular, and inclined mine shafts. A hydraulic version of the unit has also been developed and is powered by a self-contained hydraulic power pack and a longer boom. Figure 19.16 shows a grab.

The Eimco 630 rocker shovel is a miniature backhoe and front-end loader. It is pneumatically operated and digs the muck with the force of the machine moving forward and manipulation of the muck bucket. Figure 19.17 is a Eimco 630 mucker. Eimco 630 is also can be used in tunnel mucking (see Fig. 19.1).

Fig. 19.18 Steel kibble for shaft sinking

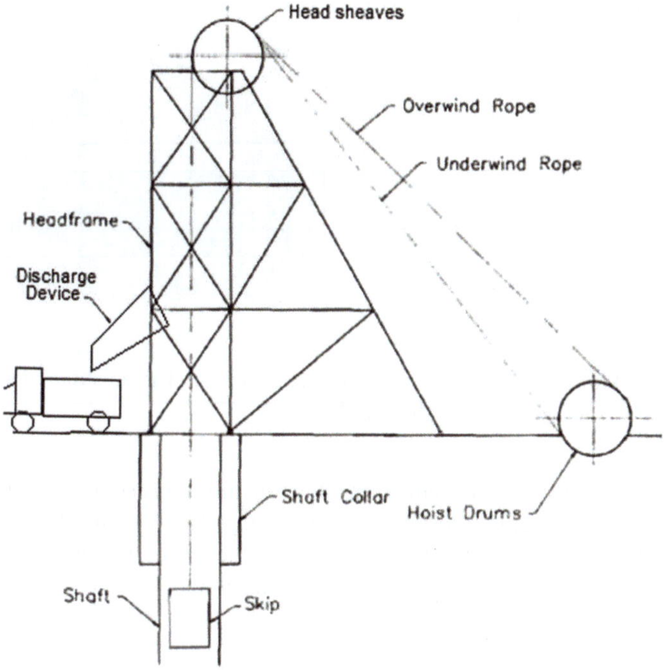

Fig. 19.19 Hoisting system for shaft transportation

Fig. 19.20 Head frame and discharge device

Fig. 19.21 Hoisting machine of the shaft

The steel kibbles are usually used as the container for hoisting mucked rock. Figure 19.18 is an example of the kibble.

Hoisting

The hoisting system for shaft excavation mainly consists of the head frame, headsheaves, hoist drums (hoisting machine), skip (kibble), and discharge device (see Fig. 19.19). Figures 19.20 and 19.21 are the examples of the head frame (including discharge device) and the hoisting machine.

References

1. Hemphill GB (2013) Practical tunnel construction. Wiley, Hoboken. ISBN: 978-0-470-64197-2
2. Atlas Copco (2015) Underground constrction—talking technically, 1st edn
3. Heiniö M (ed) (1999) Rock excavation handbook for civil engineering. Sandvic & Tamrock, Stockholm
4. Construction Projects around the world—Tunnel and Underground Civil Engineering—Leighton Contractors (Asia) Limited, Hong Kong, 2013
5. Kenyon P (2014) Norway advances Ryfast Mega-Project, TunnelTalk, July 2014
6. 赵兴东 (2012) 井巷工程,冶金工业出版社,北京,2012 (in Chinese). Zhao X (2012) Shaft & drift engineering. Metallurgical Industry Press, Beijing. ISBN: 978-7-5024-5238-4

Chapter 20
Ventilation for Underground Excavation

20.1 Requirement of Ventilation for Underground Excavation

The object of ventilation in a tunnel or shaft is to provide fresh air or to extract pollutants, in order to achieve an acceptable environment. The quality of fresh air supplied is usually determined not only by breathing requirements but by the need to dilute pollutants and to provide cooling.

This requires mechanical ventilation of all areas of tunnels, shafts, and other underground workings with clean, breathable, non-recirculated, outside air. The ventilation system must be in place and in operation before employees enter any underground workings, and the system must be kept in operation until all personnel have left the area serviced by the system.

20.1.1 Requirement of Air Quality

The air in all underground areas shall contain not less than 19.5 % oxygen and shall not contain a concentration of contaminants such as gases, vapors, and dusts greater than is safe having regard to the effects of exposure time, temperature, humidity and the combined effects of several contaminates. The concentration of inflammable contaminants shall not exceed 7 % of the lower explosive limit (minimum explosive concentration).

This issue will be discussed in detail in Chap. 23, Health and Safety, and Risk Management in Underground Excavation.

D. Zou, *Theory and Technology of Rock Excavation for Civil Engineering*,
DOI 10.1007/978-981-10-1989-0_20

20.1.2 Fresh Air Supply Quantities

To design the method and capacities (m^3/minute) of the ventilation system should take account of all factors such as follows:

- maximum number of workers at the work face at any one time;
- tunnel size, length or volume of the cavern and gradient of drive;
- ambient conditions;
- the presence of water, dust, fumes such as methane;
- amount and type of plant to be used at any one time.

Particular care is essential where dangerous dust or toxic gases are present or foreseeable. Actual pollution levels should be measured systematically.

According to British Standard, BS 6164:2011 (refer to [1]), "a minimum fresh air supply of 0.3 m^3/min per person is normally sufficient. Additional ventilation should be provided where plant (particularly diesel-powered) is to be used. An additional supply of at least 3.0 m^3/min per working kilowatt is recommended for machines with stringent emission controls." "Recommended minimum velocities in the tunnel are 0.5 m/s to contain dust and 2.0 m/s to prevent layering of methane."

20.2 Ventilation Equipment

20.2.1 Ventilation Machine: Fans

Axial fans are mostly used in tunnel ventilation. An example is shown in Fig. 20.1.

20.2.2 Requirements to Ventilation Equipment

According to Reclamation Safety and Health Standards of the US Department of the Interior Bureau of Reclamation, July 2014 [3], only class 1, division 1[1] electric motors, fans, drives, and auxiliary equipment, including wiring, starters, and controls can be used for the ventilation of tunnels and shafts during construction and the noise levels of ventilation fans should not exceed 90 dB when measured at the closest point of employee exposure.

[1](*Class 1, division 1—Explosion proof electric equipment—Electrical equipment for use in class 1 hazardous (classified) locations, as defined in the National Electric Code, is tested with respect to acceptability of the operation in the presence of flammable and explosive mixtures of specific vapors and gases with air).

Fig. 20.1 Atlas Copco's axial fun for tunnel ventilation

20.2.3 Type of Fans for Underground Ventilation System

The ventilating air current is drawn around an underground tunnel or cavern through pressure created by a fan. Different types of fans include as follows [6]:

a. Main fans, which create the primary ventilating pressure, either forcing or exhausting that generates the ventilation circuit around the underground working space.
b. Auxiliary fans, which generate an airflow beyond the circuit created by the main and booster fans. These fans are used in headings and developments.
c. Booster fans, which are located underground and reduce the pressure that should be generated by the main fans in underground space with complex underground workings.
d. Scrubber fans, which remove dust by passing air through a filter. Scrubber fans may have arrangements where the capacity of the fan is greater than the volume available in the workplace. This is planned recirculation, which removes dust through filtration.

20.3 Design of Ventilation System

20.3.1 Calculation of Discharge Volume of Ventilation (Refer to [4])

Calculation of discharge volume of ventilation

The discharge volume of ventilation in tunnels is calculated as follows.

$$Q = Q_1 + Q_2 + Q_3 + Q_4 \quad \text{(Unit:Volume per time)} \tag{20.1}$$

Table 20.1 Volume of generated noxious gas (quoted from standard specification for tunneling (mountainous tunnels) issued by Japan Society of Civil Engineers) Ref. [4]

Explosive/diesel	Classification	Noxious gas	Volume (m^3/kg) for CO (m^3/(min/piece) for NO_x
Explosive	Enoki-dynamite No. 2	Carbon monoxide	8×10^{-3}
	Other dynamite	Carbon monoxide	11×10^{-3}
	Slurry type	Carbon monoxide	2×10^{-3}
	Emulsion type	Carbon monoxide	5×10^{-3}
	ANFO	Carbon monoxide	30×10^{-3}
Diesel	Shovel	Nitrogen oxide	55×10^{-6}
	Dump truck	Nitrogen oxide	20×10^{-6}
	Others	Nitrogen oxide	20×10^{-6}

where

Q Total discharge volume of ventilation;
Q_1 Discharge volume of ventilation necessary for laborers and engineers in the tunnel $Q_1 = q_1 \times N_1$ and q_1—discharge volume of ventilation per one person, N_1—maximum number of laborers and engineers in the tunnel;
Q_2 Discharge volume of ventilation for dust caused by blasting:

$$Q_2 = (V_{21}/T)\{1 - (K_2 \times V_{21})/V_{22}\}$$

where

V_{21} Volume of tunnel where ventilation is necessary,

$$V_{21} = A_2 \times L_2,$$

where A_2—area of tunnel and L_2 = Tunnel length where ventilation is necessary;

T Time during ventilation;
K_2 Allowable density of noxious gas (Refer to Table 20.3.)
V_{22} Volume of poison gas generated by blasting (Refer to Table 20.1).
Q_3 Discharge volume of ventilation for dust generated by shotcreting,

$$Q_3 = q_3/K_3,$$

where

q_3 Dust weight per time generated by shotcreting.
K_3 Allowable density of dust (Refer to Table 20.2).
Q_4 Discharge volume of ventilation for noxious gas generated by vehicles to transport excavated materials,

Table 20.2 Allowable density of dust (refer to [4])

Category	Kind of dust	Allowable density (mg/m^3)	
		Absorbent dust	Total dust
I	Talc, soapstone, kieselguhr, aluminum, bentonite, etc.	0.5	2
II	Mineral dust, iron oxide, coal, Portland cement, limestone, etc.	1	4
III	Other organic or inorganic dust	2	8
Asbestos	Actinolite, etc.	0.12	

$$Q_4 = Q_{41} \times \alpha \times N_4 / K_4,$$

where

Q_{41} discharge volume of exhaust gas generated by one vehicle

$$Q_{41} = \beta \times V_{41} \times N_{41};$$

where

β Coefficient decided by type of engine, Q_{41} = 0.4–1.2 for meter-minute unit;
V_{41} engine displacement (volume);
N_{41} Number of rotation of engine (rpm);
α content of noxious gas in exhaust gas (volume ratio: noxious gas divided by exhaust gas);
N_4 Number of vehicles; and
K_4 Allowable density of noxious gas (refer to Table 20.3).

If we can do blasting, shotcreting and usage of vehicles separately without concurrent works, and ventilate the tunnel during each interval, Eq. (20.1) can be replaced with the following Eq. (20.2):

$$Q = Q_1 + \text{Max}(Q_2, Q3, Q_4)(\text{Unit:Volume per time}) \qquad (20.2)$$

20.3.2 Calculation of Ventilation Pressure (Refer to [2])

Ventilation pressure requirements can be calculated using the following equation:

$$\boldsymbol{P} = \boldsymbol{R}\boldsymbol{Q}^2 \quad \text{or} \quad \boldsymbol{P} = \frac{\boldsymbol{ksQ}^2}{\boldsymbol{A}^3} \qquad (20.3)$$

Table 20.3 Allowable density of noxious gas (refer to [4])

Gas	Allowable density (ppm)
Carbon monoxide	100
Nitrogen oxide	25

where

- ***P*** ventilation pressure (Pa);
- ***R*** Equivalent resistance ($Ns^2\ m^{-8}$);
- ***Q*** Airflow (m^3/s);
- ***k*** Friction factor (see Table 20.4);
- *s* Rubbing surface (m^2) (length × perimeter); and
- ***A*** Cross-sectional area (m^2).

The pressure drop is influenced by the cross-sectional area of the roadway, so roadway size is of great importance in ventilation planning.

20.3.3 Design of Fan and Ventilation Duct [4]

The power of fan and the size of ventilation duct are designed by the application of Bernoulli's theorem as follows:

$$\boldsymbol{V} = \boldsymbol{Q}/\boldsymbol{A} \tag{20.4}$$

$$\boldsymbol{h} = \lambda\left(\frac{\boldsymbol{L}}{\boldsymbol{D}}\right) \times \left(\frac{\boldsymbol{V}^2}{2\boldsymbol{g}}\right) \times \gamma \tag{20.5}$$

$$\boldsymbol{N} = \boldsymbol{Q} \times \boldsymbol{g} \times \boldsymbol{h} \times \rho_w \tag{20.6}$$

$$\boldsymbol{B} = \left(\frac{\boldsymbol{N}}{\eta}\right) \times \alpha \tag{20.7}$$

Table 20.4 k values for calculating airflow resistance based on roadway (Quoted from Appendix A2 of [2])

Condition of lining				
Smooth concrete all round	Concrete slabs or timber lagging between flanges and spring line	Concrete slabs, timber or bricks between flanges and spring line	Lagging behind arches—good straight airways	Rough conditions with irregular roof, sides and floor
$k = 0.0037$	$k = 0.0074$	$k = 0.0093$	$k = 0.0121$	$k = 0.0158$

Table 20.5 Duct characteristics (quoted from [5])

Duct type	Friction coefficient, λ	Active leakage surface, f^*
Flexible forced, class S	0.015	5 mm^2/m^2
Flexible forced, class A	0.018	10 mm^2/m^2
Flexible forced, class B	0.024	20 mm^2/m^2
Flexible exhaust, reinforced with helical-wound spring steel	0.025	5–20 mm^2/m^2
Sheet metal duct	0.010	2 mm^2/m^2

where

- V Velocity of air;
- Q Discharge volume of ventilation, refer to (20.2) or (20.3);
- A Area of ventilation duct;
- h Friction loss of air head;
- λ Coefficient of friction loss (refer to Table 20.5);
- D Diameter of ventilation duct;
- L Length of ventilation duct;
- g Acceleration of gravity;
- γ Specific gravity of air, $\gamma = 0.0012$;
- N Theoretical power of fan;
- ρ_w Density of water;
- B Actual necessary power of fan;
- η Efficiency of power of fan;
- α Safety factor, $\alpha = 1.15{-}1.2$

20.3.4 Choose of Van

Usually, the manufacturer of the fans has a ventilation system optimization program developed according to hydromechanics principles, which will output the fan characteristic chart based on the site conditions, required air quantity and pressure and the laboratory test. Figure 20.2 is an example of fan characteristic curve for a tunnel project in Hong Kong offered by Atlas Copco.

20.3.5 Ventilation Duct

There are three types of ducts used for tunnel ventilation [7]:

a. Flexible smooth ducts made of plastic fabric, see Fig. 20.3. They are used for the forcing (supply) ventilation (see the next section).

		1xAVH140	Ø 1400 mm	50 Hz	n = 1480 rpm
ß	kW	Type	ß	kW	Type
36°	1x37	AVH140.37.4.8/50Hz	52°	1x90	AVH140.90.4.8/50Hz
40°	1x45	AVH140.45.4.8/50Hz	56°	1x110	AVH140.110.4.8/50Hz
44°	1x55	AVH140.55.4.8/50Hz	60°	1x132	AVH140.132.4.8/50Hz
48°	1x75	AVH140.75.4.8/50Hz			

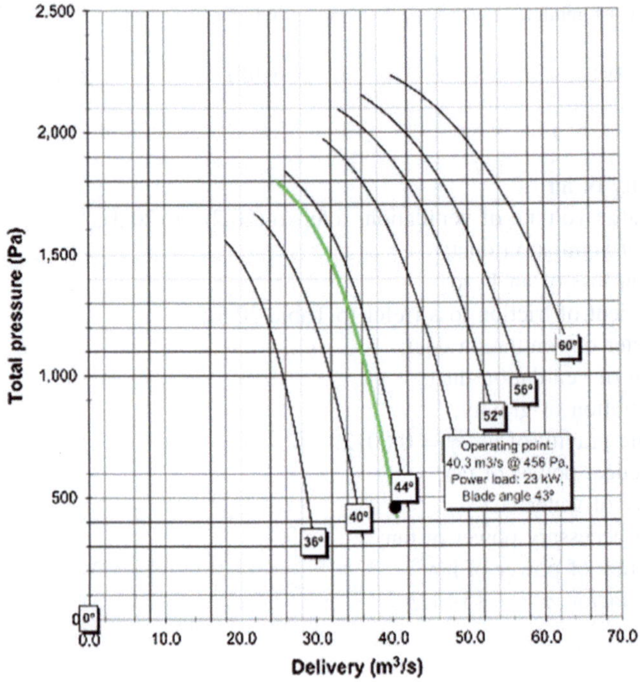

Fig. 20.2 Characteristic curve of axial fan of model AVH140 for a tunnel project in Hong Kong in 2013 by Atlas Copco

b. Flexible ducts reinforced with helical-wound spring steel (see Fig. 20.4). They are used for both forcing and exhaust (extraction) ventilation.
c. Solid ducts made of sheet metal ducts which are used for exhaust ventilation and places in the tunnel where the ducts are at risk of collision or scratch, especially near the working face and the tunnel corners.

Fig. 20.3 Flexible smooth ducts made of plastic fabric for tunnel ventilation. *Source* www.hpcproducts.com

Fig. 20.4 Flexible ducts reinforced with helical-wound spring steel

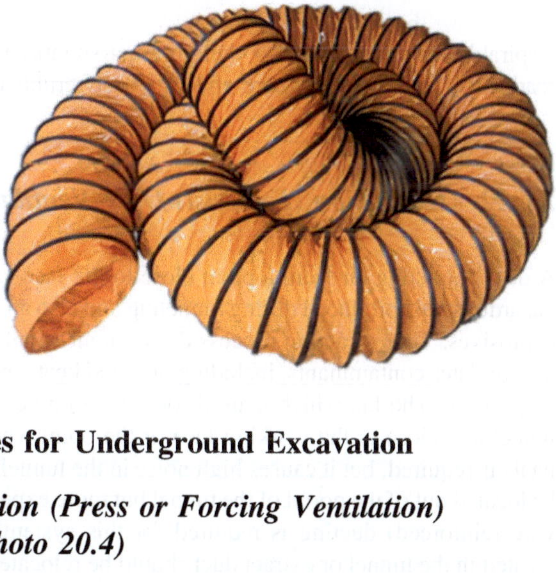

20.4 Ventilation Types for Underground Excavation

20.4.1 Supply Ventilation (Press or Forcing Ventilation) (Fig. 20.5a) (Photo 20.4)

Ducted forced ventilation to the working face provides fresh air for face workers where dust levels are not a priority. As the air passes back along the tunnel, it becomes progressively more contaminated. Where explosives are in use, it carries back a plug of heavily polluted air, which gradually diffuses into the main body of air, but which may still be dangerous. When the tunnel is longer than 1000 m, supply-only ventilation is not enough and it has significant limitations in terms of

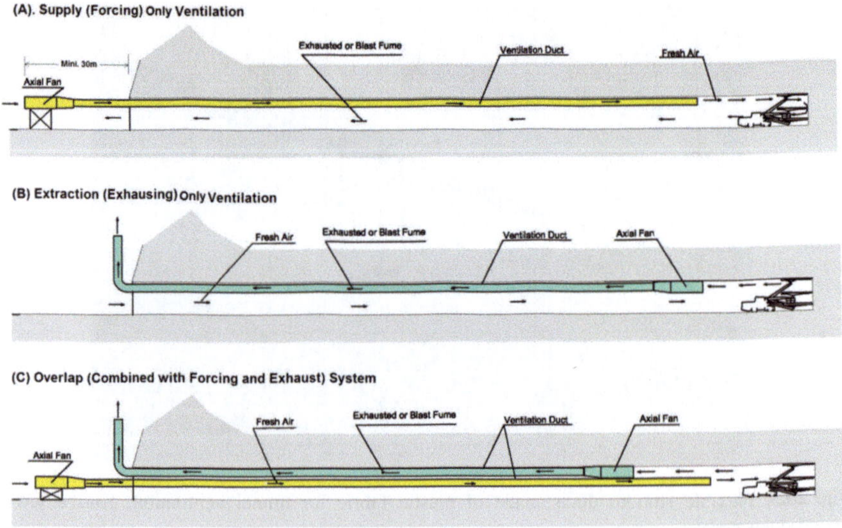

Fig. 20.5 Three types of tunnel ventilation system

respirable dust control and is therefore not suitable in machine cut headings (e.g., roadheader excavation) where effective dust scrubbers are not being used.7

20.4.2 Extraction (Exhaust)-Only Ventilation (Fig. 20.5b)

A duct extracting air from a point close to the face can be used directly to remove hazardous dust produced during tunneling and also any fumes arising from the use of explosives. The extraction process draws in air supply along the tunnel, but this air accumulates contaminants, including dust and heat, and increases in humidity during its passage. The fan which is used for extracting ventilation can be installed in the tunnel and close to the working face or the point where the extraction of contaminants is required, but it causes high noise in the tunnel. As an alternative, the fan can be located out of the portal of the tunnel but more expensive solid or semirigid (spiral wire reinforced) ducting is required for this system. The extractive fan which is located in the tunnel or extract duct should be relocated as frequently as is necessary to keep them close to the source of contamination and close to the face.

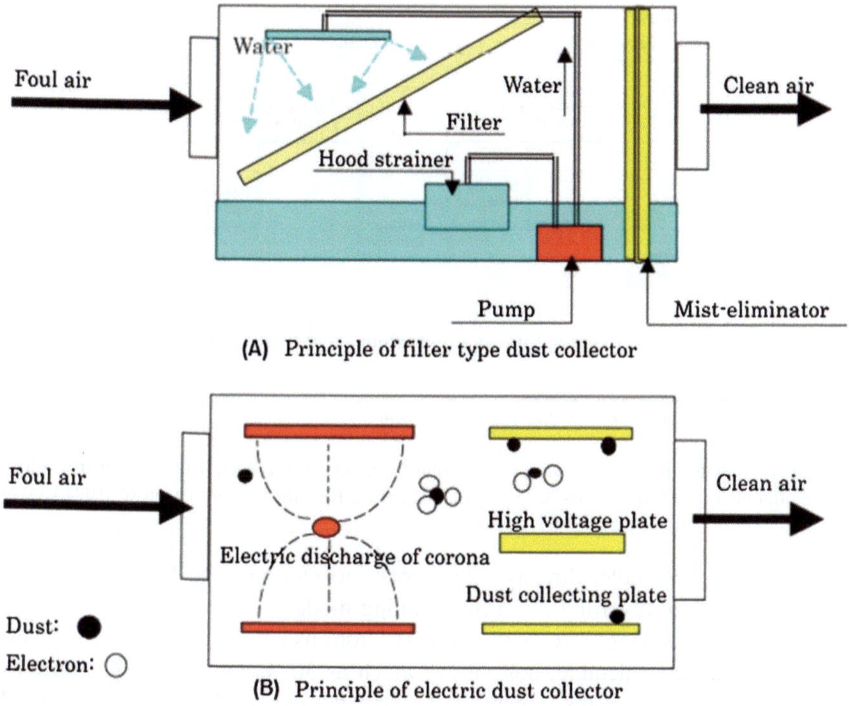

Fig. 20.6 Principles of dust collectors (*Source* [4])

20.4.3 Compound Type of Ventilation (Overlap System) (Fig. 20.5c)

An overlap system combined supply and extraction ventilation is sometimes necessary which combines the positive aspects and minimizes the weaknesses of the systems referred to above. The duct layout should be designed to maintain a circulating flow at all workstations. Filters to remove dust from the extract duct can be necessary. Dust collectors are classified into the wet type and the dry type. The filter type dust collector belongs to the former. The electric dust collector and the centrifugal dust collector belong to the latter. Figure 20.6 shows the principle of dust collectors of these two types (quoted from [4]).

20.5 Ventilation Optimization

In recent years, frequency conversion technology has been applied to the ventilation fans by some manufactures. By using a frequency inverter, the speed of the fan will not be higher than necessary and the airflow during excavation will be optimized,

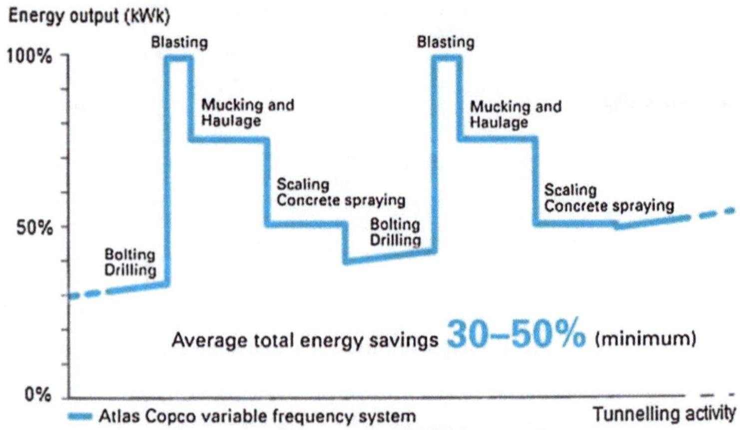

Fig. 20.7 Variable frequency enables the output to be adapted to each activity, which results in substantial energy saving of 30–50 % (reproduced from Ref. [8] by permission of Atlas Copco)

increasing, for example, directly after a blast to evacuate the fumes as soon as possible, and then reverting to normal running mode.

The potential saving on energy and costs from using the frequency conversion technology are considerable, possibly as much as 50 % compared with some traditional ventilation systems. Figure 20.7 shows the idea of Atlas Copco's variable frequency system (courtesy of Atlas Copco [8]).

References

1. United Kingdom (2011) British Standard, BS6164:2011, Code of Practice for Safety in Tunnelling in the Construction Industry
2. New Zealand (2014) Ventilation in underground mines and tunnels, approved code of practice, New Zealand Government, February, 2014
3. USA (2014) Section 23, Tunnel and Shaft Construction, Reclamation Safety and Health Standards, U.S. Department of the Interior Bureau of Reclamation, July 2014
4. Takano Y (2001) Ventilation system in tunnel during construction works. In: Proceedings of the Sino-Japanese modern engineering and technology symposium, 2001
5. Drost U et al (2006) Tunnel construction site ventilation and cooling: an integrated flow and heat load solver applied to the lyon-turin high-speed railroad tunnel project. In: TMI international conference: tunnel protection and security against fire and other hazards, 15–16. May 2006, Torino, Italy
6. Hemphill GB (2013) Practical tunnel construction. Wiley, USA. ISBN: 978-0-470-64197-2
7. Tatiya RR (2013) Surface and underground excavations—method. technique and equipment, 2nd edn. CRC Press, London. ISBN: 978-0-415-62119-9
8. Greenwood J (ed) (2015) Underground construction—talking technically. 1st edn. Atlas Copco ROck Drill AB, Lars Senf, Örebro, Sweden

Chapter 21
Ground Reinforcement and Support

21.1 Effects of the Stability of Rock Mass to Underground Excavation

21.1.1 Concept of Ground Pressure and Stress Pattern Around Underground Excavation

One important factor in underground construction is rock pressure—the in situ state of stress in a rock mass. In practice, the result of this stress is also called rock pressure.

Primary rock pressure is the summary of stresses in a rock mass before influencing it, for example, by excavating underground openings. Primary rock pressure is the result of overburden and residual or tectonic stresses.

Secondary rock pressure occurs when the primary stress field is altered by the excavation process. The secondary stress field can show considerable changes throughout the excavation process, thus indicating an unbalanced state of equilibrium.

The goal of the excavation process is to achieve a balanced state while avoiding any intermediate condition that may endanger the excavation itself, and the people and equipment working there. In practice, the stress itself does not form the critical factor, but the reactions of the rock mass caused by it.

The following characteristics of rock pressure take place in tunneling:

- Stress field is relocated resulting in the elastic deformation of the face and roof of the tunnel without fracture.
- Stress release occurs by sudden rock failure ranging in intensity from spalling to rock burst.
- Fracture and consecutive deformation of rock in the tunnel face and roof take place in the rock mass with originally elastic or quasi-elastic behavior.

D. Zou, *Theory and Technology of Rock Excavation for Civil Engineering*,
DOI 10.1007/978-981-10-1989-0_21

- Deformation and consecutive failure take place in a rock mass with originally plastic-viscous behavior.

All the above-mentioned reactions are time-dependent. The type of reaction that takes place also depends on the original state of stress and the rock mass behavior. It is also highly influenced by the mode and sequence of the excavation operations and the size and shape of the openings.

There are lots of theories and methods to estimate the ground pressure on the underground spaces. Among them, Terzaghi's method and Bierbäumer's theories are the most recognized empirical methods and will be introduced briefly as follows.

- Terzaghi's Rock Load Theory [1, 2, 15]

Terzaghi (1946) applied the failure mechanism shown in Fig. 21.1 to calculate rock load where a tunnel lining should support when the tunnel was excavated in cohesionless dry coarse soil. Vertical rock load (P_{roof}) was suggested by Eq. (21.1) for shallow tunnels.

$$P_{\text{roof}} = \frac{\gamma B}{2K \tan \phi}\left(1 - e^{K \tan \phi \frac{2H}{B}}\right) \tag{21.1}$$

where B is the relaxed range ($=2[b/2 + m \tan(45 - \phi/2)]$), v is the unit weight of rock mass, K is coefficient of lateral earth pressure, ϕ is friction angle, b, m, and H are the width, height, and depth of a tunnel, respectively.

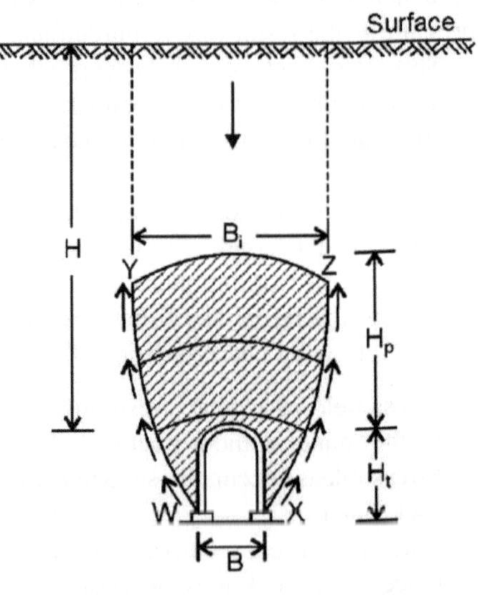

Fig. 21.1 Terzaghi's rock load concept in tunnel (Courtesy of Tsinghua University Press, Ref. [15])

For deep tunnels, the vertical rock load was also suggested as follows:

$$P_{\text{roof}} = \frac{\gamma B}{2K \tan \phi} \text{ (const.)} \tag{21.2}$$

This method is considered to be conservative, as it was established for very poor rock quantity in which the tunnels were supported by steel ribs and wooden blockings. Terzaghi's ground arch action concept and experiments with loose granular materials found that the arch thickness of a tunnel's roof is directly proportional to the excavation dimensions. This concept is shown in Fig. 21.1.

Singh et al. (1992) and Goel et al. (1995) [1] studied the effect of the tunnel size on the support pressure using the Q-system. (Q-system will be illustrated in the late Sects. 21.1.3.2 and 21.1.3.4) Based on the correlations between the measured and observed tunnel support pressures for various tunnel sizes in both squeezing and non-squeezing ground conditions, they concluded that none of rock mass classifications are applicable for squeezing ground conditions. An empirical equation relates the support pressure and the rock quality (Q).

$$P_{\text{roof}} = \frac{200}{J_{\text{r}}} Q^{1/3} \tag{21.3}$$

where P_{roof} is the roof support pressure (kPa), and J_{r} is the joint roughness in the Q-system.

Bhasin and Grimstad (1996) [1] further revised Eq. (21.3) according to the data from Norweginian tunnels:

$$P_{\text{roof}} = \frac{40D}{J_r} Q^{1/3} \tag{21.4}$$

where D represents the diameter (span) of the tunnel (m).

- Bierbäumer's Theory (Reproduced from Ref. [2] by permission of Taylor and Francis Group)

Bierbäumer's theory was developed during the construction of the great Alpine tunnels. The theory states that a tunnel is acted upon by the load of a rock mass bounded by a parabola relaxed zone of height, $h = \alpha H$, and width of B (see Fig. 21.2). ∅ is the internal friction angle of rock mass.

As shown in Fig. 21.2, the upper relaxed zone acts on the tunnel along $45 + \phi/2$ inclined plane as a vertical load. The height of the relaxed zone (h) is assumed to be proportional to the depth of the tunnel (H); $h = \alpha H$ (α is reduction factor). At tunnel crown, the vertical relaxed load (P_{roof}) can be given as follows:

$$P_{\text{roof}} = \alpha H \gamma \tag{21.5}$$

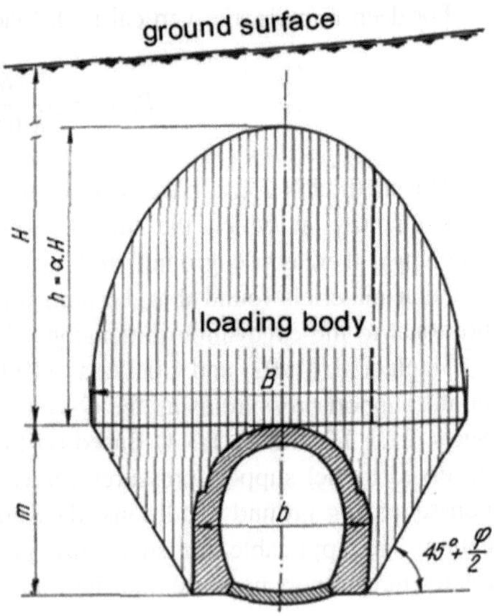

Fig. 21.2 The assumed relaxed zone of Bierbäumer (Reproduced from Ref. [2] by permission of Taylor and Francis Group)

where v is the unit weight of rock mass, and the reduction factor can be obtained as follows:when H is very small:

$$\alpha = 1 \tag{21.6}$$

when $H \leq 5B$:

$$\alpha = 1 - \frac{\tan\phi\tan^2(45 - \phi/2)H}{b + 2m\tan(45 - \phi/2)} \tag{21.7}$$

when $H \geq 5B$:

$$\alpha = \tan^4(45 - \phi/2) \tag{21.8}$$

- Approach based on the Q-system (Reproduced from Ref. [2] by permission of Taylor and Francis Group)

Barton et al. (1974) [2] proposed an empirical equation for the estimation of vertical rock load ($P_{roof)}$ based on Q values (Q-system will be illustrated in the late Sects. 21.1.3.2 and 21.1.3.4) as follows:

If the number of joint sets ≥3:

$$P_{roof} = 2Q^{-1/3}J_r^{-1}(\mathrm{kg/m^2}) \tag{21.9}$$

Table 21.1 Transformed Q value

Range of Q value	Q'
$10 < -$	$5Q$
$0.1 \leq Q \leq 10$	$2.5Q$
$Q < 0.1$	Q

If the number of joint sets <3:

$$P_{\text{roof}} = \frac{2}{3}\sqrt{J_n}Q^{-1/3}J_r^{-1}\ (\text{kg/m}^2) \tag{21.10}$$

where Q is Q value, J_n is the number of joint sets, and J_r is the joint roughness coefficient.

Also, horizontal rock load (P_{wall}) can be calculated from the same equations applied for P_{roof} with the transformed Q value (Q') according to Table 21.1.

- Results of In Situ Stress Measurement

E. Hoek and E.T. Brown [3] collected lots of in situ measured data (they were also involved in the development of stress measuring equipment in South Africa) of ground stresses from various locations around the world. The results are shown in Figs. 21.3 and 21.4 (transfered from [15]).

Figure 21.3 shows that the measured vertical stresses are in fair agreement with the simple prediction given by calculating the vertical stress due to the overlying weight of rock at a particular depth from the equation:

$$\sigma_z = \gamma z \tag{21.11}$$

where γ is the unit weight of the rock (usually in the range 20–30 kN/m^3), and z is the depth at which the stress is required [15].

Figure 21.4 shows a plot of k, the ratio of the average horizontal to vertical stress, against depth below surface. It will be seen that for most of the values plotted, the value of k lies within the limits defined by:

$$\frac{100}{z} + 0.3 < k < \frac{1500}{z} + 0.5 \tag{21.12}$$

The plot shows that at depth of less than 500 m, the horizontal stresses are significantly greater than the vertical stresses. For depth in excess of 1 km (3280 feet), the average horizontal stress and the vertical stress tend to equalize [15].

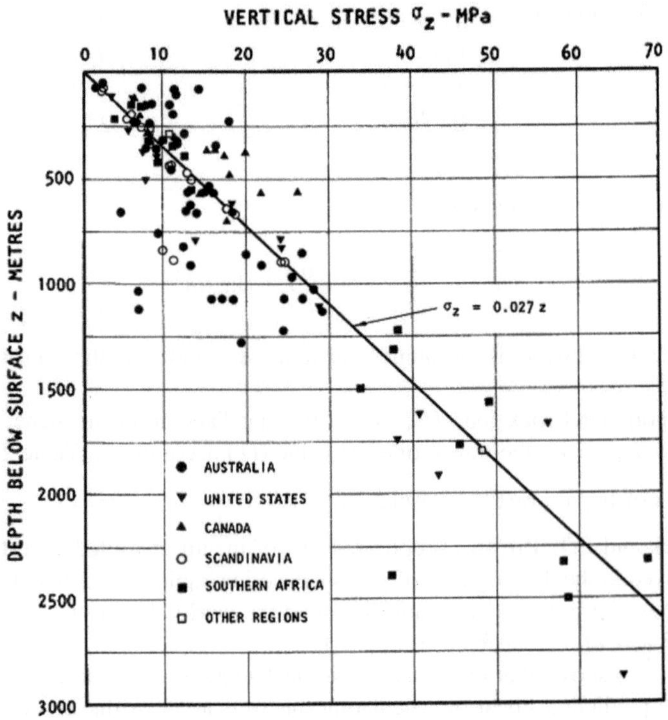

Fig. 21.3 Plot of vertical stress against depth below surface (Courtesy of Tsinghua University Press, Ref. [15])

21.1.2 Effects of Ground Conditions on Underground Excavation

Please refer to Chap. 1: Sect. 1.4. Properties of Rock Mass and their effects on Rock Excavation.

21.1.3 Classification of Stability of Rock Mass for Excavation

In Chap. 1, Sect. 1.5, the general or comprehensive classification of rock that is defined as the classification of rock sturdiness was discussed. In this section, the rock mass classification schemes or classification of stability of rock mass will be further discussed as they are closely connected with rock excavation, especially the underground excavation.

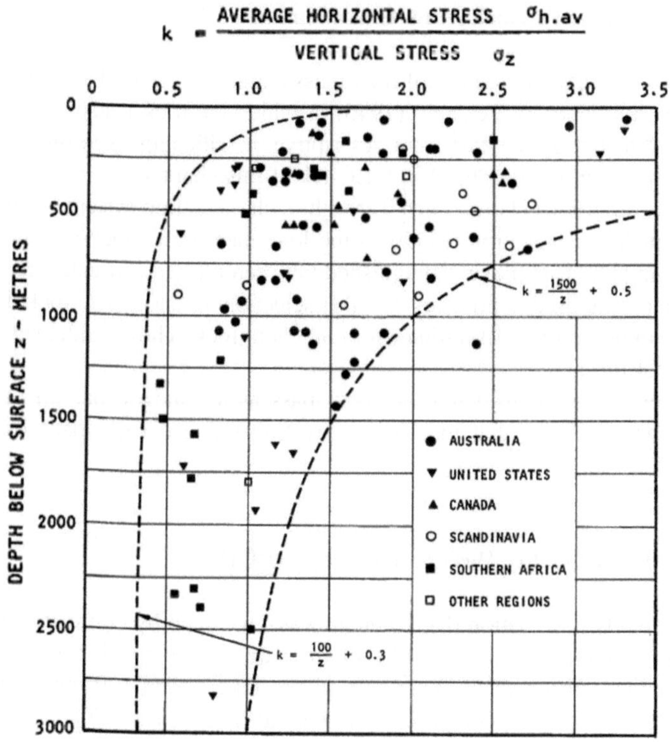

Fig. 21.4 Variation of ratio of average horizontal stress to vertical stress with depth below surface (Courtesy of Tsinghua University Press, Ref. [15])

During the feasibility and preliminary design stages of a rock excavation project, when very little detailed information is available on the rock mass and its stress and hydrological characteristics, the use of a rock mass classification scheme can be of considerable benefit. At its simplest, this may involve using the classification scheme as a checklist to ensure that all relevant information has been considered. At the other end of the spectrum, one or more rock mass classification schemes can be used to build up a picture of the composition and characteristics of a rock mass to provide initial estimates of support requirements and to provide estimates of the strength and deformation properties of the rock mass.

Rock mass classification schemes have been development for over 100 years. Summaries of some important classification systems are presented in Chap. 1 and in this chapter.

21.1.3.1 Terzaghi's Rock Mass Classification

The earliest reference to the use of rock mass classification for the design of tunnel support is in a paper by Terzaghi (1946) in which the rock loads, carried by steel sets, are estimated on the basis of a descriptive classification according to his rock load theory which is introduced in Sect. 1.1 of this chapter.

Deere et al. (1970) [1] modified Terzaghi's classification system by introducing the rock quality designation (RQD) as the lone measure of rock quality (Refer to Table 5.3 in [1]). They have distinguished between blasted and machine excavated tunnels and proposed guidelines for the selection of steel set, rock bolts, and shotcrete supports for 6–12-m-diameter tunnels in rock. These guidelines are given in Table 5.4 of [1].

Deere et al. also considered the rock mass as an integral part of the support system. They assumed that machine excavation reduced rock loads by approximately 20–25 %.

21.1.3.2 Rock Quality Designation Index (RQD)

The rock quality designation (RQD) index was developed by Deere et al. (1967) [1] to provide a quantitative estimate of rock mass quality from drill core logs. RQD is defined as the percentage of intact core pieces longer than 100 mm (4 in.) in the total length of core. The core's diameter should be at least 54.7 mm (or 2.15 in.) and should be drilled with a double-tube core barrel. The correct procedures for the measurement of the length of core pieces and the calculation of RQD are summarized in Fig. 21.5.

Palmström (1982) [1] suggested that when no core is available but discontinuity traces are visible in surface exposures or exploration adits, the RQD may be estimated from the number of discontinuities per unit volume. The suggested relationship for clay-free rock masses is as follows:

$$\mathrm{RQD} = 115 - 3.3J_v \tag{21.13}$$

where J_v is the sum of the number of joints per unit length for all joint (discontinuity) sets known as the volumetric joint count.

RQD is a directionally dependent parameter, and its value may change significantly, depending upon the borehole orientation. The use of the volumetric joint count can be quite useful in reducing this directional dependence.

RQD is intended to represent the rock mass quality in situ. When using diamond drill core, care must be taken to ensure that fractures, which have been caused by handling or the drilling process, are identified and ignored when determining the value of RQD.

When using Palmström's relationship for exposure mapping, blast-induced fractures should not be included when estimating J_v.

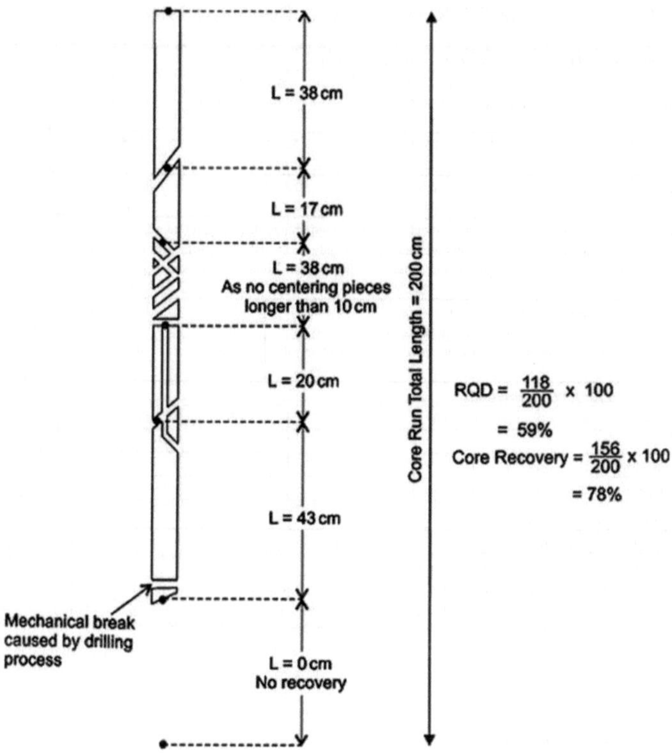

Fig. 21.5 Procedure of measurement and calculation of rock quality designation (RQD)

Deere's RQD was widely used, particularly in North America, after its introduction. Deere (1988) attempted to relate RQD to Terzaghi's rock load factors and to rock bolt requirements in tunnels (see Tables 21.2, 21.3). In the context of this discussion, the most important use of RQD is as a component of the RMR and Q rock mass classifications illustrated later in this chapter.

21.1.3.3 RMR System: Geomechanics Classification

The rock mass rating system is a rock mass quality classification developed by South African Council for Scientific and Industrial Research (CSIR). Bieniawski (1973, 1974) published the details of a rock mass classification called the geomechanics classification or the rock mass rating (RMR) system.

The following six parameters are used to classify a rock mass in the RMR system:

1. Uniaxial compressive strength of rock material.
2. RQD value.

Table 21.2 Rock mass rating (RMR) system (After Bieniawski 1989)

(a) Five basic rock mass classification parameters and their rating

Parameter	Description							
1. Strength of intact Rock material	Point load strength index (MPa)	>10	4-10	2-4	1-2			
	Uniaxial compressure strength (MPa)	>250	100-250	50-100	25-50	5-25	1-5	<1
Rating		15	12	7	4	2	1	0
2. RQD (%)		90-100	75-90	50-75	25-50	<25		
Rating		20	17	13	8	3		
3. Joint spacing (m)		>2	0.6-2	0.2-0.6	0.06-0.2	<0.06		
Rating		20	15	10	8	5		
4. Condition of Joints		not continuous. very rough surface, unwea -thed, no seperation	Slightly rough surface, slightly weathered, separation<1 mm	Slightly rough surface, highly weathered, separation<1 mm	continuous, slickensided surface, or gouge<5mm thick, or seperation 1-5mm	Continues joints, soft gouge>5mm thick, or seperation>5mm		
Rating		30	25	20	10	0		
5.Groundwater	inflow per 10m tunnel length (l/min), or	None	<10	10-25	25-125	>125		
	Joint water pressure/major in situ stress, or	0	0-0.1	0.1-0.2	0.2-0.5	>0.5		
	General conditions at excavation surface	Completely dry	Damp	Wet	Dripping	Flowing		
Rating		15	10	7	4	0		

(b) Rating adjustment for joint orientations

Strike and dip orientation of joints		Very favourable	favourable	fair	unfavourable	Very unfavourable
Rating	Tunnels	0	-2	-5	-10	-12
	Foundations	0	-2	-7	-15	-25
	slopes	0	-5	-25	-50	-60

(c) Effects of joint orientation in tunnelling

Strike perpendicular to tunnel axis				Strike parallel to tunnel axis		Dip 0°-20°
Drive with dip		Drive against dip				
Dip 45°-90°	Dip 20°-45°	Dip 45°-90°	Dip 20°-45°	Dip 45°-90°	Dip 20°-45°	Irrespective of strike
Very favourable	favourable	fair	unfarourable	Very favourable	fair	fair

3. Spacing of discontinuities.
4. Condition of discontinuities.
5. Groundwater conditions.
6. Orientation of discontinuities.

Table 21.2 shows the RMR classification updated in 1989.

Table 21.2b includes a rating adjustment that takes into account joint orientation and has to be read in conjunction with Table 21.4c where the descriptive terms of Table 21.4b are explained. The total rating for the rock is obtained by adding the rating for each parameter and the adjustment for joint orientations. The RMR rock classification is shown in Table 21.3. The table also gives the meaning of rock mass classes in terms of stand-up time, equivalent rock mass cohesion, and friction angle.

Example of Using RMR System A granite rock mass contains 3 joint sets: average RQD is 88 %, average joint spacing is 0.24 m, and joint surfaces are generally stepped and rough, tightly closed, and unweathered with occasional stains observed. The excavation surface is wet but not dripping. Average rock material uniaxial compressive strength is 160 MPa. The tunnel is excavated to 150 m below the ground where no abnormal high in situ stress is expected.

Selection of RMR parameters and calculation of RMR are shown below:

Rock material strength	160 MPa	Rating	12
RDQ (%)	88 %	Rating	17
Joint spacing (m)	0.24	Rating	10
Condition of joints	Very rough, unweathered, no separation	Rating	30
Groundwater	wet	Rating	7
RMR			76

The calculated basic RMR is 76. It falls in rock class II which indicates that rock mass is of good quality.

Bieniawski (1989) (refer to [1]) published a set of guidelines for the selection of support in tunnels in rock for which the value of RMR has been determined. Table 21.4 gives a guideline for the excavation and support of 10-m span rock tunnel in accordance with the RMR system.

It should be noted that Table 21.4 has not had a major revision since 1973. In many mining and civil engineering applications, steel fiber-reinforced shotcrete may be considered in place of wire mesh and shotcrete.

21.1.3.4 *Q*-System, NGI Tunneling Quality Index

The *Q*-system was developed by NGI (Norwegian Geotechnical Institute) (Barton et al. 1974) and updated in 1993 by Grimstad and Barton (refer to [1]). The system was based on the evaluation of a large number of case histories of underground excavation stability and is an index for the determination of the tunneling quality of a rock mass. The numerical value of this index Q is defined by:

$$Q = \left(\frac{\text{RQD}}{J_n}\right) \times \left(\frac{J_\text{r}}{J_\text{a}}\right) \times \left(\frac{J_\text{w}}{\text{SRF}}\right) \tag{21.14}$$

Table 21.3 Rock mass classification determined from total rating and meaning

RMR rating	81–100	61–80	41–60	21–40	<20
Rock mass class	I	II	III	IV	V
Description	Very good rock	Good rock	Fair rock	Poor rock	Very poor rock
Average stand-up time	10 years for 15 m span	6 months for 8 m span	1 week for 5 m span	10 h for 2.4 m span	30 min for 0.5 m span
Rock mass cohesion (KPa)	>400	300–400	200–300	100–200	<100
Rock mass friction angle	>45°	35°–45°	25°–35°	15°–25°	<15°

Table 21.4 Guidelines for excavation and support of 10-m span rock tunnels in accordance with the RMR system (After Bieniawski 1989)

Rock mass class	Excavation	Rock bolts shotcrete steel sets (20 mm diameter, fully grouted)	Shotcrete	Steel sets
I-Very good rock RMR 81–100	Full face, 3 m advance	Generally no support required except spot bolting		
ll-Good rock RMR 61–80	Full face. 1–1.5 m advance. Complete support 20 m from face	Locally, bolts in crown 3 m long, spaced 2.5 m with occasional wire mesh.	50 mm in crown where required	None
III-Fair rock RMR 41–60	Top heading and bench 1.5–3 m advance in top heading. Commence support after each Blast. Complete support 10 m from face	Systematic bolts 4 m long. Spaced 1.5–2 m in crown and walls with wire mesh in crown.	50–100 mm in crown and 30 mm in sides	None
IV-Poor rock RMR 21–40	Top heading and bench 1.0–1.5 m advance in top heading. Install support concurrently with excavation, 10 m from face	Systematic bolts 4–5 m long. Spaced 1.5–2 m in crown and walls with wire mesh.	100–150 mm in crown and 100 mm in sides	Light-to-medium ribs where required
V-Very poor rock RMR < 20	Multiple drifts 0.5–1.5 m. Advance in top heading. Install support concurrently with excavation after blasting	Systematic bolts 5–6 m Long, spaced 1–1.5 m. In crown and walls With wire mesh. Bolt invert.	150–200 mm in crown, 150 mm in sides, and 50 mm on face	Medium-to-heavy ribs spaced 0.75 m with steel lagging and forepoling if required. Close invert

where RQD is the rock quality designation measuring the fracturing degree which is defined in formula (21.1–21.4) above,

J_n is the joint set number accounting for the number of joint sets,
J_r is the joint roughness number accounting for the joint surface roughness,
J_a is the joint alteration number indicating the degree of weathering, alteration, and filling,
J_w is the joint water reduction factor accounting for the problem from groundwater pressure, and
SRF is a stress reduction factor indicating the influence of in situ stress.
Q value is considered as a function of only three parameters which are crude measures of:

(a) Block size RQD/J_n,
(b) interblock shear strength J_r/J_a, and
(c) Active stress J_w/SRF.

The parameters and rating of *Q*-system are given in Tables 21.5, 21.6, 21.7, 21.8, 21.9, and 21.10.

Q value is applied to estimate the support measure for a tunnel of a given dimension and usage as shown in Fig. 21.6. Equivalent dimension D_e is used in the figure, and the excavation support ratio (ESR value) is given in Table 21.13.

Hence,

$$D_e = \frac{\text{Excavation Span, Diameter or Height (m)}}{\text{Excavation Support Ratio (ESR)}} \tag{21.15}$$

The use of *Q*-system requires detailed engineering geological mapping and the analysis of all the geological features encountered. To simplify the classification of

Table 21.5 Rating for *Q*-system parameters: RQD

1. Rock quality designation	RQD
A Very poor	0–25
B Poor	25–50
C Fair	50–75
D Good	75–90
E Excellent	90–100

(1) where RQD is reported or measured as ≤10, (including 0), and a nominal value of 10 is used to evaluated Q in Eq. (21.14)
(2) RQD intervals of 5, i.e., 100, 95, and 90 are sufficiently accurate

Table 21.6 Rating for *Q*-system parameter, joint set number J_n

2. Joint set number	J_n
A. Massive, no, or few joints	0.5–1.0
B. One joint set	2
C. One joint set plus random joints	3
D. Two joint sets	4
E. Two joint sets plus random joints	6
F. Three joint sets	9
G. Three joint sets plus random joints	12
H. Four or more joint sets, random, heavily jointed, "sugar cube," etc.	15
J. Crushed rock, earth-like	20

(1) For intersections use (3 × J_n)
(2) For portals use (2 × J_n)

Table 21.7 Rating for Q-system parameters: joint roughness number J_r

3. Joint roughness number	J_r
Rock wall contact	
Rock wall contact before 100-mm shear	
A. Discontinuous joints	4
B. Rough and irregular, undulating	3
C. Smooth undulating	2
D. Slickensided undulating	1.5
E. Rough or irregular, planar	1.5
(Cont'd)F. Smooth, planar	1
G. Slickensided, planar	0.5
No rock wall contact when sheared	
H. Zones containing clay minerals thick enough to prevent rock wall contact	1
J. Sandy, gravely, or crushed zone thick enough to prevent rock wall contact	1

(1) Descriptions refer to small-scale features and intermediate scale features, in that order
(2) Add 1.0 if the mean spacing of the relevant joint set is greater than 3 m
(3) $J_r = 0.5$ can be used for planar, slickensided joints having lineations, provided that the lineations are oriented for minimum strength

the rock mass quality and the rock support evacuation, it is a common practice to divide the Q value range into classes as indicated in Fig. 21.6 and Table 21.12 (Table 21.11).

Example of Using Q-System A sandstone rock mass, fractured by two joint sets, plus random fractures: average RQD is 70 %, and average joint spacing is 0.11 m. Joint surface is slightly rough, highly weathered with stains, and weathered surface but no clay found on surface. Joints are generally in contact with apertures generally less than 1 mm. Average material uniaxial compressive strength is 85 MPa. The tunnel is to be excavated at 80 m below ground level, and the groundwater table is 10 m below the ground surface. Selection of Q parameters and calculation of Q value are shown below:

RQD	70 %	RQD	70
Joint set number	2 sets plus random fractures	J_n	6
Joint roughness number	Slightly rough (≥rough Planar)	J_r	1.5
Joint alteration number	Highly weathered only stain (altered non-softening mineral coating)	J_a	2
Joint water factor	70-m water head – 7 kg/cm^2 – 7 bars	J_w	0.5

(continued)

(continued)

RQD	70 %	RQD	70
Stress reduction factor	$\sigma_c / \sigma_1 = 85 / (80 \times 0.027) = 39.3$	SRF	1
Q	$(70/6) \times (1.5/2) \times (0.5/1)$		4.4

The calculated Q value is 4.4, and the rock mass is classified as fair quality.

Bieniawski (1976) shows that the relationship between the RMR rating and the equivalent Q values is adequately described by the equation:

Table 21.8 Rating for Q-system parameters: joint alteration number J_a

4. Joint alteration number	J_a	$\varnothing_r$ (approx)
Rock wall contact		
A. Tightly healed, hard, non-softening, impermeable filling, e.g., quartz or epidote	0.75	–
B. Unaltered joint walls, surface staining only	1.0	25–35°
C. Slightly altered joint walls, non-softening mineral coatings, sandy particles, clay-free disintegrated rock, etc.	2.0	25–30°
D. Silty-, or sandy-clay coatings, small clay fraction (non-softening)	3.0	20–25°
E. Softening or low-friction clay mineral coatings, i.e., kaolinite and mica and also chlorite, talc, gypsum, and graphite, and small quantities of swelling clays. (Discontinuous coatings, 1–2 mm or less)	4.0	8–16°
Rock wall contact before 100-mm shear		
F. Sandy particles, clay-free, disintegrating rock, etc.	4.0	25–30°
G. Strongly overconsolidated, non-softening clay mineral fillings (continuous < 5 mm thick)	6.0	16–24°
H. medium or low overconsolidation, softening clay mineral fillings (continuous < 5 mm thick)	8.0	12–16°
J. Swelling clay fillings, i.e., montmorillonite (continuous < 5 mm thick). Values of J_a depend on percent of swelling clay-size particles and access to water.	8–12	6–12°
No rock wall contact when sheared (Cont'd)		
K, L. Zones or bands of disintegrated or crushed rock	6 8 or	
M. and clay (See G, H, J for clay condition)	8–12	
N. Zones or bands of silty- or sandy-clay, small clay fraction, non-softening	5.0	6–24°
O, P. Thick continuous zones or bands of clay (see G.H R. and J for clay conditions)	10, 13 or 13–20	

Table 21.9 Ratings for Q-system parameters: joint water reduction factor J_w

5. Joint water reduction factor	J_w'	Approx water pressure KPa
A. Dry excavation or minor inflow, i.e., <5 l/m locally	1.0	<100
B. Medium inflow or pressure, occasional outwash of joint fillings	0.66	100–200
C. Large inflow or high pressure in competent rock with unfilled joints	0.5	250–1000
D. Large inflow or high pressure, considerable outwash of joint filling	0.33	250–1000
E. Exceptionally high inflow or water pressure at blasting, pressure decaying with time	0.2–0.1	>1000
F. Exceptionally high inflow or pressure continuing without noticeable decay	0.1 0.05	>1000

(1) Factors C to F are crude estimates; increase J_w if drainage measures are installed
(2) Special problems caused by ice formation are not considered

$$\mathrm{RMR} = 9\mathrm{Log}_e Q + 44 \tag{21.16}$$

21.2 Ground Prereinforcement for Excavation

21.2.1 Ground Freezing

21.2.1.1 Ground Freezing Method

Ground freezing is a construction technique that has been used for over one hundred years to provide temporary earth support and groundwater control. Applications are found in the underground civil, mining, and environmental remediation industries.

The ground freezing process uses a series of drilled freeze pipes and large refrigeration plants to convert existing pore water in the soil into ice, creating a strong, watertight frozen earth material similar to rock or concrete. The frozen ground acts as an excavation support system requiring no bracing, tiebacks, or additional shoring. It has been used extensively for shafts extending over a thousand feet deep. Its impermeable characteristic eliminates the need for dewatering, making the technique practical for large groundwater barriers. These barriers can be used to reduce or eliminate flow into excavations and contain contaminants or restrict the flow of contaminated plumes.

Most common method is by circulating brine (Calcium Chloride). Chilled brine is pumped to the bottom of the freeze pipe and flows up drawing heat from the soil. The other method is using liquid nitrogen (LN2). Cost per unit of heat extracted

Table 21.10 Rating for Q-system parameters: stress reduction factor SRF

6. Stress reduction factor	SRF
Weakness zones intersecting excavation, which may cause loosening of rock mass when tunnel is excavated	
A. Multiple occurrences of weakness zones containing clay or chemically disintegrated rock, very loose surrounding rock (any depth)	10
B. Single weakness zones containing clay, or chemically disintegrated rock (excavation depth <50 m)	5.0
C. Single weakness zones containing clay, or chemically disintegrated rock (excavation depth >50 m)	2.5
D. Multiple shear zones in competent rock (clay free), loose surrounding rock (any depth)	7.5
E. Single shear zone in competent rock (clay free). (depth of excavation <50 m)	2.5
F. Single shear zone in competent rock (clay free). (depth of excavation >50 m)	5.0
G. Loose open joints, heavily jointed or 'sugar cube' (any depth)	
Competent rock, rock stress problems σ_c / σ_1 σ_t / σ_1	
H. Low stress, near surface >200 >13	2.5
J. Medium stress 200–10, 13–0.66	1
K. High stress, very tight structure (usually favorable to stability, may be unfavorable to wall stability) 10–5 0.66–0.33	0.5–2
L. Mild rockburst (massive rock) 5–2.5 0.33–0.16	5–10
M. Heavy rockburst (massive rock) <2.5 <1.6	10–20
Squeezing rock, plastic flow of incompetent rock under influence of high rock pressure	
N. Mild squeezing rock pressure	5–10
O. Heavy squeezing rock pressure	10–20
Swelling rock, chemical swelling activity depending on the presence of water	
P. Mild swelling rock pressure	5–10
R. Heavy swelling rock pressure	10–15

(1) Reduce these values of SRF by 25–50 % if the relevant shear zones only influence but do not intersect the excavation
(2) For strongly anisotropic virgin stress field (if measured): when $5 \leq \sigma_1/\sigma_3 \leq 10$, reduce σ_c and σ_t to 0.8 σ_c and $0.8\sigma_t$. When $\sigma_1/\sigma_3 > 10$, reduce σ_c and σ_t to $0.6\sigma_c$ and $0.6\sigma_t$. σ_c = unconfined compression strength, and σ_t = tensile strength (point load), and σ_1 and σ_3 are the major and minor principal stresses
(3) Few case records available where depth of crown below surface is less than the span width. Suggest SRF increase from 2.5 to 5 for such cases (see H)

using LN2 is much higher, but may be competitive for small, short-term projects. Figure 21.7 shows a movable refrigeration unit.

Ground freezing may be used in any soil or rock formation regardless of structure, grain size, or permeability. However, it is best suited for soft ground rather than rock conditions. Freezing may be used for any size, shape, or depth of excavation, and the same cooling plant can be used from job to job. As the

Table 21.11 Excavation support ratio (ESR) for various tunnel categories

Excavation category		ESR
A	Temporary mine opening	3–5
B	Permanent mine openings, water tunnels for hydroelectric projects, pilot tunnels, drifts, and headings for large excavations	1.6
C	Storage rooms, water treatment plants, minor road and railway tunnels, surge chambers, and access tunnels in hydroelectric project	1.3
D	Underground power station caverns, major road and railway tunnels, civil defense chamber, tunnel portals and intersections	1.0
E	Underground nuclear power stations, railway stations, sports and excavation public facilities, underground factories	0.8

Table 21.12 Rock mass quality rating according to *Q* values

Q value	Class	Rock mass quality
400–1000	A	Exceptionally good
I00–400	A	Extremely good
40–100	A	Very good
10 to −40	B	Good
4–10	C	Fair
1–4	D	Poor
1–0.1	E	Very poor
0.1–0.01	F	Extremely poor
0.01–0.001	G	Exceptionally poor

impervious frozen earth barrier is constructed prior to excavation, it generally eliminates the need for compressed air, dewatering, or the concern for ground collapse during dewatering or excavation.

21.2.1.2 Principles of Freezing

- The effectiveness of freezing depends on the presence of water to create ice, cementing the particles and increasing the strength of the ground to the equivalent of soft or medium rock.
- If the ground is saturated or nearly so, it will be rendered impermeable.
- If the moisture does not fill the pores, it may be necessary to add water.
- The strength achieved depends on freeze temperature, moisture content, and the nature of the soil (Fig. 21.8).
- Freezing can be particularly effective in stabilizing silts, which are too fine for injection of any ordinary grouts.
- On freezing, water expands in volume by about 9 % which does not itself impose any serious stresses and strains on the soil unless the water is confined within a restricted volume. With water content up to about 30 %, the direct soil

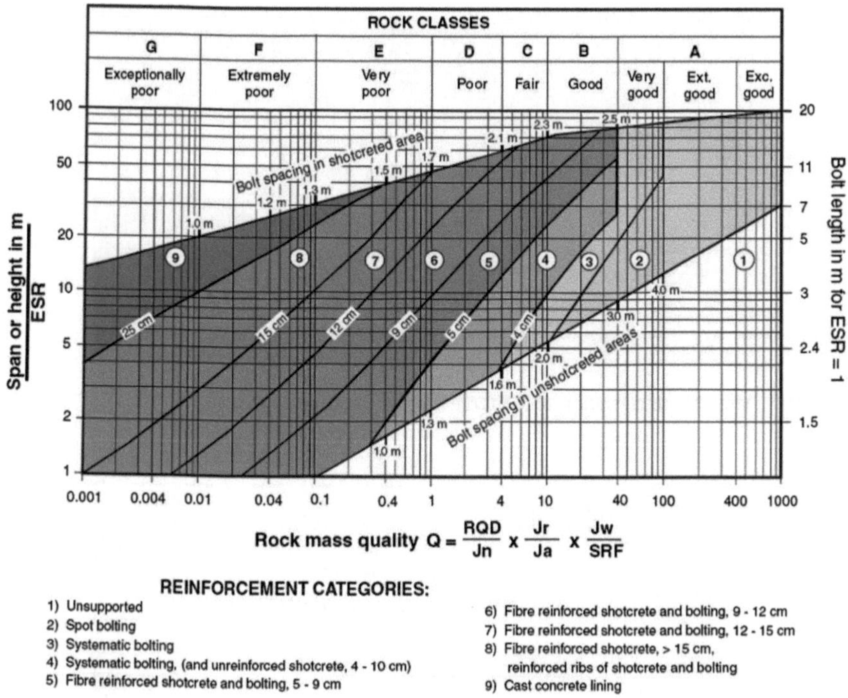

Fig. 21.6 Q–diagram with support recommendation

expansion may be about 3 %. Frost heave which may occur in fine silts and clays is a slightly different phenomenon.

- In rock and clay, ice lenses may build up and enlarge fine fissures so that causing an increase in permeability after thawing.
- If there is a flow of water through the ground to be frozen, the freezing time will be increased by reason of the continuing supply of heat energy, and if the flow is large and the water temperature is high, freezing may be completely inhibited.

21.2.1.3 Application of Freezing for Underground Excavation

- Shaft (Fig. 21.9)

Groundwater cutoff in conjunction with excavation support for the sinking of vertical shafts was the earliest application of ground freezing and remains the most common. For deep mines, no better method of sinking shafts through deep, water-bearing ground has yet been established. However, over the past 20 years, ground freezing has become a versatile tool for both soft ground and hard rock excavation and tunneling.

Fig. 21.7 Movable refrigeration unit *Source* www.mmrefrigeration.com

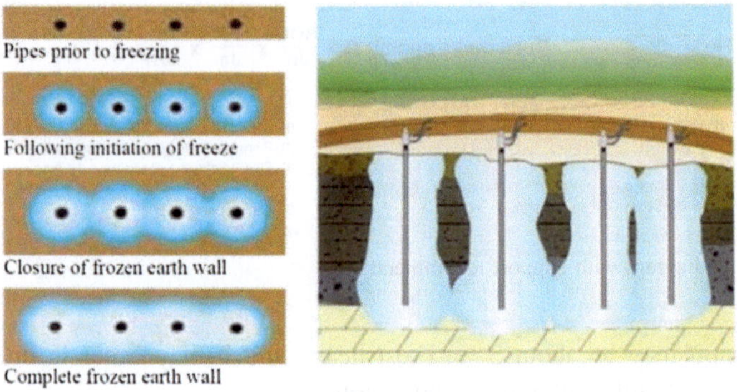

Fig. 21.8 Steps of freezing

There are several advantages of ground freezing unique to the construction of vertical shafts. The freeze can be implemented perfectly through the soil/rock interface, which is often the most difficult geology in which to increase a groundwater cutoff by other methods. Proper instrumentation can provide assurance of the integrity of the full depth freeze prior to excavation.

- Open-Face Tunneling

Ground freezing can be utilized to assure stability of tunnel excavation where the groundwater is the major problem concerned. The freeze pipe may be installed vertically or near vertically from the surface, or they may be installed horizontally from an access shaft to create a frozen envelope parallel to the tunnel alignment (see Figs. 21.10 and 21.11). With either freeze pipe orientations, a full-face freeze may be an advantageous option, depending on the excavation method, so the excavation is carried out under homogeneous conditions.

Fig. 21.9 Ground freezing for shaft shinking *Source* www.moretrench.com/

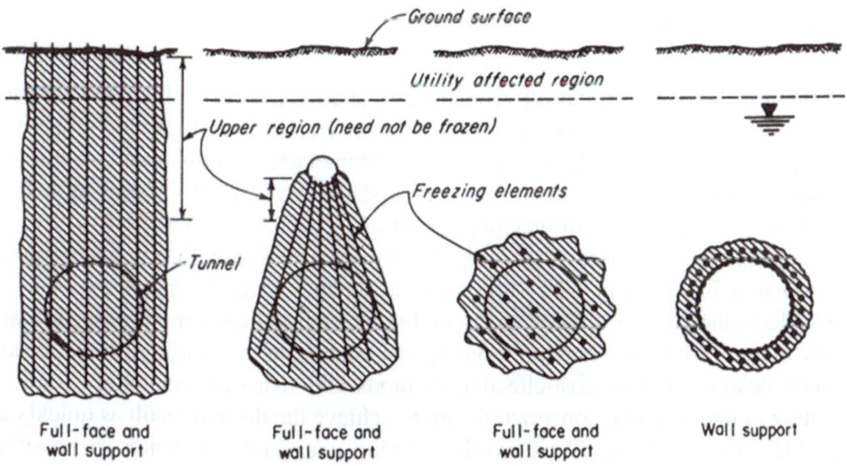

Fig. 21.10 Ground freezing methods for tunnel excavation

- TBM Tunnels

Mixed-face conditions for a hard rock tunnel boring machine (TBM) can be alleviated by creating a frozen mass that encompasses the runner face, presenting a

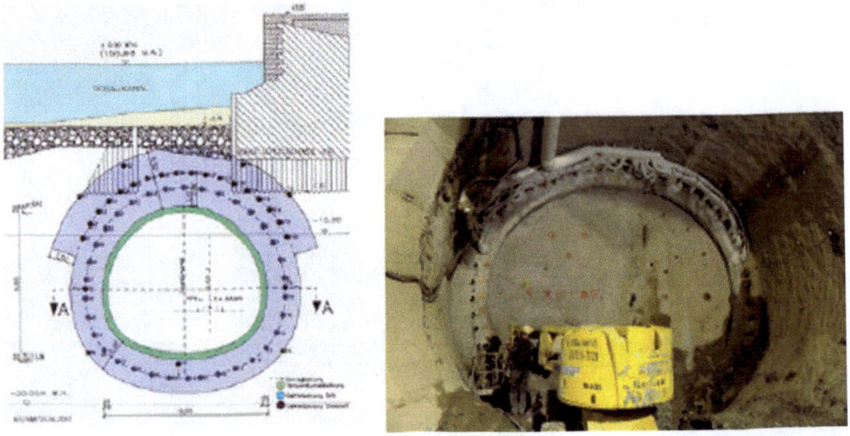

Fig. 21.11 Artificial ground freezing is applied in the excavations extending under the Danube River *Source* http://www.nap.edu/read/14670/chapter/8)

homogenous zone to allow tunneling to proceed safely. Mass ground freezing of a designated area along the tunnel alignment allows for a "safe haven" for the tunnel boring machine to undergo inspection and cutter head changes.

21.2.2 Grouting

Grouting techniques can cut off groundwater, strengthen soft soils in situ for tunneling support, remediate settlement of structures caused by soft ground tunneling, underpin structures, densify granular soils for liquefaction mitigation, construct or repair retaining walls, and construct access shafts, so grouting of the ground is widely used for ground reinforcement (see Figs. 21.12, 21.13 and 21.14).

The pregrouting method or grouting ahead of the excavation face in underground construction in rock can offer significant advantages in many situations. This is particularly the case in difficult rock conditions such as water ingress or mechanically difficult rock, in which pregrouting can contribute to avoiding problems and serious delays. Modern cost-effective methods and material technology for pregrouting in underground construction aim to achieve the desired result as quickly as possible, hence reducing the downtime during excavation as much as possible. Collapses at the tunnel face or unexpected high water inrushes are not uncommon experiences when tunneling in geologically difficult rock conditions such as fault zones in alpine terrain or tunnels with shallow location influenced by weathering or low rock stresses. Tunneling in urban areas often involve shallow location of tunnels, proximity to existing underground structures, as well as establishing connections between underground structures. The consequences of groundwater drawdown, or deformations in the ground caused by collapses are unacceptable due

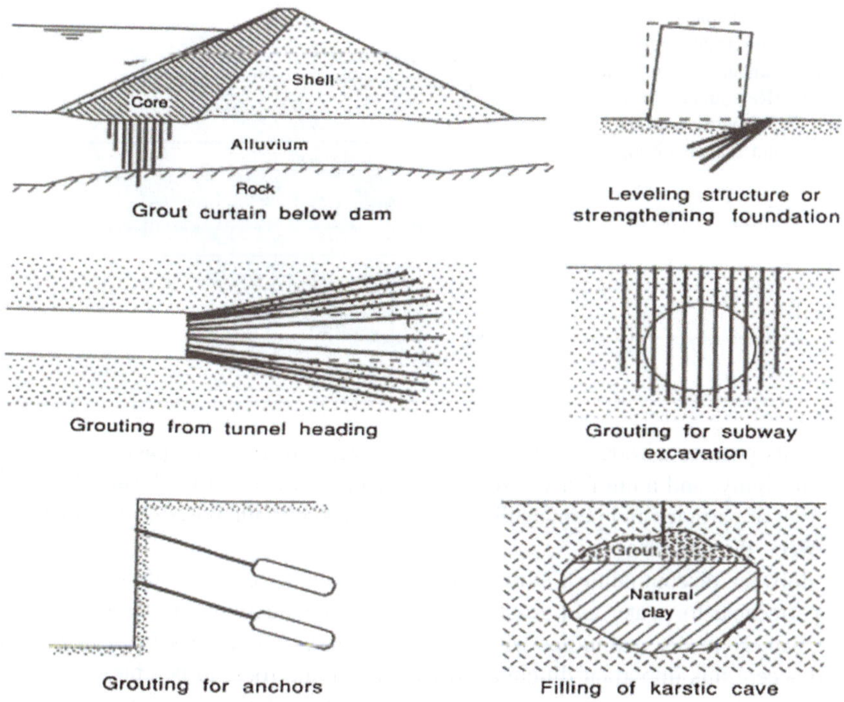

Fig. 21.12 Typical application of grouting

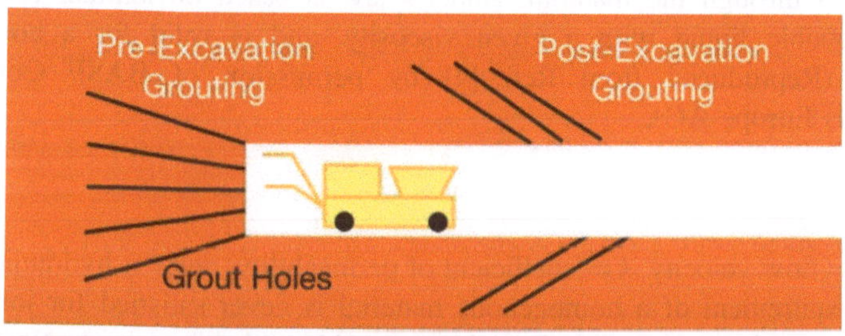

Fig. 21.13 Pre- and postexcavation grouting *Source* [6]

to the possible impact on buildings with sensitive foundations. Grouting of the ground often is the effective tool to solve all these problems.

21.2.2.1 Grouting in Rock [4]

Rock mass and soil are fundamentally different in terms of the behavior of water flow and the effect of injecting any kind of grout into the ground.

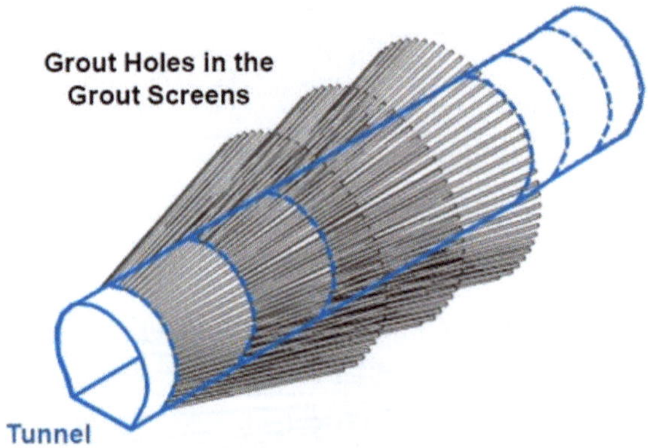

Fig. 21.14 Typical overlapping grout screen along a tunnel for stop water ingress (Reproduced from Ref. [7] by permission of Taylor and Francis Group)

Soils possess a wide variation in particle sizes, layering, compaction, porosity, permeability, and a number of other parameters. However, at the basic level, soils consist of particles and permeability is directly linked to the pores (spaces or voids) between the particles.

Between discontinuities, most rock materials, on the other hand, are practically impermeable for water and grouts. Leakage and conductivity are therefore linked exclusively to discontinuities within the rock mass. It is necessary to understand and accept this important difference between soil and rock, to be able to correctly evaluate all aspects of pressure grouting in rock tunneling and to understand why the approach has to be different to soil injection techniques.

The permeability of a material expresses how readily a liquid or a gas can be transported through the material. Darcy's law is based on laminar flow, for an incompressible liquid with a given viscosity, and is valid for a homogenous material (Reproduced from Ref. [4] by permission of BASF Construction Chemicals Europe AG):

$$v = ki \tag{21.17}$$

where v = flow velocity, k = coefficient of permeability, and i = hydraulic gradient

The requirement of a homogenous material is never satisfied for jointed rock materials, and then only when the volume considered is big enough. Normally, the term joint permeability or even better conductivity should be used.

The coefficient of permeability can be measured in the laboratory, using the above given formula of Darcy:

$$q = kAi \tag{21.18}$$

where q = liquid flow rate (m^3/s), k = coefficient of permeability (m/s), A = area of sample across flow path (m^2), and i = hydraulic gradient

The absolute permeability of a material, for liquids of varying viscosity, can be found according to the following formula:

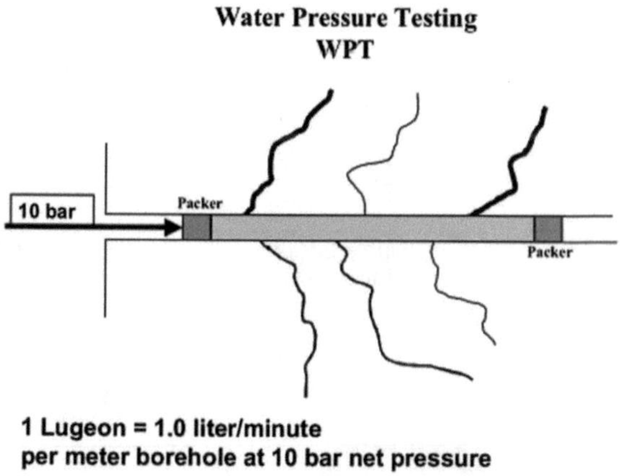

Fig. 21.15 Water pressure testing of borehole (Reproduced from Ref. [4] by permission of BASF Construction Chemicals Europe AG)

$$K = k\,(\mu \,/\, \gamma) = k\,(\nu \,/\, g) \tag{21.19}$$

where K = absolute permeability (m^2), k = coefficient of permeability (m/s), μ = dynamic viscosity (mPa) or cP, ν = kinematic viscosity (m^2/s) g 9.81 m/s^2, and γ = volume weight of the liquid (N/m^3)

For testing of rock mass conductivity through boreholes, the unit Lugeon is the most frequently used. Lugeon (L) is defined as the volume of water in liters that can be injected per minute and per meter of borehole at a net overpressure of 10 bar (see Fig. 21.15).

The Lugeon value needs interpretation and cannot be considered in isolation. If measurement has taken place over a borehole length of, say, 10 m, then there is in principle always the chance that all the water has escaped through a single leakage location. This means that if the same borehole had been measured in 0.5 m increments, nineteen of these would have had a L-value of zero, while one would be 20 times the above measured average.

To avoid possible extreme differences between Lugeon values resulting from a single measurement over a long borehole (10–30 m) and the real value over shorter segments (like 1 m), technical specifications sometimes require that the Lugeon value calculation length is set to 5 m for all borehole measuring lengths longer than 5 m.

21.2.2.2 Categories of Grouting

As shown in Fig. 21.16, the grouting technique can be categorized as the following five groups:

- Cement grouting,
- Chemical grouting,
- Compaction grouting,

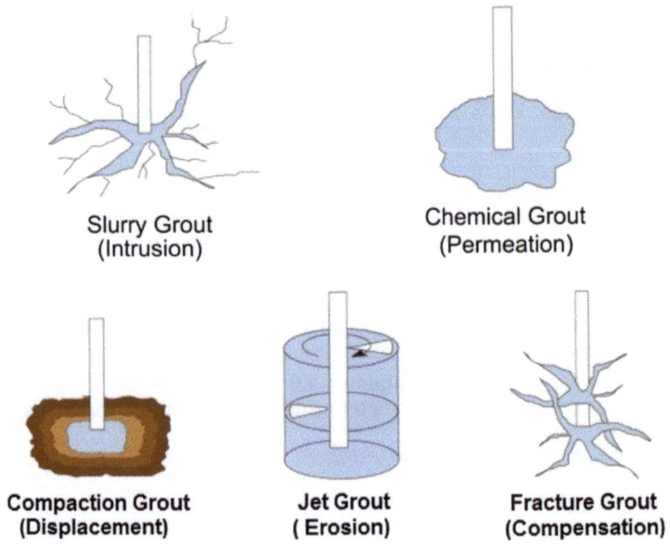

Fig. 21.16 Categories of grouting

- Fracture grouting, and
- Jet grouting.

Cement Grouting

Cement grouting, also known as slurry grouting or high mobility grouting, is a grouting technique that fills pores in granular soil or voids in rock or soil, with flowable particulate grouts.

Two important parameters governing the permeation capability of cement are the particle size and particle size distribution. The average particle size can be expressed as the specific surface of all cement particles in a given quantity. The finer the grinding, the higher is the specific surface, or Blaine value (m^2/kg).

The typical cement types available from most manufacturers, without asking for special cement qualities, are shown in Table 21.13 (Refer to [4]).

The cements with the highest Blaine value will normally be the most expensive, due to more fine grinding. The finer cements will give better penetration into fine cracks and openings.

Table 21.13 Fineness of normal cement types (largest particle size 40–150 μm) (Reproduced from Ref. [4] by permission of BASF Construction Chemicals Europe AG)

Cement type\Specific surface	Blaine (m°/kg)
Low heat cement for massive structures	250
Standard Portland cement (CEM 42.5)	300–350
Rapid hardening Portland cement (CEM 52.5)	400–450
Extra fine rapid hardening cement (limited availability)	550

Depending on the application, Portland cement or extra fine cement grout is injected under pressure at strategic locations through either single port or multiple port pipes. The grout particle size and soil/rock void size must be properly matched to permit the cement grout to enter the pores or voids. The grouted mass has an increased strength and stiffness and reduced permeability. The technique has been used to reduce water flow through rock formations beneath dams and to cement granular soils to underpin foundations or provide excavation support.

Bentonite, which is a natural clay from volcanic ashes, has traditionally been used on a routine basis in combination with cement for grouting of soil and rock, and its main mineral is montmorillonite. The reason to do so was the strong tendency of standard cement to separate when suspended in water, enhanced by the normal use of water cement ratio >1.0. Bentonite can be used to reduce the bleeding in such grouts, and a standard dosage of 3–5 % of the cement weight has a strong stabilizing effect. The final strength of the grout is not important in most cases. However, at high ground water head, or when a ground stabilization effect is valuable, the use of Bentonite at normal dosage will reduce the grout strength by 50 % and more. This is avoided when using modern admixtures in microcement grouts, without sacrificing stability or penetration.

Chemical Grouting

Chemical grouting is a grouting technique that transforms granular soils into sandstone-like masses, by permeation with a low viscosity grout. Chemical grouts available include silicates, phenolic resins, lignosulphonates, acrylamide and acrylates, sodium carboxymethylcellulose, amino resins, epoxy, polyurethane, and some other exotic materials. Without displacing the soil or changing the soil structure, the grout flows into the pores of the soil. The grout hardens, or gels, and depending on the ground characteristic, grout can be permanent or temporary. It increases the strength and cohesion of granular soils and, in tunneling, increases stand-up time and inhibits the movement of structures above or adjacent to the tunnel or shaft. A common application of chemical grouting is to provide both excavation support and underpinning when an excavation is planned immediately adjacent to an existing structure. Usually, chemical grouting can be accomplished without disrupting normal facility operations. Chemical grouting equipment is well-suited for tunneling applications in urban environments, whether for stabilizing soil around break-ins or breakouts, or for mitigating settlement of overlying structures within the influence of the tunnel alignment.

Compact Grouting

Compaction grouting is also known as low mobility grouting. Compaction grouting is a ground treatment technique that involves injection of a thick-consistency soil-cement grout under pressure into the soil mass, consolidating and thereby densifying surrounding soils in place. The injected grout mass occupies void space created by pressure densification. Typically, an injection pipe is first advanced to the maximum treatment depth. The low mobility grout is then injected as the pipe is slowly extracted in lifts, creating a column of overlapping grout bulbs. The expansion of the low mobility grout bulbs displaces surrounding soils. When performed in granular soil, compaction grouting increases the surrounding soil density, friction angle, and stiffness. Compaction grouting has been used to increase bearing capacity and decrease settlement and liquefaction potential for planned and existing structures. In karst geologies, compaction grouting has been used to treat existing sinkholes or to reduce the sinkhole potential in sinkhole-prone areas.

Fracture Grouting

Fracture grouting, also known as compensation grouting, is a grouting technique that hydrofractures in situ soil, using neat fluid grout. A sleeve port pipe is grouted into a predrilled hole beneath a foundation. The grout is injected under pressure at strategic locations through the ports in the pipe. Once the hydrofracture pressure of the soil is exceeded, fractures open up in the soil and are immediately expanded by the subsequent influx of grout. The process results in controlled heave of the overlying soils and structures. The technique has been used to relevel structures or to protect structures from settlement while a tunnel machine passes below.

Jet Grouting

Jet grouting, sometimes referred to as "jet mixing," is a technique that uses high pressure of about 300–600 bars (4350–8700 psi) and high-velocity jets of grout to hydraulically erode, mix, and partially replace the in situ soil or weak rock with cementitious grout slurry.

Jet grouting is mainly a complex procedure of grout injection inside soft soils that is divided into three methods: single, double, and triple fluid systems. However, in early 1990s, a new method that used focused jets was invented to increase the diameter of soilcrete columns. This procedure is called super jet grouting method.

The sequence of jet grouting method and different jet grouting systems are, respectively, shown in Figs. 21.17 and 21.18 quoted from [5].

Basically, creating soilcrete column, the high-velocity jet erodes the underground soil and replaces some or all of them with grout material. Considering the equal condition for jet grouting and ground factors, super jet system is able to provide a considerably larger treatment of soil since it takes the advantage of the state-of-the-art injecting system. As the equipment provides significantly higher flow rates at higher pressures, this method could improve volumes of soil 20 times as large as the previous conventional systems. This enabled jet grouting to obtain a column with a diameter in excess of 5 m, or even 9 m in softer ground. However, triple fluid system still is more frequently used among all other systems.

21.2.2.3 Grouting Equipment

The basic function of equipment in the grouting progress is to mix the grouting materials, pump it to the intended location, and control and record the operation.

The main components of grouting equipment include mixer, agitator, pump, and recorder (Fig. 21.19). Figure 21.20 shows a mobile grouting equipment.

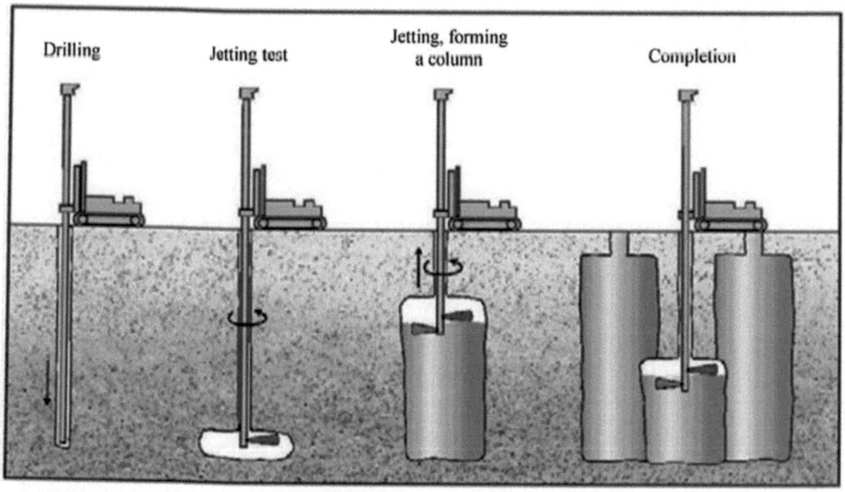

Fig. 21.17 Sequence of jet grouting method. Reproduced from [5] by permission of the Author

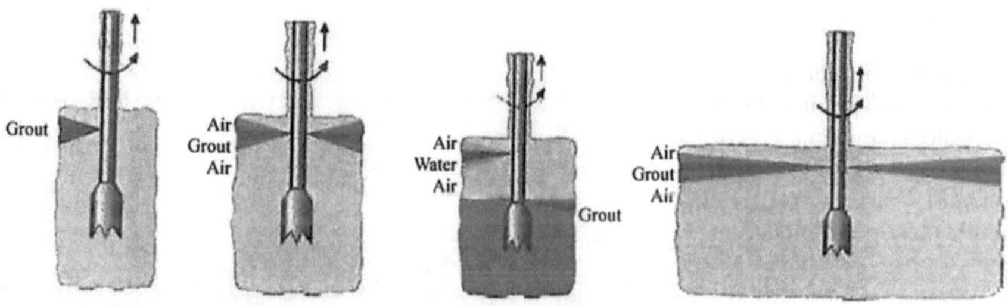

Fig. 21.18 Single, double, and triple super jet grouting. Reproduced from [5] by permission of the Author

21.3 Initial Support of Newly Excavated Space

As indicated in the above section, initial rock support assessment should be made on the basis of a rock classification such as the *Q*-system or RMR method. With the "design as you go" concept, rock support is only installed where the observation of the rock surface shows that it is necessary, often with the aid of the *Q*-system or RMR method. Thus, the amount of initial support and the support method are decided when the rock conditions are evaluated at the site of installation and as the excavation proceeds.

Rock reinforcement and rock mass act as a complex interactive system, where the individual elements always have to be seen in view of their interaction and interdependence. The overall strength of a reinforced rock mass with a joint system is governed by the characteristics of the joints (roughness, fill, rock material,

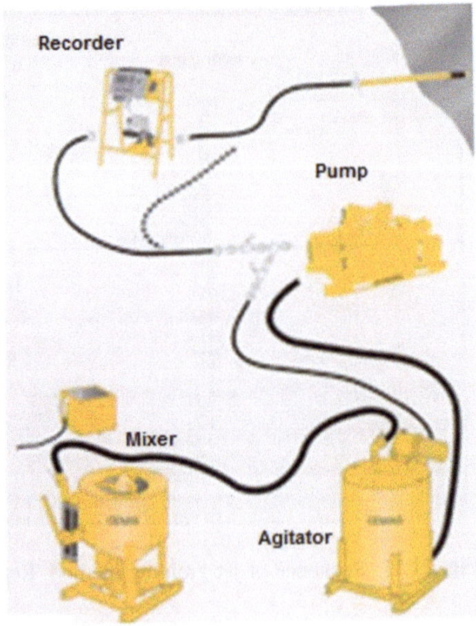

Fig. 21.19 Grouting equipment for one grout line *Source* [6]

Fig. 21.20 Dwupompowy cementing unit equipped with two Caterpillar engines, two pumps SPM TW 600S, and continuous mixing systems *Source* http://polskiproducent.pl/)

orientation) and the contribution provided by the reinforcement elements. For the design of rock mass reinforcement systems, sufficient appreciation of the expected ground conditions and experience is of fundamental importance.

Initial support is required to secure, as soon as possible, safe working conditions for the construction crews. Normally, the contractor is responsible for determining this support, based on the various supporting methods described in the contract or designed in advance by the contractor. For cost-effectiveness, initial support should, if possible, form part of the final support.

The often used initial support is a combination of steel arches, rock bolts, anchors, steel mesh, and sprayed concrete.

For some worse ground quality, further support before and after excavation may be required, such as spiling (forepoling), lattice arch girder, and steel arch ribs, in particular for the portal area of the tunnel.

All of these support techniques will be illustrated in this section.

21.3.1 Rock Dowels, Rock Bolts, and Rock Anchors

Rock dowels, bolts, and anchors are the most common methods of rock reinforcement. The main principle of them is to reinforce loosened rock or fractured in situ rock to prevent caving or spalling and to assist the rock mass to form its own self-supporting structure.

Rock Dowel—It is a kind of reinforcing element with no installed tension. It consists of a rod (usually steel bar), faceplate, and nut (or sometimes without). The rod is grouted in the hole with cement mortar, or resin.

Rock Bolt—It is similar to the rock dowel, but it is tensioned during installation. It consists of a rod and mechanical or grouted anchorage coupled with some means of applying and retaining the tension.

Rock Anchor—It is also a kind of reinforcing element for rock mass. It is tensioned following installation and is of higher capacity and generally of greater length than rock bolt.

Here, we call rock dowels and rock bolts are same as Rock Bolts.

21.3.2 Types of Rock Bolts

According to their working mechanism, rock bolts also can be classified into three types: mechanically anchored, frictionally anchored, and grouted bars.

21.3.2.1 Mechanically Anchored Bolts

Mechanically anchored bolts are usually wedge or expansion-shell bolts that are point-anchored at the bottom of the hole (see Fig. 21.21).

The bolt has an expanding anchor at its end. After insertion, the bolt is either rotated or pressed/hammered against the bottom of the hole. This expands the wedge end and anchors the bolt firmly to the sides of the hole. To install anchored bolts successfully, the hole size must be accurate and the rock must be relatively solid.

Wedge or expansion-shell bolts are typical means for temporary rock support.

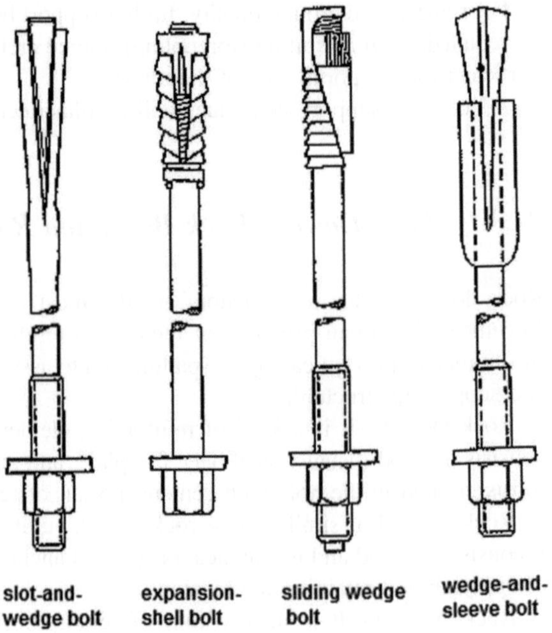

Fig. 21.21 Different types of mechanically anchored bolts

21.3.2.2 Friction-Type Bolts

Typical examples of friction-type bolts are the split-set and swellex bolts. Both are quick and easy to install and give instantaneous support. They cannot, however, be used for long-term reinforcement.

Split sets are rock bolts with only two parts, a tube with a slot cut along its length and a matching dome bearing plate (see Fig. 21.22). The split-set bolt is hammered into the hole, which has a slightly smaller diameter than the bolt causing the "tube" to be compressed by the wall of the hole. Using the correct hole size for a specific bolt, diameter is essential for successful installation. split-set bolts are very suitable for layered formations. The split-set bolt provides immediate support but only for a fairly short period of time.

The swellex bolt has a longer life span than the split set. It comprises a steel tube that has been deformed to a small diameter. It is installed by applying high-pressure water to the bolt after inserting it into the hole (see Fig. 21.23). High pressure applied into the center of the tube expands the bolt to its final dimensions in the hole, therefore enabling it to utilize the roughness and fractures in the bolt-hole surface. Same as the split-set bolt, poor corrosion protection limits this bolt for long-term support.

Fig. 21.22 The split-set bolt

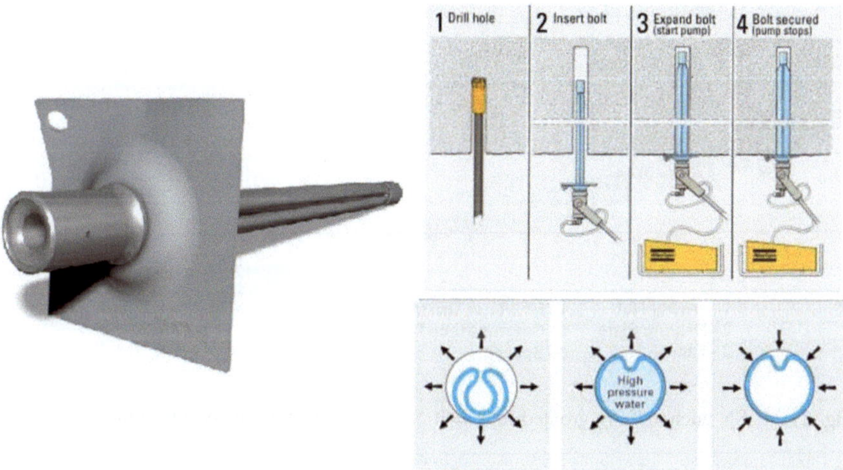

Fig. 21.23 The swellex expandable rock bolt, installation, and work mechanism (Reproduced by permission of Atlas Copco)

21.3.2.3 Grouted Bolts

Cement-grouted rebar—Cement-grouted rebar is still the most inexpensive and widely used rock bolt, because it is simple and quick to install and can be used with or without mechanized equipment. Correctly installed, a cement-grouted bolt gives rock support for years. The grout cement provides protection from corrosion. Special galvanized and/or epoxy-coated bolts can be used in extremely severe conditions. The major disadvantage of the cement-grouted bolt is its relatively long hardening period. The grout takes between 15 and 25 h to harden; therefore, it does not provide immediate support. When immediate support and/or pretensioning is needed, a grouted wedge-type or expansion-shell bolt can be used. Mixing additives in the grout can reduce the hardening time, but it also increase bolting cost (Fig. 21.24).

Resin-grouted bolts—Resin-grouted bolts give the required support relatively quickly due to a short hardening time (Fig. 21.25). When correctly installed with full-length grouting, the resin-grouted bolt is considered to give permanent support with a life span of 20–30 years. By using resins with two different hardening times, with one faster at the bottom of the hole and another that is slower at the stem, the bolts can be pretensioned. The same can be done for short-term support by only bottom-grouting the bolt.

Hollow-core self-drilling bolt—The self-drilling hollow bar anchor system is comprised of a hollow threaded bar with an attached drill bit that performs drilling, anchoring, and grouting in a single operation (Fig. 21.26). The hollow bar allows air and water to freely pass through the bar during drilling to remove debris and

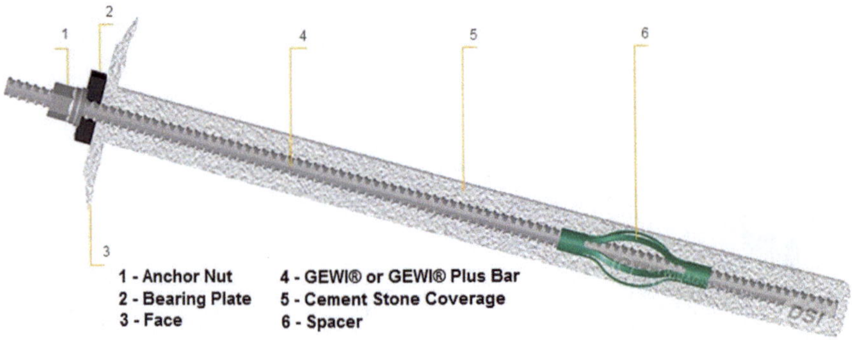

Fig. 21.24 SN anchors/fully grouted rock bolt *Source* www.dywidag-system.com

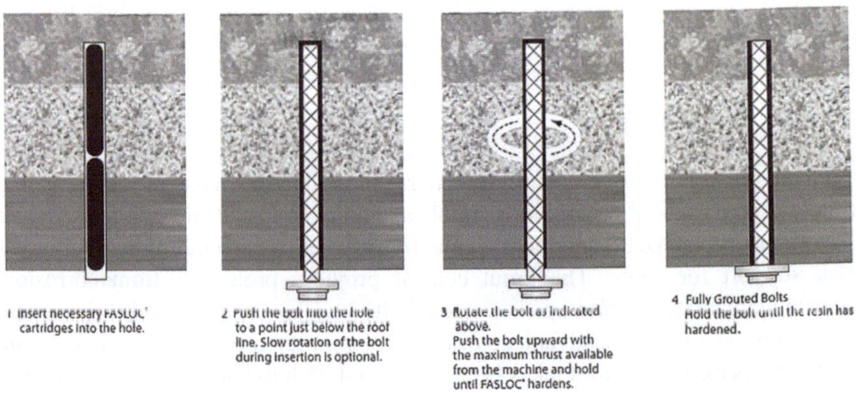

Fig. 21.25 Installation of resin-grouted bolt with FASLC resin cartridge *Source* www.dsiunderground.com

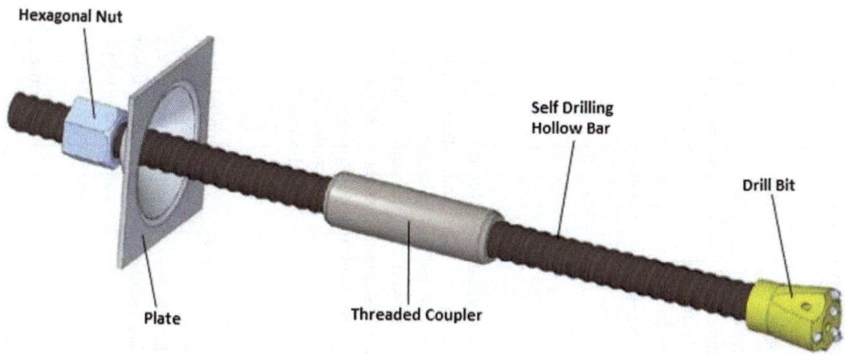

Fig. 21.26 Hollow-core self-drilling bolt *Source* www.jennmar.com/

then allows grout to be injected immediately after drilling is completed. Grout fills the hollow bar and completely covers the entire bolt. Couplers can be used to join hollow bars and extend the bolt length while nuts and plates are used to provide the required tension. The drill bit is abandoned in the hole with the bar and generally is not appropriated for very hard rock.

21.3.3 *Application Guideline and Equipment for Bolt Installation*

21.3.3.1 Guideline of Rock Bolting

Table 21.14 summarizes commonly used rock reinforcement elements and application considerations for the installation as part of initial support in tunneling in rock, which is quoted from the "Technical Manual for Design and Construction of Road Tunnels—Civil Elements" published by US Department of Transportation, Federal Highway Administration, Publication No. FHWA-NHI-10-034 December 2009, refer to [9], for reference only.

Rock bolting is not an exact science as a large number of physical and geometrical parameters are dictated by the ground conditions prevailing on site, that is what we say "design as we go." However, three common methods can be grouped under the concept of radial bolting, which refer to the spacing of bolts in a radial arch. These methods are spot bolting and two types of systematic bolting—pretensioned and untensioned bolting (see Fig. 21.27).

Spot bolting involves the use of a few bolts in spot locations in the tunnel. Their location and length are determined on site by the geotechnical engineer after making an assessment of the bolting requirements related to limited loose blocks. The size of these blocks can be determined by observing the position and directions

Table 21.14 Commonly used rock reinforcement elements and application considerations for tunneling in rock (Reproduced with no restrictions from Ref. [9])

No.	Name	Material*	Anchorage	Tensioned	Installation	Ground**	Advantages	Limitations
1	Steel rebar dowel	Deformed (solid) steel rebar	Fully bonded using cement grout or resin	No	Rebar inserted into predrilled and grout filled hole; Rebar inserted in predrilled hole together with grouting hose and grouted subsequently	Massive to highly jointed rock mass	Low cost; Availability; if properly installed, high-performance and heavy-duty support	Requires skilled and experienced installation personnel; collapsing boreholes hamper installation
2	Glass fiber dowel	Deformed fiber glass bar	Fully bonded using cement grout, more frequently with resin	No	Rebar inserted into predrilled and grout filled hole; Rebar inserted in predrilled hole together with grouting hose and grouted subsequently	Massive to highly jointed rock mass; frequently used in areas to be excavated subsequently (e.g., face bolting, breakout areas)	High-performance heavy-duty support; can be easily removed during subsequent excavations within reinforced rock mass	Requires skilled and experienced installation personnel; limited shear resistance; collapsing boreholes hamper installation
3	Split set	Longitudinally split steel pipe	Friction over entire length generated by spring action of pipe	No	Forced into predrilled borehole of slightly smaller diameter than outer diameter of split set	Massive to jointed rock mass	Immediate support action; simple installation; no grouting required	Very limited shear resistance; light support only; very corrosion sensitive; cannot be used in collapsing borehole

(continued)

Table 21.14 (continued)

No.	Name	Material*	Anchorage	Tensioned	Installation	Ground**	Advantages	Limitations
4	Swellex	Folded, inflatable steel pipe	Friction over entire length generated by inflation of tube	No	Inserted into predrilled borehole and inflated with highly pressurized water	Massive to jointed rock mass	Immediate support action; can achieve significant support capacity	Limited shear resistance and durability; cannot be retightened; requires special equipment for inflation; higher material cost; collapsing boreholes hamper installation
5	Grouted pipes	Perforated steel pipe	Fully bonded with cement or resin grout	No	Inserted into predrilled borehole (or rammed into soft ground with thick walled pipes) and grouted through pipe and perforation holes	Jointed to heavily fractured ground (soil like)	Simple installation; availability; more controllable embedment results	Limited shear resistance (depending on wall thickness); collapsing boreholes hamper installation
6	Self-drilling dowels	Thick-walled steel pipes with disposable drill bit	Fully bonded with cement or resin grout	No	Reinforcement element functions as drill rod and drill bit, and dowel remains in ground after drilling and is grouted through flushing openings	Jointed to heavily fractured rock mass	Installation steps limited to two steps (fast installation); high-performance heavy-duty support	More expensive than bar reinforcement; may become trapped in collapsing boreholes as it does not have reverse cutting tools

(continued)

Table 21.14 (continued)

No.	Name	Material*	Anchorage	Tensioned	Installation	Ground**	Advantages	Limitations
7	Rammed Dowels	Steel rebar or thick-walled steel tube	Shear resistance generated between ground and element (friction, adhesion)	No	Rammed into ground	Decomposed rock, soil	Least ground disturbance during installation; immediate support action	Relies on shear resistance generated between ground and element; requires ramming equipment; limited to soft ground conditions
8	Steel rebar bolt	Deformed steel rebar	a. End anchored: cement grout or resin; b. fully bonded: two-phase resin	Yes	a. Grouting behind grout seal through grouting hose (aeration hose); b. resin grout with two different setting times	Massive to highly jointed rock mass	Low cost; availability; if properly installed, high-performance heavy-duty support	Requires skilled and experienced installation personnel; collapsing boreholes hamper installation a. Requires grout seal; b. Resin is more expensive than grout; requires different types of resin

(continued)

Table 21.14 (continued)

No.	Name	Material*	Anchorage	Tensioned	Installation	Ground**	Advantages	Limitations
9	Glass fiber bolt	Deformed glass fiber bar	a. End anchored: cement grout, resin; b. Fully bonded: two-phase resin	Yes	a. Grouting behind grout seal through grouting hose (aeration hose); b. resin grout with two different setting times	Massive to highly jointed rock mass	High-performance heavy-duty support; can be easily removed due to limited shear resistance	Requires skilled and experienced installation personnel; collapsing boreholes hamper installation a. requires grout seal; b. resin is more expensive than grout; requires different types of resin
10	Expansion-shell bolt	Steel rebar	Mechanically end anchored	Yes	Inserted in predrilled borehole, shell at end expanded by tightening the bolt	Massive to jointed rock mass; requires competent rock material	Immediate support effect; can provide high support capacity	Relatively expensive; Slip or rock crushing may occur; tends to lose tension due to vibration (blasting) and ground deformation

*Reinforcement material

**Ground conditions described are typical application examples; reinforcement elements may also be used in other ground conditions

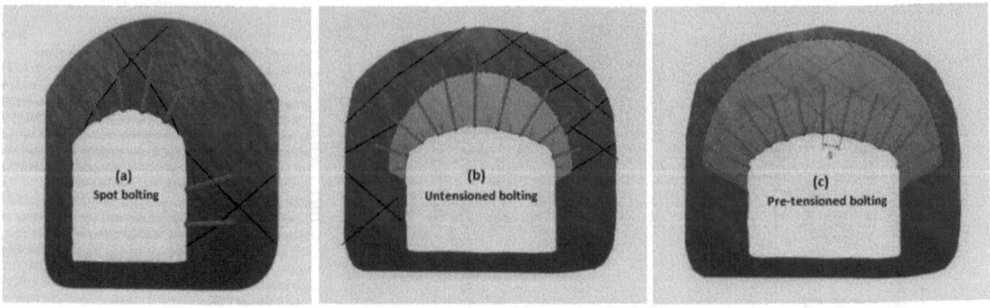

Fig. 21.27 Three groups of bolting methods (Reproduced from Ref. [10] by permission of Atlas Copco)

of the fracture planes that define the block. An inclinometer compass may be useful to register the dip of these planes, especially for large blocks.

When the block dimensions are known, the weight can be obtained by multiplying the volume with the density (normally 2.6–2.8 t/m^3). The volume must be estimated from the location, size, and orientation of structures that define the outline of the block. In order to determine the shape and weight, as well as the potential sliding direction of blocks or wedges in the roof and walls, the stereographic projection technique (Hoek and Brown 1980, [3]) is also recommended for which 3D computer modeling is available. Stereographic projection provides a mapping function whereby a sphere is projected onto a plane, which is useful in geotechnics and provides rock bolting engineers with a visualization of the loose block.

In addition to calculating the dimensions of the loose block, it is important to check that sufficient anchoring length for the chosen bolt type is to avoid rock bolt failure. Altogether, the accumulated data will help to decide on the length and number of bolts and the bolting pattern that will hold the block. A rule of thumb, however, is that bolts should always extend 1–2 m into solid rock.

If systematic bolting is chosen, the rock will most likely be heavily fissured, weakened by groundwater effects or other factors, which means that a systematic installation of rock bolts is required. As indicated in the gray area in Fig. 21.23b, the formation of a natural arch is the result of stress redistribution in the rock as the opening is created. The rock in the arch is subjected primarily to compressive stresses. In this example, untensioned rock bolts have been anchored in the natural arch to maintain stability. As most tunnels are permanent, the space between the bolt and the rock can be filled with cement or resin grout.

Untensioned bolts are generally preferred in moderately jointed rock where the lower boundary of the natural arch is relatively close to the roof of the opening. This type of bolting is adapted to the natural movement of the rock mass. The length of the bolts is estimated by calculating the block volume, weight, and density against the span of the operating measured in meters. The number of bolts and spacing between them is determined by the joint density. Movement within the rockmass as the tunnel is advanced further may allow tension to develop within the dowel.

In less competent rock structures where the lower boundary of the natural arch is further away, tensioned rock bolts are often preferred. As shown in Fig. 21.23c, these form an artificial arch near the ceiling of the opening and are used to increase both the shear resistance of the joints and the normal stress across joints. Various formulas can be adopted to determine the length and spacing between the bolts. All bolts, however, should be of the same length.

By using pretensioned bolts to create an arch or beam over a tunnel, the rock can be given a compressive stress of approximately 0.5 kg/cm^2, provided that tensioning is performed accurately. Any existing stresses in the rock must be superimposed on this value. Pretensioning allows the force structural capacity of the bolt to be developed with less ground movement, when compared with untensioned dowels.

Hoek and Brown 1980 [3] recommended the following empirical rules as a useful check for proposed bolt lengths and spacing:

Minimum bolt length—greatest of:

a. Twice the bolt spacing.
b. Three times the width of critical and potentially unstable rock blocks defined by average joint spacing in the rock mass.
c. For spans of less than 6 m (20 feet), bolt length is one-half of the span. For spans of 18–30 m (60–100 feet), bolt length is one-quarter of span in roof. For excavations higher than 18 m (60 feet), sidewall bolts is one-fifth of wall height.

Maximum bolt spacing—least of:

a. One-half the bolt length.
b. One and one-half times the width of critical and potentially unstable rock blocks defined by the average joint spacing in the rock mass.
c. When weld-mesh or chain-link mesh is to be used, bolt spacing of more than 2 m (6 feet) makes attachment of the mesh difficult (but not impossible).

The following equation for bolt lengths has been suggested by Palmström (Reproduce from Refs. [11, 12] by permission of Oslo Norwegian Group for Rock Mechanics 2000) for bolting of single loose block:

$$\mathrm{LB}_{\mathrm{roof}} = 1.4 + 0.16D_{\mathrm{t}}\left(1 + \frac{0.1}{D_{\mathrm{b}}}\right) \tag{21.20}$$

$$\mathrm{LB}_{\mathrm{wall}} = 1.4 + 0.08(D_{\mathrm{t}} + 0.5W_{\mathrm{t}}\left(1 + \frac{0.1}{D_{\mathrm{b}}}\right) \tag{21.21}$$

where D_{b} = the block diameter (in meters), W_{t} = the tunnel wall height (in meters), and D_{t} = the diameter or span of the tunnel (in meters)

21.3.3.2 Equipment for Bolt Installation

Development of mechanized equipment began as early as the 1970s. Today, there is a wide selection of fully mechanized equipment and a wide variety of different methods for bolt installation. The main factors affecting the choice of method are usually tunnel size, amount of bolts to be installed, and work cycle arrangement at the site.

Manual operation, the handheld drilling, and installation of bolts are mainly used in small drifts and tunnels where drilling is also performed by handheld equipment, and there is a limited amount of bolting work (Fig. 21.28).

Semi-mechanized installation is usually used in tunneling work sites. The drilling jumbo is used for drilling bolt holes, and bolt installation is performed from the jumbo's basket boom or from a separate utility carrier or truck.

With today's fully mechanized equipment, one operator can handle the entire bolting process from drilling to grouting and bolt installation. The operator is positioned away from the unbolted area under a safety canopy that protects him from falling rock.

The first fully mechanized bolting unit, such as Tamrock Roblt and Cabolt (Fig. 21.29), Atlas Copco's Boltec and Cabletec, was introduced in 1979.

Mechanization initially involved cement-grouted rebar bolts, but extended quickly to other bolt types. Today, all commonly used bolt types can be installed with fully mechanized bolting units. It has been proven that mechanization and automation of rock bolting process offer improved quality and safety in underground work.

Fig. 21.28 Installation of bolts manually

Fig. 21.29 Fully mechanized bolting rig (Reproduced by permission of Sandvik)

21.3.4 Installation of Wire Mesh on Rock Face

Screening, which is the installation of wire mesh, is most typically used in underground mining, but also at construction sites together with bolting and/or sprayed concrete. Screening is primarily performed manually by applying the wire mesh together with bolting of the tunnel. It can also be done by mechanized equipment, such as by having a screen manipulator on the bolting or shotcreting unit, or on a dedicated screening machine (Fig. 21.30). Wire mesh by itself has no corrosion protection and can only be used for short-term temporary support. Wire mesh should be of the welded type and not chain-link mesh.

21.3.5 Shotcrete

Sprayed concrete, or simply called shotcreting, is a widely used support method in construction. It is used for temporary or long-term support, lining, and backfilling. Usually, shotcrete is used together with bolting to obtain the best support or reinforcement.

Sprayed concrete is a form of concrete that is pneumatically projected onto the rock surface at high velocity using concrete pumps and compressed air. The sprayed concrete is sprayed via nozzles mounted on specialized equipment and sticks to the rock by adhesion. This has a stabilizing effect on the rock.

Fig. 21.30 Sandvik Robolt 320 with screen manipulator (Reproduced from Ref. [8] by permission of Sandvik)

More than a century of scientific research and product development lies behind today's modern sprayed concrete technology. First used in tunnels and mines in 1907 and patented as Gunite in 1911, it consisted of a simple blend of sand, aggregates, cement, and then injected water at the nozzle as it was released. Today, there is a multitude of different compositions and an equally wide variety of equipment designed to suit an ever-increasing number of applications.

The most common forms of shotcreting are the dry-mix and wet-mix methods (Fig. 21.31).

In the dry-mix method, the aggregate (typically 8 mm diameter stones), cement, and accelerators are mixed together and propelled by compressed air. Water is added last through a control valve on the spray nozzle. The dry method is suitable for manual shotcreting because the required equipment is usually inexpensive and small. On the other hand, the dry method can pose health hazards as it creates considerably more dust and rebound than the wet method. The quality also depends heavily on the shotcreting crew and may vary differently.

In the wet-mix method, introduced in the 1970s, the aggregate, cement, additives, and water are measured and mixed before transport. Today, wet mix is more widely used because it is easy to mechanize, and the capacity can easily outdo the dry method. Rebound rate is low, and the quality produced is even.

Shotcrete can be reinforced by adding steel fiber or polypropylene fiber to the concrete (see Fig. 21.32). The fibercrete has considerable flexural strength and has the ability to span between bolts, giving continuous rock support.

In general, thin layers are applied that are rarely more than 5 cm in thickness per round. A total layer thickness of 15 cm is common. Due to the adhesion capacity of the sprayed concrete layers, the blocks of rock around the tunnel opening become linked to each other, and the compressive load of the rock becomes more evenly distributed.

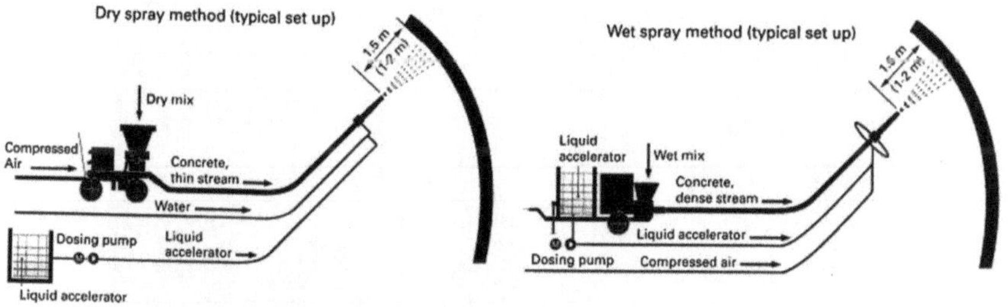

Fig. 21.31 The principles of shotcrete using either the dry or wet spraying methods (Reproduce from Ref. [10] by permission of Atlas Copco)

Fig. 21.32 Steel fiber added to the spray concrete

Manual shotcreting has been largely replaced by mechanized shotcreting machines. With mechanized equipment, much higher capacities per hour can be reached, together with consistent and even quality of the concrete layer. Safety, ergonomic, and environmental conditions are other important aspects of shotcreting. These factors are efficiently improved with mechanized shotcreting units (Fig. 21.33).

21.3.6 Steel Arch Ribs and Steel Lattice Arch Girders

In contrast to the function of rock bolting (including shotcrete), which is a kind of passive support as it becomes part of the rock and is used to help the rock support itself, steel arches and lattice girders are an active support to the rock. They are placed against the rock to be supported without penetrating the rock and just carry

Fig. 21.33 Shotcrete manipulator works in tunnel

the weight of rock that, if left unsupported, would fall into the excavation. Steel arch ribs and steel lattice girders are also commonly referred to as "steel sets" in some literatures, like Terzaghi's rock mass classification illustrated in an above section.

The preformed steel arch ribs and steel lattice girders are usually installed in the tunnel immediately after each round at the same time as rock bolting for weak rock formation. They are also commonly installed during shotcreting to give temporary support before casting the final concrete lining of the tunnel.

Steel arch ribs are not used as much now as they were even a couple of decades ago. However, there are still applications where their use is appropriate, such as unusual shapes, intersections, tunnel portal, poor rock formation in drill-blast tunnel, short starter tunnels for TBM, and reaches of tunnel where squeezing or swelling ground may occur. In today's applications, steel ribs are often installed with shotcrete being used instead of wood for the blocking (lagging) material.

The left two pictures in Fig. 21.34 show the steel arch ribs and their installation in the Sha Tin Heights tunnel (STH) of Hong Kong.

Lattice girders are used in the same way as steel arch ribs. They are lighter and thus easier to install, and their strength-to-weight ratio is higher. But they typically provide lower moment capacity due to lower total steel content. They are often used with shotcrete. Because of the open lattices, they can be covered with shotcrete with little or no voids. The right two pictures in Fig. 21.34 show the lattice girders and their installation in the same tunnel of the left pictures.

Fig. 21.34 Steel arches (*left*) and lattice girders (*right*) and their installation in STH tunnel of Hong Kong

There are two type of lattice girders: three-bar and four-bar lattice girders (see Fig. 21.35). Four-bar lattice girders have higher normal and bending moment resistance than three-bar ones.

Blocking and lagging are necessary when installing steel sets. Steel sets are held in place by the use of wood blocks and wedges.

Blocks are placed between the rib and the rock. The blocking is used to distribute the load of rock across the set to mitigate differential loading. Blocking should be distributed as evenly as possible to ensure even distribution of the load.

The members of the tunnel support that span the spaces between the primary supporting ribs are called "lagging." Lagging provides protection from spalling or falling rock and transfers the ground load to the ribs. Lagging materials can be wood or steel (see lower left picture of Fig. 21.34).

21.3.7 Spiling (Forepoling) and Pipe Roofing

Spiling, also called forepoling or prebolting, is normally considered a temporary support measure for weak rock mass during tunneling. Spiles or forepoles are used to describe the support elements consisting of pipes or pointed boards or rods driven ahead of the steel sets or lattice girders. These elements (herein called spiles) provide temporary overhead protection while excavation for and installation of the next set or girder is accomplished. Typically, spiles are driven in an overlapping arrangement as shown in Fig. 21.36 so that there is never a gap in coverage. Design of spiles is best described as "intuitive" as it must be kept flexible and constantly

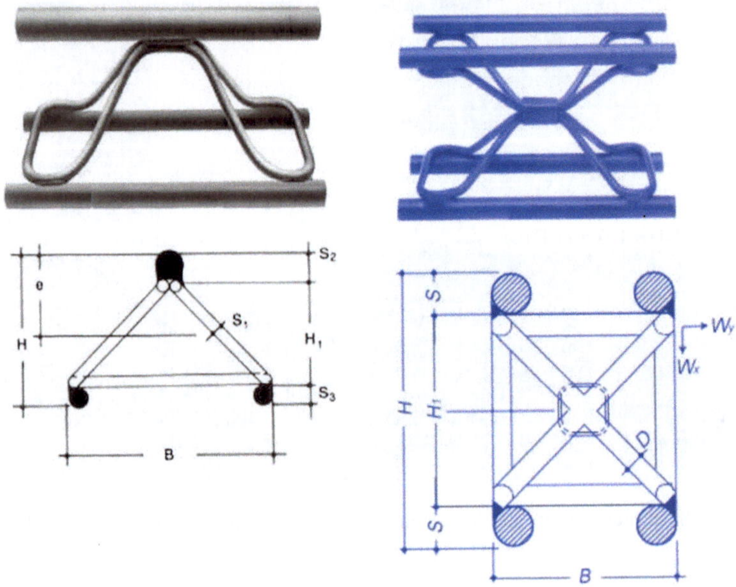

Fig. 21.35 Structure of 3-bar and 4-bar lattice girders *Source* www.dsiunderground.com/

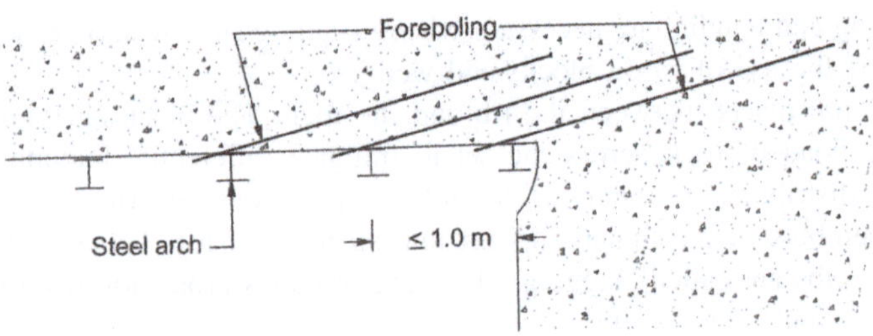

Fig. 21.36 Spilling (forepolling) method of supporting ahead of the working face

adjusted in the field as the ground behavior is observed during the construction. A working first approximation of design load might be a height of rock equal to 0.1B–0.25B, where B is the width of the opening. The recommended spile angle to the tunnel axis is usually 10–15°. A key consideration in this technique is to achieve safe anchoring at the rear end of the spile before each new blasting round. A typical procedure will be to use steel straps, radial bolts, and fiber-reinforced sprayed concrete as back anchorage. The spiling method may be combined with sprayed concrete to create a temporary support solution. In these cases, bolts with corrosion protection must be used. The bolts are installed in ground so that they adapt to the rock mass.

Another method for providing support ahead of the tunnel face is pipe screening (pipe roofing or pipe canopy), which has traditionally been used when tunneling in loose material. Today, it is applied when excavating a tunnel in rock mass with wider weakness zones and the tunnel portal excavation in soft rock. The method involves the installation of a screen of steel pipes in front of the tunnel face, over the entire roof or part of it, in the tunnel profile. The dimensions of the pipes used are typically 76.2–139.7 mm. The steel thickness in the pipes is 5–7 mm (see Figs. 21.37 and 21.38).

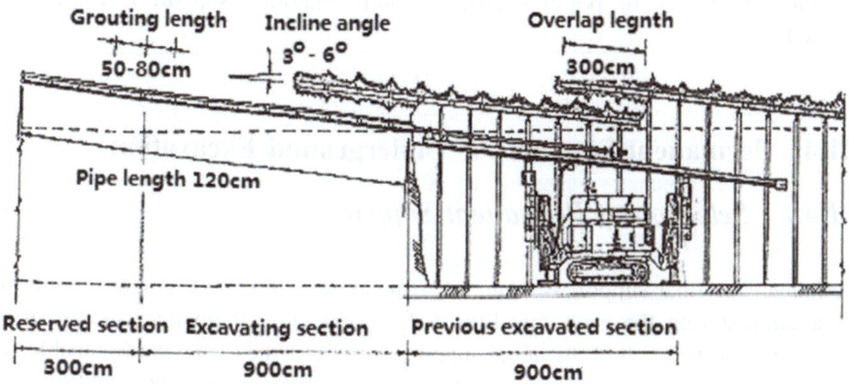

Fig. 21.37 Installation of pipe screen for tunnel excavation

Fig. 21.38 Pipe screening for tunnel portal excavation *Source* http://mitac.sptc.com.tw/

21.3.8 Portal Excavation and Temporary Support

Portal excavation and support are the most complicated and difficult work for tunneling as the worse ground condition is usually encountered.

In the worst ground situation, such as very fractured rock mass, serious ground water inflow, or very loose material such as sand, prereinforcement techniques, such as ground freezing or grouting, may need to be carried out to ensure the stability of the ground in the excavation.

In most situations, the temporary support techniques described above are used according to the actual ground condition. Figure 21.39 shows an example of the working procedure of portal excavation and temporary support measures for reference.

21.4 Permanent Support for Underground Excavation

21.4.1 Selection of Permanent Support

Permanent or final support is required to maintain stable rock conditions in the excavation during the economic life of the excavation. Permanent support may incorporate some or all of the initial support. Normally, the owner is responsible for the decision on the type and amount of support to achieve these safe conditions. As indicated above, the rock support methods and amount should be determined when the rock conditions can be studied and mapped on the actual site of installation during or after the excavation.

The objective of the lining system is to stabilize ground movement, not to carry ground loads. The most efficient tunnel stabilization and lining system are one that mobilize the strength of the ground by permitting controlled ground deformation.

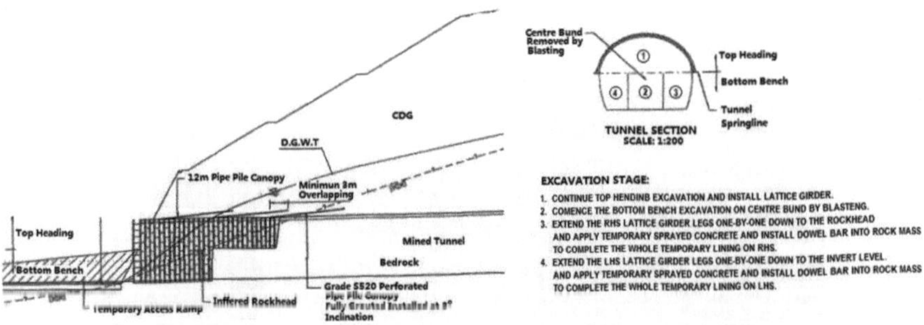

Fig. 21.39 Portal excavation and temporary support method and working sequence for a horseshoe-shaped tunnel in soft ground condition

Selection and design of permanent support of underground space should mainly consider the following factors:

- The most important factor is the ground that surrounds the underground space. It includes the following:
  - ground load which will apply on to the lining;
  - groundwater. Water sealing measures may be required in the lining design; and
  - rock characteristics and rock mass structures, including its anti-weathering ability.
- The excavation methods are suited to the ground characteristics, of which stand-up time is usually most significant. Timing of lining installation can substantially affect the magnitudes of ground deformation and lining loads.
- The purpose, function, and service life of the underground construction.

21.4.2 Types of Permanent Support [13]

21.4.2.1 Unlined Rock Tunnel

Many old mountain road tunnels have served for years without any lining at all or with linings limited to portals and weak rock zones. The Yosemite Tunnel (Fig. 21.40) in California, USA, has served for more than 60 years with large sections of unlined rock. Unlined rock tunnel is generally limited to massive, stable rock formations.

Fig. 21.40 Yosemite tunnel in California, USA, by Rennett Stowe *Source* www.flickr.com/

21.4.2.2 Sprayed Concrete (Shotcrete)

Shotcrete is widely used for stabilization of rock tunnels excavated by drill and blast methods. The shotcrete may cover lattice girders for additional strength. Shotcrete is usually used as the temporary support as its great attraction is that it can provide early construction support in rock with limited "stand-up" time. Shotcrete is sometimes used as a permanent lining. Figure 21.41 shows shotcrete-lined underground railway tunnel in London. Permanent shotcrete linings are usually built up in layers. The surface layer may contain wire mesh to provide long-term ductility or fire resistance. Alternatively, reinforced shotcrete containing randomly oriented steel or synthetic fibers may be used where toughness and ductility are desirable. Figure 21.42 displays a typical shotcrete final lining section with waterproofing system, welded wire fabric (WWF), lattice girder, and grouting hoses for contact grouting and a final shotcrete layer with PP fiber addition.

21.4.2.3 Ribbed System

Ribbed system as a traditional tunnel support system has been used for decades. This technique involves rolled steel sections being placed around the circumference of the excavated tunnel profile at specified intervals (Fig. 21.43).

It is common these days to combine steel ribs with sprayed concrete. The legs of the steel arches are often set into concrete blocks to help distribute the loads into the ground and prevent settlement.

Lattice girders, rather than rolled steel section, combined with sprayed concrete are also commonly used these days.

Fig. 21.41 A second mass-pass-transit cross-rail line in London lined with shotcrete *Source* www.theguardian.com/uk/crossrail

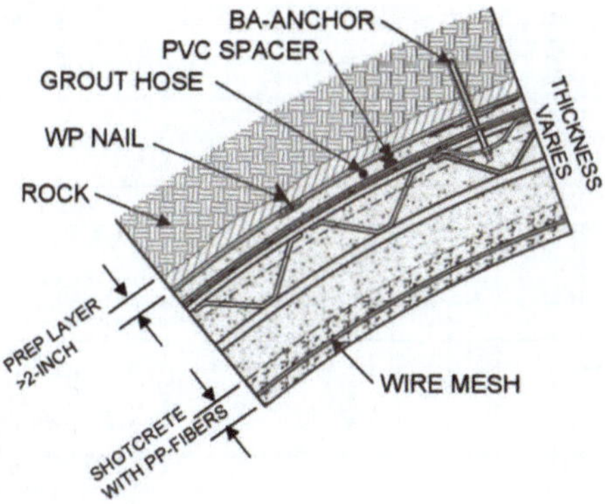

Fig. 21.42 Typical shotcrete with lattice girder and PP fiber addition lining detail *Source* [14]

Fig. 21.43 The old steel arches of Holme Tunnel were replace with new arches *Source* www.railmagazine.com

21.4.2.4 Segmental Lining

Precast segmental linings support the ground with a structure made up of a number of preformed interlocking structural elements. It is usually used in the tunnel excavated by a tunnel boring machine (TBM). They can be used in both soft and hard ground. Several curved precast elements or segments are assembled inside the tail of the tunnel boring machine to form a complete circle. The segments are relatively thin, 8–12 in. (20–30 cm) and typically 40–60 in. (1–1.5 m) wide measured along the length of the tunnel, and for large tunnels, this can now be up to 2 m or more.

Fig. 21.44 Segment lining system support the ground with a structure made up of a number of preformed interlocking elements

Precast segmental linings can be used as initial ground support followed by a cast-in-place concrete lining (the "two-pass" system) or can serve as both the initial ground support and final lining (the "one-pass" system) straight out of the tail of the TBM.

Segments used as initial linings are generally lightly reinforced, are erected without bolting them together, and have no waterproofing.

Precast segmental linings used as both initial support and final lining are built to high tolerances and quality. They are typically heavily reinforced, fitted with gaskets on all faces for waterproofing, and bolted together to compress the gaskets after the ring is completed but prior to advancing the TBM (see Figs. 21.44 and 21.45).

Segmental Lining is also used as the permanent support of shafts (Fig. 21.46).

21.4.2.5 In Situ Concrete Lining

Cast-in-place concrete can be used in any tunnel with any tunneling method. It requires some form of initial ground support to maintain the excavated opening while the lining is formed, placed, and cured. Cast-in-place concrete is usually used in hard ground tunnels mined using drill and blast excavation, and cast-in-place concrete is used in soft ground tunnels mined using sequential excavation. Cast-in-place concrete can be formed into any shape so that the lining shape can be optimized to the required opening requirements.

In wet ground, particularly where end usage may make water leakage objectionable, it may be advisable to install a waterproofing membrane layer between the initial stabilization system and the inner lining.

Fig. 21.45 Precast concrete segment for tunnel support

Fig. 21.46 Precast segments support shaft

As an example shown in Fig. 21.47, the typical section for the cast-in-place lining is used for the Cumberland Gap Tunnel. The Cumberland Gap Tunnel is a highway tunnel excavated in rock by the drill and blast method. Initial ground support is untreated rock, shotcrete, and rock bolts. The initial ground support varied along the length of the tunnel due to varying ground condition.

Usually, the in situ concrete lining requires pouring the invert first; then, the form is placed on the invert and secured. See Fig. 21.48.

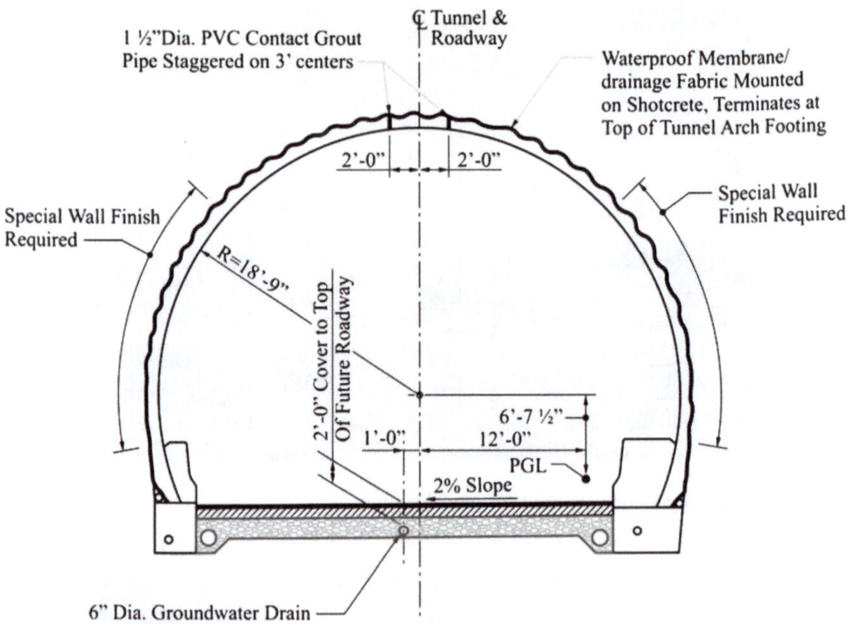

Fig. 21.47 Cumberland Gap Tunnel Lining (Unfinished) Source [9]

Fig. 21.48 Pouring the invert of the tunnel

Figures 21.49 and 21.50 show the works of installation of a waterproofing membrane layer on the tunnel inner face and the lining form for inner lining construction, respectively.

Fig. 21.49 Laying the waterproofing membrane layer on tunnel inner face

Fig. 21.50 Steel arch self-propelled traveling lining form for in situ concrete lining

21.5 New Austrian Tunneling Method (NATM)

The New Austrian Tunneling method (NATM), also known as Sequential Excavation Method (SEM), describes a popular method of modern tunnel design and construction. It was developed between 1957 and 1965 in Austria. It was given its name in Salzburg in 1962 to distinguish it from the old Austrian tunneling approach. The main contributors to the development of NATM were Ladislaus von Rabcewicz, Leopold Miiller, and Franz Pacher. The main idea is to use the geological stress of the surrounding rock mass to stabilize the tunnel itself.

Müller, L. and Fecker, E. published 22 principles to fully describe the NATM. The main principles can be summarized as follows:

1. Exploitation of the strength of rock mass—The method relies on the inherent strength of the surrounding rock mass being conserved as the main component of tunnel support. Primary support is directed to enable the rock to support itself.

2. Shotcrete protection—Loosening and excessive rock deformation must be minimized. This is achieved by applying a thin layer of shotcrete immediately after face advance.
3. Measurements—Every deformation of the excavation must be measured. NATM requires installation of sophisticated measurement instrumentation. It is embedded in the lining, ground, and boreholes.
4. Flexible support—The primary lining is thin and reflects recent strata conditions. Active rather than passive support is used, and the tunnel is strengthened not by a thicker concrete lining but by a flexible combination of rock bolts, wire mesh, and steel ribs.
5. Closing of invert—Quickly closing the invert and creating a load-bearing ring are important. It is crucial in soft ground tunnels where no section of the tunnel should be left open even temporarily.
6. Contractual arrangements—Since the NATM is based on monitoring measurements, changes in support and construction method are possible. This is possible only if the contractual system enables those changes.
7. Rock mass classification determines support measures—There are several main rock classes for tunnels and corresponding support systems for each. These serve as the guidelines for tunnel reinforcement.

Figure 21.31 shows the working procedure of NATM (Fig. 21.51).

When NATM works as a construction method, the key features are as follows:

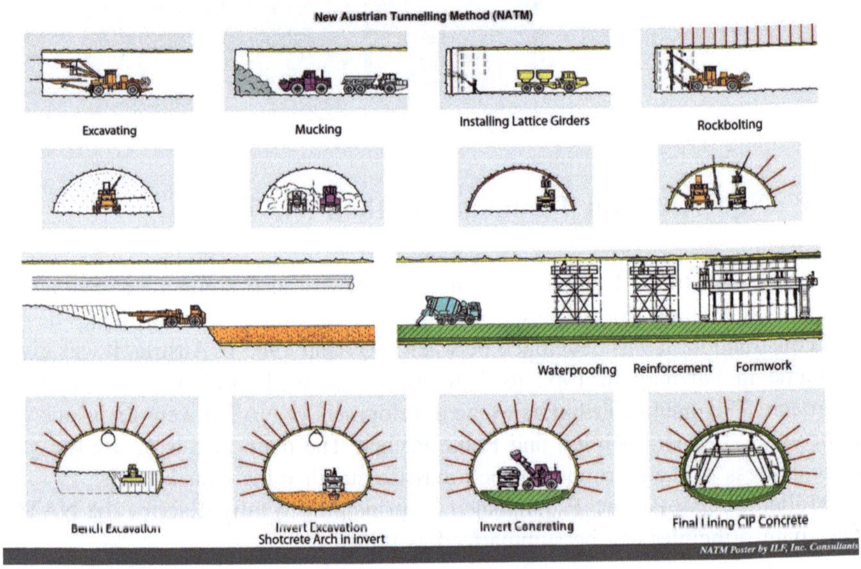

Fig. 21.51 Graphic explanation of working procedure of New Austrian Tunnel Method (Reproduced by permission of ILF Consulting Engineers Austria)

- The tunnel is sequentially excavated and supported, and the excavation sequences can be varied.
- The initial ground support is provided by shotcrete in combination with fiber or welded wire fabric reinforcement, steel arches (usually lattice girders), and sometimes ground reinforcement (e.g., soil nails and spiling).
- The permanent support is usually (but not always) a cast-in-place concrete lining.

References

1. Singh B, Goel RK (2011) Engineering rock mass classification: tunnelling, foundations and landslides, Elsevier, Amsterdam • Boston • Heidelberg • London New York • Oxford • Paris • San Diego San Francisco • Singapore • Sydney • Tokyo, 2011, ISBN: 978-0-12-385878-8
2. You K, et al (2007) Estimation of rock load for the design of 2-arch tunnel lining, underground space—the 4th dimension of metropolises—Proceedings of the World Tunnel Congress 2007 and 33rd ITA/AITES Annual General Assembly, Prague, May 2007 Edited by Ivan Hrdina, CRC Press 2007, London, ISBN:978-0-415-40807-3
3. Hoek E, Brown ET (1980) Underground excavation in rock, Institution of Mining and Metallurgy, London, 1980. ISBN: 9780415461658
4. Garshol KF (2003) Pre-excavation grouting in rock tunnel, MBT International Underground Construction Group, Division of MBT (Switzerland) Ltd.
5. Ehsanzadeh B, Ahangari K (2014) A novel approach in estimation of the soilcrete column's diameter and optimization of the high pressure jet grouting using adaptive neuro fuzzy inference system (ANFIS), Open J Geol 4:386–398. Published Online August 2014 in SciRes. http://www.scirp.org/journal/ojg; doi:10.4236/ojg.2014.48030
6. Garshol KF, Tam J (2013) Pre-excavation grouting of underground construction in hard rock —site practice, 2013. http://www.ags-hk.org/tcg/TCG06.pdf. Accessed 17 June 2016
7. Garshol KF et al (2013) Deep subsea rock tunnels in Hong Kong. World Tunnel Congress 2013 Geneva, Underground—the Way to the Future. Taylor & Francis Cop. ISBN: 9781138000940 1138000949
8. Heiniö M, Editor in Chief (1999) Rock excavation handbook for civil engineering, Sandvik & Tamrock, 1999
9. Hung CJ et al (2009) Technical manual for design and construction of road tunnels—Civil Elements, National Highway Institution, U. S. Department of Transportation Federal Highway Administration, Washington, December 2009
10. Atlas Copco (2015) Underground construction—a global review of tunnelling and subsurface installations, 1st Edition, 2015
11. Palmström A, Nilson B (2000) Engineering geology and rock engineering, Oslo Norwegian Group for Rock Mechanics 2000. ISBN: 8291341338 9788291341330
12. Hjalmarsson EH (2011) Tunnel support, use of lattice girders in sedimentary rock, master's thesis, University of Iceland, Reykjavik, Iceland, 2011
13. Bickel JO et al (1996) Tunnel engineering handbook, 2nd edn. Chapman & Hall, Boston
14. RailSystem.Net (2015) New Austrian Tunneling Method (NATM). Source: http://www.railsystem.net/natm. Accessed April 14, 2016
15. Shang Y-Q et al (2006) Engineering geology (in Chinese). Tsinghua University Press, Beijing. ISBN: 7-302-12540-6. 尚岳全等 (2006), 地质工程学, 清华大学出版社, 北京

Chapter 22
Monitoring and Instrumentation for Underground Excavation

22.1 Introduction [5–7]

All underground constructions require monitoring by direct observation or by instrumentation.

The purposes of monitoring are as follows:

- preventing any significant disruption to the public and environment, noticeable effect on buildings, structures, and utilities in the vicinity of underground excavation;
- verification of design to reveal developing adverse conditions of the underground space during and after excavation and construction, thus allowing timely corrective action, especially the large caverns and tunnels or complex ground conditions. In tunneling, using NATM methods monitoring inside the tunnel as construction proceeds is integral part of the construction process.

In this chapter, the following types of measurements typically will be discussed:

- Ground movement around the tunnel,
- Building movement for structures within the zone of influence,
- Tunnel movement of the tunnel being constructed or adjacent tubes,
- Groundwater movement and pressure due to changes in the water percolation pattern, and
- Dynamic ground movement and air overpressure from drill and blast.

The first three items comprise quasi-static changes in position, and the fourth is also concerned with long-term effects. In contrast, dynamic ground movement covers response due to vibration and air overpressure caused by the shock waves generated by explosive charges used to excavate rock.

D. Zou, *Theory and Technology of Rock Excavation for Civil Engineering*,
DOI 10.1007/978-981-10-1989-0_22

Regularly monitoring all the changes caused by underground excavation is one of the important works during the whole progress of the project.

(a) Initial readings shall be taken immediately after the instruments have been installed and after effects of installation have been subsided.
(b) All of the monitoring needs to be coordinated to fit with the tunnel construction schedule.
(c) Alert, Action, and Alarm levels shall be established in response to the instrumentation findings. Necessary action shall be taken when breaching Alert, Action, and Alarm levels.

22.2 Settlement Monitoring

22.2.1 Ground Settlement Monitoring

Ground settlement or subsidence caused by tunneling is monitored by leveling. It is recommended to plot not only the settlement troughs of several points or sections but also contour maps. Leveling is also carried out inside a tunnel to determine, e.g., the crown settlement (the leveling rod can be suspended from the tunnel crown).

Several types of instrumentation are used to monitor the ground settlement:

- Deep benchmarks,
- Survey points,
- Borros points, and
- Telltales or roof monitors.

Figure 22.1 shows the survey point, and Fig. 22.2 shows the section of the deep benchmark.

Figure 22.3 shows the section of triple telltale or roof monitor, and Fig. 22.4 shows the picture of it.

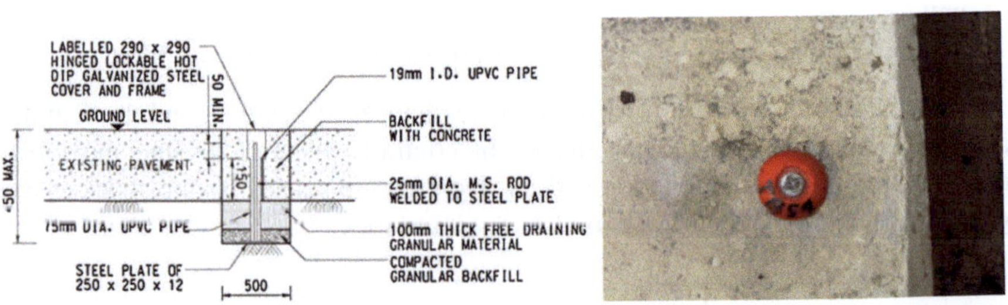

Fig. 22.1 Survey point

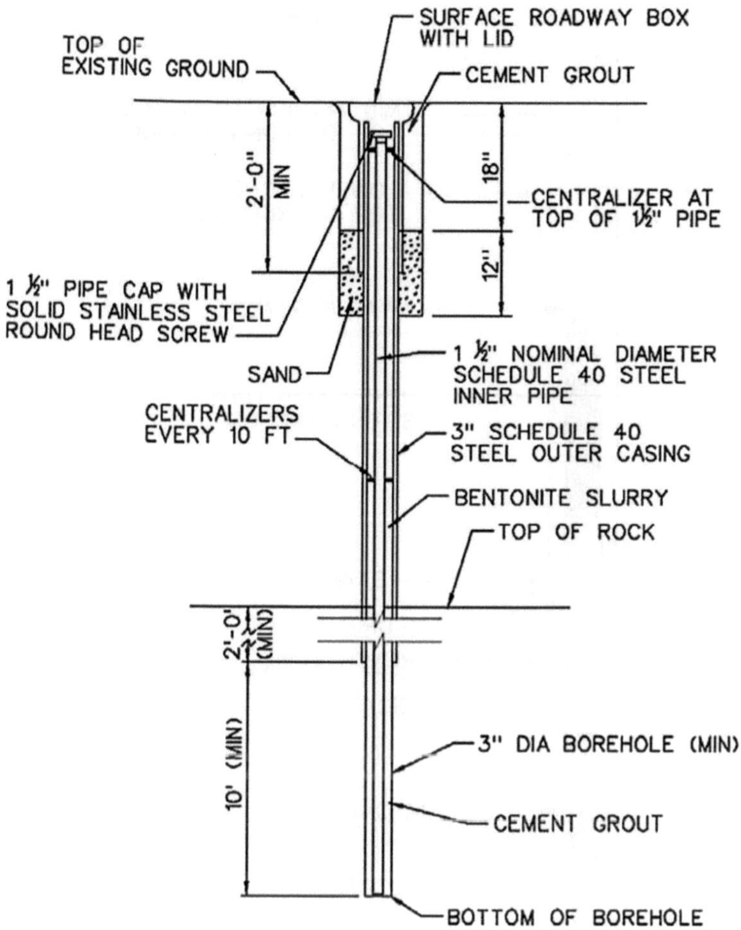

Fig. 22.2 Deep benchmark

22.2.2 *Utility Settlement and Building Settlement Monitoring*

Ground settlement caused by underground excavation may result in damage due to the nearby buildings, structures, or utilities, especially by the uneven settlement.

There are different instrumentations that can be used to monitor the settlement of buildings, structures, and utilities, and the commonly used ones are as follows:

- Structure Monitoring Points

Structure monitoring points are survey points that are placed directly on the structures of concern, most often being installed on a vertical wall of a building or a structural element (see Figs. 22.5 and 22.6).

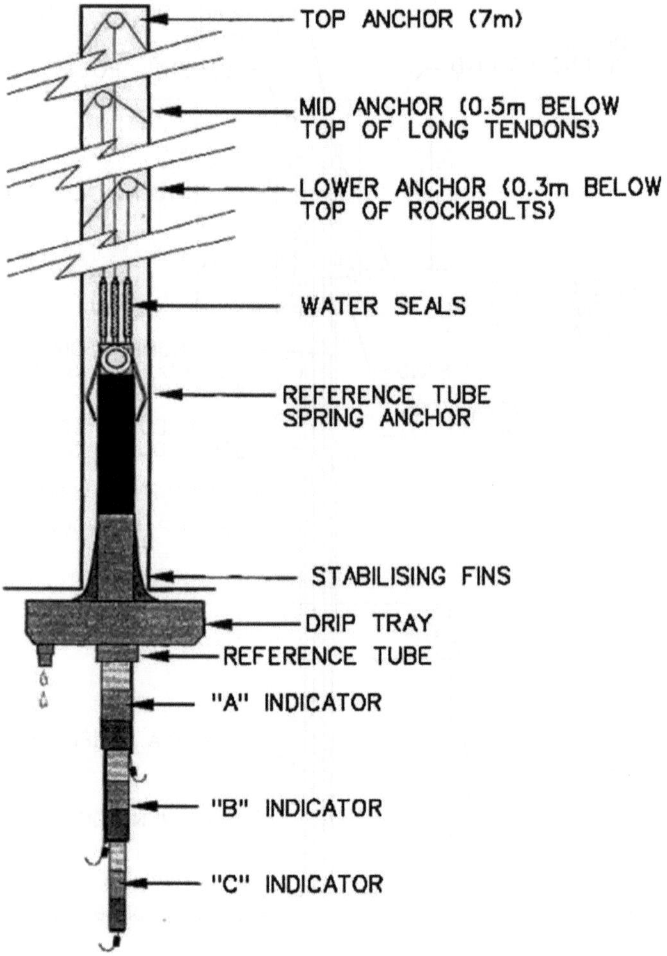

Fig. 22.3 Triple height telltale. *Source* [1]

- Robotic Total Stations [1]

Robotic total stations are used for obtaining almost real-time data on movements in three dimensions when it is not feasible to continually mobilize survey crews to collect data. The operation of a total station instrument (theodolite) is based on an electronic distance meter (EDM), which uses electromagnetic energy to determine distances and angles with a small computer built directly into the instrument. Accuracy is generally much greater than that achievable with the use of classical optical surveying. Moreover, the equipment based on EDMs is capable of detecting target movements along all three possible plotting axes, the *x*, the *y*, and the *z*. Total stations used in geotechnical and structural monitoring are electro-optical, and either lasers or infrared light is used as the signal generator. As shown in Fig. 22.7,

Fig. 22.4 Typical installation of triple height telltale

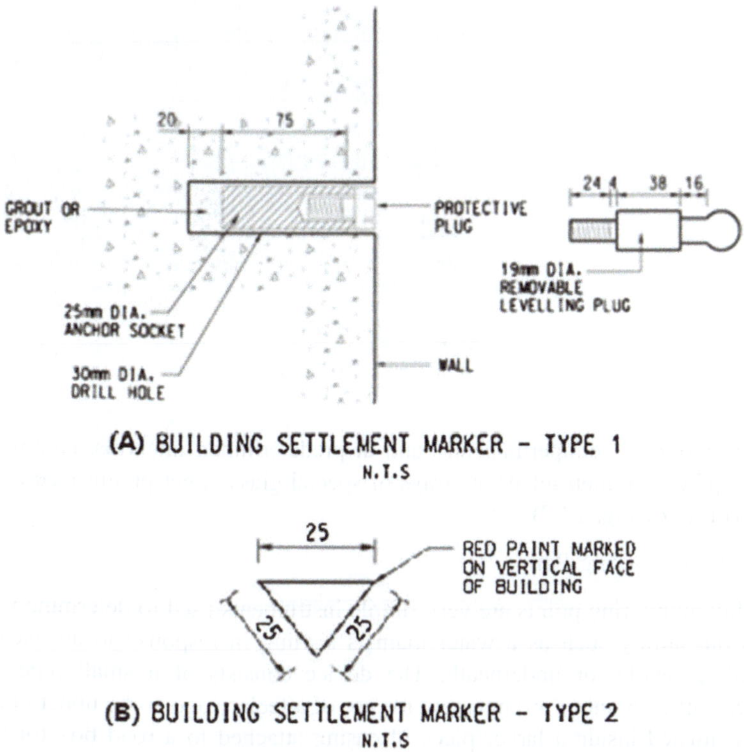

Fig. 22.5 Building and structure settlement mark

Fig. 22.6 Building and structure monitoring

they are mounted semipermanently and at predetermined intervals, automatically "wake up" to aim themselves at arrays of special glass target prisms (see the right lower corner of Fig. 22.7).

- Utility Monitoring Points

Utility monitoring points are very simple instruments used to determine whether an existing utility such as a water main is settling in response to an excavation proceeding nearby or underneath. The device consists of a small pipe with a rounded survey point or arrangement for use of a feeler gage at the upper end. This pipe is situated inside a larger piece of casing attached to a road box for surface protection. The lower end of the small pipe is attached to the top of the utility to be

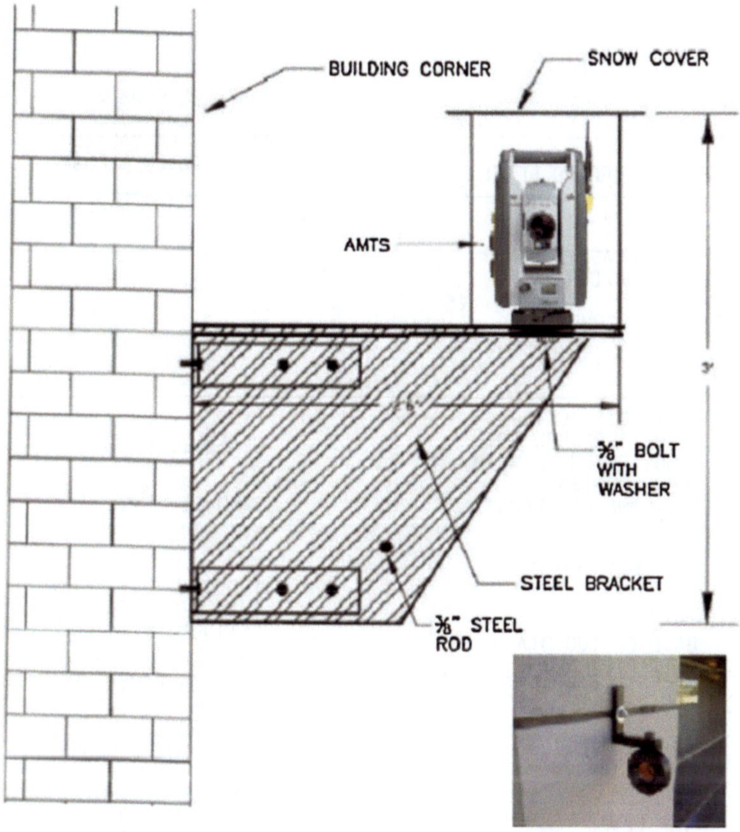

Fig. 22.7 Robotic total station instrument. *Source* [1]

monitored and data collected by determining whether the top seems to be moving downward (see Figs. 22.8 and 22.9).

22.3 Displacement and Deformation Monitoring

22.3.1 Inclinometer Monitoring

Vertical inclinometers are instruments for measuring relative horizontal displacements affecting the shape of a guide casing embedded in the ground or structure. Inclinometer probes usually measure displacement in two perpendicular planes; therefore, displacement magnitudes and directions (vectors) can be calculated. The bottom end of the guide casing serves as a stable reference (datum) and must be embedded beyond the displacement zone. Relative displacement over time is

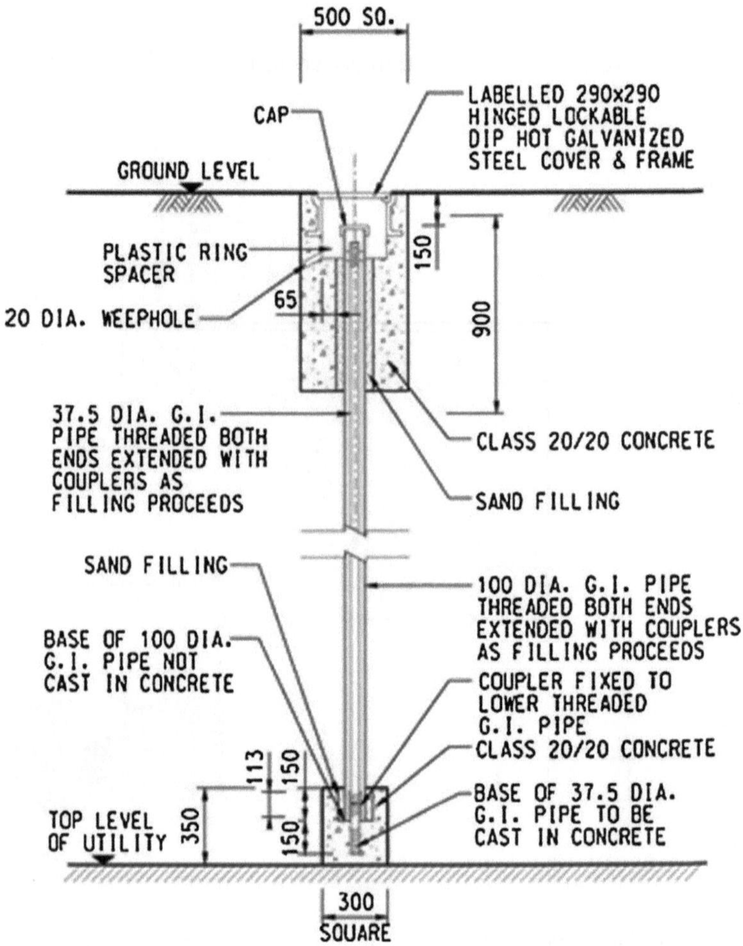

Fig. 22.8 Utility settlement mark. N.T.S

determined by repeating measurements at the same depths and comparing data sets. The guide casing is installed vertically for most applications in order to measure horizontal ground movements. Figures 22.10 and 22.11 show the vertical inclinometer.

Horizontal inclinometers are used to obtain high-resolution profiles of settlement or heave. The digital horizontal inclinometer system consists of inclinometer casing, a horizontal probe, control cable, pull cable, and a readout unit.

Inclinometer casing is installed in a horizontal trench or borehole. As shown in the Fig. 22.12, the casing can be open at both ends, or closed at the far end. When the casing is closed at the far end, a dead-end pulley and cable-return pipe are also installed. The probe, control cable, pull cable, and readout unit are used to survey

Fig. 22.9 Settlement mark for utility monitoring

the casing. The initial survey establishes the profile of the casing, and subsequent surveys will reveal changes in the profile if ground movement has occurred.

22.3.2 Magnetic Extensometer and Monitoring

Magnet extensometer is a system for measuring either settlement or heave at various depths in soil, embankments, earthfill dams, and dikes. The system consists of an access tube with external corrugated pipe, magnet rings, telescopic bottom section with datum ring, and suspension head. Magnet rings (targets) are fixed, externally to the access tube, in the ground where movement may occur. Magnet rings move together with the surrounding soil along the axis of the access tube (see Fig. 22.13 together with Fig. 22.14).

Readings are obtained with a portable readout, lowering the reed switch probe through the access tube. Comparison of surveys taken over time provides profiles of ground settlement or heave.

22.3.3 Tiltmeter

A tiltmeter is a sensitive inclinometer designed to measure very small changes from the vertical level, either on the ground or in structures.

Manual tiltmeters generally consist of reference points on plates attached to the surface of interest (Fig. 22.15) and monitored by means of a portable readout unit, the functioning of which is based on an accelerometer transducer. The modern electronic tiltmeter uses a simple bubble-level principle, as used in the common carpenter level. As shown in Fig. 22.16, an arrangement of electrodes senses the

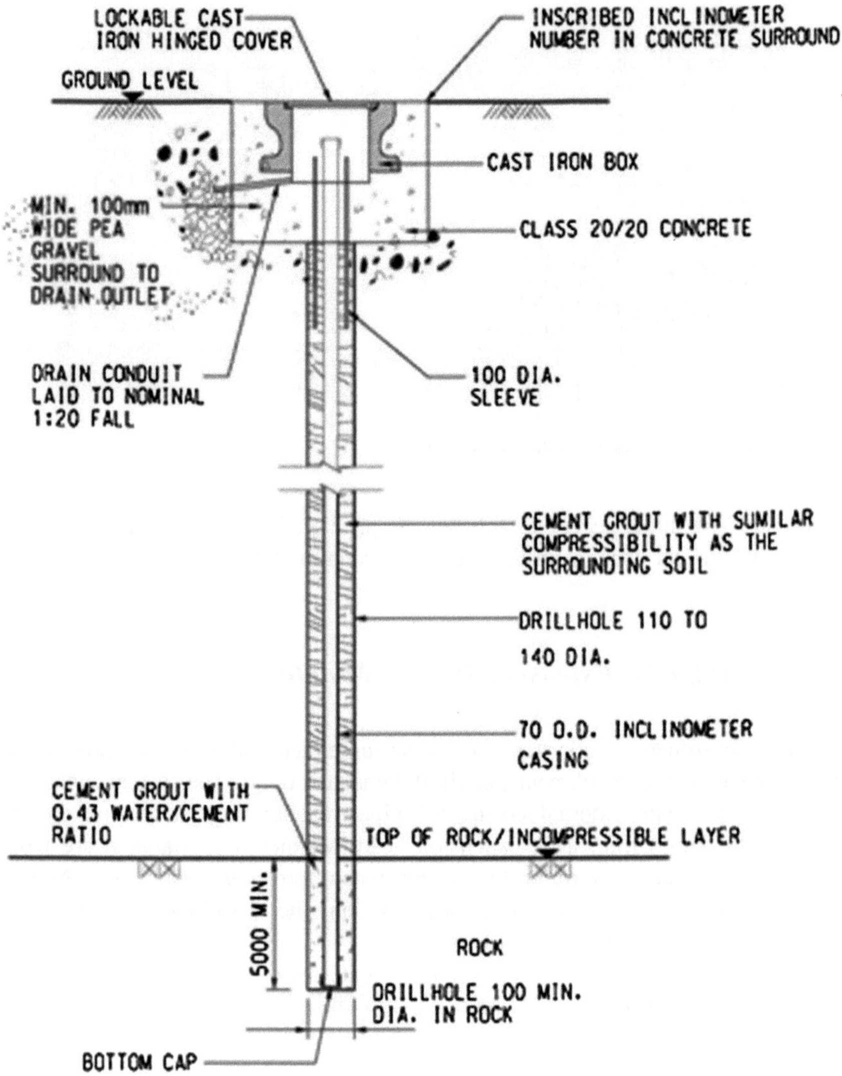

Fig. 22.10 Vertical inclinometer (in rock and soil) N.T.S

exact position of the bubble in the electrolytic solution, to a high degree of precision. Any small changes in the level are recorded using a standard datalogger. This arrangement is quite insensitive to temperature and can be fully compensated, using built-in thermal electronics.

Fig. 22.11 Vertical inclinometer and inclinometer measurement

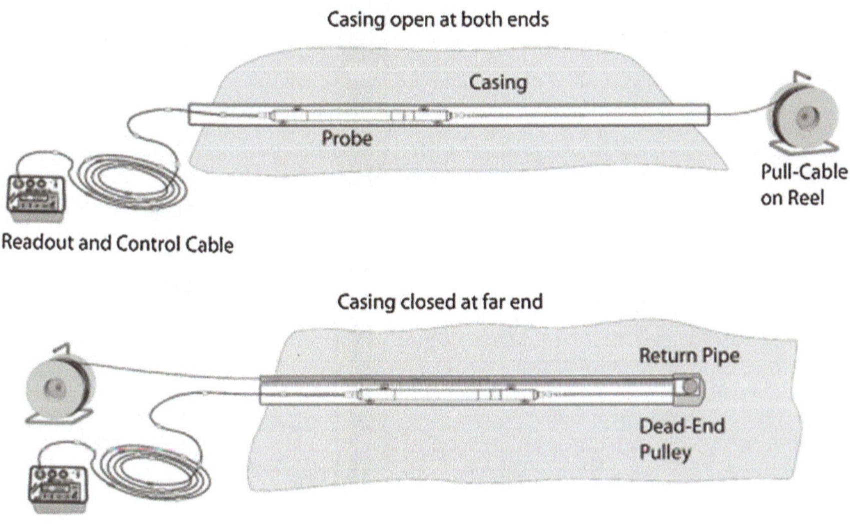

Fig. 22.12 Installation of horizontal inclinometer probe (reproduced from Ref. [8] with the permission from ©2006 DGSI)

22.3.4 Crack Monitoring: Telltales

Telltales consist of two plates which overlap for part of their length. One plate is calibrated in millimeters, and the overlapping plate is transparent and marked with a hairline cursor. As the crack width opens or closes, one plate moves relative to the other. The relationship of the cursor to the scale represents the amount of movement occurring. These gauges typically have a range of ±20 mm and resolution of 1 mm. The telltales are fixed with screws and adhesive across the crack to be monitored. The standard telltale is produced in durable acrylic plastic and is used for monitoring movement across cracks in vertical and horizontal directions on flat surfaces. Figure 22.17 shows a telltale.

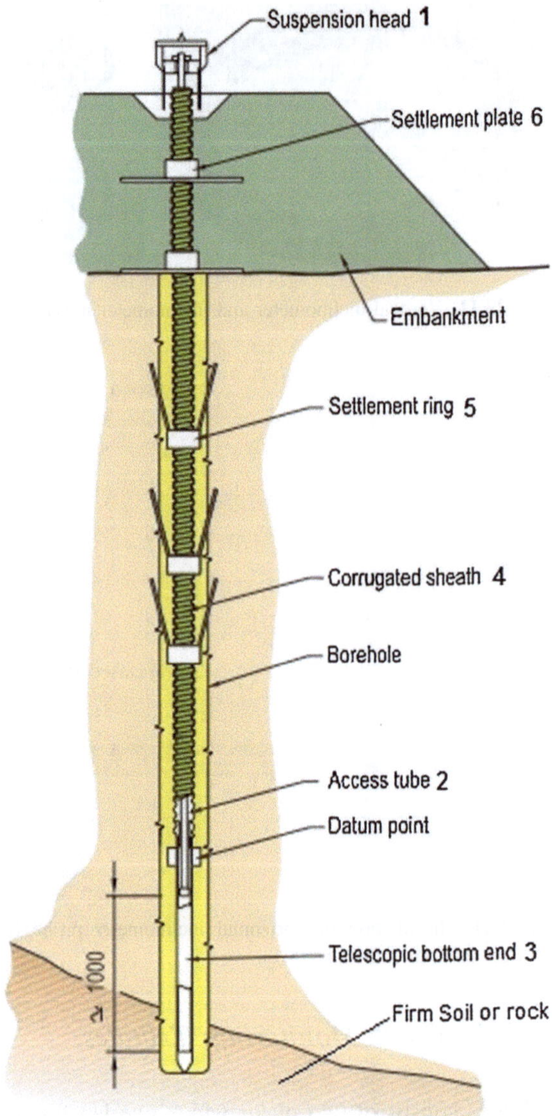

Fig. 22.13 Installation scheme of magnetic extensometer (reproduced with the permission from SISGEO SRL—Italy)

22.3.5 Convergence Array Monitoring

For monitoring the deformation of the tunnel lining, convergence array is carried out with the help of pins or target plates mounted onto the tunnel wall. To determine convergences, the readings between individual pins of the same cross section should be plotted overtime. Figure 22.18 shows the traditional method of convergence array for temporary lining with tape extensometer (TEX) (Fig. 22.19).

Fig. 22.14 Components of magnetic extensometer (reproduced by permission of SISGEO SRL—Italy)

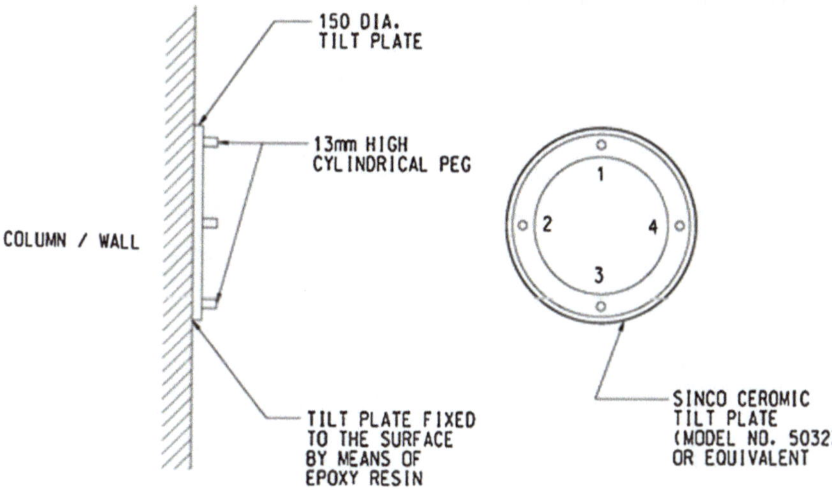

Fig. 22.15 Manual tiltmeter (N.T.S.)

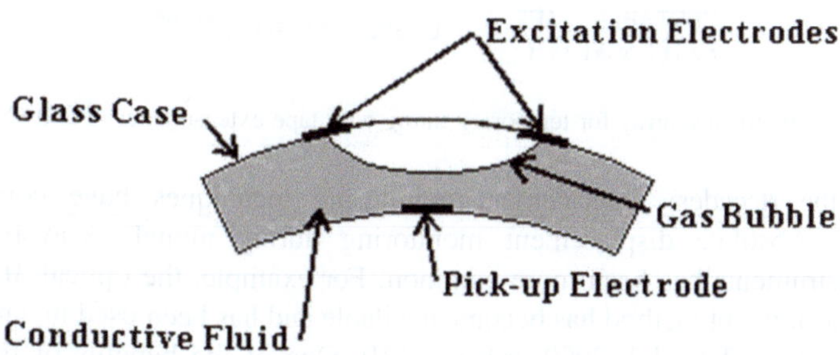

Fig. 22.16 Principle of electronic tiltmeter

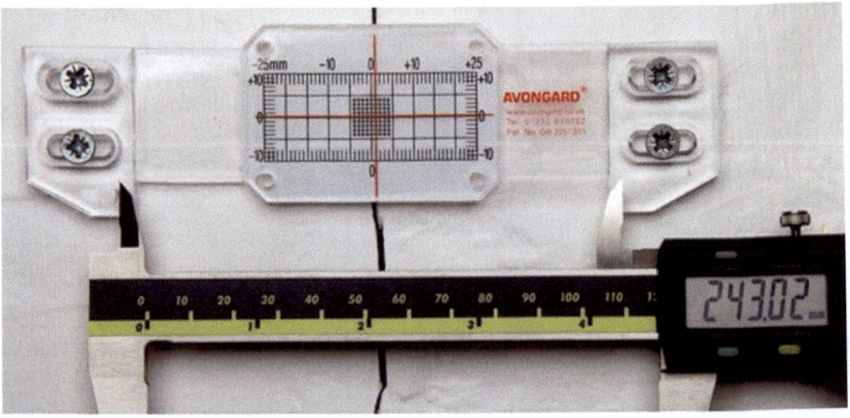

Fig. 22.17 Telltales—crack monitoring

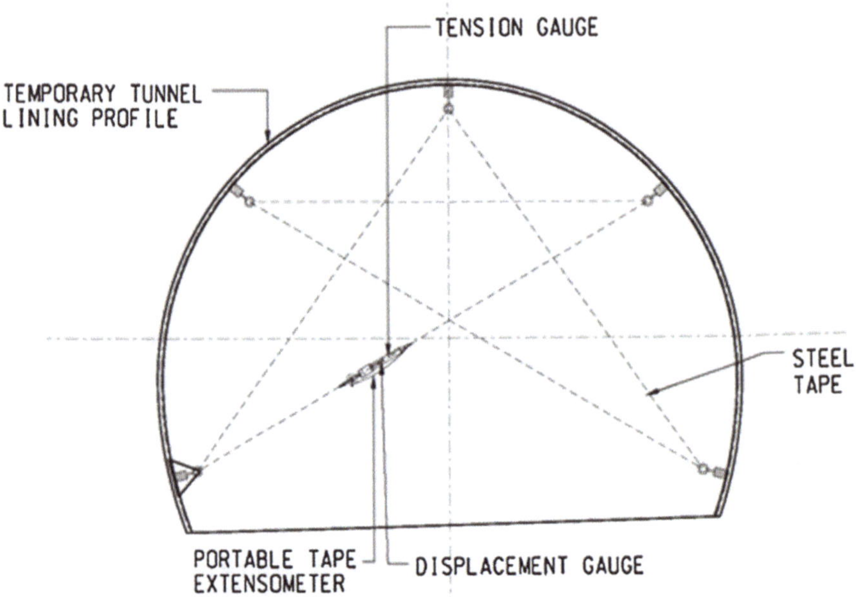

Fig. 22.18 Convergence array for temporary lining with tape extensometer (TEX) (N.T.S.)

Over the decades, displacement monitoring techniques have considerably improved. Absolute displacement monitoring during tunnel excavation using optical instruments has been more common. For example, the optical 3D convergence measurement method has become available and has been used in a number of tunneling sites (ITA-CET 2009, refer to [2]). One of the benefits of the optical method is that the monitoring system does not impair the tunnel construction process.

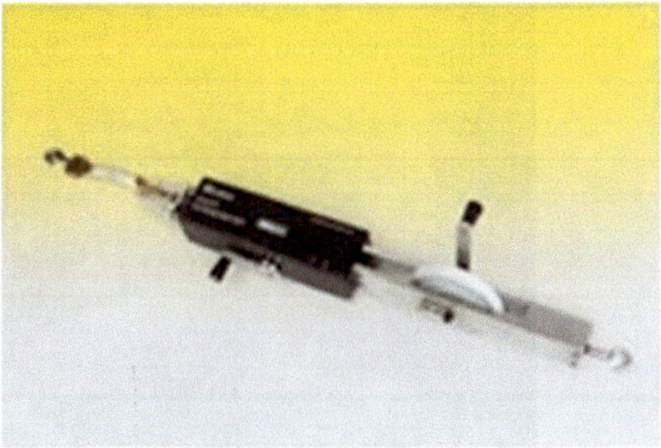

Fig. 22.19 Tape extensometer

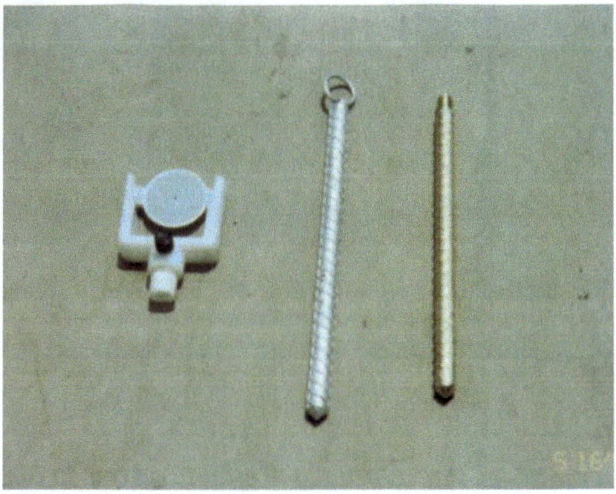

Fig. 22.20 Bireflex target, convergence, and settlement pins. *Source* [9]

The 3D optical measuring system consists of the following major components: electronic theodolite with an integrated coaxial electro-optical distance-measuring system (total station); special reflective targets (Fig. 22.20); hardware requirements including: PC or notebook, interface for transferring data from theodolite to PC, output devices, and comprehensive software system including: data base management, geodetic calculations, graphical evaluation, utilities, and drivers. The 3D optical measuring system instead of conventional displacement monitoring methods can assist automatic data logging, to make proper judgment by rapidly safety evaluation from monitoring raw data (Fig. 22.21). By adding fixed wire or wireless connection to a server, this monitoring can now be carried out in locations where routine man access is difficult or impossible.

Fig. 22.21 Illustration of optical 3D convergence measurement, measurement results are absolute, accuracy 1–2 mm for monitoring the deformation of cavity deformation of the tunnel walls surface (reproduced from Ref. [2] with the permission from AITES/ITA)

22.3.6 Time-Domain Reflectometer Probe: Slope Movement Detector

A time-domain reflectometer (TDR) probe is an electronic instrument that uses time-domain reflectometry to characterize and locate faults in metallic cables (e.g., twisted pair wire or coaxial cable). It can also be used to locate discontinuities in a connector, printed circuit board, or any other electrical path. The equivalent device for optical fiber is an optical time-domain reflectometer.

The following Fig. 22.22 shows the instrument of time-domain reflectometer probe, which is powered by solar energy. It can detect slope movement, through a coaxial cable or optical fiber which is fixed on the slope surface, and report the domain of slope movement wirelessly to the control center.

Fig. 22.22 Time-domain reflectometer probe for detecting slope movement

22.4 Groundwater Monitoring

22.4.1 Standpipe and Piezometer [3]

Standpipe piezometers are used to monitor piezometric water levels. Typical applications include the following:

- Monitoring pore water pressure to determine the stability of slopes, embankments, and landfill dikes.
- Monitoring the effectiveness of dewatering schemes.
- Monitoring seepage and groundwater movements in embankments, landfill dikes, and dams.

The standpipe piezometer consists of a filter tip joined to a riser pipe. The filter has 60–70-micron pores and is made from polyethylene or porous stone. The riser pipe is typically made from PVC plastic pipe. A water level indicator is used to monitor the piezometric water level (Figs. 22.23 and 22.24).

22.4.2 Vibrating Wire Piezometer [4]

Vibrating wire (VW) piezometers are used to monitor pore water pressure. They can also be used to monitor water levels. Typical applications include the following:

- Monitoring pore water pressures to determine safe rates of fill or excavation.
- Monitoring pore water pressures to determine slope stability.
- Monitoring the effects of dewatering systems used for excavations.
- Monitoring the effects of ground improvement systems such as vertical drains and sand drains.
- Monitoring pore pressures to check the performance of earth fill dams and embankments.

Fig. 22.23 Standpipe and piezometer

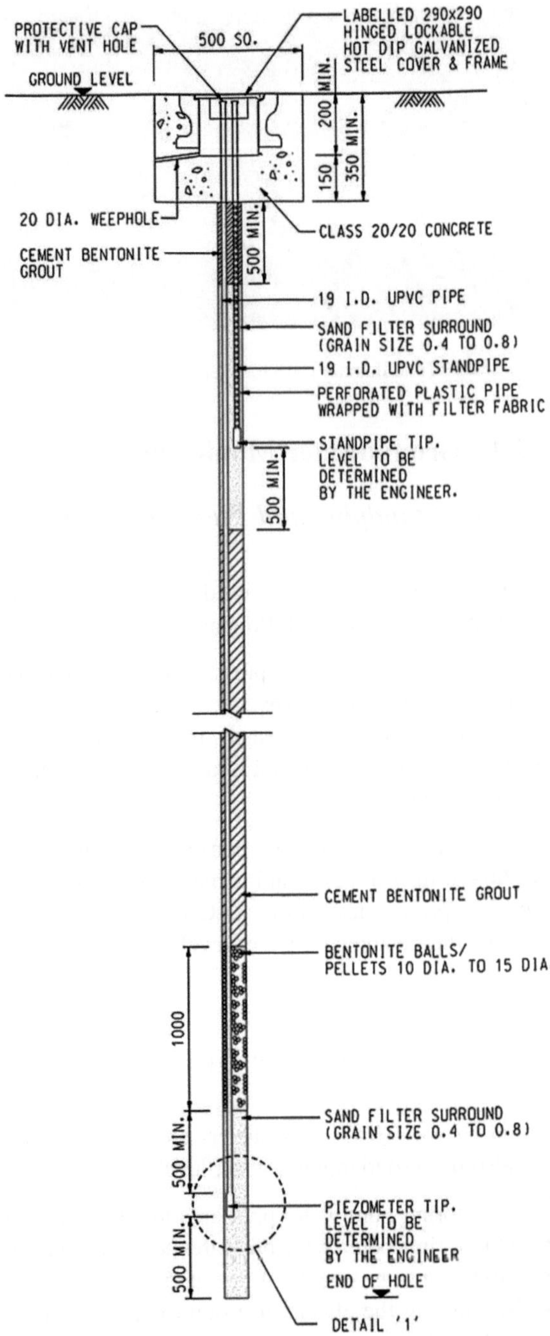

Fig. 22.24 Standpipe and piezometer

- Monitoring pore pressures to check containment systems at landfills and tailings dams.
- Monitoring water levels in stilling basins and weirs.

The VW piezometer converts water pressure to a frequency signal via a diaphragm, a tensioned steel wire, and an electromagnetic coil. The piezometer is designed so that a change in pressure on the diaphragm causes a change in tension of the wire. An electromagnetic coil is used to excite the wire, which then vibrates at its natural frequency. The vibration of the wire in the proximity of the coil generates a frequency signal that is transmitted to the readout device. The readout or data logger stores the reading in Hz. Calibration factors are then applied to the reading to arrive at a pressure in engineering units. VW piezometers can be installed in fully grouted boreholes and do not require sand filter zones. This greatly simplifies the installation of multiple sensors in the same borehole. It also makes it possible to install piezometers with inclinometer casing within the same borehole. Figure 22.25 shows a VW piezometer installed in the tunnel lining. Figure 22.26

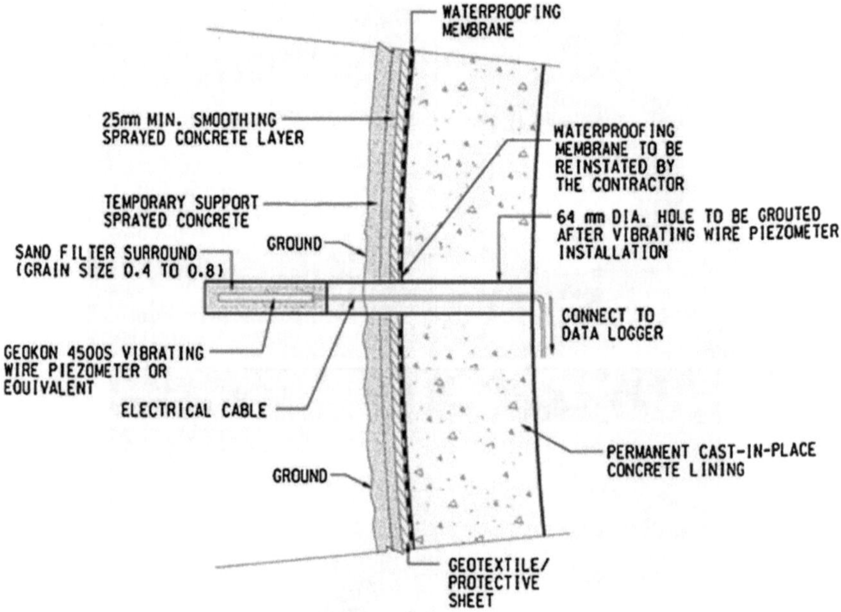

Fig. 22.25 Vibrating wire piezometer (N.T.S.)

shows a vibrating wire piezometer. VW piezometers provide a resolution of 0.025 % FS. There are several types of VW piezometers available on the market. Fig. 22.26 shows two types of them.

22.4.3 *Pore Water Pressure Gauge*

The difference compared to VW piezometers is that the water pressure gauges are not fully grouted in the rock, but instead are connected to a drain tube which is installed in the concrete lining and collects the water from rock (see Figs. 22.27 and 22.28).

22.5 Ground Vibration and AOP Monitoring

The instruments for monitoring ground vibration and air overpressure are same as that used in surface blasting monitoring and are discussed in Chap. 11.

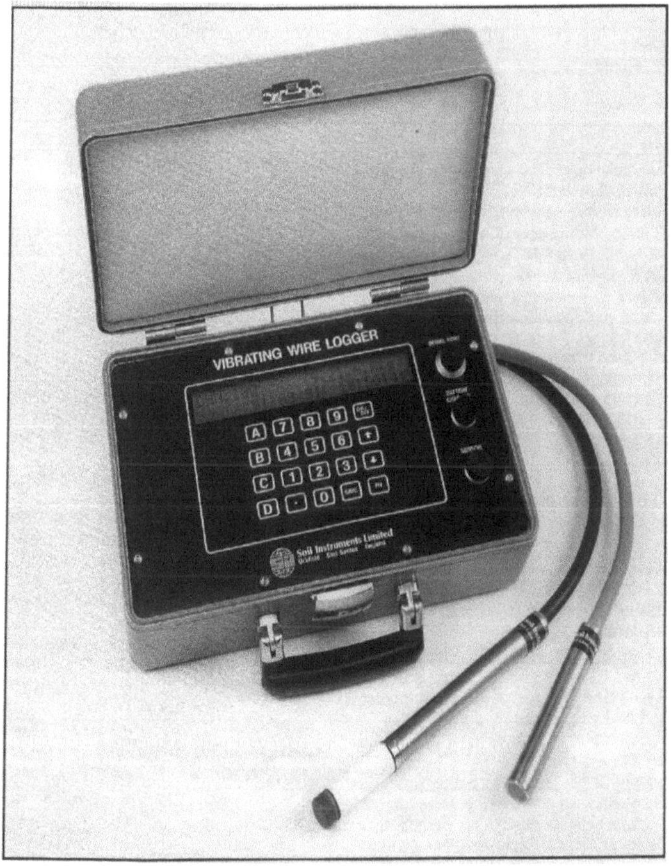

Fig. 22.26 Vibrating wire piezometer. *Source* http://jatiluhurdam.wordpress.com

22.6 Instrumentation Management

As noted in the introduction to this chapter, the primary function of most instrumentation programs is to monitor performance of the construction process in order to avoid or mitigate problems.

22.6.1 Instrumentation Selection

Generally, the physical parameters monitored include strains, relative displacements, changes in curvature (in the tunnel lining), stresses in the lining and in the rock mass, rock or earth pressures on the tunnel lining, groundwater pressure, forces in rock anchors, and piezometric levels (ground and tunnel settlement). ITA Report

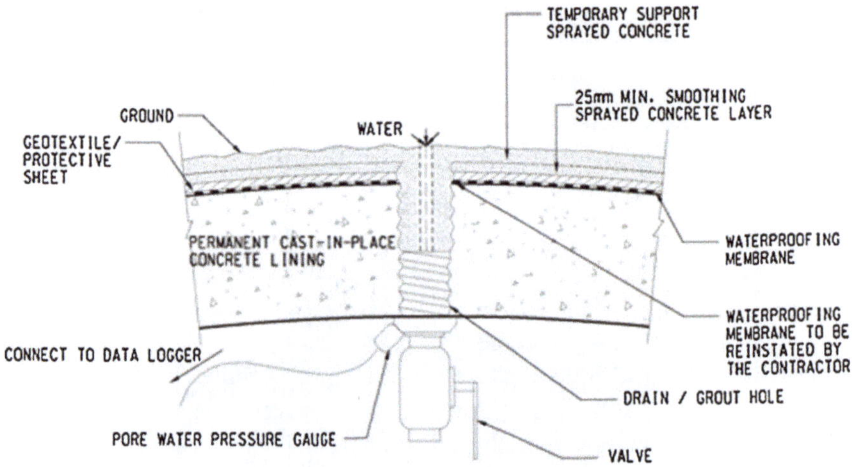

Fig. 22.27 Pore water pressure gauge (N.T.S.)

Fig. 22.28 Pore water pressure gauge. *Source* www.controls-group.com/

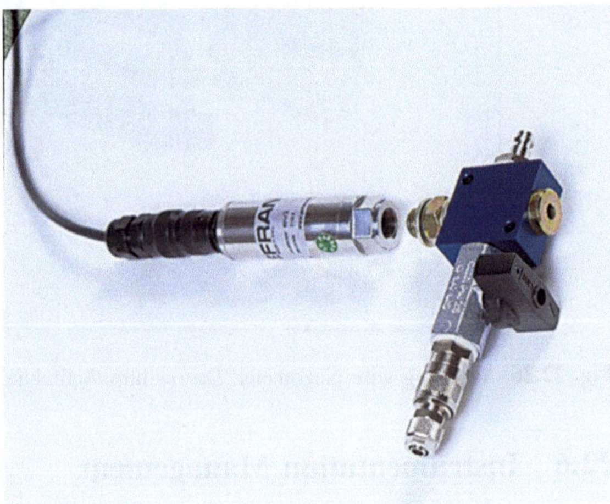

No. 009 [2] gives the typical measurement items for conventional tunneling and TBM tunneling, and they are shown in Table 22.1:

ITA Report No. 009 [2] also gives the typical instrumentation equipments used for routine instrumentation in Table 22.2 below:

22.6.2 Frequency of Monitoring Readings

The frequency of monitoring readings can vary according to the monitoring phases as follows:

Table 22.1 Leading parameters to be monitored in four tunnel configurations (modified AITES 2005) (reproduced from Ref. [2] with the permission from ITA/AITES)

	Conventional method (low overburden in urban areas)	Closed-face TBM (urban areas)	Operational tunnel creeping ground
0. Visual inspection (face and sidewall)	●		●
1. Geometrical parameters			
Face extrusion (horizontal displacement)	●		
Surface settlement	●	●	
Surface rotation	○	○	
Extrusion of the ground ahead of tunnel face (extrusion meter)	■		
Displacement in borehole (extensometer, inclinometer)	○	○	×
Convergence at sidewall	●		●
Crack monitoring	○		●
Deformation of permanent lining	×	×	●
2. Mechanical parameters			
Face (arch base, anchoring rod, rock bolt, etc.)	●		×
Stress in ground			○
Stress in support/lining	○	×	○
3. Hydraulic parameters			
Pumped out water rate	○		○
Surface rainfall	×		×
Piezometric levels in ground	●	●	●
Temperature of leakage			×
4. Other parameters			
Tunnel air temperature		○	×
Tunnel air pressure		×	
Tunnel hygrometer	×	×	○
Date and time	●	●	●
Vibration from blasting	●		

×Usually Secondary parameter
○Frequently important parameter
●Essential parameter, always monitored
■Essential parameter, always monitored when advance full face with preconfinement

Table 22.2 Types of instrumentation equipment (modified Tunnel Lining Design Guide, 2004) (reproduced from Ref. [2] with the permission from ITA/AITES)

Objective	Instrumentation	Range	Resolution	Accuracy
Extrusion of ground ahead of face	Increx probe	0.1 mm	0.01 mm	±0.003 mm/m
	Sliding micrometer	1 m	0.01 mm	0.002 mm/m
	Sliding deformeter		0.01 mm	0.02 mm/m
Relative Vertical movement	Precise leveling pins installed on structures, settlement points, geodetic surveying targets in structures or tunnel linings	Any	0.1 mm	0.5–1.0 mm
	Precise liquid level settlement gauges with LVDTs installed in surface structures	100 mm	0.01–0.02 mm	±0.25 mm
	Borehole magnet extensometer	Any	±0.1 mm	±1–5 mm
	Borehole rod or invar tape extensometer	100 mm	0.01 mm	±0.01–0.05 mm
	Satellite geodesy	Any	To ±50 mm	To ±1 mm
Lateral displacement	Surface horizontal invert wire extensometer	0.01 %	0.001–0.005 %	0.01–0.05 mm
Change in inclination	Borehole electrolevels	Any	±10 mm	±10–20 mm
	Electrolevel beams on structures and in tunnels tilt meters	(To 175 mm/m)	(To 0.3 mm/m)	
	Horizontal borehole deflectometer	±50 mm	±0.02 mm	±0.1 mm
	Borehole inclinometer Probes	±53° from vertical	0.04 mm/m	±5 mm/25 m
Change in earth pressure	"Push-in" total pressure dells	Up to 1 MPa	Up to 0.1 % FS	Up to 1.0 % FS
Change in Water Pressure	Standpipe piezometers	Any	±10 mm	±10–20 mm
	Pneumatic Piezometer (pore pressures are balanced by applied pneumatic pressure); Electronic (vibrating wire type) piezometric sensors	0.20 bar	0.01 bar	0.5 % FS ±0.02 bar

(continued)

Table 22.2 (continued)

Objective	Instrumentation	Range	Resolution	Accuracy
Crack or joint movement	Telltales	±20 mm	0.5 mm	±1 mm
	Calliper pins/micrometer or mechanical strain gauges	Up to 150 mm	0.02 mm	±0.02 mm
	Vibrating wire jointmeters	Up to 100 mm	Up to 0.02 % FS	Up to 0.15 % FS
Strain in structural member or lining	Vibrating wire gauges	Up to 3000με	0.5–1.0 με	±1–4 με
	Fiber optics	To 10,000 με (1 % strain)	5 με	20 με
Tunnel lining diametric distortion	Tape extensometers across fixed chords	Up to 30 mm	0.01–0.05 mm	±0.003–0.5 mm
	3D geodetic optic leveling ('retro' or 'bioflex') targets, leveling diodes or prisms	Any	0.1–1.0 mm	0.5–2.0 mm
	Strain gauged borehole extensometers installed from within tunnel	100 mm (3000 με)	0.01 mm (0.5 με)	±0.01–0.05 mm (±1–10 με)
	Convergence system	±50 mm	0.01 mm	±0.05 mm
Lining Stress	Total pressure (or 'stress') cells	2–20 MPa	0.025–0.25 % FS	0.1–2.0 % FS
Lining leakage	Flow meter	Any	1 l/min	2 l/min
Vibration	Triaxial vibration monitor/seismograph	250 mm/s	0.01 0.1 mm/s	3 % at 15 Hz

Tunnel Lining Design Guide (2004), British Tunnelling Society, Institution of Civil Engineers, Thomas Telford Ltd

- *Instrument installation phase*: It is important to record the sign of the measurements, to check the numbering of the instrumented points and data channels, and to detect any anomalous behavior.
- *Initial reading phase*: The initial reading of all instrumentation equipment is the "baseline reading."
- *Routine monitoring phase*: In this phase, the reading frequency must be chosen with due consideration of the rate of change in the measured quantity and monitoring stage, i.e., active monitoring and closeout monitoring. It should also be periodically reviewed in light of observed results. In addition, time synchronization of various data acquisition system and consideration of seasonal variation of reading are also important.

Table 22.3 Monitoring frequencies of used instrumentation equipment

Instrument type	Depth	Purpose	Minimum frequency of monitoring		
			Background monitoring	Standard monitoring	Active monitoring
Standpipes/piezometers in critical areas	At as-built depth	To monitor change in groundwater level or piezometric pressure heads in soil/rock	Monthly	Weekly	Daily
Other standpipes/piezometers	At as-built depth	To monitor change in groundwater level or piezometric pressure heads in soil/rock	Monthly	Weekly	Weekly
Building settlement marker	On selected buildings and structures	To monitor building settlement	Monthly	Weekly	Daily
Settlement plate/ground settlement marker	On ground surface	To monitor ground settlement	Monthly	Weekly	Daily
Utility settlement marker	On selected utilities	To monitor utility settlement	Monthly	Weekly	Daily
Tiltmeter	On selected buildings and structures	To monitor tilting of structures	Monthly	Weekly	Daily
Telltale5	On selected buildings and structures	To monitor any cracks	Monthly	Weekly	Daily
Inclinometer/extensometer	Below Selected GROUND SURFACE	To Monitor Subsurface lateral/vertical ground movement	Monthly	Weekly	Daily

Table 22.4 Alert, Action, and Alarm levels for tunnel excavation monitoring for reference only

Type	Alert level		Action level		Alarm level	
	Settlement (mm)	Slope of settlement trough	Settlement (mm)	Slope of settlement trough	Settlement (mm)	Slope of settlement trough
Existing building/structures	13	1/600	20	1/375	25	1/300
Existing MTR structures	10	1/2000	16	1/1500	20	1/1000
Existing historic building and existing fresh water reservoir	6	1/2000	8	1/1500	10	1/1000
Existing road	13	1/500	20	1/312	25	1/250
Existing/proposed slopes and retaining walls	15	1/600	24	1/375	30	1/300
Existing utilities	13	1/600	20	1/375	25	1/300
Level ground	15	1/140	24	1/112	30	1/70
Groundwater drawdown in piezometers[b]	1000[a]	–	1600[a]	–	2000[a]	–
Groundwater drawdown in piezometers[b]	1500[a]	–	2400[a]	–	3000[a]	–
Groundwater drawdown in piezometers[b]	2500[a]	–	4000[a]	–	5000[a]	–

[a]The groundwater drawdown shall be referred to the lowest measured water levels on site

[b]The allowable drawdown must be checked in advance, and different 3A levels may be assigned depending on the sensitivity of the location

The following table is a table of typical monitoring requirements for a tunnel in Hong Kong. This tunnel is excavated in granite and partly beneath the urban area (Table 22.3).

Table 22.5 Action to be taken when breaching Alert, Action, and Alarm levels

	Minimum course of action
Alert level	1. The engineer shall be informed immediately
	2. The contractor shall submit an investigation report to describe works being undertaken. to review the instrument responses and to study the cause of undue response
	3. The contractor shall review and increase the instrumentation monitoring and reporting frequency, if applicable
	4. The contractor shall submit a detailed plan of action describing the measures to be taken when the concerned instrument reach the action level to the engineer for approval
Action level	1. The engineer shall be informed immediately
	2. The active construction works may require to be suspended subject to the engineer's review of monitoring data
	3. The contractor shall immediately implement the measures as defined in the detailed plan of action to prevent further ground movement and groundwater drawdown, etc
	4. The contractor shall prepare a detailed investigation report to study the cause of the exceedance
	5. The contractor shall propose a contingency plan for the engineer's approval in the event that alarm value is reached or exceeded
	6. The contractor shall develop an emergency plan for the engineer's approval in the event the applied contingency measures cannot control the situation
	7. The contractor shall meet the engineer to discuss the instrument response and review the effectiveness of the implemented measures
	8. The contractor shall carry out design review of the works
Alarm level	1. Consideration shall be given to suspend all active construction works and the engineer shall be informed immediately
	2. The contractor shall immediately implement the measures defined in the contingency plan
	3. The contractor shall implement the measures defined in the emergency plan in the event that the applied contingency measures are found inadequate
	4. The contractor shall provide a complete report to examine the construction method and review the response of the instruments with full history of the monitoring data and construction activities and necessary design update
	5. To resume the suspended activities, the contractor shall demonstrate to the engineer's satisfaction that it is safe to do so with approval from the engineer

Engineer is the representative of the owner of the project

22.6.3 *Warning Mechanism: 3A Levels*

Measurement must be part of the risk management process and used as a possible warning mechanism enabling preventive measures to be introduced in acceptable time. It is normal practice to establish 3A—Alert, Action, and Alarm levels for key indicator parameters (such as displacement, strain, or pressure), which determine appropriate actions in response to these values being exceeded. As an example for reference, the following table gives typical 3A levels (Table 22.4).

When setting up any of the 3A levels, the following have to be specified:

- Procedures for passing on information.
- Allocation of responsibilities between the owner, supervisor, designer, and contractor.
- Time allowed for each person to pass on information or make decision.
- Remedial actions for dealing with foreseeable situations.

As an example for reference, Table 22.5 lists the action to be taken when breaching Alert, Action, and Alarm levels.

References

1. Hung CJ et al (2009) Technical manual for design and construction of road tunnels—civil elements. U. S. Department of Transportation, Federal Highway Administration, publication no. FHWA-NHI-10-034, December 2009
2. AITES/ITA WG2-Research (2011) Monitoring and control in tunnel construction. ITA report no. 009/November 2011. ISBN 978-2-9700776-3-3
3. DGSI (2013) Standpipe piezometer and water level indicator. http://www.slopeindicator.com/pdf/standpipe%20piezo%20datasheet.pdf. Accessed 15 Apr 2016
4. DSGI (2014) Vibrating wire piezometers. http://www.slopeindicator.com/pdf/vw-piezometer-datasheet.pdf. Accessed 15 Apr 2016
5. Chapman D et al (2010) Introduction to tunnel construction. Spon Prest, London
6. Hoek E, Brown ET (1980) Underground excavations in rock. Institution of Mining and Metallurgy, London
7. Kolymbas D (2008) Tunnelling and tunnel mechanics. Springer, Berlin
8. DGSI (2006) Horizontal digitilt inclinometer probe, DGSI (Durham Geo Slope Indi-cator). http://www.slopeindicator.com/pdf/manuals/digitilt-horizontal-probe-manual-2006.pdf. Accessed 15 Apr 2016
9. Wang I-T et al (2000) 3D-optical deformation measurement and its automation applications in tunnelling. China Engineering Consultants, Inc., Taiwan, ISARC. http://www.iaarc.org/publications/fulltext/isarc2000-192_MD4.pdf. Accessed 13 May 2016

Chapter 23
Health and Safety, and Risk Management in Underground Excavation

23.1 Introduction

Compared to the surface excavation, the working conditions and working environment of the underground excavation are more complex [10].

- In an underground situation, the working space is inherently tight, distorted, congested, isolated and inaccessible, of poor quality, and deteriorating. These adverse conditions endanger personnel, damage mobile equipment, and affect all activities.
- It is not only the confined space underground, but also adverse working conditions such as darkness, heat, humidity, gassy, and watery conditions that make the worker's job most difficult and risky.
- Variable rockhead and mixed ground conditions, such as presence of weak or compressible ground, drifting sand, or karst cave filled by water-bearing rock and soil, may be encountered.
- The workers are also liable to occupational diseases, such as asbestosis, silicosis, and some others. In addition, the risk of fire, explosion, inundation, and ground failure is not rare in underground construction.

There have been serious failures and accidents in underground construction worldwide in the past decades [1]. These have resulted in fatalities, damage to property, and other socioeconomic consequences. Most of these are due to inadequacies in the risk management and safety control. The achievement of good health and safety performance on site, and the avoidance of any "adverse events", such as accidents, cases of ill health, incidents, and losses, arises from the effective identification and control of risks.

The risk management approach is now widely adopted to control risks in construction projects. Codes and technical guidance have been published on safety in tunneling and risk management of tunnel works, for example, BS 6164: 2011 [2], ITA Report No. 001 [3], and A Code of Practice by the ITIG (2012) [4].

D. Zou, *Theory and Technology of Rock Excavation for Civil Engineering*,
DOI 10.1007/978-981-10-1989-0_23

A structured approach of safe system and risk management includes the following stages:

(a) Hazard identification.
A situation or condition that has the potential for unwanted consequences:

- Human injury;
- Damage to property;
- Damage to environment;
- Economic loss;
- Delay to project completion.

(b) The risks arising from the identified hazards.

It is a combination of the frequency of occurrence of a defined hazard and the consequences of the occurrence.

Table 23.1 gives an overview of the main geological hazards, their effects or potential consequences, and preventive actions. (Refer to [5]). The table is directed mainly to drill and blast (D&B) tunneling, but applies in principle also to tunnel boring machine (TBM) tunneling.

Table 23.2 indicates the large variety of potential hazards that present dangers of personal injuries in tunnels, based on risk analyses made by domestic contractors. It is of importance that personnel working underground recognize the special risk factors.

(c) Risk acceptance criteria—A qualitative or quantitative expression defining the maximum risk level that is acceptable or tolerable for a given system.
(d) Risk analysis—A structured process which identifies both the probability and extent of adverse consequences arising from a given activity. Risk analysis includes identification of hazards and descriptions of risks, which may be qualitative or quantitative.
(e) An assessment of the identified risks.
Risk assessment is the formalized process of identifying hazards and evaluating their consequence and probability of occurrence together with strategies as appropriate for preventative and contingent actions.
(f) Risk elimination.
The first form of risk control is to eliminate the hazards and risk completely; if this is possible, then this can be achieved by making engineering choices in the project, or perhaps by exercising design choices during the planning and specification stage.
(g) Risk evaluation—comparison of the results of a risk analysis with risk acceptance criteria or other decision criteria.
(h) Risk mitigation measure and risk-control measures for residual risks—action to reduce risk by reducing consequences or frequency of occurrence. The risk

Table 23.1 Overview of main geological hazards with potential for accidents (swelling material may be involved in 2nd, 4th, and 5th column) (Reproduced from Ref. [5] by permission of courtesy of Norwegian Tunnelling Society)

Hazard	Water under pressure	Unconsolidated zones	High rock stresses	Poor confinement	Crushed or blocky rock mass	Gas, methane
Effects or potential consequences	Flooding, cave-in, and dangerous drill rod changing	Immediate cave-in cannot be controlled at face	Rock spalling or bursting and slab or block fails	Block fails	Block falls and cave-in	Explosion Delay of work activities
Warning signals	Water in probe or blastholes, Inflow through joints in the face, and Karstic features	Water, mud, and sand in probe or blast holes	Drilling problems in stress release cracks, noises; crackling "shots", visible deformations, and *may be lacking!*	Drilling problems in open joints *May be lacking!*	Drilling problems in rock, drizzling continues with time, and *may be Lacking* in blocky rock	Bubbles in seepage water and Rotten smell of associated gas
Preventive actions	Probe drill to localize potential inflow, pregrouting and/or drainage, and do not blast until treatment is done	As for "Water under pressure" and ground freezing ahead of face	Scaling, bolting, sprayed concrete and drill stress release holes	Prebolting "spilling" scaling, bolting, and spray concrete ribs	For intact contour: sprayed concrete and bolting and for lost contour, water present: cast-in-place concrete lining	Probe drill Increased ventilation for dilution and circulation, and measurements and monitoring

Table 23.2 Risk factors concerning personal injures in tunneling (The Shortfirer's Manual, Norweigian Tunnelling Society, NFF 2003) (Reproduced from Ref. [6] by permission of courtesy of Norwegian Tunnelling Society)

Item	Block fall	Explosion	Fire	Rock burst or fly rock	Objects into eye	Gas, dust	Injury due to electricity	Toppling, fail injuries	Traffic	Pinch injury
Drilling	×	×	×	×	×		×			×
Charging holes	×	×			×			×		
Blasting	×	×		×		×				
Loading	×		×	×		×	×		×	×
Scaling from pile	×				×	×		×		
Machine scaling	×	×			×					×
Watering pile	×						×		×	×
Transportation	×		×	×		×			×	×
Scaling from basket	×					×		×		×
Storing of explosives		×	×							
Placement on site of explosives		×	×							
Internal transport of explosives		×	×							
Storing of fuel and oil		×	×							
Storing of gas		×	×			×				
Ventilation work					×			×		×
Grouting of bolts[a]	×				×	×		×		×
Electric work	×		×		×		×			
Hot work		×	×		×					

[a]Some chemical additives for grouting may cause skin injuries

control strategies are needed for those risks that remain to be addressed during the construction stage.

(i) Specified safe methods of working.
The ultimate aim of this process is to devise project-specific ways of safety undertaking all work activities.

(j) There is a special need to draw up specific plans for both contingencies and emergencies that might be foreseen on the project.

In this chapter, the major hazards and risks involved in the underground excavation and the elimination measures will be illustrated in Sect. 23.3.

23.2 General Requirements for Underground Excavation

Except the general requirements of health and safety for all civil engineering works, the following requirement should be addressed herewith for the underground excavation and construction.

23.2.1 Employee Identification System

Entrances to all underground facilities must have a check-in and checkout system that provides the contractor with an accurate record of each person underground. The system must be able to identify each individual and general location. General locations include heading, train crew, track crew, maintenance area, storage area, and survey stations. Figure 23.1 shows a tally system (or called tag system) which is an effective check-in and checkout system for all personnel in underground construction.

23.2.2 Illumination

Underground lighting and illumination intensities must adhere to the relevant regulations on industry lighting (including emergency) and power equipment. The regulations require that all workplaces and their approaches, where there is insufficient natural light, are lit by artificial means. In addition, they require the provision of emergency lighting in workplaces where failure of the main lighting system would result in danger.

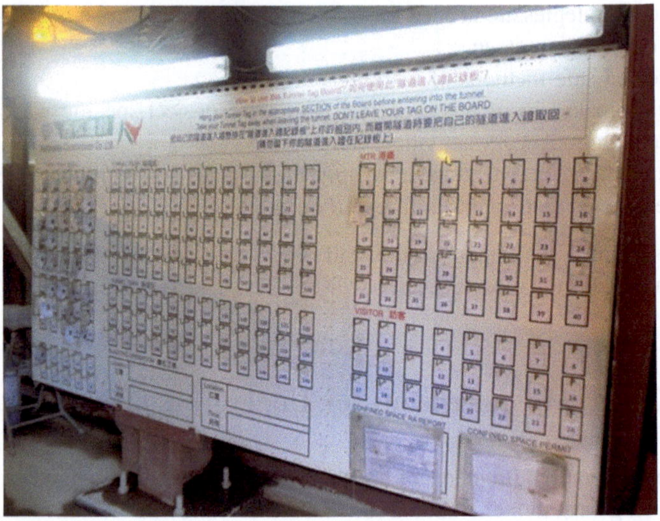

Fig. 23.1 Board of tally system which is located at the entrance of underground construction for check-in and checkout of all personnel

Table 23.3 Mean lighting levels

Area	Lighting level
Walkways and tracks	10 lux at walkway level
General working areas	100 lux at working surfaces
Tunnel face Excavation areas Crane lifting points	100 lux illuminated from at least two widely separated sources to avoid shadows

Good lighting contributes greatly to safety in underground construction. General lighting levels should be such that any hazards on walkways and tracks can readily be seen. Table 23.3 sets out the recommended mean lighting levels (quoted from BS 6164:2011 [2]).

23.2.3 Communication

Good communications throughout the site are fundamental to the safety and efficiency of all aspects of a tunnel project, in particular to the passing of information and instructions, the monitoring of systems, the control of lifting operations, the transportation persons, materials, and plant, and in the management of emergencies.

Install a telephone system or an equivalent powered communication system between the tunnel heading and the portal, the shaft bottom and shaft head, and the first-aid station. Keep the powered communication systems independent of the tunnel or shaft power supply and install the powered communication systems so that failure or disruption of any one station will not disrupt the operation of any other station.

23.2.4 Signals

Safety signals including audible or visual signals should be provided where a significant risk to health and safety cannot be avoided by the implementation of engineering controls.

23.3 Major Hazards and Risks and the Elimination Measures

23.3.1 Ground Risk

Ground settlement and ground collapses are the serious failures in tunnel construction usually caused by geotechnical risk. They may result in fatalities, damage to property, and other socioeconomic consequences. Figure 23.2 shows a serious accident of one people dead and four people injured caused by the ground settlement. Ground collapse in a tunnel not only seriously affects the progress of the project and causes financial loss, but also results in fatalities. Figure 23.3 shows an accident of rock collapse during the tunnel construction.

Elimination Measures:

(a) Adequate site investigation is essential to identify ground parameters, discontinuities, water, gas, and contamination;
(b) Closely monitoring the ground settlement at both ground surface and in the tunnel;
(c) Closely monitoring the variation of ground water;
(d) Closely monitoring the stability of the surrounding rock mass and the support system (including rock bolts, rock anchor, shotcrete, and permanent lining);
(e) Carefully inspecting the rock surface of the whole underground void after each blast and carefully scaling all loosened rock from the tunnel faces. Necessary reinforcement measures for rock stability must be taken as early as possible.

Fig. 23.2 Ground settlement during underground excavation for construction of Metro Line 7 of Shenzhen in June 2015. This accident caused the death of one person and four injured

Fig. 23.3 Michael O'Brien (not in picture) died after a rock hit his head while he was working on the East Side Access Project in New York, 2011 (*Source* www.dailymail.co.uk/news/)

23.3.2 Ground Vibration Produced by Blasting

Ground vibration caused by blasting in the tunnel construction may cause damage and disturbance to nearby buildings, structures, and utilities. The features and propagation of seismic waves through ground, the restriction on ground vibration in different countries, the monitoring instruments, and the measures to control ground vibration produced by blasting have been illustrated in previous chapters.

According to the experience of the author of this book, the propagation of seismic waves produced from the underground blasting has the following characteristics:

- When the underground work is at shallow depth in overburden soil or heavily weathered poor quality rock, the characteristics of the propagation of seismic wave produced by underground blasting have no obvious difference from that by the surface blasting.
- When blasting is carried out in hard rock deep beneath the ground overburden, the intensity of seismic wave produced may be greater than that produced by surface blasting for the same scaled distance, especially when monitored in other nearby underground spaces, e.g., Metro and Subway tunnels. Figure 23.4 is an

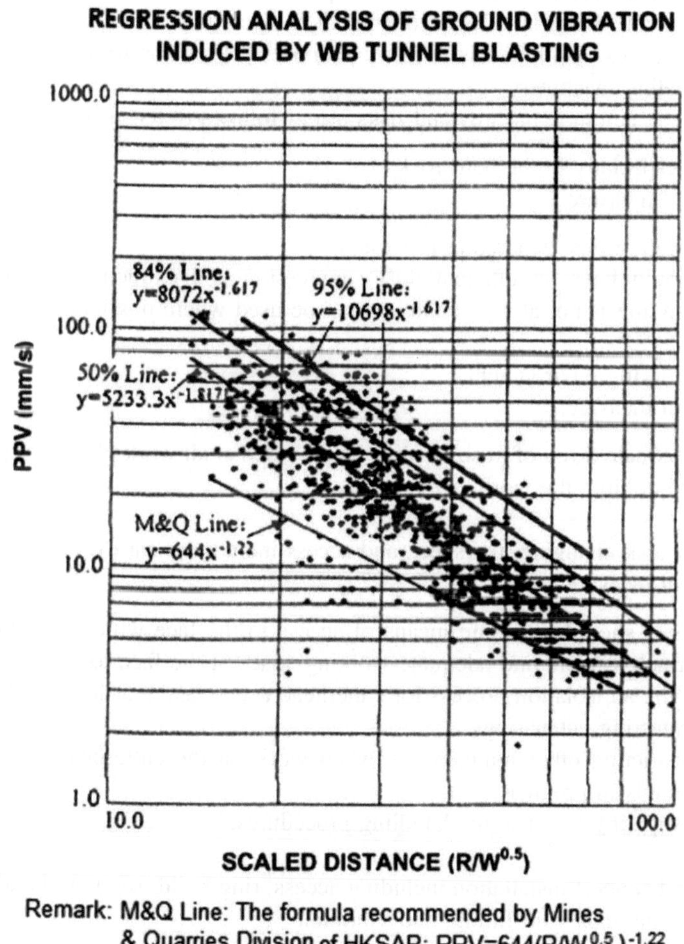

Fig. 23.4 Analysis chart of ground vibration for WB tunnel in Contract MTR 612, June 2000 in Hong Kong

example which was plotted according to the vibration records collected from monitoring points by the author in June 2000 in Hong Kong. The chart shows that the PPV of induced vibration by the tunnel blasting in hard granite are higher than the data calculated according to the M&Q recommended formula which was conducted using more than one hundred data collected from both surface (most) and underground blasts in Hong Kong (Chap. 6, Ref. [14]).

23.3.3 Tunnel Boring Machine (TBM)

Tunnel boring machines have become sophisticated and complex machines, and there are many hazards associated with their operation. There are a number of standards, codes, and manuals in many countries [7] relating to the safe operation with tunneling machinery.

The most common hazards and risks are as follows (refer to [8, 9]):

- Dust, especially in the hard rock.
  Control measure:
  - Isolating dust generating processes.
  - Should have an efficient dust suppression and extraction system (water spraying is not always sufficient but be used where possible).
- Hot working environment.
  Control measure:
  - Adequate flows of fresh air shall be provided at all work places for personnel and cooling the work environment.
- Cutter disk changing, transport, and access including heat exposure.
  Control measures:
  - There should be a safe means of access to the face which may include an airlock system, and adequate working space at the face to work safely.
  - Using an isolation process for cutterhead access and rear loading cutters and protecting cutter rings.
  - Restricting other maintenance when work on the cutterhead or cutterhead entry is in progress.
  - Instigating safe manual handling procedures.
- Ground support installation including access, ring build, segment handling with cranes, annulus void filling, and vibration
  Control measures:

- Using remote control bolting.
- Correctly operating segment feeder and handling devices including having operators visually check area before use.
- Ensuring:

  workers are competent in the use of segment handlers and follow manufacturer's instructions;
  the ring builder can visually see rams when moving them;
  use of exclusion zones so no work is done under unsupported ground;
  limbs are kept well clear of plant.

- Inspecting installed ground support.

- Tunnel collapse including ground and rock fall near shields and fingers; lining failure leading to potential ground and rock falls and degradation of excavated ground through drying, flaking, and support failure
  Control measures:
  - using geotechnical assessment during TBM design;
  - ongoing assessment of ground conditions and ground support with geologists and designers during excavation;
  - using finger shields including hoods;
  - mapping ground conditions immediately behind shields;
  - designing tunnel lining using geotechnical data and assessment, including faults (shearing), and seismic activity including liquefaction;
  - having quality assurance programs for lining and annulus void filling.

- Water inrush including flooding
  Control measures:
  - Using:

    flood doors, tunnel portals, and shafts designed to prevent inundation;
    water spray barriers;
    bulkheads between face and workers with sealable doors;
    sealable spoil screw conveyor tube;
    automated or remote operated pumping systems.

  - Designing dewatering systems through geotechnical modeling.
  - Implementing emergency exit plan.
  - Monitoring groundwater inflow.
  - Probing under sea or river.
  - Estimated hydrostatic pressure input into tunnel lining design.
  - Alignment design incorporates known geotechnical data.

  - Having quality assurance programs for lining and annulus void filling.
  - Designing the tunnel to prevent flotation.
  - Calculating pressure for groundwater control.

- Fire
  Control measures:
  - Providing fire suppression.
  - Having individual electrical cabinet fire detection and suppression systems.
  - Installing aqueous film forming foam systems at locations where grease, oil and fuel lines and tanks are present.
  - Designing tunnel lining for fire durability.
  - Substituting equipment to reduce diesel and oils.
  - Having detailed safe work method statements (SWMS) for hot works, like oxy cutting and welding.
  - Using:
    - Fire resistant hydraulic fluid and fire resistant power cables for high voltage supply;
    - Fire retardant tail shield grease.

- Electricity
  Control measures:
  - All safety critical electrical equipment should be explosion protected;
  - using cutoff switches and lock out systems;
  - implementing procedures for power failure;
  - installing warning signs.

23.3.4 Tunnel Transport

Tunnel transportation for men and plants, especially the muck transportation, is always very busy. A tunnel is a confined space in which visibility is often poor due to lack of lighting. Consequently, there is a high risk of collision between men and machines which has resulted in a number of fatal and serious injury accidents in recent years. Control measures should be implemented to eliminate or minimize, so far as is reasonably practicable, the risks associated with tunnel transportation.

- Pedestrian access is generally provided by walkways. Every tunnel should have a safe pedestrian route from the face to the entrance of the tunnel. Pedestrian should only be allowed to walk on one side of the tunnel. Adequate separation between pedestrian and vehicular traffic should be considered in the layout of the transport infrastructure.

- Passengers should be transported within specially enclosed vehicles which should be constructed to prevent any part of the passenger's body from reaching or leaning out the vehicle while it is in motion.
- Using a traffic management system, roadways and rail tracks shall be carefully laid out and properly maintained. Vehicles and rolling stock shall be chosen according to the gradients to be traversed and properly maintained.
- Trains and free-steered vehicles for use in underground should be designed, assembled, and equipped, so that the operator can fully observe the danger zone on his vehicle in both directions. All rail or free-steered vehicles underground should be equipped with two white lights in the direction of travel and with red rear lights, and an audible warning signal.
- Special transport vehicles (explosives and emergency vehicles) should be equipped with a special signal (revolving orange lights for example).

23.3.5 Shaft Under Construction

Shafts are constructed to provide entry for people, materials, equipment, and ventilation to a tunnel. Shaft construction methods and excavation techniques vary depending on conditions and their purpose. Shafts may be vertical or inclined and lined or unlined of various shapes.

The following three main problems (hazards) are of special concern during shaft construction:

1. People working in the bottom of the shaft which is a small space and in most instances need to be working together with some moving mechanics or underneath the hoisting materials;
2. Failure of the hoisting system; and
3. Some object falling down from the top or the wall of the shaft when people are working in the bottom of the shaft.

Control measures should be implemented to eliminate or minimize the risks associated with shaft sinking:

1. The number of persons in the shaft bottom area should be kept to a minimum while operations are in progress.
2. Procedures should be set up to avoid persons being underneath suspended loads wherever possible and persons should be alerted to any loads being sent down.
3. When a hydraulic excavator or grab is working in the shaft, personnel should be either protected within the shaft or removed from the shaft before working commences.

4. When handling loads with a crane or hoist, precaution should be taken to ensure that:

   (a) the load or skip does not swing or twist causing it to strike the lining of the shaft or other structure;
   (b) the load or skip does not catch a ledge, either in lowering or in hoisting, causing it to tip over and spill out its contents (whether persons or materials);
   (c) the rope does not become slack when the load is resting on the bottom or on a stage and catch in some part of the shaft structure, with resultant damage when tightened.

As a standard procedure in lifting, the load should be lifted a short distance then stopped, steadied, and inspected before hoisting continues.

5. The layout and detail at the top of the shaft should be designed to prevent the accidental fall of persons, plant, spoil, or material into the shaft.
6. Cranes at shaft

The cranes most commonly utilized for access shafts are crawler crane, mobile cranes, gantry cranes, and tower cranes for shallow shafts. Special hoists can be required in deeper shafts (50 m or deeper).

Rope length should be checked to confirm that at least two full turns of rope remain on the hoist drum when the hook is at the shaft bottom at full depth.

Special caution should be paid for hoisting long or difficult loads.

A competent person must inspect the crane at the beginning of each shift, and each time it is set up at the work site.

7. Personnel Access

Personnel access in shaft should be by fixed access equipment such as ladders, a mast climbing hoist, or man-riding crane. The secondary means should be available in the event that the main personnel access is out of work (Figs. 23.5, 23.6).

When a crane is used for the carriage of persons, it should conform to the relevant regulations.

8. Communications

Good communication between the surface and the working level is essential both for control of hoisting and lowering and for exchange of information on loads to be handled. The emergency channel should be kept clear at anytime.

At least two systems should be provided, one of which should be a voice system, and another should be capable of working in an emergency. As a general rule, operation should always be controlled from the shaft bottom.

Slingers/signalers should be nominated and should have undergone suitable training. No fully automatic lifting equipment is allowed to be used in shafts.

Fig. 23.5 Personnel accesses to both ladders and a mast climbing hoist

Fig. 23.6 Cage used to hoist personnel with the shaft primary hoisting system

23.3.6 Tunnel Atmosphere and Air Quality Control

The quality of the atmosphere is very important, and contaminants in the tunnel atmosphere affect everyone working in it.

- Requirements of Air Quality
  According to the US Reclamation Safety and Health Standards (RSHS) (2014) [11] and British Standard BS 6164:2011 [2], Underground air quality must meet the following requirements:
  - Oxygen concentrations must be between 19.5 and 22.0 %. The tunnel atmosphere should be considered as oxygen-deficient when the concentration of oxygen falls below 19 %.
  - Carbon monoxide concentrations must not exceed 25 ppm (RSHS)/30 ppm for long-term[a] and 200 ppm for short-term[b] (BS).
  - Carbon dioxide concentration must not exceed 5000 ppm (RSHS)/5000 ppm for long-term and 15,000 ppm for short-term (BS).
  - Nitrogen dioxide concentration must not exceed 3 ppm (RSHS)/3 ppm for long-term and 5 ppm for short-term (BS).
  - Hydrogen sulfide must not exceed 10 ppm (RSHS)/10 ppm for long term and 15 ppm for short term (BS).
  - Methane gas must not exceed 20 % (RSHS) of the lower explosive limit (4.4 %).
  - Other flammable gases or vapors must not exceed 10 % of the lower explosive limit (for instance, approx. 1.0 % for petrol/diesel vapor).
  - Other airborne contaminants, including dust, must not exceed the relevant regulations or standards on occupational health. According to BS 6164:2011, the occupational exposure limit for any respirable dust of unknown specific hazard, including dusts containing less than 1 % crystalline silica, is 5 mg/m^3. The long-term maximum exposure limit (MEL) for respirable crystalline silica dust is 0.3 mg/m^3.

  (Note: long-term: 8 h; time weighted average; Short-term: 15 min.)
  In addition, wherever possible, the web-bulb temperature in any working area should not be allowed to exceed 27 °C (BS).

- Air quality control
  - Well-designed and maintained ventilation system for the tunnel works is the most effective measure to contribute greatly to comfort and efficiency.
  - The concentration of oxygen, dust, toxic or potentially explosive fumes, or harmful gases in the tunnel atmosphere should be routinely monitored and steps taken as necessary to ensure contaminant levels do not exceed those laid down by national legislation or guidance. Where a specific work activity known to generate significant contamination, such as welding, is being carried out, local monitoring should be undertaken.

23.3.7 Fire and Rescue System

Among the most significant safety hazards of tunneling, to which the personnel and plants are exposed, are fire and smoke. In particular, it is the rapid spread of smoke through the whole tunnel space, rather than radiant heat generated by a fire, which can lead to fatalities.

In most tunnels under construction, the main source of fuel for fire is the large quantities of plastic, rubber, and other flammable materials found on plant, and equipment, along with the significant quantity of hydraulic fluid and possibly diesel fuel kept underground.

There was a fire incident in a tunnel in Hong Kong in 2012, which was under construction at the time (Fig. 23.7).

Elimination Measure:

(a) Never store any flammable materials, particularly fuel or hydraulic fluid, in the tunnel, and gasoline or liquid petroleum gases shall not be permitted underground;
(b) Reduce the quantity of flammability hydraulic fluid in plants and use flame retardant grease in all underground plant including the TBM. All hydraulic system should be well engineered. Only fire-resistant hydraulic fluids approved by a recognized authority can be used underground unless the equipment is protected by a fire protection system;
(c) An effective fire-fighting system must be built up and continually extended along with the tunnel advancing. The system must include both the fixed

Fig. 23.7 Fire in a tunnel

onboard fire suppression system on all plant and equipment and movable handheld extinguishers and fire mains with hydrants and hose reels;

(d) An evacuation and fire-fighting training program for all underground workers should be undertaken in order to prepare for possible cave-ins, flood, gas, explosion, fire, or other disasters; and

(e) Good housekeeping is another vital precaution in minimizing the build-up of flammable rubbish.

Rescue System

- Unrestricted and unobstructed access to all areas including all tunnel areas should be provided at all time for FSD appliances and personnel and maintained to the satisfaction of the FSD.
- An emergency control center at ground level incorporating an emergency roll call system should be established near each controlled access point to the tunnel. All other points of entry to the tunnels are secured to prevent unauthorized access.
- An updated plan showing the layout of the tunnels which includes the locations of refuge, locations of firefighting installations, and designated emergency assembly points should be provided at the emergency control center for reference by FSD personnel.
- Where a tunnel extends greater than 200 m, a suitable vehicle should be provided for conveyance of FSD equipment and personnel inside the tunnel during emergency. Trolleys should be available at the tunnel entrance to assist conveying equipment to the scene of an incident.
- A tally system should be provided at all access points to control and record the number of persons in the tunnel at any time.
- A dedicated intercommunication system, in the form of direct telephone shall be provided to the emergency control center and locations along the tunnels at not greater than 60 m intervals.
- Manual alarm points with audible and visual warning devices should be provided. The radio communication system (TDTRS) should be also provided. If alternative communication system should be adopted, it should be accepted by FSD. The alarm system should be connected to the emergency control center.
- Self-luminous exit signs/directional exit signs and emergency lighting should be provided.
- Gas detectors which are capable of detecting combustible gas, oxygen concentration level, and hazardous poisonous gas should be provided and readily available for emergency use at the emergency control center and at each tunnel portal.
- A secondary power supply or back-up generators should be provided to support the operation of tunnel lighting and ventilation systems and all emergency services in the event of power failure.
- A separate compartment/refuge chamber should be provided in the close proximity of the tunnel entrances designated by FSD to serve as a buffer zone to safeguard fire service operation and breathing apparatus entry control point.

Table 23.4 Provision of fire extinguishing equipment

Location of fire	Extinguishing medium				
	Water (jet)	Water (spray)	Foam	Inert gas	Powder
Tunnel—general	F		P	P	P
TBM—general			P	P	P
TBM—hydraulics			F		F
TBM—electrics				F	F
Diesel plants			F		F
Battery locomotive				F	F
Fuel store			P		P
Battery charging					P
Compressed-air working	F	F	P		
Timber headings, break-outs, etc.	F		P		

F fixed, *P* portable

- A comprehensive evacuation plan showing the emergency rescue procedures for workers should be produced, reviewed monthly, and updated by the safety officer. A copy should be forwarded to FSD for their record. An up-to-date copy should be shown in the emergency control center.
- Close contact, communication, and consulting with FSD should be established prior to the commencement and during the whole period of tunnel construction.

Tables 23.4 and 23.5 (quoted from BS 6164:2011 [2]) list the suitable fire-fighting equipment.

23.3.8 Explosives and Blasting Works

For drill–blast tunneling, the main hazards are dust, noise, vibration, and the risk associated with storing and using explosives. The main risks from using explosives include premature detonation and atmospheric contamination from the dust and blast fume released by the blast.

Table 23.5 Portable fire extinguishing equipment

Class of materials involved	Extinguishing medium
Fire involving solid usually of an organic nature, in which combustion normally takes place with the formation of glowing embers	Water extinguisher
Fire involving liquids or liquefiable solids	Foam extinguisher, CO_2, and dry powder
Fire involving gases	Water spray to cool cylinder, foam to extinguish any fire when valve has been closed
Fire involving metals	Dry powder, dry sand
Electrical equipment (if live)	Inert gas, dry powder, and dry sand

23.3.8.1 General Requirement for Explosives and Blasting Works

- All explosives delivered to the site will be moved directly to the blasting location and charging commenced immediately. No temporary storage within the tunnel will be allowed.
- Explosives will be handled and used only by the registered shotfirers.
- Only non-electric (shock tube) or electronic detonators will be used for tunnel blasting. If non-electric initiation system is used, one (and one or two for back-up) electric detonator can be used for initiating the non-electric initiation system. These electric detonators must be locked in a wooden case and located in a dry place and visible by the shotfirer at all times. The shotfirer must keep the key of the case all the time. Only at the time when the evacuation procedure has been completed and the shot is to be fired, the electric detonator can be connected to the non-electric initiation system by the registered shotfirer.
- Nobody except the shotfirers and their assistants can stay in the tunnel when the explosive charging is underway.
- After charging, an approved evacuation procedure will be carried out. Firstly, the shotfirer will clear the tunnel, including the adjacent underground work site if any, and ensure all people have been evacuated out of the tunnels.
- A tally system (or called tag system, see Sect. 23.2.1) will be used to ensure all people have left the tunnels prior to firing; all people must leave the portal area and stay in designated safe places.
- The blast door will be closed and the guarders must not let any people enter the tunnel before blasting is completed, and entering tunnel is permitted only after sufficient ventilation.
- If there is a misfire, the shotfirer must clear the tunnels again, and the evacuation procedure will remain effective. The shotfirer must handle the misfire personally according to the safety procedure.
- After drilling and blasting, the rock face of the tunnel roof and walls must be carefully checked and scaled. Any loosened rock must be barred down and temporary support must be carried out immediately if necessary.

23.3.8.2 Blasting Door

A blasting door made with at least 5 mm thick steel plate must be constructed at the entrance of the Tunnel, and the minimum distance from the blasting face must be 5 m (Fig. 23.8).

Fig. 23.8 Blasting door for tunnel blasting

23.3.8.3 Reduce Blasting Impact and Noise

In order to reduce the noise produced by blasting and the impact of rock fragments to the door, some buffer materials, such as waste rubber belts (see Fig. 23.9), and sound insulation materials are recommended to be set up behind the door.

Fig. 23.9 Hanging waste rubber belt and wire mesh between the blasting door and the blasting face to reduce the impact to the steel door

23.4 Explosives Stored in Site Magazine and Site Transportation

23.4.1 Site Explosives Magazine

For underground blasting works, a site magazine usually needs to be set up to temporary store explosive products. The location, structure, storage quantities, protective and precaution measures, and the daily management should strictly meet the regulations of the local government. A valid license for storage of explosive must be approved by the local authority.

23.4.1.1 Types of Explosive Magazines

In USA, according to the regulation (ATF) there are five types of magazines for the storage of explosive materials [12]:

- Type 1 magazines are permanent magazines for the storage of high explosives. Other classes of explosive materials may also be stored in Type 1 magazines.
- Type 2 magazines are mobile or portable indoor and outdoor magazines for the storage of high explosives.
- Type 3 magazines are portable outdoor magazines for the temporary storage of high explosives while attended (a day box, for example).
- Type 4 magazines are for the storage of low explosives. Blasting agents, Class C detonators, safety fuses, squibs, igniters, and igniter cords may also be stored in Type 4 magazines.
- Type 5 magazines are for the storage of blasting agents.

In Australia, there are only divided explosives magazines as above-ground magazine and portable magazine [13] (Figs. 23.10, 23.11).

In Hong Kong, there are also two types of explosive magazines: mode A store for high explosives and mode B store for non-blasting explosives [15].

In China, according to the Safety Regulation for Blasting, GB6722-2014, there are five types of explosive storage: permanent above-ground explosive magazines, small scale explosive magazines, permanent underground and soil-covered explosive magazines, mobile explosive magazines, and underground mine explosive magazines [14].

23.4.1.2 Safety Requirements for Site Explosives Magazine

- Location of Magazine

The explosive regulations enacted by the local government require explosives storage magazines to be located certain minimum distances from inhabited

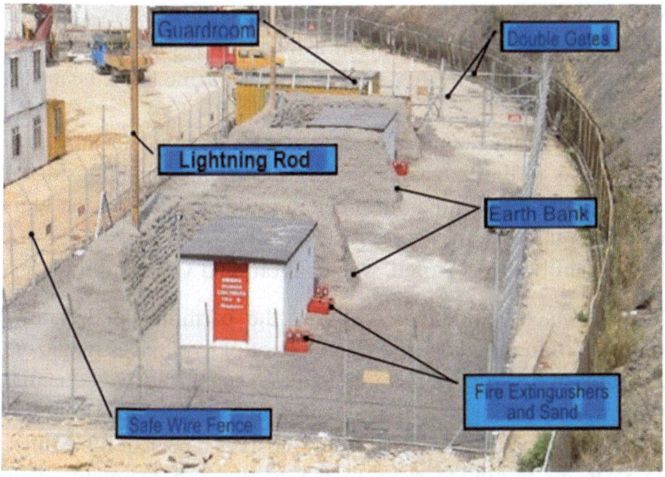

Fig. 23.10 A typical site explosive magazine in Hong Kong

Fig. 23.11 A typical type 2 portable explosive outdoor magazine (*Source* http://usexplosivestorage.com/)

buildings, public highways, passenger railways, and other magazines based on the quantity of explosive materials in each magazine. Tables of distance were adopted to protect the public in the event of a magazine explosion, e.g., Table 3.2.3.2 of AS 2187.1-1998 [13], Schedule 2 of UK ER 2014 No. 1638 [16], China GB 50089-2007 [17], Table 555.218 of US ATF [12].

- Tables of distances apply to the outdoor storage of explosive materials.
- When determining the distance from a magazine to a highway, an individual should measure from the nearest edge of the magazine to the nearest edge of the highway.
- If any two or more magazines are separated by less than the specified distance, then the weights in the magazines must be combined and considered as one.
- Barricading (earth bank mounding) can significantly reduce the required minimum distances under some tables of distances.
- Permitted Quantity to be Stored, Segregation, and Compatibility

Some regulations or guidelines specified the permitted maximum explosives quantity to be stored in the magazine. The license of explosives magazine approved by the local authority also specify the maximum net explosives quantity (NEQ).

In Hong Kong, according to the "Guidance Note on How to Apply for a Mode A Store Licence for Storage of Blasting Explosives" issued by Mines Division, CEDD, HKSAR, to minimize risks to the public and maintain security of the explosives, the storage capacity of all Mode A stores should be kept to the minimum required for the project (normally 2–3 days' supply). In Hong Kong, it is also necessary to carry out a quantitative risk assessment (QRA) to show that the hazard to life (probability of fatality) is within an acceptable range, for both storage and transport of explosives.

- Segregation and Compatibility

Any kinds of detonators, detonating relays, and capped fuses shall not be stored with blasting explosives, detonating cords, and boosters.

Fireworks shall not be stored with blasting explosives or detonators.

23.4.1.3 Construction of Outdoor Site Explosive Magazine

- General

The basic considerations in the construction of magazines for the storage of blasting explosives are as follows:

1. to maintain the explosives in good condition;
2. to keep the explosives from unauthorized persons;
3. to reduce the risks and consequences of accidental explosion.

The construction must allow for adequate ventilation but must resist illegal entry and interference and be bullet-resisting. The location of the magazine should take advantage of any natural features of the area which may help to reduce fire or explosion risks. All magazines are required to be marked "EXPLOSIVES" or "DETONATORS" in red letters on a white background for the advice of persons in their vicinity.

- Construction

In most locations, magazines must be of substantial construction i.e., steel, reinforced concrete, or reinforced brickwork. In some circumstances, approval may be given for magazines of light construction to be used (converted freight containers for use as magazine—refer to Appendices E of AS 2187.1-1998 [13]).

The latest standard of Hong Kong Mode A magazine construction drawings are block-work walls with timber framed roof. The structure is designed to disintegrate to small pieces, to avoid hazard of large flying fragments in the event of explosion.

Magazines shall be provided with all-weather access and good drainage system. For reduction of fire risk, magazine surrounds shall be cleared of vegetation, including trees, to form a firebreak not less 5 m wide or a clearing not less than 8 m wide.

Barricading (mounding) is the effective screening of a magazine containing explosive materials from another magazine, a building, a railway, or a highway, either by a natural barricade or by an artificial barricade. The detailed requirements of the mounding can refer to AS 2187.1-1998 [13].

Fencing is required for prevention of access by unauthorized persons and animals. It shall be 2500 mm high and a minimum distance of 600 mm from the outside of the mound measured from its base and not less than 3000 mm from magazines with no mound.

Lighting in magazines may be either natural or artificial. Electrical fittings and wiring shall comply with relevant regulations for electrical equipment in hazardous locations.

Lightning protection shall be provided and maintained in accordance with the requirement of relevant regulations.

All hinges and locks should be made of non-ferrous metal. No ferrous metal should be left exposed in the interior of the magazine.

A warning signboard with prohibited articles and substances painted in red and black, shown in symbols should be posted at the gate. Each symbol should be at least 100 mm in diameter. A typical warning signboard is shown in Fig. 23.12 (quoted from [15])

A guardhouse should be provided. A separate outer security fence should be installed to protect this guardhouse (Fig. 23.10).

23.4.1.4 Management of Site Explosives Magazine

- Explosives entering the magazine at different dates must be placed at different piles and the principle of "first in, first out" must be complied with to avoid product deteriorating due to overdue storage.
- Any explosive products to be taken out from the magazine must be recorded in the register book and signed by the registered shotfirer, contractor's staff, and staff of the Engineer. All explosive products taken from the magazine must be used in the site area where a valid blasting permit has been issued by local authority (Mines Division in Hong Kong).

Fig. 23.12 Warning sign board should be posted at the gate of the explosive magazine. (Reproduced from Ref. [15] by permission of Mines Division, GEO, CEDD, and HKSAR)

- The surplus explosive products after blasting must be returned to the magazine and recorded in the register book and then signed by the registered shotfirer, contractor's staff, and staff of the Engineer. Any damaged explosives (including the packing materials of cartridge explosive that is torn up or cut open) cannot be returned to the magazine and must be destroyed with the safe method recommended by the manufacturers.
- Any kindling (matches, lighters), radio communication device, mobile phone, alcohol, or any articles which may produce spark, including shoes with iron nails are prohibited to be brought into the explosive magazine. A placard must be posted at the entrance gate listing all prohibited articles and those articles which would require prior approval for bringing into the magazine.
- Explosives containers must not be placed directly against an interior wall for ventilation reasons.

- Explosives containers must be stored so that identifying marks are visible and easy to read.
- Tools for opening and closing containers must be non-sparking and may not be stored in magazines.
- Housekeeping
  - Interior must be clean, dry, and free of grit, paper, and empty packages and containers.
  - Floors must be regularly swept.
  - Exterior must be clear of rubbish, brush, dry grass, or small trees.

23.4.2 *Explosives Transport (Relevant Contents from Ref. [18] by Permission of Mines Division, GEO, CEDD, Hong Kong SAR)*

- The vehicle used for delivery of explosives must be powered by a diesel engine and equipped with the facilities according to the requirements of relevant regulation or specified by local authorities and checked and approved by relevant local authorities.
- Detonators and other types of blasting explosives shall not be loaded or transported within the same cargo compartment of the vehicle, unless the cargo compartment fulfills the additional requirements as specified by the authorities (refer to Annex B of [18]). Care must be taken when moving, loading, and unloading any explosive products. Throwing or dropping of any explosive products is prohibited.
- The 'permissible laden weight' of explosives allowed to be carried by the vehicle shall not exceed the carrying capacity of the vehicle. The 'permissible laden weight' will be calculated according to the following formula (refer to [18]):

$$\mathrm{PLW} = \mathrm{PGVW} - (\mathrm{VNW} + 75 \times \mathrm{N\,kg}) \tag{23.1}$$

Where

PLW (kg)	permissible laden weight;
PGVW (kg)	permitted gross vehicle weight;
VNW (kg)	vehicle net weight;
N	maximum number of persons permitted in the vehicle.

- Detonators must be loaded in the vehicle after other explosives loaded, and in the blasting site, they must be unloaded first and placed in a safe location on level ground prior to unloading other explosives.

- One armed security guard and the registered shotfirer must accompany each trip of explosives delivery.
- A vehicle loaded with detonators must go ahead and be followed by vehicle loaded with explosives. A suitable distance (about 30 m) between the two vehicles should be kept. Radio communication should be provided in any vehicle loaded with explosives and maintain communication at any emergency time with the site manager, police, FSD, and other authorities (Mines Division in Hong Kong). The radio communication device must be turned off in the vehicle loaded with detonators during the trip.
- Always maintain the vehicle in good condition with regular maintenance. Enough fuel should be filled before loading any explosive products for each trip. The driver must check the vehicle condition again before loading, including engine, electric circuit, water tank, and braking system.
- The vehicle is prohibited to stop for shopping, refilling of fuel, or stop at any public area. The fixed delivery route cannot be changed without prior approval from the authorities.
- No smoking in the vehicle or nearby during explosives loading, transporting, and unloading is allowed.
- The vehicle loaded with explosive products must run safely on the fixed route, comply with the traffic regulations, and no overspeeding and dangerous driving is allowed.

23.4.3 *Fire Extinguisher for Site Magazine and Explosives Trucks and Its Use*

- The following fire extinguishers must be provided in the site magazine:

Four fire extinguishers, including two of water type and two of CO_2 cartridge type, and 4 buckets of sand (Fig. 23.13).

- The above materials must be placed on two shelves and located between the security fence and the magazine.
- The fire extinguishers must be renewed and checked regularly and possess a valid test certificate.
- For a typical vehicle with gross vehicle weight of 9 tonnes or above, four fire extinguishers, comprising two 2.5 kg dry powder and two 9-L foam fire extinguishers of an approved type, with certificates, shall be provided. They shall be mounted in front and on both sides of the rear body in easily accessible positions with securely mounted brackets and quick release clamps. For a vehicle with gross vehicle weight of less than 9 tonnes, the required number of fire extinguishers shall be agreed with local authorities (in Hong Kong, the Mines Division).

Fig. 23.13 Fire extinguishers for site magazine

- All storekeepers, security guards, and drivers must be trained for use of the above extinguishers by the qualified safety officer.
- Water type extinguisher should only be used to extinguish wood, paper, and cloth fires and should never be used to extinguish flammable liquid (i.e., oil) or electrical fires. Oil and electrical fire should be extinguished by Carbon dioxide (CO_2) extinguishers.
- Always aim at the base of the fire when the fire extinguisher is being used.

23.4.4 *Emergency Plan*

- The contact telephone list for emergency (including site supervisor, safety officer, local authority (Mines Division in Hong Kong), police, FSD, and ambulance station) should be posted on the security room next to the telephone and the cab of the vehicle.
- If there is a fire in the magazine, the following procedure must be complied with:

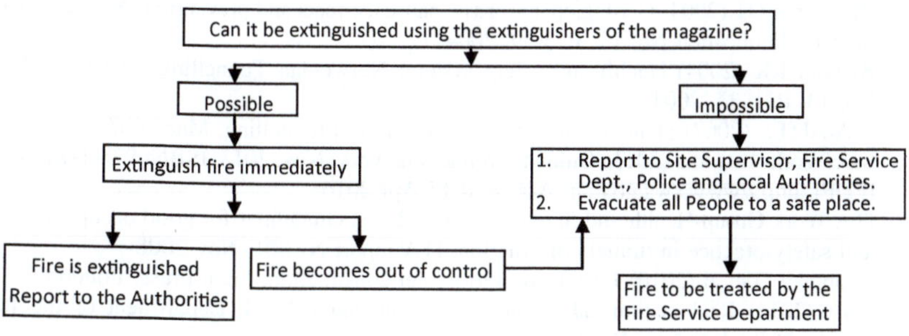

- If a fire occurs during the delivery trip:

(a) If there is a fire on the vehicle loaded with detonators, all people including the driver, car, crew, and all nearby pedestrians should be evacuated immediately to a safe place, close the scene, and report the incident to the police, FSD, local authorities (Mines Division in Hong Kong), and site manager.
(b) If there is a fire on the vehicle loaded with explosives, estimate whether the fire can be extinguished. If it is possible to control, extinguish it immediately with the extinguishers provided in the vehicle. If the fire cannot be controlled, all people including nearby pedestrians should be evacuated to a safe place and close off the scene of the incident. Report the incident to the police, FSD, local authorities (Mines Division in Hong Kong), and site manager.

- If a loss occurs in the magazine, secure the scene immediately and report to the police, local authority (Mines Division in Hong Kong), and site manager.
- If a traffic accident occurs during explosive delivery, evacuate all pedestrians and other vehicles nearby first, and then report to the police, local authorities (Mines Division in Hong Kong), and site manager.
- When there is a thunderstorm affecting wide local areas or partly over the site area, stop all operations at magazine and evacuate all people to a safe place until the thunderstorm passes away.

References

1. Geotechnical Engineering Office (2015) Catalogue of notable tunnel failure—case histories (up to April 2015), GEO, CEDD, HKSAR, 2015
2. United Kingdom (2011) British Standard, BS 6164:2011, Code of practice for safety in tunnelling in the construction industry, 2011. ISBN: 978 0 580 71086 5
3. ITA Report No. 001 (2008) Guidelines for good occupational health and safety practice in tunnel construction, Nov 2008
4. ITIG (2012) A code of practice for risk management of tunnel works, 2nd edn. The International Insurance Group, Aug 2012
5. Blindheim OT (2004) Geological hazards—causes, effects and prevention, Norwegian Tunnel Society, Publication No. 13, P23-29, 2004
6. Bjerkan RK (2004) Health and safety system, Norwegian Tunnelling Society, Publication No. 13, P31–34, 2004
7. ITA/AITES (2007) Standards/manuals of mechanized tunnelling, Mar 2007
8. Safe Work Australia (2013) Guide for tunnelling work, Nov. 2013. ISBN: 978-1-74361-239-2 [PDF]. http://www.swa.gov.au. Accessed 15 Apr 2016
9. ITA Work Group-Health and Safety in Work (2008) Guidelines for good occupational health and safety practice in tunnel construction, ITA report N° 001, Nov 2008
10. Chapman D et al (2010) Introduction to tunnel construction. Spon Press, London
11. USA (2014) Reclamation safety and health standards. U. S. Department of the Interior, Bureau of Reclamation, Washington

12. USA (2011) ATF—explosive magazine construction requirements, U. S. Department of Justice, Bureau of Alcohol, Tobacco, Firearms and Explosives, Office of Enforcement Programs and Services. ATF Publication 5400.17, October 2011. http://www.atf.gov. Accessed 15 Apr 2016
13. Australia (1998) AS 2187.1—1998, Australia standard, explosives—storage, transport and use, part 1. Storage. 5 July 1998
14. 中华人民共和国国家标准 (2014) GB 6722-2014: 爆破安全规程,中华人民共和国国家质量监督检验检疫总局, 中国国家标准化管理委员会, 2015.7.1. P. R. China National Standard (2014) GB6722-2014: Safety regulation for blasting. Standardization Administration of China, 1 July 2015
15. Mines Division (2015) Guidance note on how to apply for a mode a store licence for storage of blasting explosives, mines division, GEO, CEDD, HKSAR, March 2015. http://www.cedd.gov.hk/eng/services/mines_quarries/index.html. Accessed 15 Apr 2016
16. United Kingtom (2014) The explosives regulations 2014, Statutory Instrument 2014 No.1638, United Kingdom (see also Explosives Regulations 2014—Safety Provisions)
17. 中华人民共和国国家标准 (2007), GB 50089-2007: 民用爆破器材工程设计安全规范, 中华人民共和国国家质量监督检验检疫总局, 2007.8.1. P. R. China National Standard (2007), GB 50089-2007: Safety Code for Design of Engineering of Civil Explosives Materials. The State Administration of Quality Supervision, Inspection and Quarantine People's Republic of China, 1 August 2007
18. Mines Division (2013) Guidance note on requirements for approval of an explosives delivery vehicle, Mines Division, GEO, CEDD, HKSAR, June 2013. http://www.cedd.gov.hk/eng/services/mines_quarries/index.html. Accessed 15 Apr 2016

MIX
Papier aus verantwortungsvollen Quellen
Paper from responsible sources
FSC® C105338

If you have any concerns about our products,
you can contact us on
ProductSafety@springernature.com

In case Publisher is established outside the EU,
the EU authorized representative is:
Springer Nature Customer Service Center GmbH
Europaplatz 3, 69115 Heidelberg, Germany

Printed by Libri Plureos GmbH
in Hamburg, Germany